T0328818

LONDON MATHEMATICAL SOCIETY LECTURE NOTE SERIES

Managing Editor: Professor N.J. Hitchin, Mathematical Institute,
University of Oxford, 24–29 St Giles, Oxford OX1 3LB, United Kingdom

The titles below are available from booksellers, or from Cambridge University Press at www.cambridge.org/mathematics

185 Representations of solvable groups, O. MANZ & T.R. WOLF
186 Complexity: knots, colourings and counting, D.J.A. WELSH
187 Surveys in combinatorics, 1993, K. WALKER (ed)
188 Local analysis for the odd order theorem, H. BENDER & G. GLAUBERMAN
189 Locally presentable and accessible categories, J. ADAMEK & J. ROSICKY
190 Polynomial invariants of finite groups, D.J. BENSON
191 Finite geometry and combinatorics, F. DE CLERCK *et al*
192 Symplectic geometry, D. SALAMON (ed)
194 Independent random variables and rearrangement invariant spaces, M. BRAVERMAN
195 Arithmetic of blowup algebras, W. VASCONCELOS
196 Microlocal analysis for differential operators, A. GRIGIS & J. SJÖSTRAND
197 Two-dimensional homotopy and combinatorial group theory, C. HOG-ANGELONI *et al*
198 The algebraic characterization of geometric 4-manifolds, J.A. HILLMAN
199 Invariant potential theory in the unit ball of $\mathbf{C}^n$, M. STOLL
200 The Grothendieck theory of dessins d'enfant, L. SCHNEPS (ed)
201 Singularities, J.-P. BRASSELET (ed)
202 The technique of pseudodifferential operators, H.O. CORDES
203 Hochschild cohomology of von Neumann algebras, A. SINCLAIR & R. SMITH
204 Combinatorial and geometric group theory, A.J. DUNCAN, N.D. GILBERT & J. HOWIE (eds)
205 Ergodic theory and its connections with harmonic analysis, K. PETERSEN & I. SALAMA (eds)
207 Groups of Lie type and their geometries, W.M. KANTOR & L. DI MARTINO (eds)
208 Vector bundles in algebraic geometry, N.J. HITCHIN, P. NEWSTEAD & W.M. OXBURY (eds)
209 Arithmetic of diagonal hypersurfaces over infinite fields, F.Q. GOUVÉA & N. YUI
210 Hilbert C^*-modules, E.C. LANCE
211 Groups 93 Galway/St Andrews I, C.M. CAMPBELL *et al* (eds)
212 Groups 93 Galway/St Andrews II, C.M. CAMPBELL *et al* (eds)
214 Generalised Euler-Jacobi inversion formula and asymptotics beyond all orders, V. KOWALENKO *et al*
215 Number theory 1992–93, S. DAVID (ed)
216 Stochastic partial differential equations, A. ETHERIDGE (ed)
217 Quadratic forms with applications to algebraic geometry and topology, A. PFISTER
218 Surveys in combinatorics, 1995, P. ROWLINSON (ed)
220 Algebraic set theory, A. JOYAL & I. MOERDIJK
221 Harmonic approximation, S.J. GARDINER
222 Advances in linear logic, J.-Y. GIRARD, Y. LAFONT & L. REGNIER (eds)
223 Analytic semigroups and semilinear initial boundary value problems, KAZUAKI TAIRA
224 Computability, enumerability, unsolvability, S.B. COOPER, T.A. SLAMAN & S.S. WAINER (eds)
225 A mathematical introduction to string theory, S. ALBEVERIO, *et al*
226 Novikov conjectures, index theorems and rigidity I, S. FERRY, A. RANICKI & J. ROSENBERG (eds)
227 Novikov conjectures, index theorems and rigidity II, S. FERRY, A. RANICKI & J. ROSENBERG (eds)
228 Ergodic theory of $\mathbf{Z}^d$ actions, M. POLLICOTT & K. SCHMIDT (eds)
229 Ergodicity for infinite dimensional systems, G. DA PRATO & J. ZABCZYK
230 Prolegomena to a middlebrow arithmetic of curves of genus 2, J.W.S. CASSELS & E.V. FLYNN
231 Semigroup theory and its applications, K.H. HOFMANN & M.W. MISLOVE (eds)
232 The descriptive set theory of Polish group actions, H. BECKER & A.S. KECHRIS
233 Finite fields and applications, S.COHEN & H. NIEDERREITER (eds)
234 Introduction to subfactors, V. JONES & V.S. SUNDER
235 Number theory 1993–94, S. DAVID (ed)
236 The James forest, H. FETTER & B.G. DE BUEN
237 Sieve methods, exponential sums, and their applications in number theory, G.R.H. GREAVES *et al*
238 Representation theory and algebraic geometry, A. MARTSINKOVSKY & G. TODOROV (eds)
240 Stable groups, F.O. WAGNER
241 Surveys in combinatorics, 1997, R.A. BAILEY (ed)
242 Geometric Galois actions I, L. SCHNEPS & P. LOCHAK (eds)
243 Geometric Galois actions II, L. SCHNEPS & P. LOCHAK (eds)
244 Model theory of groups and automorphism groups, D. EVANS (ed)
245 Geometry, combinatorial designs and related structures, J.W.P. HIRSCHFELD *et al*
246 p-Automorphisms of finite p-groups, E.I. KHUKHRO
247 Analytic number theory, Y. MOTOHASHI (ed)
248 Tame topology and o-minimal structures, L. VAN DEN DRIES
249 The atlas of finite groups: ten years on, R. CURTIS & R. WILSON (eds)
250 Characters and blocks of finite groups, G. NAVARRO
251 Gröbner bases and applications, B. BUCHBERGER & F. WINKLER (eds)
252 Geometry and cohomology in group theory, P. KROPHOLLER, G. NIBLO, R. STÖHR (eds)
253 The q-Schur algebra, S. DONKIN
254 Galois representations in arithmetic algebraic geometry, A.J. SCHOLL & R.L. TAYLOR (eds)
255 Symmetries and integrability of difference equations, P.A. CLARKSON & F.W. NIJHOFF (eds)
256 Aspects of Galois theory, H. VÖLKLEIN *et al*
257 An introduction to noncommutative differential geometry and its physical applications, 2nd edn, J. MADORE
258 Sets and proofs, S.B. COOPER & J. TRUSS (eds)
259 Models and computability, S.B. COOPER & J. TRUSS (eds)
260 Groups St Andrews 1997 in Bath, I, C.M. CAMPBELL *et al*
261 Groups St Andrews 1997 in Bath, II, C.M. CAMPBELL *et al*
262 Analysis and logic, C.W. HENSON, J. IOVINO, A.S. KECHRIS & E. ODELL
263 Singularity theory, B. BRUCE & D. MOND (eds)

264 New trends in algebraic geometry, K. HULEK, F. CATANESE, C. PETERS & M. REID (eds)
265 Elliptic curves in cryptography, I. BLAKE, G. SEROUSSI & N. SMART
267 Surveys in combinatorics, 1999, J.D. LAMB & D.A. PREECE (eds)
268 Spectral asymptotics in the semi-classical limit, M. DIMASSI & J. SJÖSTRAND
269 Ergodic theory and topological dynamics, M.B. BEKKA & M. MAYER
270 Analysis on Lie Groups, N.T. VAROPOULOS & S. MUSTAPHA
271 Singular perturbations of differential operators, S. ALBEVERIO & P. KURASOV
272 Character theory for the odd order function, T. PETERFALVI
273 Spectral theory and geometry, E.B. DAVIES & Y. SAFAROV (eds)
274 The Mandelbrot set, theme and variations, TAN LEI (ed)
275 Descriptive set theory and dynamical systems, M. FOREMAN *et al*
276 Singularities of plane curves, E. CASAS-ALVERO
277 Computational and geometric aspects of modern algebra, M.D. ATKINSON *et al*
278 Global attractors in abstract parabolic problems, J.W. CHOLEWA & T. DLOTKO
279 Topics in symbolic dynamics and applications, F. BLANCHARD, A. MAASS & A. NOGUEIRA (eds)
280 Characters and automorphism groups of compact Riemann surfaces, T. BREUER
281 Explicit birational geometry of 3-folds, A. CORTI & M. REID (eds)
282 Auslander–Buchweitz approximations of equivariant modules, M. HASHIMOTO
283 Nonlinear elasticity, Y. FU & R.W. OGDEN (eds)
284 Foundations of computational mathematics, R. DEVORE, A. ISERLES & E. SÜLI (eds)
285 Rational points on curves over finite fields, H. NIEDERREITER & C. XING
286 Clifford algebras and spinors, 2nd edn, P. LOUNESTO
287 Topics on Riemann surfaces and Fuchsian groups, E. BUJALANCE *et al*
288 Surveys in combinatorics, 2001, J. HIRSCHFELD (ed)
289 Aspects of Sobolev-type inequalities, L. SALOFF-COSTE
290 Quantum groups and Lie theory, A. PRESSLEY (ed)
291 Tits buildings and the model theory of groups, K. TENT (ed)
292 A quantum groups primer, S. MAJID
293 Second order partial differential equations in Hilbert spaces, G. DA PRATO & J. ZABCZYK
294 Introduction to the theory of operator space, G. PISIER
295 Geometry and integrability, L. MASON & YAVUZ NUTKU (eds)
296 Lectures on invariant theory, I. DOLGACHEV
297 The homotopy category of simply connected 4-manifolds, H.-J. BAUES
298 Higher operads, higher categories, T. LEINSTER
299 Kleinian Groups and Hyperbolic 3-Manifolds, Y. KOMORI, V. MARKOVIC & C. SERIES (eds)
300 Introduction to Möbius Differential Geometry, U. HERTRICH-JEROMIN
301 Stable modules and the D(2)-problem, F.E.A. JOHNSON
302 Discrete and Continuous Nonlinear Schrödinger Systems, M.J. ABLOWITZ, B. PRINARI & A.D. TRUBATCH
303 Number Theory and Algebraic Geometry, M. REID & A. SKOROBOGATOV (eds)
304 Groups St Andrews 2001 in Oxford Vol. 1, C.M. CAMPBELL, E.F. ROBERTSON & G.C. SMITH (eds)
305 Groups St Andrews 2001 in Oxford Vol. 2, C.M. CAMPBELL, E.F. ROBERTSON & G.C. SMITH (eds)
306 Peyresq lectures on geometric mechanics and symmetry, J. MONTALDI & T. RATIU (eds)
307 Surveys in Combinatorics 2003, C.D. WENSLEY (ed)
308 Topology, geometry and quantum field theory, U.L. TILLMANN (ed)
309 Corings and Comodules, T. BRZEZINSKI & R. WISBAUER
310 Topics in Dynamics and Ergodic Theory, S. BEZUGLYI & S. KOLYADA (eds)
311 Groups: topological, combinatorial and arithmetic aspects, T. W. MÜLLER (ed)
312 Foundations of Computational Mathematics, Minneapolis 2002, FELIPE CUCKER *et al* (eds)
313 Transcendental aspects of algebraic cycles, S. MÜLLER-STACH & C. PETERS (eds)
314 Spectral generalizations of line graphs, D. CVETKOVIC, P. ROWLINSON & S. SIMIC
315 Structured ring spectra, A. BAKER & B. RICHTER (eds)
316 Linear Logic in Computer Science, T. EHRHARD *et al* (eds)
317 Advances in elliptic curve cryptography, I. F. BLAKE, G. SEROUSSI & N. SMART
318 Perturbation of the boundary in boundary-value problems of Partial Differential Equations, D. HENRY
319 Double Affine Hecke Algebras, I. CHEREDNIK
321 Surveys in Modern Mathematics, V. PRASOLOV & Y. ILYASHENKO (eds)
322 Recent perspectives in random matrix theory and number theory, F. MEZZADRI & N.C. SNAITH (eds)
323 Poisson geometry, deformation quantisation and group representations, S. GUTT *et al* (eds)
324 Singularities and Computer Algebra, C. LOSSEN & G. PFISTER (eds)
325 Lectures on the Ricci Flow, P. TOPPING
326 Modular Representations of Finite Groups of Lie Type, J. E. HUMPHREYS
328 Fundamentals of Hyperbolic Manifolds, R. D. CANARY, A. MARDEN & D. B. A. EPSTEIN (eds)
329 Spaces of Kleinian Groups, Y. MINSKY, M. SAKUMA & C. SERIES (eds)
330 Noncommutative Localization in Algebra and Topology, A. RANICKI (ed)
331 Foundations of Computational Mathematics, Santander 2005, L. PARDO, A. PINKUS, E. SULI & M. TODD (eds)
332 Handbook of Tilting Theory, L. ANGELERI HÜGEL, D. HAPPEL & H. KRAUSE (eds)
333 Synthetic Differential Geometry, 2nd edn, A. KOCK
334 The Navier–Stokes Equations, P. G. DRAZIN & N. RILEY
335 Lectures on the Combinatorics of Free Probability, A. NICA & R. SPEICHER
336 Integral Closure of Ideals, Rings, and Modules, I. SWANSON & C. HUNEKE
337 Methods in Banach Space Theory, J. M. F. CASTILLO & W. B. JOHNSON (eds)
338 Surveys in Geometry and Number Theory, N. YOUNG (ed)
339 Groups St Andrews 2005 Vol. 1, C.M. CAMPBELL, M.R. QUICK, E.F. ROBERTSON & G.C. SMITH (eds)
340 Groups St Andrews 2005 Vol. 2, C.M. CAMPBELL, M.R. QUICK, E.F. ROBERTSON & G.C. SMITH (eds)
341 Ranks of Elliptic Curves and Random Matrix Theory, J.B. CONREY *et al* (eds)
342 Elliptic Cohomology, H.R. MILLER & D.C. RAVENEL (eds)
343 Algebraic Cycles and Motives Vol. 1, J. NAGEL & C. PETERS (eds)
344 Algebraic Cycles and Motives Vol. 2, J. NAGEL & C. PETERS (eds)
345 Algebraic and Analytic Geometry, A NEEMAN
346 Surveys in Combinatorics, A HILTON & J. TALBOT (eds)

Algebraic and Analytic Geometry

Amnon Neeman
Australian National University

CAMBRIDGE UNIVERSITY PRESS
Cambridge, New York, Melbourne, Madrid, Cape Town, Singapore, São Paulo

Cambridge University Press
The Edinburgh Building, Cambridge CB2 8RU, UK

Published in the United States of America by Cambridge University Press, New York

www.cambridge.org
Information on this title: www.cambridge.org/9780521709835

First published 2007

Printed in the United Kingdom at the University Press, Cambridge

A catalogue record for this book is available from the British Library

ISBN 978-0-521-70983-5 paperback

Contents

Preface

This book came out of a course I taught, twice, at the Australian National University. I taught it first in the Fall of 2004, and then again, because of interest from some students and colleagues, in the Fall of 2005. The course was a one-semester affair, and the students were fourth-year undergraduates.

Given that these were undergraduate students in their final year, this could be one of the last few mathematics courses they would ever see. They might well decide to pursue interests having nothing to do with mathematics; they could, for all I know, choose to become doctors, or lawyers, or bankers, or politicians. My task was to present to them an overview of algebraic geometry. It would be premature to give them a thorough grounding in the field; a broad, panoramic picture seemed far more appropriate, and if possible the panorama should include glimpses into a wide assortment of pretty vistas, into more specialized areas, each of which is beautiful in its own right. I tried to cover interesting topics, without delving into too much detail on any one of them.

The first order of business was to choose the subject matter for the course. I had the option of teaching classical algebraic geometry; there are several excellent textbooks to choose from, written specifically for students at this level. But I wanted to teach modern algebraic geometry, and there really are no undergraduate treatments of the field. The consensus seems to be that this topic is beyond undergraduates, suitable only for courses at the graduate level.

So here I was, soon to face a class of math majors in their final year, and I had decided to teach them some modern algebraic geometry, even though there was no available textbook. I had to assemble the material myself. In so doing, I had to take into account the mathematics the students are likely to have seen in their first three years at university. Usually this would include some background on point-set topology, maybe a course on analytic functions in one complex variable, possibly a course on functional analysis, which would probably cover the Hahn–Banach Theorem and the Open Mapping Theorem, possibly a little about manifolds, maybe a rudimentary course on algebraic topology, and perhaps some basic algebra—groups, rings, fields, modules, if I were lucky maybe even the Hilbert Basis Theorem and the Nullstellensatz.

Algebraic geometry is a meeting place in which all the previous strands of knowledge magically converge; why not present it this way? This was my guiding philosophy in planning the course, and later the book. Bearing in mind that only an unusual math major will have seen every one of the topics listed, I tried not to lean too heavily on any one of them. But strong math majors should have met a large portion of these subjects, and my hope was that I was building on familiar ground.

Whenever we teach a course, especially a wide-ranging survey course, there will be the keen students, the enthusiastic ones who want to go a little beyond the discussions presented in class, who might even wish to pursue the subject further some day. This book was written for them. It covers the same topics treated in the course, and it tries to remain at the same level, demanding from the reader no prerequisites beyond what was assumed in my course. But the book has far more detail than the course, with many of the proofs included. I mention this because, if you decide to use the book to teach a course, and the course you have in mind is to be modeled approximately on the one I gave, then it would be a mistake to follow the book slavishly. In your lectures you will be presenting only selected portions of the book; you will have to pick and choose, deciding which parts of the material to present, and what to leave out.

I do not presume to make these decisions for you; the author has no authority to direct the reader how to use a book. As a rough guide let me, nevertheless, tell you what I did. Perhaps even more helpful: let me tell you what I would do if I come to teach the course again. The Introduction is worth presenting, just to give the students an overview; I spent the first class doing that. Chapter 2 rated two classes; one for an overview, and one for the proof of Theorem 2.3.2. I spent a few weeks on Chapter 3; it contains the definition of schemes, and a thorough understanding is worthwhile. In the case of Chapters 4, 5 and 6, I stated the results and mostly skipped the proofs. The results are about the complex topology and the sheaf of holomorphic functions on schemes of finite type over $\mathbb{C}$. Everything is plausible enough and, in my opinion, sketchy arguments sufficed. Now that the book is available, the need to cover this material thoroughly is even less than it was when

I was teaching the course; the students would be able fill in the details as carefully as they want, of any of the parts which you choose to omit in class.

I spent most of the semester on Chapters 7 and 8. Chapter 7 introduces coherent sheaves, both algebraic and analytic, while Chapter 8 gives a little window on geometric invariant theory, and uses it to study the properties of projective space. My feeling was that these were topics well worth treating a little more completely.

By the time you have finished with Chapter 8 you should have a reasonable idea how much time you have left. Based on that, I would decide how much of Chapter 9 to cover; most of the chapter can be omitted. Time permitting, I would present a little of the material on representations of linear algebraic groups, treat the special case of the multiplicative group $\mathbb{G}_m$, and present the proof of Hilbert's theorem on the finite generation of rings of invariants. Just how much is presented, and how many of the proofs, would very much depend on how pressed I were for time. The chapter is very skippable.

If the ground is thoroughly prepared, then Chapter 10 should not demand much time. It is the punchline of the course and is worth doing well. I do have to admit, however, that in both of my attempts to teach the material I skipped the computational parts; I stated, without proof, the results of Lemma 10.6.4. This meant, among other things, that I basically skipped all of Section 10.6. This is a much more significant omission than it may seem; if you choose to dispense with the proofs of Section 10.6 then you can also leave out all the preparatory material, which takes up several earlier sections in the book. For example there would be no compelling need to say much about the Fréchet topology, of the vector space of sections of a coherent analytic sheaf. And the various computations, of what I tend to refer to as the "elementary" or "concrete" examples, all become optional. My rough estimate is that it renders about 15% of the book dispensable.

Let me explain that I did not pull this figure out of thin air; the book used to be approximately this much shorter. At the suggestion of the reviewer I expanded the book to include Section 10.6, together with all the preparatory material. I am grateful to the reviewer for proposing this improvement; it certainly makes the treatment far more self-contained.

The first draft of the book was completed in 2005. Then, in 2006, I had an unusual fourth-year, undegraduate student. Michael Carmody wanted to do his senior thesis with me, but he told me, in advance, that this would be his last mathematical year. He had decided that his passion was for philosophy. This meant that, after his fourth year finished, his intention was to start a PhD in philosophy.

This made him another student ideal for this type of book. Giving him a solid grounding in the field, the sort that would prepare him for research, was not a priority; it seemed far more appropriate to present him with a panoramic view. I therefore gave him the manuscript of this book to read. He took about four months to get through it. We met every couple of weeks. At these meetings he would present me with long lists of misprints, as well as with some points in the mathematics which he found unclear. I took his comments extremely seriously; whenever he found anything confusing, I would rewrite the text to elucidate the point. I owe him a tremendous debt for his help. Anyway, I was pleased that he managed to plough through the book, almost unaided, in about four months. It meant that the book is accessible to its intended audience.

Before I end the Preface I should thank the many people whose help has been invaluable. Let me begin with the students who took the courses; I have already thanked Michael Carmody, but special mention goes also to Joanne Hall, Jason Lo and Kester Tong, whose questions helped inform what I wrote. I would like to thank my colleagues Eugene Lerman and Shahar Mendelson for making me teach the course a second time, only one year after the first, and for many comments on the manuscript as it was being written. I am grateful also to my (graduate) student Daniel Murfet, for pointing out several notational inconsistencies. Thanks go also to Boris Chorny, Jonathan Manton and Greg Stevenson, who kept sending me corrections, as well as ideas for more substantial improvements, right up until the very last minute. I wish to thank the anonymous reviewer for some wonderful suggestions, and my editor, Diana Gillooly, for her patient help and good humor. Speaking of patience and good humor: I would like offer my warmest thanks to my family, for putting up with me during the months when this manuscript was being written. Thank you Terry (my wife), and Ted, Joe and Jeremy (our sons). Jeremy was always ready with a good joke. For example: when Cambridge University Press and I were in the process of choosing a title for the book, it was Jeremy who finally said that, if we really wanted the book to sell, then we should name it *Harry Potter and Algebraic Geometry.* Diana Gillooly improved this to *Harry Potter and the Proof of GAGA,* which sounds more mysterious. And Jeremy also proposed that, if I wanted to lighten the mood by having a joke somewhere in the book, it could start out with: "Three algebraic varieties walk into a bar..."

Enough of the Preface; let the Quidditch match begin.

1
Introduction

Algebraic geometry is an old subject. There are many introductory books about it, at various levels. There are even some introductory texts which, like the present one, are addressed primarily to advanced undergraduates or beginning graduate students. The idea of these elementary introductions is to sell the subject. We do not yet attempt to train people in the field, only to convince them that it is fascinating and well worth the effort required to learn it.

There is no doubt that learning algebraic geometry entails substantial effort. The modern way of approaching the subject makes use of several technical machines, and a well-trained algebraic geometer needs to master at least one of these machines, preferably more than one. The very elementary introductions to the field try to avoid the machinery. They are generally very classical, using mathematics from the nineteenth century and the first half of the twentieth, before algebraic geometry underwent the Industrial Revolution and became so mechanized. The classical introductory books talk a great deal about curves, using the Riemann-Roch theorem to study them. They also might deal a little with simple singularities and their resolutions.

There are also many excellent books which do a thorough job teaching the foundations. These are for the serious graduate student, who already knows that this is the subject in which she wants to write her PhD. The current book is addressed to the uncertain graduate student, who is trying to decide if she really wants to spend the next four years of her life learning how to use sheaf cohomology to solve problems in algebraic geometry. The idea of the book is not to avoid the machinery, but rather to give an impressive illustration of its power.

The current book goes right for the mechanical apparatus, and tries to persuade the beginner of its value. We chose a theorem whose proof needs the machine. We chose an interesting, powerful theorem, and present a proof of it. Since we want to impress the reader with the value of the machine we explain the proof fairly completely, developing along the way the parts of the machine needed in the proof. But, bearing in mind that we want the book to be accessible to beginners, we keep the mechanical parts minimal. We only develop those parts of the machine which are unavoidable in the proof.

Let us therefore immediately make our disclaimer: this is not the right book for a serious graduate student, preparing herself for a research career in algebraic geometry. The treatment we give here, of the foundations of the subject, is much too patchy and has far too many glaring holes. For a thorough introduction to the foundations the student is referred to the many excellent books pitched at a higher level.

The theorem we chose to prove in this book is Serre's GAGA theorem. The GAGA stands for *Géométrie algébrique et géométrie analytique,* the title of [7], the 1956 paper by Jean-Pierre Serre containing the proof. In the remainder of the introduction we will try to explain what the theorem is about, and why it is surprising and important. There is, however, a problem: we do not yet have the language to state the theorem in the generality in which Serre proved it. Instead we will state three theorems, all of which are essentially immediate consequences of Serre's GAGA. We can state the consequences already, and will try to explain their importance. As the book progresses we will develop the language necessary for the more general theorem, and most of the machinery needed in the proof. We should make one more disclaimer: there will be some small parts of the proof we will not fully explain. We will give references and say something about the ideas.

Before we proceed any further we should say something about the prerequisites for reading the book. Let us start with the algebraic prerequisites: the reader is assumed to know what are commutative rings and ideals, what are ring homomorphisms, and the relation between ideals and kernels of ring homomorphisms; in particular given a ring R and an ideal $I \subset R$ the reader should know how to form the ring homomorphism $R \longrightarrow R/I$. The reader should also know what it means for an ideal to be either prime or maximal. Given a field k, our most important example of a ring will be the polynomial ring $k[x_1, x_2, \ldots, x_n]$; the reader is assumed to know this ring. The reader should also know about modules, homomorphisms between modules and exact sequences of modules. When we need anything more sophisticated we will state the results we need and give references. We made sure to appeal only to facts in commutative algebra which are contained in Atiyah and Macdonald's book [1], and the vast majority of what we need may be found in the first few chapters.

We also assume a basic familiarity with homological algebra. Atiyah and Macdonald's book [1] does not cover this, but there are many excellent accounts of the subject; see for example Weibel [9]. In most of the book we use very little; we assume a familiarity with exact sequences, an acquaintance with the 5–lemma, and an ability to do simple diagram chasing to prove sequences exact. Towards the end of the book the level of sophistication goes up a notch; starting in Section 10.4 we will look at the homology of chain complexes, we will appeal to the result that homotopic maps of chain complexes induce equal maps in homology, and we will rely on the fact that a short exact sequence of chain complexes gives a long exact sequence in homology.

The reader is assumed to know some basic point-set topology; we will freely refer to topological spaces, open and closed sets and continuous maps. We expect the reader to know what a homeomorphism is. Given a topological space X and a subset U, the reader should know what is the subspace topology on U. Given a topological space X, a set Y and surjective map $X \longrightarrow Y$, the reader should know how to form the quotient topology on Y. We will also feel free to talk about connectedness and compactness of topological spaces, and about Hausdorff spaces, and we will assume the reader knows what all these concepts mean.

And finally, while it is not absolutely indispensable, it would help to have seen at least one course on functions of one or more complex variables. We will freely refer to "holomorphic functions", and occasionally also to "meromorphic functions". For a reader who has never met them before let us briefly introduce them. Let $U \subset \mathbb{C}^n$ be an open set. A holomorphic function on U is a function $f : U \longrightarrow \mathbb{C}$ so that, for every point $p \in U$, the Taylor series of the function f at the point p converges near p to the function f. More formally this means that for every point $p \in U$ there exists a real number $\delta > 0$ so that, on the ball of radius δ centered at $p = (p_1, p_2, \ldots, p_n)$, the function f is given by a convergent power series

$$f(z_1, z_2, \ldots, z_n) = \sum_{(i_1,i_2,\ldots,i_n)\in\mathbb{N}^n} a_{i_1,i_2,\ldots,i_n}(z_1-p_1)^{i_1}(z_2-p_2)^{i_2}\cdots(z_n-p_n)^{i_n} .$$

This defines holomorphic functions. The ratio f/g of two holomorphic functions f and g, where g does not vanish on any open subset of U, is called a meromorphic function. Taken very literally it is not a function; if f and g are holomorphic on U then f/g is only defined on the subset of U where g does not vanish; in an abuse of language we say that f/g is meromorphic on U. In general, a meromorphic function h on U is only assumed to be locally of the form f/g. That is, every point $p \in U$ is contained in a ball $V \subset U$ so that, on the ball V, there exist holomorphic functions f_V and $g_V \neq 0$, and f_V/g_V agrees with the restriction to V of h.

1.1 Algebraic and analytic subspaces

Consider the closed subsets of $\mathbb{C}^n$, the n–dimensional complex space. There are many closed subsets. We are particularly interested in two classes of closed subsets: the closed algebraic and the closed analytic subspaces. Closed algebraic subspaces of $\mathbb{C}^n$ are the closed sets on which a finite number of polynomials vanish. Closed analytic subspaces are, at least locally, the closed subsets on which finitely many holomorphic functions vanish. Here are the definitions, given more precisely.

Definition 1.1.1. *A closed subset $X \subset \mathbb{C}^n$ is called a* closed algebraic subspace *if there are finitely many polynomial functions on $\mathbb{C}^n$, let us say $f_1, f_2, \ldots, f_r$, so that*

$$X = \{x \in \mathbb{C}^n \mid f_i(x) = 0 \quad \forall\, 1 \leq i \leq r\}.$$

The definition of analytic subspaces is slightly more delicate, being local.

Definition 1.1.2. *A closed subset $X \subset \mathbb{C}^n$ is called a* closed analytic subspace *if, for every point $x \in X$, there exists an open neighborhood U of x in $\mathbb{C}^n$, and finitely many holomorphic functions $f_1, f_2, \ldots, f_r$ on U, so that*

$$X \cap U = \{y \in U \mid f_i(y) = 0 \quad \forall\, 1 \leq i \leq r\}.$$

Note that, in the definition of a closed analytic subset, we do not require the existence of finitely many global holomorphic functions. The definition is local. Furthermore it is immediate, from the definitions, that any closed algebraic subspace of $\mathbb{C}^n$ must automatically be a closed analytic subspace.

It is also very clear that there are many closed analytic subspaces of $\mathbb{C}^n$ which are not algebraic. The easiest is to consider the case $n = 1$. If $f \neq 0$ is a polynomial function in one variable (that is a polynomial function on $\mathbb{C}^1$), then the set of points $\{x \in \mathbb{C}^1 \mid f(x) = 0\}$ is finite. This just says that a polynomial in one variable has finitely many roots. If X is a closed algebraic subset of $\mathbb{C}^1$, Definition 1.1.1 tells us that there is a finite set of polynomial functions $f_1, f_2, \ldots, f_r$, and X is the set of points $p \in \mathbb{C}$ at which all the f_i vanish. If the f_i are all zero then $X = \mathbb{C}^1$. Otherwise one of the f_i must be non-zero, it vanishes only at finitely many roots, and X is a subset of this finite set of roots. We conclude that a closed algebraic subspace $X \subset \mathbb{C}^1$ is either finite or is equal to $\mathbb{C}^1$.

Now consider the subset $X \subset \mathbb{C}^1$ given by

$$X = \{x \in \mathbb{C}^1 \mid \sin(x) = 0\}.$$

Since $\sin(x)$ is a holomorphic function of $x \in \mathbb{C}^1$, the set X is a closed analytic subspace of $\mathbb{C}^1$. On the other hand we know that $\sin(x)$ vanishes whenever x is a multiple of π; even better, it vanishes no place else. That is

$$X = \{n\pi \mid n \in \mathbb{Z}\}.$$

The set X is neither finite nor all of $\mathbb{C}^1$; it therefore is not algebraic.

In Definitions 1.1.1 and 1.1.2 we defined closed algebraic and analytic subspaces of $\mathbb{C}^n$. For the definitions to make sense we needed to have a clear notion of which functions on $\mathbb{C}^n$ are polynomial, and which are holomorphic. In the definitions we can replace $\mathbb{C}^n$ by any space P provided that P has, at least locally, well defined classes of polynomial and holomorphic functions.

There is a family of such spaces, which come equipped with classes of polynomial and holomorphic functions. In the first few chapters of the book we will define them and talk a little about their elementary properties. Such spaces are called *schemes of finite type over* $\mathbb{C}$. For the remainder of the introduction we ask the reader to accept, without seeing the formal definitions, that schemes of finite type over $\mathbb{C}$ are some topological spaces which have a well-defined notion of which functions are polynomial and which are holomorphic. For every P, a scheme of finite type over $\mathbb{C}$, it makes sense to ask which closed subsets $X \subset P$ are algebraic subspaces, and which are analytic. The first important corollary of GAGA is a theorem which predates Serre's paper, a theorem whose first proof was due to Chow:

Theorem 1.1.3. *Let P be a scheme of finite type over $\mathbb{C}$. Assume P is compact. Then any closed analytic subspace of P is algebraic.*

We saw above that not all closed analytic subspaces of $\mathbb{C}^n$ are algebraic. This does not contradict Theorem 1.1.3, since $\mathbb{C}^n$ is not compact. One very interesting space to which Theorem 1.1.3 applies is the complex projective space $\mathbb{CP}^n$. I do not want to define $\mathbb{CP}^n$ in the introduction;

the readers who know it do not need such a definition, the readers who do not yet know $\mathbb{CP}^n$ will learn much about it in the rest of the book.

It turns out that the general statement of Theorem 1.1.3 is easy enough to reduce to the special case where $P = \mathbb{CP}^n$. There is a lemma due to Chow that says an arbitrary compact P, which is a scheme of finite type over $\mathbb{C}$, admits a surjective, polynomial map $\pi : X \longrightarrow P$, with X a closed algebraic subset of $\mathbb{CP}^n$. Suppose $Y \subset P$ is a closed analytic subspace. The special case of Theorem 1.1.3, where the space is $\mathbb{CP}^n$, tells us that $\pi^{-1}Y \subset X \subset \mathbb{CP}^n$ is closed and algebraic, and then fairly easy, standard arguments imply that $Y = \pi(\pi^{-1}Y)$ is an algebraic subspace of P. In this book we will confine ourselves to proving the special case of Theorem 1.1.3 with $P = \mathbb{CP}^n$; the reader is expected to remember that the more general fact is an easy consequence.

The next consequence of GAGA is about vector bundles. Given a topological space P it is possible to define vector bundles on P. We will not remind the reader of the definition of a vector bundle in the introduction; there will be a great deal said about vector bundles, and more general sheaves, later in the book. The only thing we want to recall here is that one way to construct a vector bundle on P is in terms of transition functions, as follows. Take an open cover $\{U_i, i \in \mathscr{I}\}$ of the topological space P. For each pair U_i, U_j of open sets in the cover give a function

$$\varphi_{ij} : U_i \cap U_j \longrightarrow \mathrm{GL}(n, \mathbb{C}).$$

The functions φ_{ij} are called the *transition functions.* If the φ_{ij} satisfy the identities

$$\varphi_{ij}\varphi_{ji} = 1, \qquad \varphi_{ij}\varphi_{jk}\varphi_{ki} = 1$$

then the data is enough to specify a vector bundle on P. If P happens to be a scheme of finite type over $\mathbb{C}$ it makes sense to speak of polynomial functions and it makes sense to speak of holomorphic functions. It therefore makes sense to speak of vector bundles where the transition functions are polynomial, and of vector bundles where the transition functions are holomorphic. The ones with polynomial transition functions are called *algebraic vector bundles,* while the ones with holomorphic transition functions are called *analytic vector bundles.* The next theorem says

Theorem 1.1.4. *Let P be a scheme of finite type over $\mathbb{C}$. Assume P is compact. Let $\mathscr{V}$ be an analytic vector bundle on P. Then $\mathscr{V}$ is isomorphic to an algebraic vector bundle.*

And the last theorem asserts

Theorem 1.1.5. *Let P be a scheme of finite type over $\mathbb{C}$. Assume P is compact. Let $\mathscr{V}$ and $\mathscr{V}'$ be algebraic vector bundles on P, and let $\varphi : \mathscr{V} \longrightarrow \mathscr{V}'$ be an analytic map of vector bundles. Then φ is algebraic.*

Now that we have stated the theorems we should explain their import. To the extent that algebraic geometry concerns itself with schemes of finite type over $\mathbb{C}$, and with vector bundles over these schemes, we have learnt that, as long as the scheme P is compact, we can study it either by algebraic or by analytic methods. This is actually a very powerful, important observation. There are theorems we know how to prove by analytic methods without having an algebraic proof, and theorems for which the only known proof is algebraic. It is a little strange: we end up proving theorems in analysis using commutative algebra, and theorems in algebra using partial differential equations.

The most beautiful, intriguing parts of mathematics are those which lie at the confluence of different fields. Algebraic geometry is one of these. To be a good algebraic geometer one needs to be aware of both the algebraic and the analytic approaches to the subject. It also does not hurt to know some number theory; we will not, in this book, describe the interaction between algebraic geometry and number theory.

To be completely honest I should tell the reader that, even before Serre, it was known that algebraic geometry was a subject which lay at the intersection of algebra and analysis. Riemann knew this in the nineteenth century. Let me very briefly remind the reader of the relation between the algebraic and analytic approaches to elliptic curves; everything I say, in the remainder of the introduction, was known in the nineteenth century. None of the remainder of the introduction will be used in the rest of the book; the only purpose is to give the reader a very old, explicit example of the way algebra and analysis interact in algebraic geometry.

1.2 Elliptic curves

Let us give ourselves two complex numbers a and b, in general position (meaning not real multiples of each other). We want to consider functions periodic with the two periods a and b, that is

$$f(z+a) = f(z) = f(z+b).$$

These functions go by the name of *doubly periodic functions on* $\mathbb{C}$. Doubly periodic functions are determined by their values on a fundamental domain. There is a parallelogram in the complex plane with sides a and

b, and any doubly periodic function f is determined by its value on this parallelogram. The picture is

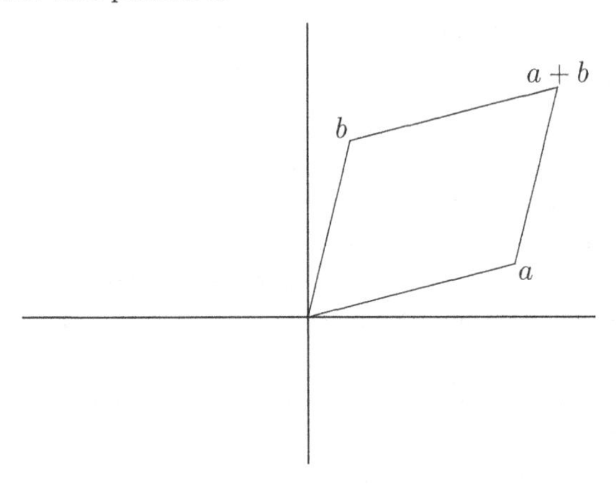

If f were holomorphic it would be bounded, and Liouville's theorem tells us that all bounded holomorphic functions are constant. A little more subtle, but also easy, is the proof that if the function f has a single simple pole, in the fundamental parallelogram above, then it must also be constant. To be interesting f would have to have a double pole.

The simplest idea on how to construct f would be to take the function $1/z^2$, which has a double pole at the origin, and add up all its translates. That is, we might be tempted to define

$$f(z) = \sum_{(m,n)\in\mathbb{Z}^2} \frac{1}{(z - ma - nb)^2} \quad .$$

The only problem with this definition is that the sum does not converge. To make it converge we have to subtract a constant from each of the summands. The definition which works is

$$\mathfrak{p}(z) \;=\; \frac{1}{z^2} + \sum_{(m,n)\neq(0,0)} \left(\frac{1}{(z - ma - nb)^2} - \frac{1}{(ma + nb)^2} \right) \quad .$$

Differentiating term by term we get

$$\mathfrak{p}'(z) \;=\; -2 \sum_{(m,n)\in\mathbb{Z}^2} \frac{1}{(z - ma - nb)^3} \quad .$$

These two infinite sums give doubly periodic meromorphic functions on $\mathbb{C}$ whose only poles are at lattice points, that is points of the form $ma+nb$ with m and n integers. It turns out not to be difficult to show that the function $\mathfrak{p}$ satisfies the differential equation

$$(\mathfrak{p}')^2 = 4\mathfrak{p}^3 - g_2\mathfrak{p} - g_3,$$

where g_2 and g_3 are constants; that is they are complex numbers. They depend on a and b, in a way which can be made explicit. In other words the map which takes $z \in \mathbb{C}$ to the pair $(\mathfrak{p}(z), \mathfrak{p}'(z)) \in \mathbb{C}^2$ is a map from $\mathbb{C}$ (modulo periods) to the set of solutions of the equation

$$y^2 = 4x^3 - g_2 x - g_3 \ . \tag{1.1}$$

Of course for z in the lattice, that is $z = ma + nb$ with m and n integers, the functions $\mathfrak{p}$ and $\mathfrak{p}'$ have poles and the map is not well defined. It turns out that, outside of the poles, we get a parametrization of solutions to the equation. In other words every solution (x, y) to Equation 1.1 is expressible in the form $(\mathfrak{p}(z), \mathfrak{p}'(z))$, and z is unique up to translation by the lattice.

This means that, if we take two solutions to Equation (1.1), we can "add" them to obtain a third solution. After all, if (x, y) and (x', y') are solutions to Equation 1.1, then there exist complex numbers z and z' with

$$(x, y) = (\mathfrak{p}(z), \mathfrak{p}'(z)) \qquad \text{and} \qquad (x', y') = (\mathfrak{p}(z'), \mathfrak{p}'(z')) \ .$$

We can form the complex number $z + z'$. As long as the number $z + z'$ is not in the lattice, that is we do not have $z + z' = ma + nb$ with m and n integers, we can form the new point $(\overline{x}, \overline{y}) = (\mathfrak{p}(z + z'), \mathfrak{p}'(z + z'))$, which is a solution to Equation 1.1. Since z and z' are well defined up to translation in the lattice, so is $z + z'$. Therefore the point $(\overline{x}, \overline{y}) = (\mathfrak{p}(z+z'), \mathfrak{p}'(z+z'))$ is unambiguously well defined. We have that the set of solutions to Equation 1.1 almost forms an abelian group. To make it an abelian group we have to formally adjoin one more "infinite" solution to the equation.

It becomes interesting to understand the group law. At the moment, in order to "add" the solutions (x, y) and (x', y'), we need to first express them as

$$(x, y) = (\mathfrak{p}(z), \mathfrak{p}'(z)) \qquad \text{and} \qquad (x', y') = (\mathfrak{p}(z'), \mathfrak{p}'(z')) \ ,$$

then form the complex number $z + z'$, and finally define the sum by the formula

$$(\overline{x}, \overline{y}) = (\mathfrak{p}(z + z'), \mathfrak{p}'(z + z')) \ .$$

The next question is can one find a simple description of the group law, one which does not explicitly use the parametrization $(x, y) = (\mathfrak{p}(z), \mathfrak{p}'(z))$?

It turns out that the group law is simple: the points (x, y), (x', y') and (x'', y'') add to zero in the group if and only if they are colinear in

$\mathbb{C}^2$. Perhaps we could rephrase this a little more explicitly. We know that there exist complex numbers z, z' and z'' with

$$(x, y) = \big(\mathfrak{p}(z), \mathfrak{p}'(z)\big)\,, \qquad (x', y') = \big(\mathfrak{p}(z'), \mathfrak{p}'(z')\big)\,,$$
$$(x'', y'') = \big(\mathfrak{p}(z''), \mathfrak{p}'(z'')\big)\,.$$

What we are asserting is that the points (x, y), (x', y') and (x'', y'') are colinear if and only if $z + z' + z''$ is in the lattice; that is

$$z + z' + z'' = ma + nb$$

with m and n integers.

So far we have described the analytic theory, which is very classical and proceeds by way of elliptic functions; the reader can find the proofs of all the assertions we made about elliptic functions, and much more, in Hurwitz and Courant's book [4]. Next we can generalize to the more modern, algebraic approach. Take Equation 1.1, but now with coefficients g_2, g_3 lying in your favorite field k. One can look at the solutions of Equation 1.1 over k, and formally adjoin one more "infinite" solution. And one can attempt to define a group law as above, using colinearity in k^2.

If the characteristic of the field k is not 2 or 3 it turns out that we get an abelian group. This is the group of points of an elliptic curve, and has been studied extensively, mainly by number theorists, for more than a century. An excellent account of the algebraic version of the theory, and its relation with number theory, may be found in Silverman [8].

1.3 Notation

It is always best to agree in advance on our notation. In this entire book all rings are commutative rings with 1. A ring homomorphism $f : R \longrightarrow S$ is a homomorphism of (commutative) rings with $f(1) = 1$. Whenever we speak about inclusions of sets, and write $A \subset B$, we mean weak inclusion; we allow the possibility that $A = B$. But when we write an inequality of real numbers in the form $a < b$ the inequality is strict. Weak inequalities will be written $a \leq b$. Given two sets A and B, we denote by $A - B$ the set of all elements of A not belonging to B.

2
Manifolds

This chapter is intended as a gentle introduction to Chapter 3. In Chapter 3 we will define schemes; by way of preparation, before we begin the technicalities, it might be helpful to take a close look at a related concept that could already be somewhat familiar, that of a manifold. The reader might be outraged, and complain that differential geometry was not listed among the prerequisites for this book. How dare I presume that the reader will find manifolds familiar?

My answer is twofold: first, this chapter was written in such a way that it should be readable, even by the reader who has never before met a manifold. And second, we do only a very minimal amount of differential geometry. In this chapter we will not go beyond the definition of a manifold, and the definition of $\mathcal{C}^k$–functions on $\mathcal{C}^k$–manifolds. Even a reader without much background in differential geometry might have seen this much.

The idea of a scheme, which will occupy us from Chapter 3 on, mimics that of a manifold; but to make the parallel transparent it helps to start with the right definition of a manifold. With the right definitions the formalisms really are precisely the same, not just similar. In this chapter we treat manifolds. We will start with the traditional definition of a manifold, then modify it slightly. In Chapter 3 we then proceed to give the parallel definition of a scheme.

2.1 Manifolds defined in the traditional way

The usual way to define a manifold is as a topological space, locally homeomorphic to $\mathbb{R}^n$. We remind the reader.

Definition 2.1.1. *An n–dimensional topological manifold, with an at-*

las, *is a topological space M, with an open cover $M = \bigcup_{i\in I} U_i$, such that each U_i is homeomorphic to $\mathbb{R}^n$.*

When we just say "topological manifold" we mean an n–dimensional topological manifold for some unspecified positive integer n. Topological manifolds are interesting, but one frequently wants to consider more special manifolds with extra structure. As the reader might already know, manifolds come in many delightful versions, such as $\mathcal{C}^k$–manifolds or $\mathcal{C}^\infty$–manifolds. In this section we remind the reader of the usual definitions. In the remainder of the chapter we then modify them slightly to make the generalization to schemes work more smoothly. But first a little reminder.

Definition 2.1.2. *Let $k > 0$ be an integer, and let $U \subset \mathbb{R}^n$ be an open set. A function $\varphi : U \longrightarrow \mathbb{R}^m$ is $\mathcal{C}^k$ if and only if all the partial derivatives, up to order k, exist and are continuous.*

Now a $\mathcal{C}^k$–manifold is first of all a topological manifold, that is there exists an open cover as in Definition 2.1.1. In order for a topological manifold to be a $\mathcal{C}^k$–manifold an open cover must be given, that is $M = \bigcup_{i\in I} U_i$, for which the transition functions are $\mathcal{C}^k$. More formally, we have

Definition 2.1.3. *An* n–dimensional $\mathcal{C}^k$–manifold, with an atlas, *is a topological space M, together with an open cover $M = \bigcup_{i\in I} U_i$, such that*

(i) *Each U_i comes with a homeomorphism $\varphi_i : \mathbb{R}^n \longrightarrow U_i$.*
(ii) *The maps φ_i and φ_j give two homeomorphisms of the open set $U_i \cap U_j$ with open subsets of $\mathbb{R}^n$. The induced homeomorphism from $\varphi_j^{-1}(U_i \cap U_j) \subset \mathbb{R}^n$ to $\varphi_i^{-1}(U_i \cap U_j) \subset \mathbb{R}^n$ must be $\mathcal{C}^k$; see Definition 2.1.2.*

Remark 2.1.4. Perhaps a picture might help. We have a diagram of spaces

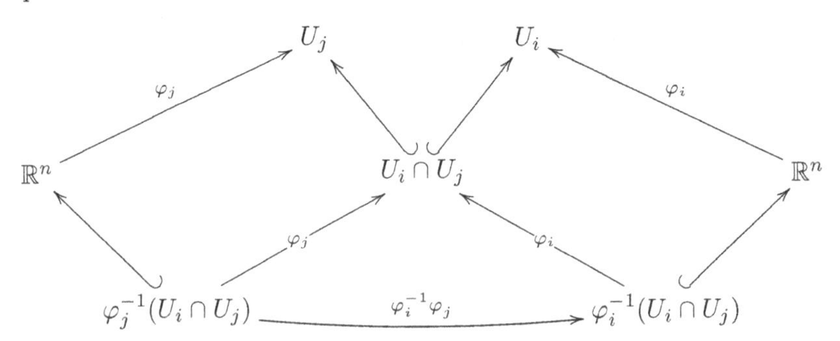

The functions $\varphi_i^{-1}\varphi_j$, which take the open set $\varphi_j^{-1}(U_i \cap U_j) \subset \mathbb{R}^n$ to the open set $\varphi_i^{-1}(U_i \cap U_j) \subset \mathbb{R}^n$, are called the "transition functions", and are often denoted φ_{ji}. What is being asserted is that all the maps $\varphi_{ji} = \varphi_i^{-1}\varphi_j$ are k–times continuously differentiable.

Remark 2.1.5. The usual definition of an n–dimensional $\mathcal{C}^k$–manifold is as a $\mathcal{C}^k$–isomorphism class of n–dimensional $\mathcal{C}^k$–manifolds with atlases. In Example 2.2.13 we will remind the reader what is a $\mathcal{C}^k$–mapping $f : M \longrightarrow N$, of $\mathcal{C}^k$–manifolds with atlases; once we know what such a map is, a $\mathcal{C}^k$–isomorphism is just a $\mathcal{C}^k$–map $f : M \longrightarrow N$ with a $\mathcal{C}^k$–inverse $f^{-1} : N \longrightarrow M$.

Now we come to a couple more variants. We have

Definition 2.1.6. *An* n–dimensional $\mathcal{C}^\infty$–manifold, with an atlas, *is a topological space* M, *together with an open cover* $M = \bigcup_{i\in I} U_i$, *such that*

(i) *Each* U_i *comes with a homeomorphism* $\varphi_i : \mathbb{R}^n \longrightarrow U_i$.
(ii) *The transition functions* $\varphi_{ji} = \varphi_i^{-1}\varphi_j$ *of Remark 2.1.4 are all continuously differentiable to infinite order; that is, arbitrarily high partial derivatives exist and are continuous.*

Definition 2.1.7. *An* n–dimensional real analytic manifold, with an atlas, *is a topological space* M, *together with an open cover* $M = \bigcup_{i\in I} U_i$, *such that*

(i) *Each* U_i *comes with a homeomorphism* $\varphi_i : \mathbb{R}^n \longrightarrow U_i$.
(ii) *The transition functions* $\varphi_{ji} = \varphi_i^{-1}\varphi_j$ *of Remark 2.1.4 are all real analytic, that is the Taylor series converges to the function in a neighborhood of every point.*

For a complex (instead of real) variant we have

Definition 2.1.8. *An* n–dimensional complex manifold, with an atlas, *is a topological space* M, *together with an open cover* $M = \bigcup_{i\in I} U_i$, *such that*

(i) *Each* U_i *comes with a homeomorphism* $\varphi_i : B \longrightarrow U_i$, *where* $B \subset \mathbb{C}^n$ *is the unit ball in complex* n*–space* $\mathbb{C}^n$.
(ii) *The transition functions* $\varphi_{ji} = \varphi_i^{-1}\varphi_j$ *of Remark 2.1.4 are all holomorphic.*

2.2 Sheaves of rings and ringed spaces

As we have said, at the beginning of the chapter, we want to slightly modify the definition of manifolds, to make the analogy with schemes more transparent. The starting point of our modification is that we want a definition which is not explicitly in terms of transition functions. Transition functions depend on a choice of an open cover for M and we want a cleaner, more intrinsic formulation. To achieve this we have to introduce a new concept, that of a sheaf.

So our first project is to understand sheaves; this section is devoted to developing the concept. Once we have understood sheaves we can then turn around and use them to define manifolds, which we will do in Section 2.4. Before all else it might be best to start with an example of a sheaf, so we can have in mind something concrete to relate to. The example we will consider is built from manifolds, but do not worry; the argument is not circular. If you prefer to treat manifolds as something which has not yet been defined, and will only come in Section 2.4, then in Example 2.2.2 let M be an open set in $\mathbb{R}^n$, rather than an arbitrary $\mathcal{C}^k$–manifold with an atlas.

We begin with a little reminder, about differentiable functions.

Reminder 2.2.1. If M is a topological space and $f : M \longrightarrow \mathbb{R}$ is a function, then in general we have no idea how to differentiate f; to form a directional derivative in the direction h one would have to look at something like

$$\lim_{t \longrightarrow 0} \frac{f(x+th) - f(x)}{t} ,$$

and, unless there is a special feature to the topological space M, we have no idea what is meant by $x + th$. If $M = \mathbb{R}^n$ then there is, of course, no problem. If M is a $\mathcal{C}^k$–manifold with an atlas then it is locally homeomorphic to $\mathbb{R}^n$, allowing us to overcome the difficulty. To make our life a little easier let us not try to obtain the derivative of a function f, but confine ourselves to the easier task of deciding whether f is differentiable.

Therefore let M be a $\mathcal{C}^k$–manifold with an atlas, let $U \subset M$ be an open set, and let $f : U \longrightarrow \mathbb{R}$ be a function. Definition 2.1.3 gives us an open cover $M = \bigcup_{i \in I} U_i$, homeomorphisms $\varphi_i : \mathbb{R}^n \longrightarrow U_i$, and we are given that the transition functions $\varphi_i^{-1}\varphi_j$ are all $\mathcal{C}^k$. We say that the function $f : U \longrightarrow \mathbb{R}$ is k–times continuously differentiable at a point $x \in U$ if, when we choose a U_i in the given cover of M with $x \in U_i$, then the composite map

$$\mathbb{R}^n \xrightarrow{\ \varphi_i\ } U_i \xrightarrow{\ f\ } \mathbb{R}$$

is k–times continuously differentiable at $\varphi_i^{-1}(x) \in \mathbb{R}^n$. Of course it is possible that x is contained in more than one U_i; let us suppose $x \in U_i \cap U_j$. We need to convince ourselves that if $f\varphi_i$ is k–times continuously differentiable at $\varphi_i^{-1}(x)$, then $f\varphi_j$ is k–times continuously differentiable at $\varphi_j^{-1}(x)$. For this observe the identity

$$f\varphi_j = \{f\varphi_i\}\{\varphi_i^{-1}\varphi_j\} \ ;$$

we are assuming that the function $f\varphi_i$ is k–times continuously differentiable at $\varphi_i^{-1}(x)$, while the transition function $\varphi_i^{-1}\varphi_j$ is k–times continuously differentiable everywhere, by hypothesis. The chain rule tells us that the composite is k–times continuously differentiable at $\varphi_j^{-1}(x)$; the chain rule expresses each partial derivative of the composite as a sum of products of the partial derivatives.

The function $f : U \longrightarrow \mathbb{R}$ is defined to be $\mathcal{C}^k$ if it is k–times continuously differentiable at every point $x \in U$.

With this reminder out of the way, we now give our motivating example of a sheaf of rings.

Example 2.2.2. Let M be a $\mathcal{C}^k$–manifold with an atlas. For every open subset $U \subset M$, Reminder 2.2.1 gives us a subset of the set of all functions $f : U \longrightarrow \mathbb{R}$, namely the $\mathcal{C}^k$–functions. We adopt the notation

$$\Gamma(U, \mathcal{O}_M) \quad = \quad \{f : U \longrightarrow \mathbb{R} \mid f \text{ is } \mathcal{C}^k\} \ .$$

For each U the set $\Gamma(U, \mathcal{O}_M)$ is a ring; the addition and multiplication are defined by the formulas

$$(f+g)(x) = f(x) + g(x)\,, \qquad (f \cdot g)(x) = f(x) \cdot g(x)\,.$$

Furthermore, for any pair of open subsets $U \subset V \subset M$, the functions $f : V \longrightarrow \mathbb{R}$ can be restricted to the open subset U; we send f the composite $U \longrightarrow V \xrightarrow{f} \mathbb{R}$. In the notation above we obtain a restriction map

$$\mathrm{res}_U^V : \Gamma(V, \mathcal{O}_M) \longrightarrow \Gamma(U, \mathcal{O}_M)$$

which is a ring homomorphism. What we propose to do next is give the axioms defining a sheaf of rings on a topological space M. The example to keep in mind, which might help, is the sheaf which takes an open subset $U \subset M$ to the ring $\Gamma(U, \mathcal{O}_M)$.

Before defining a sheaf we begin with the slightly easier notion of a presheaf.

Definition 2.2.3. *Let M be a topological space. A* presheaf of rings $\mathcal{O}$ *on M is*

(i) *To every open set $U \subset M$ we assign a ring $\Gamma(U, \mathcal{O})$.*
(ii) *For every inclusion of open sets $U \subset V \subset M$ there is a restriction homomorphism*

$$\mathrm{res}_U^V : \Gamma(V, \mathcal{O}) \longrightarrow \Gamma(U, \mathcal{O}) .$$

(iii) *This data satisfies*
  (a) *If $U \subset V \subset W$ then $\mathrm{res}_U^W = \mathrm{res}_U^V \circ \mathrm{res}_V^W$.*
  (b) *res_U^U is the identity map.*
  (c) *$\Gamma(\emptyset, \mathcal{O}) = 0$.*

Remark 2.2.4. It is possible to rephrase the definition of a presheaf using the language of category theory. In this remark I will give the reformulation; the readers who have never seen category theory before can safely ignore the remark.

In the language of category theory the collection of open subsets of M forms a category $\mathfrak{Open}(M)$. The objects of $\mathfrak{Open}(M)$ are the open sets in M, and the morphisms are the inclusions. A presheaf of rings is simply a contravariant functor $\mathfrak{Open}(M) \longrightarrow \mathfrak{Rings}$, taking the initial object $\emptyset$ of $\mathfrak{Open}(M)$ to the terminal object 0 in the category of rings.

Remark 2.2.5. By changing the word 'ring' in Definition 2.2.3 into group, abelian group, module, etc, we would have a presheaf of groups, abelian groups, modules, and so on. We leave the definitions to the reader.

Remark 2.2.6. There is a little bit of jargon that comes with the terrain, and it probably will do us no harm to become accustomed to it. Given a presheaf $\mathcal{O}$ (of rings, or groups, or modules...) on the topological space M, it is customary to refer to an element $s \in \Gamma(U, \mathcal{O})$ as a *section of the presheaf $\mathcal{O}$ over the open set U*. In the special case where $U = M$ is the whole space, a section $s \in \Gamma(M, \mathcal{O})$ is called a *global section of $\mathcal{O}$*.

Definition 2.2.7. *A* sheaf of rings *on a topological space M is a presheaf $\mathcal{O}$ that satisfies two additional properties*

(i) *Suppose $U = \cup U_i$ is an open cover of U. If $s \in \Gamma(U, \mathcal{O})$ and $\mathrm{res}_{U_i}^U(s) = 0$ for all U_i, then $s = 0$.*

(ii) *Suppose* $U = \cup U_i$, *and we are given* $s_i \in \Gamma(U_i, \mathcal{O})$ *so that, for all pairs* i *and* j, $\mathrm{res}^{U_i}_{U_i \cap U_j}(s_i) = \mathrm{res}^{U_j}_{U_i \cap U_j}(s_j)$. *Then there exists* $s \in \Gamma(U, \mathcal{O})$ *such that* $\mathrm{res}^U_{U_i}(s) = s_i$ *for each* i.

Remark 2.2.8. Restating Definition 2.2.7 in terms of the jargon we introduced in Remark 2.2.6, we have that a presheaf $\mathcal{O}$ is a sheaf if and only if

(i) Any section s, of the presheaf $\mathcal{O}$ over the open set U, which vanishes when restricted to an open cover U_i of U, vanishes on all of U.

(ii) Suppose $U = \cup U_i$, and suppose we are given sections s_i of the presheaf $\mathcal{O}$ over the open sets U_i. If the sections are compatible, that is if for all pairs i and j the restriction of s_i to $U_i \cap U_j$ agrees with the restriction of s_j to $U_i \cap U_j$, then they extend to a section on all of U. There exists a section s of the presheaf $\mathcal{O}$, over the open set U, restricting to the s_i.

Remark 2.2.9. Note that part (i) of Definition 2.2.7 means that the section s constructed in part (ii) is unique.

The motivation for the definition is that sheaves link the local and the global; they are globally defined, yet local in nature. Parts (i) and (ii) of Definition 2.2.7 turn out to be exactly what is needed to make sure that sections of a sheaf $\mathcal{O}$ on the open set U can be studied by replacing U by an open cover. They can be studied locally.

Let M be a topological space, V an open subset and $V = \cup_{i \in I} U_i$ an open cover. For any presheaf $\mathcal{O}$ of abelian groups, rings or modules on M we have restriction maps

$$\Gamma(V, \mathcal{O}) \longrightarrow \Gamma(U_i, \mathcal{O}) , \qquad \Gamma(U_i, \mathcal{O}) \longrightarrow \Gamma(U_i \cap U_j, \mathcal{O}) .$$

They can be assembled to

$$\Gamma(V, \mathcal{O}) \xrightarrow{\alpha} \prod_{i \in I} \Gamma(U_i, \mathcal{O}) \xrightarrow{\beta} \prod_{i,j \in I} \Gamma(U_i \cap U_j, \mathcal{O}) .$$

Here, α is the map taking $s \in \Gamma(V, \mathcal{O})$ to the product of its restrictions, and β takes the family $\{s_i \in \Gamma(U_i, \mathcal{O})\}$ to the family of

$$\mathrm{res}^{U_i}_{U_i \cap U_j}(s_i) - \mathrm{res}^{U_j}_{U_i \cap U_j}(s_j) \qquad \in \qquad \Gamma(U_i \cap U_j, \mathcal{O}) .$$

With this notation we will prove the following lemma, which is definitely useful and might be illuminating.

Lemma 2.2.10. *The presheaf $\mathcal{O}$ over the topological space M is a sheaf if and only if, for every open subset $V \subset M$ and any open cover $V = \cup_{i\in I} U_i$, the maps constructed above give an exact sequence*

$$0 \longrightarrow \Gamma(V,\mathcal{O}) \xrightarrow{\ \alpha\ } \prod_{i\in I}\Gamma(U_i,\mathcal{O}) \xrightarrow{\ \beta\ } \prod_{i,j\in I}\Gamma(U_i\cap U_j,\mathcal{O}) .$$

Proof. The statement that α is injective is equivalent to the condition of Definition 2.2.7(i), while the statement that an element of the kernel of β lies in the image of α is equivalent to Definition 2.2.7(ii). □

We now know what is a sheaf of rings on a topological space X, and we have at least one example to bear in mind; any $\mathcal{C}^k$–manifold M carries a sheaf of rings $\mathcal{O}_M$, the sheaf of k–times continuously differentiable functions, as in Example 2.2.2. A pair, consisting of a topological space X and a sheaf of rings $\mathcal{O}$ over it, will be called a ringed space. We formalize this as a definition.

Definition 2.2.11. *A* ringed space $(X,\mathcal{O})$ *is a topological space X together with a sheaf of rings $\mathcal{O}$ on X.*

Definition 2.2.12. *A* morphism of ringed spaces, *from the ringed space $(X,\mathcal{O}_X)$ to the ringed space $(Y,\mathcal{O}_Y)$, is a pair (φ,φ^*), where*

(i) *$\varphi : X \longrightarrow Y$ is a continuous map.*
(ii) *For every open set $U \subset Y$, we have an associated ring homomorphism $\varphi_U^* : \Gamma(U,\mathcal{O}_Y) \longrightarrow \Gamma(\varphi^{-1}U,\mathcal{O}_X)$.*
(iii) *The maps φ^* must be compatible with the restriction maps of the sheaves $\mathcal{O}_X$ and $\mathcal{O}_Y$. That is, given an inclusion of open sets $U \subset V \subset Y$, the following square must commute*

$$\begin{array}{ccc} \Gamma(V,\mathcal{O}_Y) & \xrightarrow{\ \varphi_V^*\ } & \Gamma(\varphi^{-1}V,\mathcal{O}_X) \\ {\scriptstyle \mathrm{res}_U^V}\downarrow & & \downarrow{\scriptstyle \mathrm{res}_{\varphi^{-1}U}^{\varphi^{-1}V}} \\ \Gamma(U,\mathcal{O}_Y) & \xrightarrow[\ \varphi_U^*\ ]{} & \Gamma(\varphi^{-1}U,\mathcal{O}_X) \end{array} .$$

Example 2.2.13. We will soon show how, starting with a $\mathcal{C}^k$–map $f : X \longrightarrow Y$ of $\mathcal{C}^k$–manifolds with atlases, one can construct an induced map of ringed spaces. But first let us briefly remind the reader what is a $\mathcal{C}^k$–map $f : X \longrightarrow Y$.

Let X and Y be $\mathcal{C}^k$–manifolds with atlases. Because X and Y are $\mathcal{C}^k$–manifolds with atlases, Definition 2.1.3 gives us open covers $X = \cup_{i\in I} U_i$ and $Y = \cup_{j\in J} V_j$, together with homeomorphisms $\varphi_i : \mathbb{R}^m \longrightarrow U_i$ and

$\psi_j : \mathbb{R}^n \longrightarrow V_j$. A continuous map $f : X \longrightarrow Y$ is called $\mathcal{C}^k$ if all the continuous maps

$$\varphi_i^{-1}(U_i \cap f^{-1}V_j) \xrightarrow{\ \varphi_i\ } U_i \cap f^{-1}V_j \xrightarrow{\ f\ } V_j \xrightarrow{\ \psi_j^{-1}\ } \mathbb{R}^n$$

are k–times continuously differentiable; this makes sense because each $\varphi_i^{-1}(U_i \cap f^{-1}V_j)$ is an open subset of $\mathbb{R}^m$, and we can differentiate maps from open subsets of $\mathbb{R}^m$ to $\mathbb{R}^n$.

Now that we have reminded the reader what the terms mean, suppose we are given a $\mathcal{C}^k$–map of $\mathcal{C}^k$–manifolds with atlases $f : X \longrightarrow Y$. Example 2.2.2 gives us two ringed spaces, $(X, \mathcal{O}_X)$ and $(Y, \mathcal{O}_Y)$, where the sheaves $\mathcal{O}_X$ and $\mathcal{O}_Y$ are the sheaves of $\mathcal{C}^k$–functions on the respective manifolds with atlases. To get a morphism of ringed spaces we need to construct a pair (f, f^*). The map $f : X \longrightarrow Y$ is given to us; it is the given $\mathcal{C}^k$–map of $\mathcal{C}^k$–manifolds with atlases. The definition of

$$f_U^* : \Gamma(U, \mathcal{O}_Y) \longrightarrow \Gamma(f^{-1}U, \mathcal{O}_X)$$

is that it takes $g \in \Gamma(U, \mathcal{O}_Y)$, that is $g : U \longrightarrow \mathbb{R}$, to the composite

$$f^{-1}U \xrightarrow{\ f\ } U \xrightarrow{\ g\ } \mathbb{R}\ .$$

We need to prove that this is an element in $\Gamma(f^{-1}U, \mathcal{O}_X)$; in other words we must show that the composite gf is k–times continuously differentiable. That is, at any point $p \in f^{-1}U$ we must show that the composite is k–times continuously differentiable at p. Using the atlases we may find a neighborhood of $p \in f^{-1}U$ which is homeomorphic to an open subset $V \subset \mathbb{R}^m$, a neighborhood of $f(p) \in U$ which is homeomorphic to an open subset $W \subset \mathbb{R}^n$, and the problem is to show that the composite

$$V \xrightarrow{\ f\ } W \xrightarrow{\ g\ } \mathbb{R}$$

is $\mathcal{C}^k$. What we are given is that $f : V \longrightarrow W$ and $g : W \longrightarrow \mathbb{R}$ are each $\mathcal{C}^k$. As in Reminder 2.2.1, the fact that the composite is $\mathcal{C}^k$ comes down to the chain rule of calculus.

2.3 There are not many maps of ringed spaces

In Examples 2.2.2 and 2.2.13 we learned how, starting with a $\mathcal{C}^k$–map of $\mathcal{C}^k$–manifolds with atlases $f : X \longrightarrow Y$, one can construct a map of ringed spaces (f, f^*) from the ringed space $(X, \mathcal{O}_X)$ to the ringed space $(Y, \mathcal{O}_Y)$. The main theorem of this section is that the maps of Examples 2.2.13 are the only maps between these ringed spaces.

The statement above is correct, but since we want to use it in the definition of manifolds we will pretend that we have no idea what a manifold is. In this section we will talk only about open subsets of $\mathbb{R}^n$. That is, we make the following trivial observation, which is a special case of Examples 2.2.2 and 2.2.13:

Remark 2.3.1. Let M be an open subset in $\mathbb{R}^n$. We can define a sheaf of rings $\mathcal{O}_M$ on M by letting

$$\Gamma(U, \mathcal{O}_M) \quad = \quad \{g : U \longrightarrow \mathbb{R} \mid g \text{ is } \mathcal{C}^k\} \ ,$$

for all open subsets $U \subset M$. Note that on open subsets $U \subset \mathbb{R}^n$ there is absolutely no difficulty in defining what it means for g to be k–times continuously differentiable.

Furthermore, if M is an open subset of $\mathbb{R}^m$, N is an open subset of $\mathbb{R}^n$ and $f : M \longrightarrow N$ is a k–times continuously differentiable function, then there is no problem at all in defining the map (f, f^*) from the ringed space $(M, \mathcal{O}_M)$ to the ringed space $(N, \mathcal{O}_N)$; for all open subsets $U \subset N$, the ring homomorphism

$$f^*_U : \Gamma(U, \mathcal{O}_N) \longrightarrow \Gamma(f^{-1}U, \mathcal{O}_M)$$

takes $g : U \longrightarrow \mathbb{R}$ to $gf : f^{-1}U \longrightarrow \mathbb{R}$. This is, of course, a special case of Example 2.2.13; but in the special case of open subsets of $\mathbb{R}^m$ and $\mathbb{R}^n$ the assertions really are trivial.

The main theorem states

Theorem 2.3.2. *Let M be an open subset of $\mathbb{R}^m$ and let N be an open subset of $\mathbb{R}^n$. Suppose*

$$(f, f^*) : (M, \mathcal{O}_M) \longrightarrow (N, \mathcal{O}_N)$$

is a map of ringed spaces, where $\mathcal{O}_M$ and $\mathcal{O}_N$ are the sheaves of k–times continuously differentiable functions, as in the first paragraph of Remark 2.3.1. Then $f : M \longrightarrow N$ is a $\mathcal{C}^k$–map, and the ring homomorphisms

$$f^*_U : \Gamma(U, \mathcal{O}_N) \longrightarrow \Gamma(f^{-1}U, \mathcal{O}_M)$$

are the homomorphisms of the second paragraph of Remark 2.3.1, taking g to gf.

Proof. We will prove this by a sequence of lemmas, which will occupy the rest of this section. Since the rest of the section is devoted to the proof of Theorem 2.3.2, our notation throughout this section is as in the statement of the theorem.

Lemma 2.3.3. *Let $U \subset N$ be an open subset. The ring homomorphism*

$$f_U^* : \Gamma(U, \mathcal{O}_N) \longrightarrow \Gamma(f^{-1}U, \mathcal{O}_M)$$

preserves order, in the weak sense that, if g and g' are elements of $\Gamma(U, \mathcal{O}_N)$ with $g < g'$, then $f_U^* g \le f_U^* g'$.

Proof. Suppose $g < g'$; then $\{g' - g\}(x) > 0$ for all $x \in U$. But $g' - g : U \longrightarrow \mathbb{R}$ has a square root: there is a function $h : U \longrightarrow \mathbb{R}$ with $h^2 = g' - g$. If $g' - g$ is k–times continuously differentiable then so is h. Now because f_U^* is a ring homomorphism $f_U^*(g' - g) = f_U^*(h^2) = f_U^*(h)^2 \ge 0$. □

Lemma 2.3.4. *Let $U \subset N$ be an open subset. The ring homomorphism*

$$f_U^* : \Gamma(U, \mathcal{O}_N) \longrightarrow \Gamma(f^{-1}U, \mathcal{O}_M)$$

is a homomorphism of $\mathbb{R}$–algebras.

Proof. Because f_U^* is a ring homomorphism, $f_U^*(1) = 1$. But then $f_U^*(n) = n$ for all $n \in \mathbb{Z}$, and if we take m, n in $\mathbb{Z}$ with $n \ne 0$, we have

$$f_U^*\left(\frac{m}{n}\right) = f_U^*(mn^{-1}) = f_U^*(m) f_U^*(n^{-1}) = \frac{m}{n} ,$$

which proves that f_U^* is a homomorphism of $\mathbb{Q}$–algebras. It remains to deduce that f_U^* acts as the identity on all real numbers. Let $a \in \mathbb{R}$. If a is rational we already know that $f_U^*(a) = a$, hence assume a irrational. For any $\ell \in \mathbb{Z}$ there exists an integer n with

$$\frac{n}{2^\ell} \quad < \quad a \quad < \quad \frac{n+1}{2^\ell} \quad .$$

Lemma 2.3.3 allows us to deduce

$$f_U^*\left(\frac{n}{2^\ell}\right) \quad \le \quad f_U^*(a) \quad \le \quad f_U^*\left(\frac{n+1}{2^\ell}\right) \quad ,$$

that is

$$\frac{n}{2^\ell} \quad \le \quad f_U^*(a) \quad \le \quad \frac{n+1}{2^\ell} \quad .$$

Letting ℓ go to infinity we conclude $f_U^*(a) = a$. □

Lemma 2.3.5. *Let $U \subset N$ be an open subset. Let g be an element of $\Gamma(U, \mathcal{O}_N)$; that is, $g : U \longrightarrow \mathbb{R}$ is a $\mathcal{C}^k$–function. Then*

$$\{f_U^* g\}^{-1}(0) \quad \subset \quad \{gf\}^{-1}(0) \, .$$

Proof. We want to prove that

$$\{\{f_U^* g\}(x) = 0\} \implies \{gf(x) = 0\} ;$$

equivalently, we will show that

$$\{gf(x) \neq 0\} \implies \{\{f_U^* g\}(x) \neq 0\} .$$

Consider the open set $U_g = g^{-1}\big(\mathbb{R} - \{0\}\big)$. That is,

$$U_g = \{y \in U \mid g(y) \neq 0\} .$$

Then $f^{-1}U_g$ is precisely the set

$$f^{-1}U_g = \{x \in f^{-1}U \mid gf(x) \neq 0\} .$$

We need to show that $f_U^* g$ does not vanish anywhere on $f^{-1}U_g$.

But now on U_g the function g has a reciprocal†; the function $1/g$ is k–times continuously differentiable where $g \neq 0$. To be careful with the notation, what we have is a function $h \in \Gamma(U_g, \mathcal{O}_N)$ so that

$$h \cdot \mathrm{res}_{U_g}^U(g) = 1 .$$

Next we apply the ring homomorphism $f_{U_g}^*$ and we discover, in the ring $\Gamma(f^{-1}U_g, \mathcal{O}_M)$, the identity

$$\{f_{U_g}^* h\} \cdot \{f_{U_g}^* \mathrm{res}_{U_g}^U(g)\} = 1 ,$$

which means that $f_{U_g}^* \mathrm{res}_{U_g}^U(g)$ vanishes nowhere on $f^{-1}U_g$. But by Definition 2.2.12(iii) we know that

$$f_{U_g}^* \mathrm{res}_{U_g}^U(g) = \mathrm{res}_{f^{-1}U_g}^{f^{-1}U} f_U^*(g) ,$$

and this tells us that the restriction of $f_U^* g$ to $f^{-1}U_g \subset f^{-1}U$ has no zero. □

Lemma 2.3.6. *Let $U \subset N$ be an open subset. Let g be an element of $\Gamma(U, \mathcal{O}_N)$; that is, $g : U \longrightarrow \mathbb{R}$ is a $\mathcal{C}^k$–function. Then*

$$f_U^* g = gf .$$

Proof. Let a be any real number. We wish to apply Lemma 2.3.5 to the function $h = g - a \in \Gamma(U, \mathcal{O}_N)$. By Lemma 2.3.4 we know that $f_U^* a = a$, and hence

$$f_U^* h = f_U^*(g - a) = f_U^* g - f_U^* a = \{f_U^* g\} - a .$$

† It would be more in keeping with tradition to use the word "inverse" rather than "reciprocal", but in this lemma the symbol g^{-1} is already being overworked

Lemma 2.3.5 asserts that

$$\{f_U^*h(x) = 0\} \quad \Longrightarrow \quad \{hf(x) = 0\}\ ,$$

or in other words

$$\{f_U^*g(x) = a\} \quad \Longrightarrow \quad \{gf(x) = a\}\ .$$

The statement being true for every $a \in \mathbb{R}$, we deduce $f_U^*g = gf$. □

To prove Theorem 2.3.2, it remains only to show that the mapping $f : M \longrightarrow N$ is $\mathcal{C}^k$; Lemma 2.3.6 already guarantees that the maps

$$f_U^* : \Gamma(U, \mathcal{O}_N) \longrightarrow \Gamma(f^{-1}U, \mathcal{O}_M)$$

are the homomorphisms of the second paragraph of Remark 2.3.1, taking g to gf.

Now we complete the proof of Theorem 2.3.2: in the special case, where $U = N$, the above tells us that, on the level of global sections (see Remark 2.2.6 for the jargon),

$$f_N^* : \Gamma(N, \mathcal{O}_N) \longrightarrow \Gamma(M, \mathcal{O}_M)$$

is the map sending g to gf. This means that, for any $\mathcal{C}^k$–function $g : N \longrightarrow \mathbb{R}$, the composite gf must be an element of the ring $\Gamma(M, \mathcal{O}_M)$; it is a $\mathcal{C}^k$–function on M. Next we will apply this observation to some particular choices of g, to deduce the fact that f is a $\mathcal{C}^k$–map.

We know that $N \subset \mathbb{R}^n$ is an open subset, and the coordinate projections $\pi_i : \mathbb{R}^n \longrightarrow \mathbb{R}$, taking the point $(y_1, y_2, \ldots, y_n) \in \mathbb{R}^n$ to $y_i \in \mathbb{R}$, are clearly $\mathcal{C}^k$–functions on $N \subset \mathbb{R}^n$. By the above the composite $f_i = \pi_i f$, of the map $f : M \longrightarrow N$ with the coordinate projection π_i, is a $\mathcal{C}^k$–function on M. But the map f is just the map taking $m \in M$ to the point $\big(f_1(m), f_2(m), \ldots, f_n(m)\big) \in \mathbb{R}^n$; the fact that the f_i are k–times continuously differentiable means that so is f. □

2.4 The sheaf theoretic definition of a manifold

In this section we give the modified definition of a manifold, which we have been promising all chapter. But first we need an easy fact.

Proposition 2.4.1. *Let X be a topological space, and let $V \subset X$ be an open subset. Let $\mathcal{O}$ be a sheaf (of rings, abelian groups...) on X. Then defining, for every open subset $U \subset V$,*

$$\Gamma\big(U, \mathcal{O}|_V\big) \quad = \quad \Gamma(U, \mathcal{O})$$

yields a sheaf $\mathcal{O}|_V$ on V.

Proof. The axioms of a sheaf are all assertions that for all open subsets something happens. There are fewer open subsets in V than in X, and hence the statements about open subsets of V follow immediately from the statements about the open subsets of X. □

Definition 2.4.2. *As in Proposition 2.4.1 let X be a topological space, and let $V \subset X$ be an open subset. Suppose $\mathcal{O}$ is a sheaf on X. Then the sheaf $\mathcal{O}|_V$ on V, constructed in Proposition 2.4.1, is called* the restriction of the sheaf $\mathcal{O}$ to V. *If $(X, \mathcal{O})$ is a ringed space and $V \subset X$ is an open subset, then the open subset has a canonical structure of a ringed space $(V, \mathcal{O}|_V)$.*

Now we are ready for the definition of a manifold.

Definition 2.4.3. *An* n–dimensional $\mathcal{C}^k$–manifold *is a ringed space $(X, \mathcal{O})$, which is locally (as a ringed space) isomorphic to $(\mathbb{R}^n, \mathcal{O}_{\mathbb{R}^n})$. Here $\mathcal{O}_{\mathbb{R}^n}$ stands for the sheaf of $\mathcal{C}^k$–functions on $\mathbb{R}^n$, as in Remark 2.3.1.*

Remark 2.4.4. It might be helpful to elaborate a little on Definition 2.4.3. In order for the ringed space $(X, \mathcal{O})$ to be a $\mathcal{C}^k$–manifold we need the existence of the following:

(i) An open cover $X = \cup U_i$.

(ii) For every open set U_i in the cover, we need an isomorphism of ringed spaces

$$(\varphi_i, \varphi_i^*) : (\mathbb{R}^n, \mathcal{O}_{\mathbb{R}^n}) \longrightarrow (U_i, \mathcal{O}|_{U_i}) \quad ,$$

where $\mathcal{O}|_{U_i}$ is the restriction of the sheaf $\mathcal{O}$, to the open set $U_i \subset X$, described in Proposition 2.4.1 and Definition 2.4.2, while the sheaf $\mathcal{O}_{\mathbb{R}^n}$ on $\mathbb{R}^n$ is the sheaf of $\mathcal{C}^k$–functions as in Remark 2.3.1.

The choice of such a cover $X = \cup U_i$ is an atlas for X. Note that the sheaf definition is cleaner in that the atlas is not part of the given data. Anyway, to see that the definition agrees with the traditional one we observe:

Proposition 2.4.5. *For the open cover $X = \cup U_i$, as in Remark 2.4.4(i) above, the transition functions are automatically $\mathcal{C}^k$.*

Proof. We have a diagram of maps of ringed spaces, where the vertical maps are isomorphisms

$$\begin{array}{ccccc} (\mathbb{R}^n, \mathcal{O}_{\mathbb{R}^n}) & & & & (\mathbb{R}^n, \mathcal{O}_{\mathbb{R}^n}) \\ {\scriptstyle (\varphi_j, \varphi_j^*)}\downarrow & & & & \downarrow{\scriptstyle (\varphi_i, \varphi_i^*)} \\ (U_j, \mathcal{O}|_{U_j}) & \longleftarrow & (U_i \cap U_j, \mathcal{O}|_{U_i \cap U_j}) & \longrightarrow & (U_i, \mathcal{O}|_{U_i}) \end{array}$$

Let $M = \varphi_j^{-1}(U_i \cap U_j)$ and $N = \varphi_i^{-1}(U_i \cap U_j)$, both of which are open subsets in $\mathbb{R}^n$. Let

$$\mathcal{O}_M = \{\mathcal{O}_{\mathbb{R}^n}\}|_M , \qquad \mathcal{O}_N = \{\mathcal{O}_{\mathbb{R}^n}\}|_N ,$$

that is the restrictions to the open sets M and N of the sheaf $\mathcal{O}_{\mathbb{R}^n}$ on $\mathbb{R}^n$. The left half of the diagram gives an isomorphism of ringed spaces

$$(M, \mathcal{O}_M) \longrightarrow (U_i \cap U_j, \mathcal{O}|_{U_i \cap U_j}) ,$$

while the right half gives an isomorphism

$$(N, \mathcal{O}_N) \longrightarrow (U_i \cap U_j, \mathcal{O}|_{U_i \cap U_j}) .$$

We deduce an isomorphism of ringed spaces

$$(\varphi_{ji}, \varphi_{ji}^*) : (M, \mathcal{O}_M) \longrightarrow (N, \mathcal{O}_N)$$

and, since M and N are open subsets in $\mathbb{R}^n$, Theorem 2.3.2 tells us that the transition function $\varphi_{ji} : M \longrightarrow N$ is C^k. □

Exercise 2.4.6. The reader can check that, as asserted in the first paragraph of Section 2.3, any map of ringed spaces between C^k–manifolds

$$(f, f^*) : (X, \mathcal{O}_X) \longrightarrow (Y, \mathcal{O}_Y)$$

is induced by a C^k–function $f : X \longrightarrow Y$. [Hint: the key is to convince yourself that the question is local. Locally $(X, \mathcal{O}_X)$ and $(Y, \mathcal{O}_Y)$ are isomorphic to $(\mathbb{R}^m, \mathcal{O}_{\mathbb{R}^m})$ and $(\mathbb{R}^n, \mathcal{O}_{\mathbb{R}^n})$ respectively, and therefore Theorem 2.3.2 applies.]

3
Schemes

In Section 2.4 we defined $\mathcal{C}^k$–manifolds, as ringed spaces $(M, \mathcal{O}_M)$ which are locally isomorphic to $(\mathbb{R}^n, \mathcal{O}_{\mathbb{R}^n})$. In this chapter we will define schemes as ringed spaces, which are locally isomorphic to certain building blocks $\left(\operatorname{Spec}(R), \widetilde{R}\right)$.

There are actually some technicalities here, which we do our best to avoid. It is true that schemes are, among other things, ringed spaces locally isomorphic to $\left(\operatorname{Spec}(R), \widetilde{R}\right)$. But this is not the customary way to view them: because not all morphisms of ringed spaces between schemes are permissible as morphisms of schemes, it is traditional to consider schemes as *locally ringed spaces*, whatever these are. It turns out that the ringed space $\left(\operatorname{Spec}(R), \widetilde{R}\right)$ is always a locally ringed space. Therefore any ringed space, built up of pieces isomorphic to $\left(\operatorname{Spec}(R), \widetilde{R}\right)$, must automatically be a locally ringed space. This means that all schemes are locally ringed spaces. The correct way to define morphisms of schemes, in glorious generality, is as morphisms of locally ringed spaces. Given two locally ringed spaces $(X, \mathcal{O}_X)$ and $(Y, \mathcal{O}_Y)$, there could be morphisms of ringed spaces between them which are not morphisms of locally ringed spaces.

In this book I refuse to include the definition of a locally ringed space. I have not given the definition of the stalk of a sheaf at a point, nor do I intend to; we will get through this entire book without ever mentioning germs of functions. Without telling the reader what a stalk is, I cannot possibly discuss locally ringed spaces. To avoid becoming bogged down in technicalities we will make the simplifying assumption, almost at the start, that all our schemes are locally of finite type over $\mathbb{C}$. This simplifying assumption sidesteps the difficulty; for schemes locally of finite type over $\mathbb{C}$, all morphisms of ringed spaces automatically respect the added structure of locally ringed spaces. Serious algebraic geometers will object that this is not the generality in which the subject ought to be developed. I remind the reader that this book is not intended to give a thorough exposition of the foundations of algebraic geometry. All we aim to do is develop the subject in the generality needed to state and (almost) prove GAGA. Let me also remind the reader that this book is meant to be accessible to undergraduates. Whenever possible we try to avoid technical distractions which might be confusing. As I have already said, we manage never to discuss stalks. Perhaps more remarkable, we get away without once mentioning the sheafification of a presheaf. It is not our aim to give a comprehensive treatment; we attempt to reach the highlights by the most efficient shortcuts.

First things first: before we worry about how to assemble together the permissible building blocks $\left(\mathrm{Spec}(R), \widetilde{R}\right)$, we had better define what they are.

3.1 The space Spec(R)

Let R be a ring, which is always assumed commutative with 1. In this chapter we will learn how to construct a ringed space $\left(\mathrm{Spec}(R), \widetilde{R}\right)$. To give such a ringed space we must give a topological space Spec(R), and on it a sheaf of rings $\widetilde{R}$. In this section we begin this project by constructing the topological space Spec(R).

Reminder 3.1.1. Recall that an ideal $\mathfrak{p} \subset R$ is called *prime* if $\mathfrak{p} \neq R$, and $xy \in \mathfrak{p}$ implies that either x or y lies in $\mathfrak{p}$.

Remark 3.1.2. Every non-zero ring has at least one prime ideal; one applies Zorn's lemma to obtain a maximal element in the collection of ideals not containing 1.

Proposition 3.1.3. *Let R be a ring. Let*

$$\mathrm{Spec}(R) = \{\mathfrak{p} \subset R \mid \mathfrak{p} \text{ is a prime ideal}\}\,.$$

That is, Spec(R) *is the set of prime ideals of R. For any ideal $\mathfrak{a}$ define $V(\mathfrak{a}) \subset \mathrm{Spec}(R)$ by*

$$V(\mathfrak{a}) = \{\mathfrak{p} \in \mathrm{Spec}(R) \mid \mathfrak{a} \subset \mathfrak{p}\}\,.$$

Declare that a subset of Spec(R) *is closed if and only if it is of the form $V(\mathfrak{a})$, for some ideal $\mathfrak{a} \subset R$. Then* Spec(R) *becomes a topological space.*

The proof is easy. Because we need to refer to parts of the proof later, we break it up into a string of lemmas. This will occupy most of the remainder of the (brief) section.

Remark 3.1.4. The topology of Proposition 3.1.3, on the set $\mathrm{Spec}(R)$, is called the *Zariski topology.*

Remark 3.1.5. In order to prove Proposition 3.1.3 we need to establish the following:

(i) $V(\mathfrak{a}) = \emptyset$ for some ideal $\mathfrak{a}$.
(ii) $V(\mathfrak{a}) = \mathrm{Spec}(R)$ for some ideal $\mathfrak{a}$.
(iii) For any two ideals $\mathfrak{a}_1$ and $\mathfrak{a}_2$, there is an ideal $\mathfrak{b}$ such that

$$V(\mathfrak{a}_1) \cup V(\mathfrak{a}_2) = V(\mathfrak{b}) .$$

(iv) For any set of ideals $\{\mathfrak{a}_\alpha\}_{\alpha \in I}$, there is an ideal $\mathfrak{b}$ such that

$$V(\mathfrak{b}) = \bigcap_{\alpha \in I} V(\mathfrak{a}_\alpha) .$$

Parts (i) and (ii) would prove that the empty set $\emptyset \subset \mathrm{Spec}(R)$ and the entire space $\mathrm{Spec}(R)$ are closed, part (iii) would prove that finite unions of closed subsets are closed, and part (iv) would prove that arbitrary intersections of closed sets are closed.

Lemma 3.1.6. *Remark 3.1.5(i) is true: there is an ideal $\mathfrak{a} \subset R$ with $V(\mathfrak{a}) = \emptyset$.*

Proof. Put $\mathfrak{a} = R$; no prime ideal contains R. □

Lemma 3.1.7. *Remark 3.1.5(ii) is true: there is an ideal $\mathfrak{a} \subset R$ with $V(\mathfrak{a}) = \mathrm{Spec}(R)$.*

Proof. Put $\mathfrak{a} = \{0\}$; every prime ideal contains 0. □

Lemma 3.1.8. *Remark 3.1.5(iii) is true: given two ideals $\mathfrak{a}_1$ and $\mathfrak{a}_2$ in R, there is an ideal $\mathfrak{b}$ with*

$$V(\mathfrak{a}_1) \cup V(\mathfrak{a}_2) = V(\mathfrak{b}) .$$

In fact, we may take $\mathfrak{b} = \mathfrak{a}_1\mathfrak{a}_2$, that is the ideal generated by all products $\alpha\beta$, with $\alpha \in \mathfrak{a}_1$ and $\beta \in \mathfrak{a}_2$.

Proof. If $\mathfrak{a}_1 \subset \mathfrak{p}$ or $\mathfrak{a}_2 \subset \mathfrak{p}$ then certainly $\mathfrak{b} = \mathfrak{a}_1\mathfrak{a}_2 \subset \mathfrak{p}$; we conclude

$$V(\mathfrak{a}_1) \cup V(\mathfrak{a}_2) \quad \subset \quad V(\mathfrak{b}) .$$

We need to prove that the containment is an equality.

Suppose therefore $\mathfrak{p} \notin V(\mathfrak{a}_1) \cup V(\mathfrak{a}_2)$. Then $\mathfrak{p}$ contains neither $\mathfrak{a}_1$ nor $\mathfrak{a}_2$. Therefore there are $x \in \mathfrak{a}_1 - \mathfrak{p}$ and $y \in \mathfrak{a}_2 - \mathfrak{p}$. But $xy \in \mathfrak{b} = \mathfrak{a}_1\mathfrak{a}_2$, while the fact that $\mathfrak{p}$ is prime means $xy \notin \mathfrak{p}$. Hence $\mathfrak{p} \notin V(\mathfrak{b})$. □

Lemma 3.1.9. *Remark 3.1.5(iv) is true: given any set of ideals* $\{\mathfrak{a}_\alpha\}_{\alpha\in I}$, *there is an ideal* $\mathfrak{b}$ *such that*

$$V(\mathfrak{b}) = \bigcap_{\alpha\in I} V(\mathfrak{a}_\alpha)\,.$$

In fact, we may take $\mathfrak{b}$ *to be the ideal generated by the union of the ideals* $\mathfrak{a}_\alpha$.

Proof. If $\mathfrak{p}$ contains $\mathfrak{b}$ then it must contain all of the ideals $\mathfrak{a}_\alpha$; hence

$$V(\mathfrak{b}) \quad \subset \quad \bigcap_{\alpha\in I} V(\mathfrak{a}_\alpha)\,.$$

For the reverse inclusion, if $\mathfrak{p} \in \cap_\alpha V(\mathfrak{a}_\alpha)$ then it contains all the ideals $\mathfrak{a}_\alpha$ and, being an ideal, it contains the ideal $\mathfrak{b}$ generated by their union. □

Remark 3.1.10. Note that the Zariski topology on $\mathrm{Spec}(R)$ is very different from most topological spaces we are accustomed to. For example it decidedly is not Hausdorff. If a closed set $V(\mathfrak{a})$ contains a point $\mathfrak{p} \in \mathrm{Spec}(R)$, that is if $\mathfrak{a} \subset \mathfrak{p}$, then $V(\mathfrak{a})$ contains all larger prime ideals. Not all points are closed. Precisely: given a point $\mathfrak{p} \in \mathrm{Spec}(R)$, its closure $\overline{\mathfrak{p}}$ is the set of primes containing $\mathfrak{p}$. The point $\mathfrak{p}$ is closed if and only if the ideal $\mathfrak{p} \subset R$ is maximal.

3.2 A basis for the Zariski topology

In Section 3.1 we described the topological space $\mathrm{Spec}(R)$. We know that the underlying set is the set of prime ideals in R, and the topology is the Zariski topology. It is very convenient, for what we will do, to note that there is a pleasant basis of open sets for the Zariski topology.

For this section we fix therefore a ring R, and we let $X = \mathrm{Spec}(R)$. The closed subsets are all of the form $V(\mathfrak{a})$, where the $\mathfrak{a} \subset R$ are ideals. Among the ideals are the principal ideals $Rf \subset R$. That is choose an element $f \in R$, and Rf is the ideal of all multiples of f.

Definition 3.2.1. *Let* f *be an element of* R. *The open set* X_f *is defined to be the complement of* $V(Rf)$. *That is*

$$X_f = \mathrm{Spec}(R) - V(Rf)\,.$$

The open sets X_f *are called the* basic open sets *for the Zariski topology.*

Remark 3.2.2. For any $f \in R$, and any prime ideal $\mathfrak{p} \subset R$, we have

$$\{\mathfrak{p} \in V(Rf)\} \quad \Longleftrightarrow \quad \{Rf \subset \mathfrak{p}\} \quad \Longleftrightarrow \quad \{f \in \mathfrak{p}\}\,.$$

In words (rather than symbols) this means that $\mathfrak{p} \in V(Rf)$ if and only if $f \in \mathfrak{p}$. For the complement $X_f = \operatorname{Spec}(R) - V(Rf)$ we conclude that X_f is the set of primes $\mathfrak{p} \subset R$ not containing f.

Remark 3.2.3. Let f be any unit in R. The proof of Lemma 3.1.6 tells us that $V(R) = V(Rf) = \emptyset$; taking the complement we have $X_f = \operatorname{Spec}(R)$. The proof of Lemma 3.1.7 establishes that $V(0) = V(R0) = \operatorname{Spec}(R)$; once again, the complement statement is $X_0 = \emptyset$.

Proposition 3.2.4. *The basic open sets X_f form a basis for the Zariski topology.*

Proof. We need to show two things.

(i) Every open set is a union of basic open sets.
(ii) The intersection of two basic open sets is basic.

Let us first show (i). The assertion, that every open set is the union of sets X_f, is equivalent to saying that every closed set is the intersection of closed sets $V(Rf)$. But for any ideal $\mathfrak{a} \subset R$ we have that $\mathfrak{a}$ is generated by all the principal ideals $\{Rf, f \in \mathfrak{a}\}$. Lemma 3.1.9 now says

$$V(\mathfrak{a}) = \bigcap_{f \in \mathfrak{a}} V(Rf) .$$

Next we have to prove (ii), that is we need to prove that $X_f \cap X_g$ is of the form X_h for some h. Since we want to remember an h that works, we label this as a separate lemma:

Lemma 3.2.5. *Let f and g be two elements of the ring R. Then*

$$X_f \cap X_g = X_{fg} .$$

Proof. Equivalently we need to show that $V(Rf) \cup V(Rg) = V(Rfg)$; but this is a special case of Lemma 3.1.8. □

This completes the proof of Proposition 3.2.4 □

Remark 3.2.6. From Lemma 3.2.5 it follows, by an easy induction, that if $f_1, f_2, \ldots, f_n$ are elements of R then

$$X_{f_1} \cap X_{f_2} \cap \cdots \cap X_{f_n} \quad = \quad X_{f_1 f_2 \cdots f_n} .$$

If we let $f = f_1 = f_2 = \cdots = f_n$ we deduce that $X_{f^n} = X_f$.

3.3 Localization of rings

We wish to remind the reader of some commutative algebra. Given a ring R and an element $f \in R$, there is a ring $R[1/f]$ in which f is formally inverted. We recall briefly.

Reminder 3.3.1. The elements of $R[1/f]$ are of the form x/f^n, with $x \in R$ and n an integer ≥ 0. Equality is defined by the rule that

$$x/f^m = y/f^n$$

if and only if there exists an integer $N \geq 0$ with

$$f^{n+N}x = f^{m+N}y \ .$$

In particular, $x/f^m = 0 = 0/1$ if and only if, for some integer $N \geq 0$, the identity $f^N x = 0$ is satisfied. Addition is given by the formula

$$(x/f^m) + (y/f^n) \quad = \quad (f^n x + f^m y)/f^{m+n} \ .$$

Multiplication is defined by the formula

$$(x/f^m)(y/f^n) = xy/f^{m+n}.$$

There is a ring homomorphism $\alpha_f : R \longrightarrow R[1/f]$, which takes $x \in R$ to $x/f^0 = x/1 \in R[1/f]$. This ring homomorphism has the following universal property:

Proposition 3.3.2. *Let $\varphi : R \longrightarrow S$ be a ring homomorphism such that $\varphi(f)$ is invertible in S. Then φ factors uniquely through the map $\alpha_f : R \longrightarrow R[1/f]$; that is, in the diagram below there exists a unique map $\varphi' : R[1/f] \longrightarrow S$ rendering the triangle commutative*

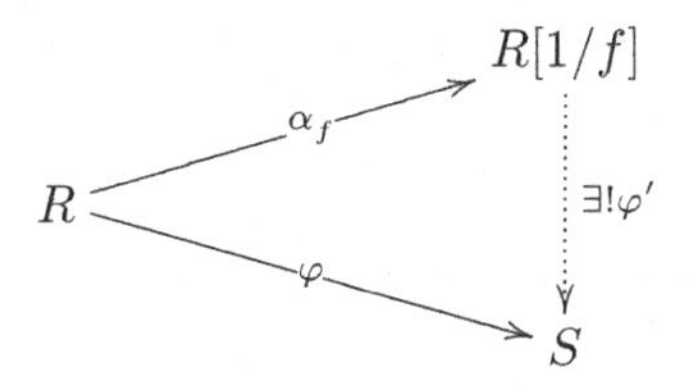

The formula for φ' is $\varphi'(x/f^n) = \varphi(f)^{-n}\varphi(x)$.

Example 3.3.3. Let $R = \mathbb{C}[t]$ be the polynomial ring in one variable t, and let $S = \mathbb{C}[t, t^{-1}]$ be the ring of Laurent polynomials. We remind the reader: the elements of $S = \mathbb{C}[t, t^{-1}]$ are finite sums

$$\sum_{i=-m}^{n} a_i t^i$$

with $a_i \in \mathbb{C}$, and the addition and multiplication rules are obvious. Let

$\varphi : R \longrightarrow S$ be the inclusion, of $R = \mathbb{C}[t]$ into $S = \mathbb{C}[t, t^{-1}]$. Clearly $\varphi(t)$ is an invertible element of S, and Proposition 3.3.2 asserts that there is a commutative triangle

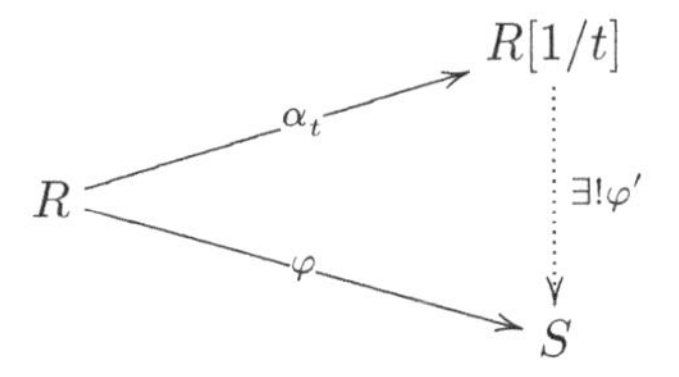

Let us prove that the map φ' is an isomorphism.

We need to show that φ' is injective and surjective. The surjectivity is clear; given the element $t^m/t^n \in R[1/t]$, we have

$$\varphi'(t^m/t^n) = t^{m-n} .$$

As m, n range over integers ≥ 0, the integer $m-n$ takes all possible values in $\mathbb{Z}$; that is, the elements t^{m-n} span the vector space $S = \mathbb{C}[t, t^{-1}]$.

Let us now show that φ' is injective. Reminder 3.3.1 tells us that every element of $R[1/t]$ is of the form a/t^m, with $a \in R = \mathbb{C}[t]$. Suppose therefore that we have, in S, the identity

$$\varphi'(a/t^m) = \varphi'(b/t^n) .$$

That is, in the ring $S = \mathbb{C}[t, t^{-1}]$ we have an identity of Laurent polynomials

$$a \cdot t^{-m} = b \cdot t^{-n} .$$

Of course this gives us, for every integer N, an identity in S

$$a \cdot t^{N-m} = b \cdot t^{N-n} .$$

If we choose $N = m + n$, we discover that

$$a \cdot t^n = b \cdot t^m ;$$

this is an equality of ordinary polynomials (as opposed to Laurent polynomials), since all the degrees are positive. It is an identity in the subring $R \subset S$, and Reminder 3.3.1 allows us to conclude that $a/t^m = b/t^n$, as elements of $R[1/t]$.

Remark 3.3.4. Later in the book, we will feel very free to confuse the ring $R[1/t] = \mathbb{C}[t][1/t]$ with the Laurent polynomial ring $S = \mathbb{C}[t, t^{-1}]$; Example 3.3.3 gives a canonical isomorphism $\varphi' : R[1/t] \longrightarrow S$.

Exercise 3.3.5. There is a generalization of Example 3.3.3 to polynomial rings in several variables. Let $R = \mathbb{C}[x_0, x_1, \ldots, x_n]$ be the polynomial ring in $(n+1)$ variables, let j be an integer in the range $0 \le j \le n$, and let

$$S = \mathbb{C}[x_0, x_0^{-1}, x_1, x_1^{-1}, \ldots, x_j, x_j^{-1}, x_{j+1}, x_{j+2}, \ldots, x_n]$$

be the ring of Laurent polynomials, as indicated. The elements of S are finite linear combinations of Laurent monomials

$$x_0^{a_0} x_1^{a_1} \cdots x_n^{a_n}$$

with $a_i \in \mathbb{Z}$, and where $a_i \ge 0$ if $i > j$. The inclusion $\varphi : R \longrightarrow S$ is a ring homomorphism, and the monomial $f = x_0 x_1 \cdots x_j \in R$ becomes invertible in S. Proposition 3.3.2 says that φ factors uniquely through α_f; there is a commutative triangle

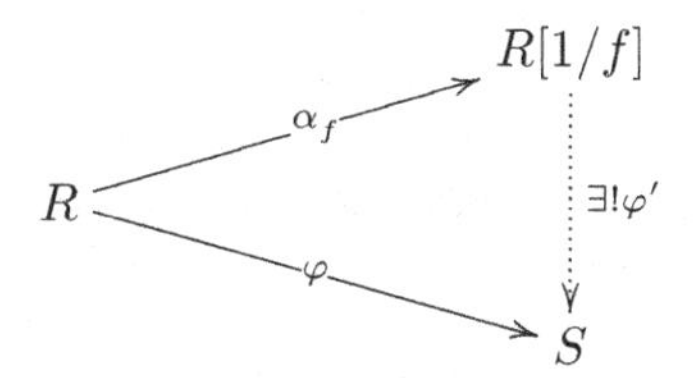

We leave it to the reader to check that $\varphi' : R[1/f] \longrightarrow S$ is an isomorphism.

Remark 3.3.6. Example 3.3.3 and Exercise 3.3.5 are related as follows. Let $\varphi_i : \mathbb{C}[x_i] \longrightarrow \mathbb{C}[x_i, x_i^{-1}]$ be the homomorphism of Example 3.3.3. Then the homomorphism φ, given in Exercise 3.3.5, is (up to isomorphism) nothing more nor less than

$$\left[\bigotimes_{i=1}^{j} \mathbb{C}[x_i]\right] \otimes \left[\bigotimes_{i=j+1}^{n} \mathbb{C}[x_i]\right] \xrightarrow{\left[\bigotimes_{i=1}^{j} \varphi_i\right] \otimes 1} \left[\bigotimes_{i=1}^{j} \mathbb{C}[x_i, x_i^{-1}]\right] \otimes \left[\bigotimes_{i=j+1}^{n} \mathbb{C}[x_i]\right]$$

where all the tensor products are over $\mathbb{C}$. To see this, let $\varphi : R \longrightarrow S$

be as in Exercise 3.3.5. Consider the following square

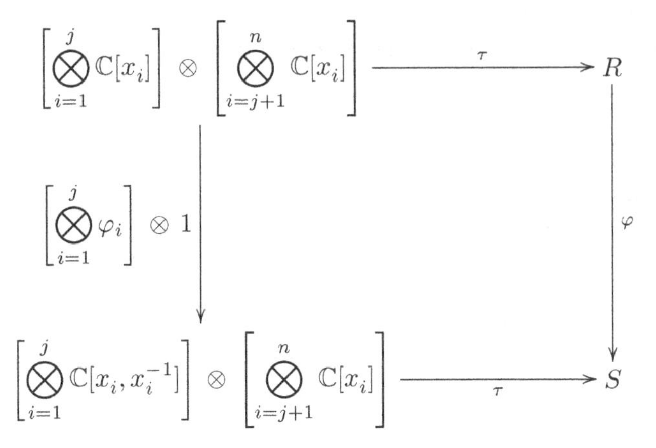

The horizontal maps τ, both of them, are given by the simple formula

$$\tau(f_0 \otimes f_1 \otimes \cdots \otimes f_n) \quad = \quad f_0 f_1 \cdots f_n \ .$$

We leave it to the reader to check that the square commutes, and that both horizontal maps τ are isomorphisms. The hint is that the monomials form a natural basis everywhere.

So much for examples. Now it is time to return to more abstract theory; we recall some basic facts, which follow immediately from the elementary properties of localization.

Lemma 3.3.7. *Let R be a ring, and assume that $f \in R$ is an element contained in every prime ideal of R. Then f is nilpotent.*

Proof. Let us prove first that the ring $R[1/f]$ must be the zero ring. Suppose it is not; by Remark 3.1.2 the ring $R[1/f]$ then has a prime ideal $\mathfrak{q}$. The homomorphism

$$R \xrightarrow{\ \alpha\ } R[1/f] \longrightarrow \frac{R[1/f]}{\mathfrak{q}} \quad = \quad S$$

is a homomorphism from R to an integral domain S. The kernel is a prime ideal of the ring R which, by assumption, must contain f. But f goes to an invertible element in $R[1/f]$, and hence also in S; this is a contradiction.

Thus $R[1/f]$ must be the zero ring. But then, in the ring $R[1/f]$, we have the identity $1 = 0$. Reminder 3.3.1 tells us that there must be an integer m with $f^m 1 = f^m = 0$. □

Lemma 3.3.8. *With the notation as in Section 3.2, suppose $X_f \subset \cup_{i \in I} X_{g_i}$. That is, we have elements $f, g_i \in R$, the sets X_f, X_{g_i} are* *the basic open sets of Definition 3.2.1, and we have an inclusion* $X_f \subset \cup_{i \in I} X_{g_i}$. *Then there is a finite subset* $\{g_{i_1}, g_{i_2}, \ldots, g_{i_n}\}$ *of the set* $\{g_i, i \in I\}$ *with*

$$X_f \quad \subset \quad \bigcup_{j=1}^{n} X_{g_{i_j}} \quad .$$

Furthermore, there exists an integer $m > 0$, *elements* $\{a_1, a_2, \ldots, a_n\}$ *of* R *and an identity*

$$f^m = a_1 g_{i_1} + a_2 g_{i_2} + \cdots + a_n g_{i_n} \, .$$

Proof. The inclusion $X_f \subset \cup_{i \in I} X_{g_i}$ can be expressed as $V(Rf) \supset \cap_{i \in I} V(Rg_i)$. By Lemma 3.1.9 we know that

$$\bigcap_{i \in I} V(Rg_i) = V(\mathfrak{b}) \, ,$$

where $\mathfrak{b}$ is the ideal generated by $\{g_i, i \in I\}$. Hence we have $V(Rf) \supset V(\mathfrak{b})$. This means that any prime ideal containing $\mathfrak{b}$ must contain f. Therefore the image of $f \in R$, under the natural ring homomorphism $R \longrightarrow R/\mathfrak{b}$, is contained in every prime ideal of $R/\mathfrak{b}$. By Lemma 3.3.7 we conclude that the image of f in the ring $R/\mathfrak{b}$ is nilpotent, i.e. there exists an $m > 0$ with $f^m \in \mathfrak{b}$.

The ideal $\mathfrak{b}$ might have lots of generators, but any one of its elements is a finite linear combination of these generators. Since f^m belongs to the ideal $\mathfrak{b}$, there must be an identity

$$f^m = a_1 g_{i_1} + a_2 g_{i_2} + \cdots + a_n g_{i_n} \, ,$$

involving only finitely many of the generators g_{i_j}. This in turn implies that every prime ideal containing $\{g_{i_1}, g_{i_2}, \ldots, g_{i_n}\}$ contains f, or

$$V(Rf) \quad \supset \quad \bigcap_{j=1}^{n} V(Rg_{i_j}) \, .$$

The complement of this statement is, as required, that X_f is contained in the union of $X_{g_{i_j}}$. □

Remark 3.3.9. In the special case $X_f \subset X_g$ of Lemma 3.3.8 (that is, the set $\{g_i, i \in I\}$ has only one element) the conclusion of the lemma asserts that there must exist an integer $m > 0$, and an element $a \in R$, with $f^m = ag$.

3.4 The sheaf $\widetilde{R}$ on Spec(R)

In Sections 3.1 and 3.2 we defined the topological space Spec(R) and produced a convenient basis of open sets to work with. As in Section 3.2 we fix our ring R and $X = \text{Spec}(R)$. We promised the reader a ringed space, and it is now time to produce a sheaf of rings $\widetilde{R}$ on Spec(R). The idea is simple enough. On the basic open sets, of Section 3.2, the sheaf should be defined by the simple formula

$$\Gamma(X_f, \widetilde{R}) = R[1/f] ,$$

with $R[1/f]$ as in Reminder 3.3.1. What we need to do, in this section, is convince ourselves that this indeed extends to a well–defined sheaf on $X = \text{Spec}(R)$.

Remark 3.4.1. Assume for a second there is a sheaf as above, with $\Gamma(X_f, \widetilde{R}) = R[1/f]$. In the special case, where $f = 1$, the subset $X_f \subset X$ is equal to X by Remark 3.2.2. This would make

$$\Gamma(X, \widetilde{R}) \quad = \quad \Gamma(X_1, \widetilde{R}) \quad = \quad R[1/1] \quad = \quad R .$$

Now we return to proving that the sheaf $\widetilde{R}$ exists. There are some things to prove. For example, it is not even clear that the ring $R[1/f]$ is independent of f. That is, if we have $X_f = X_g$, why can we be certain that $R[1/f] = R[1/g]$? Before tackling such problems we prove some formal facts about the rings $R[1/f]$, and about the way they relate to the open sets X_f. We begin by fixing the notation for the localization map.

Definition 3.4.2. *For every element $f \in R$, define $\alpha_f : R \longrightarrow R[1/f]$ to be the localization map.*

Next we prove a couple of lemmas about the rings $R[1/f]$.

Lemma 3.4.3. *Let f and g be elements of R, and suppose $X_f \subset X_g$. Then there exists a unique map $\alpha_f^g : R[1/g] \longrightarrow R[1/f]$ rendering commutative the diagram*

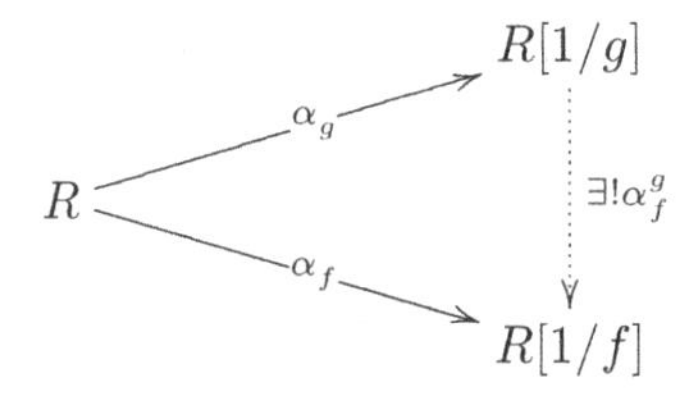

Proof. By Proposition 3.3.2 it suffices to show that $\alpha_f(g)$ is invertible in

$R[1/f]$. We are given that $X_f \subset X_g$, and hence Remark 3.3.9 says that there is an identity $f^m = ag$, for some integer $m > 0$ and some $a \in R$.

But now in the ring $R[1/f]$ we deduce the identity

$$g \cdot \left(\frac{a}{f^m}\right) = 1,$$

which means that $\alpha_f(g)$ is invertible in $R[1/f]$. □

Lemma 3.4.4. *Suppose we have three basic open sets $X_f \subset X_g \subset X_h$. Then $\alpha_f^h : R[1/h] \longrightarrow R[1/f]$ is equal to the composite*

$$R[1/h] \xrightarrow{\alpha_g^h} R[1/g] \xrightarrow{\alpha_f^g} R[1/f] .$$

Proof. The commutative diagram below

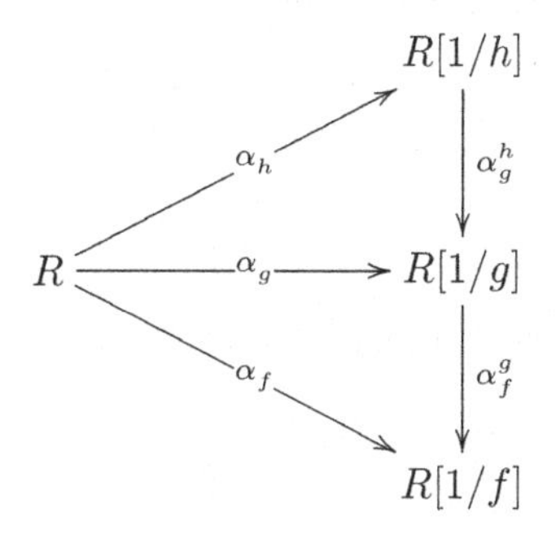

coupled with the uniqueness of the map α_f^h rendering commutative the triangle

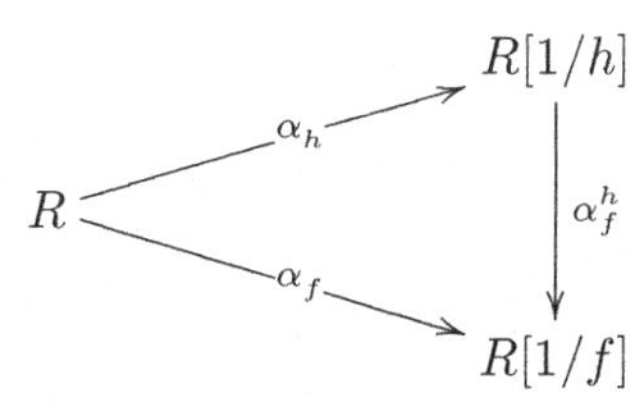

tells us that $\alpha_f^h = \alpha_f^g \alpha_g^h$, as required. □

Remark 3.4.5. If $X_f = X_g$ then $X_f \subset X_g$ and $X_g \subset X_f$, and we obtain canonical maps $\alpha_f^g : R[1/g] \longrightarrow R[1/f]$ and $\alpha_g^f : R[1/f] \longrightarrow R[1/g]$. By Lemma 3.4.4 we know that $1 = \alpha_f^f = \alpha_f^g \alpha_g^f$ and $1 = \alpha_g^g = \alpha_g^f \alpha_f^g$. That is, the maps α_g^f and α_f^g give canonical isomorphisms between $R[1/f]$ and $R[1/g]$.

Now we define $\Gamma(X_f, \widetilde{R})$ for all basic open sets X_f.

Definition 3.4.6. *For every basic open set U we choose and fix an f*

with $U = X_f$*. Note that there is usually more than one possible choice for* $f \in R$*. With this fixed choice of* f*, define*

$$\Gamma(X_f, \widetilde{R}) = R[1/f] \ .$$

By Remark 3.4.5 the choice of f, made in Definition 3.4.6, is unimportant; if $X_f = X_g$ then there is a canonical isomorphism of $R[1/f]$ with $R[1/g]$.

Definition 3.4.7. *Suppose we have basic open sets* $X_f \subset X_g$*, with* f *and* g *the arbitrary choices made in Definition 3.4.6. The restriction map*

$$\mathrm{res}_{X_f}^{X_g} : \Gamma(X_g, \widetilde{R}) \longrightarrow \Gamma(X_f, \widetilde{R})$$

is defined to be $\alpha_f^g : R[1/g] \longrightarrow R[1/f]$*.*

Lemma 3.4.8. *Given three basic open sets* $X_f \subset X_g \subset X_h$*, the composite*

$$\Gamma(X_h, \widetilde{R}) \xrightarrow{\mathrm{res}_{X_g}^{X_h}} \Gamma(X_g, \widetilde{R}) \xrightarrow{\mathrm{res}_{X_f}^{X_g}} \Gamma(X_f, \widetilde{R})$$

is equal to the map $\mathrm{res}_{X_f}^{X_h} : \Gamma(X_h, \widetilde{R}) \longrightarrow \Gamma(X_f, \widetilde{R})$*.*

Proof. This is just the statement that $\alpha_f^h = \alpha_f^g \alpha_g^h$, proved in Lemma 3.4.4. □

Remark 3.4.9. It is clear that $\mathrm{res}_{X_f}^{X_f} = 1$, and Lemma 3.4.8 tells us that restriction maps compose correctly on basic open sets. We also have that

$$\Gamma(\emptyset, \widetilde{R}) \quad = \quad \Gamma(X_0, \widetilde{R}) \quad = \quad R[1/0] \quad = \quad 0 \ .$$

In other words what we have is very much like a presheaf, the only problem being that, so far, it is only defined on the basic open subsets. We now need to understand when we can be guaranteed the existence of an extension to a sheaf on X.

The next observation is that any sheaf $\mathcal{O}$ on X is determined by its values on any basis of open sets. After all, any open set V admits a cover by elements of the basis, that is $V = \cup_{i \in I} U_i$ with U_i in the basis. Lemma 2.2.10 gives an exact sequence

$$0 \longrightarrow \Gamma(V, \mathcal{O}) \xrightarrow{\ \alpha\ } \prod_{i \in I} \Gamma(U_i, \mathcal{O}) \xrightarrow{\ \beta\ } \prod_{i,j \in I} \Gamma(U_i \cap U_j, \mathcal{O}) \ .$$

If we know $\Gamma(U_i, \mathcal{O})$ and $\Gamma(U_i \cap U_j, \mathcal{O})$ and the restriction maps between them (this is given to us, since U_i and $U_i \cap U_j$ belong to the basis) then the sequence permits us to compute $\Gamma(V, \mathcal{O})$. A little more care shows how, for open sets $V \subset V'$, we can compute the restriction maps

$$\mathrm{res}_V^{V'} : \Gamma(V', \mathcal{O}) \longrightarrow \Gamma(V, \mathcal{O}) .$$

The question before us is whether this unique extension to a sheaf on X exists in our case, where the basis of open sets are the $X_f \subset X$, and we set $\Gamma(X_f, \widetilde{R}) = R[1/f]$.

Let us return to the glorious generality of point set topology. Suppose we have a topological space X and a basis of open sets. Suppose we have a sheaf $\mathcal{O}$ on X, and we know the values of this sheaf on the open sets U belonging to the basis; that is we are given $\Gamma(U, \mathcal{O})$ for basic open sets U and the restriction maps between them. Clearly if U is a basic open set, and $U = \cup_{i \in I} U_i$ with U_i in the basis, then Lemma 2.2.10 gives an exact sequence

$$(*) \qquad 0 \longrightarrow \Gamma(U, \mathcal{O}) \xrightarrow{\ \alpha\ } \prod_{i \in I} \Gamma(U_i, \mathcal{O}) \xrightarrow{\ \beta\ } \prod_{i,j \in I} \Gamma(U_i \cap U_j, \mathcal{O}) .$$

For the extension to exist all such sequences must be exact.

There is a helpful little lemma in point set topology which says that the converse holds:

Lemma 3.4.10. *Let X be a topological space, and suppose we are given a basis for the topology. Suppose that, on every open set $U \subset X$ in the basis, we define a ring $\Gamma(U, \mathcal{O})$, and, if $U \subset U'$ are both in the basis, we define a restriction map*

$$\mathrm{res}_U^{U'} : \Gamma(U', \mathcal{O}) \longrightarrow \Gamma(U, \mathcal{O}) .$$

Suppose further that, if $U \subset U' \subset U''$ are all in the basis, then $\mathrm{res}_U^{U''} = \mathrm{res}_U^{U'} \mathrm{res}_{U'}^{U''}$, and that $\Gamma(\emptyset, \mathcal{O}) = 0$. If the sequence (*) above is exact for every open cover of a basic open set U by basic open sets U_i, then there is a (unique) extension of $\mathcal{O}$ to a sheaf on X.

Proof. We will explain the beginning of the proof, and then leave most of the details to the reader. We need to define $\Gamma(V, \mathcal{O})$ for every open set $V \subset X$. Let $V = \cup_{i \in I} U_i$ be the maximal open cover of V by open subsets in the basis. That is, I is the set of all basic open sets contained in V. Define $\widetilde{\Gamma}(V, \mathcal{O})$ by setting the following sequence to be exact (i.e. by defining $\widetilde{\Gamma}(V, \mathcal{O})$ as the kernel of the map β below):

$$0 \longrightarrow \widetilde{\Gamma}(V, \mathcal{O}) \xrightarrow{\ \alpha\ } \prod_{i \in I} \Gamma(U_i, \mathcal{O}) \xrightarrow{\ \beta\ } \prod_{i,j \in I} \Gamma(U_i \cap U_j, \mathcal{O}) .$$

If $V \subset V'$ then the maximal open cover of V by basic open sets is contained in the maximal open cover of V'; hence there is a commutative diagram with exact rows

$$\begin{array}{ccccccc} 0 & \longrightarrow & \widetilde{\Gamma}(V',\mathcal{O}) & \xrightarrow{\alpha'} & \prod_{i\in I'}\Gamma(U_i,\mathcal{O}) & \xrightarrow{\beta'} & \prod_{i,j\in I'}\Gamma(U_i\cap U_j,\mathcal{O}) \\ & & & & \downarrow & & \downarrow \\ 0 & \longrightarrow & \widetilde{\Gamma}(V,\mathcal{O}) & \xrightarrow{\alpha} & \prod_{i\in I}\Gamma(U_i,\mathcal{O}) & \xrightarrow{\beta} & \prod_{i,j\in I}\Gamma(U_i\cap U_j,\mathcal{O}) \end{array},$$

in which the vertical maps are the obvious projections. We can define $\mathrm{res}_V^{V'} : \widetilde{\Gamma}(V',\mathcal{O}) \longrightarrow \widetilde{\Gamma}(V,\mathcal{O})$ to be the unique map rendering the diagram commutative. It is clear that restrictions compose well and that $\Gamma(\emptyset,\mathcal{O}) = 0$. In other words we have a presheaf of rings on X. If U is a basic open set then $\widetilde{\Gamma}(U,\mathcal{O}) = \Gamma(U,\mathcal{O})$. We know this either from the hypothesis of the lemma, about open covers of a basic open set U by basic open sets, or else from the fact that $U \subset U$ is a member of the maximal cover of U by basic open sets. And then one needs to check:

Exercise 3.4.11. The sheaf axioms hold for the presheaf whose value, at an open $V \subset X$, is $\widetilde{\Gamma}(V,\mathcal{O})$. [Hint: you might wish to first compare the maximal cover of V with a cover that is not maximal. Also, you might want to imitate the techniques of the proof of Theorem 3.4.12 below, especially the methods of the proofs of Lemmas 3.4.14 and 3.4.16, which compare arbitrary covers with finite subcovers.] □

Now we come to the main theorem of this section:

Theorem 3.4.12. *There is a unique sheaf on $X = \mathrm{Spec}(R)$, with*

$$\Gamma(X_f,\widetilde{R}) = R[1/f]$$

as in Definitions 3.4.6 and 3.4.7.

Remark 3.4.13. By Lemma 3.4.10 we are reduced to checking the following: let $X_f = \cup_{i\in I} X_{g_i}$ be an open cover of a basic open set by basic open sets. Then the natural sequence

$$0 \longrightarrow \Gamma(X_f,\widetilde{R}) \xrightarrow{\alpha} \prod_{i\in I}\Gamma(X_{g_i},\widetilde{R}) \xrightarrow{\beta} \prod_{i,j\in I}\Gamma(X_{g_i}\cap X_{g_j},\widetilde{R})$$

of Lemma 2.2.10 is exact. We will break up the proof into three lemmas; Lemma 3.4.14 will show that α is injective, while Lemmas 3.4.15 and 3.4.16 will establish that the kernel of β equals the image of α.

Lemma 3.4.14. *The map α, of Remark 3.4.13, is injective.*

Proof. Suppose $s \in \Gamma(X_f, \widetilde{R})$ is in the kernel of α. Then the restriction of s to each $X_{g_i} \subset X_f$ must vanish. Lemma 3.3.8 tells us first of all that X_f is the union of finitely many of the X_{g_i}. Choose such a finite set. Relabeling we may assume that

$$X_f \quad = \quad \bigcup_{i=1}^{r} X_{g_i} \ ,$$

and the other elements of I are labeled any way we desire. We certainly know that the restriction of s to $\Gamma(X_{g_i}, \widetilde{R})$ vanishes for $1 \leq i \leq r$. We will show that this already implies the vanishing of s.

Now s is an element of $\Gamma(X_f, \widetilde{R}) = R[1/f]$, therefore $s = x/f^n$ for some $x \in R$ and some positive integer n. The fact that the image of s vanishes in $\Gamma(X_{g_i}, \widetilde{R}) = R[1/g_i]$ means (see Reminder 3.3.1) that there exists an integer $m_i > 0$ with

$$g_i^{m_i} x = 0 \, .$$

There are finitely many i; therefore we may choose an integer m so large that $g_i^m x = 0$ for all i; just take the maximum of the m_i. Now we recall that $X_{g_i^m} = X_{g_i}$; see for example Remark 3.2.6. We therefore have

$$X_f \quad = \quad \bigcup_{i=1}^{r} X_{g_i} \quad = \quad \bigcup_{i=1}^{r} X_{g_i^m} \ .$$

Lemma 3.3.8 now tells us that there exists an identity

$$f^\ell = a_1 g_1^m + a_2 g_2^m + \cdots + a_r g_r^m \, .$$

Multiplying this identity by x, and remembering that $g_i^m x = 0$ for all i, we deduce that $f^\ell x = 0$. This means that $s = x/f^n$ vanishes in $\Gamma(X_f, \widetilde{R}) = R[1/f]$. □

Lemma 3.4.15. *Let the notation be as in Remark 3.4.13, but assume that the set I is finite. Then the kernel of β equals the image of α.*

Proof. We are assuming that the set I is finite; relabeling, put $I = \{1, 2, \ldots, r\}$. That is

$$X_f \quad = \quad \bigcup_{i=1}^{r} X_{g_i} \ .$$

We are given an element in the kernel of β. This means that, for each $1 \leq i \leq r$, we have an element $s_i \in \Gamma(X_{g_i}, \widetilde{R})$, and these satisfy

$$\mathrm{res}^{X_{g_i}}_{X_{g_i} \cap X_{g_j}}(s_i) \quad = \quad \mathrm{res}^{X_{g_j}}_{X_{g_i} \cap X_{g_j}}(s_j) \ .$$

We need to produce a section $s \in \Gamma(X_f, \widetilde{R})$ with $s_i = \mathrm{res}^{X_f}_{X_{g_i}}(s)$. We write $s_i \in \Gamma(X_{g_i}, \widetilde{R}) = R[1/g_i]$ in the form $s_i = x_i/g_i^{n_i}$. By multiplying numerators and denominators by appropriate powers of g_i we can choose all the n_i to be equal to some positive integer n, so $s_i = x_i/g_i^n$. The assertion that $\mathrm{res}^{X_{g_i}}_{X_{g_i}\cap X_{g_j}}(s_i) = \mathrm{res}^{X_{g_j}}_{X_{g_i}\cap X_{g_j}}(s_j)$ means that they become equal in the ring

$$\Gamma(X_{g_i} \cap X_{g_j}, \widetilde{R}) \quad = \quad \Gamma(X_{g_i g_j}, \widetilde{R}) \quad \cong \quad R[1/g_i g_j] \ .$$

Note that the equality $X_{g_i} \cap X_{g_j} = X_{g_i g_j}$ follows from Lemma 3.2.5, while the canonical isomorphism $\Gamma(X_{g_i g_j}, \widetilde{R}) \cong R[1/g_i g_j]$ comes from Remark 3.4.5. This concretely means that, in the ring $R[1/g_i g_j]$, we have an identity $x_i/g_i^n = x_j/g_j^n$. Putting them over a common denominator, the identity becomes

$$\frac{g_j^n x_i}{(g_i g_j)^n} \quad = \quad \frac{g_i^n x_j}{(g_i g_j)^n} \ .$$

Reminder 3.3.1 tells us that these are equal if and only if there exists some non-negative integer N_{ij} such that

$$(g_i g_j)^{N_{ij}} g_j^n x_i \quad = \quad (g_i g_j)^{N_{ij}} g_i^n x_j \ .$$

Since there are finitely many N_{ij} we may, replacing by the maximum if necessary, assume them all equal. We get a single integer N and identities in R, for every i and j,

$$(g_i g_j)^N g_j^n x_i \quad = \quad (g_i g_j)^N g_i^n x_j \ . \tag{3.1}$$

Now recall that $X_{g_i} = X_{g_i^{N+n}}$, and

$$X_f \quad = \quad \bigcup_{i=1}^r X_{g_i} \quad = \quad \bigcup_{i=1}^r X_{g_i^{N+n}} \ .$$

Lemma 3.3.8 tells us that there exists an identity in R

$$f^\ell = a_1 g_1^{N+n} + a_2 g_2^{N+n} + \cdots + a_r g_r^{N+n} \ . \tag{3.2}$$

Let x be given by the formula

$$x = a_1 g_1^N x_1 + a_2 g_2^N x_2 + \ldots + a_r g_r^N x_r \ , \tag{3.3}$$

and put $s = x/f^\ell \in R[1/f]$. I assert that $\mathrm{res}^{X_f}_{X_{g_i}}(s) = s_i$.

To prove this, observe

$$\begin{aligned} g_i^{N+n}x &= g_i^{N+n}\sum_{j=1}^{r} a_j g_j^N x_j && \text{by equation (3.3)} \\ &= \sum_{j=1}^{r} a_j g_i^{N+n} g_j^N x_j \\ &= \sum_{j=1}^{r} a_j g_i^N g_j^{N+n} x_i && \text{by equation (3.1)} \\ &= g_i^N x_i \sum_{j=1}^{r} a_j g_j^{N+n} \\ &= f^\ell g_i^N x_i && \text{by equation (3.2).} \end{aligned}$$

In the ring $R[1/g_i]$, where both f and g_i are invertible, this gives the identity

$$\frac{x}{f^\ell} = \frac{x_i}{g_i^n} ,$$

meaning $\mathrm{res}^{X_f}_{X_{g_i}}(s) = s_i$. □

Lemma 3.4.16. *With the notation as in Remark 3.4.13, the kernel of* β *equals the image of* α*; there is no need to assume* I *finite.*

Proof. Let I be arbitrary. As we already observed, in the proof of Lemma 3.4.14, there is a finite subset of the set I which provides a cover for X_f. Relabeling I, as in Lemma 3.4.14, we may assume that

$$X_f = \bigcup_{i=1}^{r} X_{g_i} ,$$

and the other elements of I are labeled any way we wish. We are given an element in the kernel of β; this means that, for every $t \in I$, we have a section $s_t \in \Gamma(X_t, \widetilde{R})$, and furthermore, for every pair $t, t' \in I$, we have

$$\mathrm{res}^{X_{g_t}}_{X_{g_t} \cap X_{g_{t'}}}(s_t) = \mathrm{res}^{X_{g_{t'}}}_{X_{g_t} \cap X_{g_{t'}}}(s_{t'}) .$$

If we restrict attention to the subset $\{1, 2, \ldots, r\} \subset I$, then Lemma 3.4.15 already tells us that there exists a section $s \in \Gamma(X_f, \widetilde{R})$ so that, for all $1 \le i \le r$, we have

$$s_i = \mathrm{res}^{X_f}_{X_{g_i}}(s) .$$

We assert that this $s \in \Gamma(X_f, \widetilde{R})$ works for every $t \in I$; that is

$$s_t = \mathrm{res}^{X_f}_{X_{g_t}}(s) .$$

The point is the following. We know that

$$\operatorname{res}^{X_{g_i}}_{X_{g_i}\cap X_{g_t}} \operatorname{res}^{X_f}_{X_{g_i}}(s) = \operatorname{res}^{X_{g_i}}_{X_{g_i}\cap X_{g_t}}(s_i) = \operatorname{res}^{X_{g_t}}_{X_{g_i}\cap X_{g_t}}(s_t) .$$

On the other hand,

$$\operatorname{res}^{X_{g_i}}_{X_{g_i}\cap X_{g_t}} \operatorname{res}^{X_f}_{X_{g_i}}(s) = \operatorname{res}^{X_f}_{X_{g_i}\cap X_{g_t}}(s) = \operatorname{res}^{X_{g_t}}_{X_{g_i}\cap X_{g_t}} \operatorname{res}^{X_f}_{X_{g_t}}(s) .$$

Combining these equalities, we have that

$$\delta = s_t - \operatorname{res}^{X_f}_{X_{g_t}}(s)$$

is an element of $\Gamma(X_{g_t}, \widetilde{R})$ so that, for all $1 \le i \le r$,

$$\operatorname{res}^{X_{g_t}}_{X_{g_i}\cap X_{g_t}}(\delta) = 0 .$$

But the open sets $X_{g_i g_t} = X_{g_i} \cap X_{g_t}$, with $1 \le i \le r$, cover X_{g_t}; Lemma 3.4.14 allows us to conclude that δ vanishes. □

3.5 A return to the world of simple examples

We have just gone through some considerable abstraction; it might be helpful to make another small foray into the world of elementary examples.

Example 3.5.1. Let $R = \mathbb{C}[x_0, x_1, \ldots, x_n]$ be the polynomial ring in $(n+1)$ variables, and let j, k be integers in the range $0 \le j \le k \le n$. Let $g = x_0 x_1 \cdots x_j$, and let $f = x_0 x_1 \cdots x_k$. Let $X = \operatorname{Spec}(R)$. Since g divides f we have $X_f \subset X_g$. What can we say about the homomorphism $\alpha^g_f : R[1/g] \longrightarrow R[1/f]$, defined in Lemma 3.4.3?

As in Exercise 3.3.5, we let S_j and S_k be the rings

$$\begin{aligned} S_j &= \mathbb{C}[x_0, x_0^{-1}, x_1, x_1^{-1}, \ldots, x_j, x_j^{-1}, x_{j+1}, x_{j+2}, \ldots, x_n] , \\ S_k &= \mathbb{C}[x_0, x_0^{-1}, x_1, x_1^{-1}, \ldots, x_k, x_k^{-1}, x_{k+1}, x_{k+2}, \ldots, x_n] . \end{aligned}$$

In Exercise 3.3.5 we saw that the natural inclusions

$$\varphi_j : R \longrightarrow S_j , \qquad \varphi_k : R \longrightarrow S_k$$

factor, respectively, as

$$R \xrightarrow{\alpha_g} R[1/g] \xrightarrow{\varphi'_j} S_j , \qquad R \xrightarrow{\alpha_f} R[1/f] \xrightarrow{\varphi'_k} S_k ,$$

so that φ'_j and φ'_k are isomorphisms. Lemma 3.4.3 defines for us a map $\alpha^g_f : R[1/g] \longrightarrow R[1/f]$. But then $\varphi'_k \circ \alpha^g_f \circ \{\varphi'_j\}^{-1}$ is a map $S_j \longrightarrow S_k$. What is it?

We will give the proof that $\varphi'_k \circ \alpha^g_f \circ \{\varphi'_j\}^{-1}$ is simply the natural inclusion $S_j \longrightarrow S_k$. Consider the square

$$\begin{array}{ccc} R[1/g] & \xrightarrow{\ \alpha^g_f\ } & R[1/f] \\ {\scriptstyle \varphi'_j}\big\downarrow & & \big\downarrow{\scriptstyle \varphi'_k} \\ S_j & \xrightarrow[\text{inclusion}]{} & S_k \end{array} \quad ;$$

we wish to show that it commutes. To this end consider the diagrams

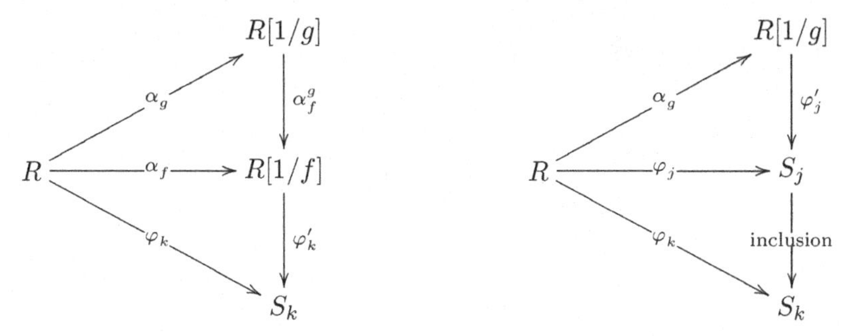

Each of the four right-angle triangles commutes. For three of the triangles, the commutativity comes from the definitions of α^g_f, φ'_j and φ'_k. For the fourth triangle the commutativity is obvious; the map φ_j (respectively φ_k) is just the natural inclusion of R into S_j (respectively, into S_k). The inclusion of S_j into S_k respects the inclusions of R into both. We conclude that the two composites

$$\begin{array}{ccc} R & & \\ {\scriptstyle \alpha_g}\big\downarrow & & \\ R[1/g] & & \\ {\scriptstyle \varphi'_j}\big\downarrow & & \\ S_j & \xrightarrow[\text{inclusion}]{} & S_k \end{array} \qquad\qquad \begin{array}{ccc} R & & \\ {\scriptstyle \alpha_g}\big\downarrow & & \\ R[1/g] & \xrightarrow{\ \alpha^g_f\ } & R[1/f] \\ & & \big\downarrow{\scriptstyle \varphi'_k} \\ & & S_k \end{array}$$

are equal; both are simply $\varphi_k : R \longrightarrow S_k$. Now the uniqueness statement in Proposition 3.3.2 asserts that the morphism $\varphi_k : R \longrightarrow S_k$ can factor through $\alpha_g : R \longrightarrow R[1/g]$ in only one way; hence the two composites in the square

$$\begin{array}{ccc} R[1/g] & \xrightarrow{\ \alpha^g_f\ } & R[1/f] \\ {\scriptstyle \varphi'_j}\big\downarrow & & \big\downarrow{\scriptstyle \varphi'_k} \\ S_j & \xrightarrow[\text{inclusion}]{} & S_k \end{array}$$

must be equal.

Exercise 3.5.2. In Example 3.5.1 we showed that the isomorphisms

$$\varphi'_j : R[1/g] \longrightarrow S_j \ , \qquad\qquad \varphi'_k : R[1/f] \longrightarrow S_k$$

fit into a commutative square

$$\begin{array}{ccc} R[1/g] & \xrightarrow{\alpha^g_f} & R[1/f] \\ {\scriptstyle \varphi'_j}\downarrow & & \downarrow{\scriptstyle \varphi'_k} \\ S_j & \xrightarrow[\text{inclusion}]{} & S_k \end{array} .$$

We leave it to the reader to check the commutativity of the square

$$\begin{array}{ccc} \left[\bigotimes_{i=1}^{j} \mathbb{C}[x_i, x_i^{-1}]\right] \otimes \left[\bigotimes_{i=j+1}^{k} \mathbb{C}[x_i]\right] \otimes \left[\bigotimes_{i=k+1}^{n} \mathbb{C}[x_i]\right] & \xrightarrow{\tau} & S_j \\ {\scriptstyle 1 \otimes \left[\bigotimes_{i=j+1}^{k} \varphi_i\right] \otimes 1}\Big\downarrow & & \Big\downarrow{\scriptstyle \text{inclusion}} \\ \left[\bigotimes_{i=1}^{j} \mathbb{C}[x_i, x_i^{-1}]\right] \otimes \left[\bigotimes_{i=j+1}^{k} \mathbb{C}[x_i, x_i^{-1}]\right] \otimes \left[\bigotimes_{i=k+1}^{n} \mathbb{C}[x_i]\right] & \xrightarrow[\tau]{} & S_k \end{array}$$

where the horizontal maps τ are the isomorphisms of Remark 3.3.6, and where the maps $\varphi_i : \mathbb{C}[x_i] \longrightarrow \mathbb{C}[x_i, x_i^{-1}]$ are the natural inclusions, also as in Remark 3.3.6. Putting these two commutative squares together, we obtain a commutative square where the rows are isomorphisms

$$\begin{array}{ccc} \left[\bigotimes_{i=1}^{j} \mathbb{C}[x_i, x_i^{-1}]\right] \otimes \left[\bigotimes_{i=j+1}^{k} \mathbb{C}[x_i]\right] \otimes \left[\bigotimes_{i=k+1}^{n} \mathbb{C}[x_i]\right] & \xrightarrow{\{\varphi'_j\}^{-1}\tau} & R[1/g] \\ \Big\downarrow {\scriptstyle 1 \otimes \left[\bigotimes_{i=j+1}^{k} \varphi_i\right] \otimes 1} & & \Big\downarrow {\scriptstyle \alpha_f^g} \\ \left[\bigotimes_{i=1}^{j} \mathbb{C}[x_i, x_i^{-1}]\right] \otimes \left[\bigotimes_{i=j+1}^{k} \mathbb{C}[x_i, x_i^{-1}]\right] \otimes \left[\bigotimes_{i=k+1}^{n} \mathbb{C}[x_i]\right] & \xrightarrow[\{\varphi'_k\}^{-1}\tau]{} & R[1/f] \end{array}$$

The discussion above can easily be generalized, but to state the generalization we need to establish some notation.

Notation 3.5.3. As above, let $R = \mathbb{C}[x_0, x_1, \ldots, x_n]$ be the polynomial ring in $(n+1)$ variables, and let $X = \mathrm{Spec}(R)$. Let $J \subset \{0, 1, \ldots, n\}$ be any subset. We define

(i) The element $f_J \in R$ is given by the formula $f_J = \prod_{j \in J} x_j$.

(ii) The basic open subset $X_J \subset X$ is simply $X_J = X_{f_J}$, with f_J as in (i).

(iii) The ring S_J will be the Laurent polynomial ring

$$S_J \quad = \quad \mathbb{C}[x_0, x_1, \ldots, x_n, x_{j_1}^{-1}, x_{j_2}^{-1}, \ldots, x_{j_\ell}^{-1}]\,,$$

where $J = \{j_1, j_2, \ldots, j_\ell\}$.

(iv) The ring C_J will be the tensor product

$$C_J \quad = \quad \bigotimes_{i=0}^{n} C_i^J\,,$$

where

$$C_i^J \quad = \quad \begin{cases} \mathbb{C}[x_i] & \text{if } i \notin J, \\ \mathbb{C}[x_i, x_i^{-1}] & \text{if } i \in J. \end{cases}$$

(v) The ring homomorphism $\tau : C_J \longrightarrow S_J$ is given by the formula

$$\tau(f_0 \otimes f_1 \otimes \cdots \otimes f_n) \quad = \quad f_0 f_1 \cdots f_n\,.$$

(vi) Suppose $J \subset J' \subset \{0, 1, \ldots, n\}$ are two subsets. The ring homomorphism $\psi^{JJ'} : C_J \longrightarrow C_{J'}$ is the tensor product

$$\bigotimes_{i=0}^{n} \psi_i^{JJ'} : \bigotimes_{i=0}^{n} C_i^J \longrightarrow \bigotimes_{i=0}^{n} C_i^{J'} ,$$

where $\psi_i^{JJ'} : C_i^J \longrightarrow C_i^{J'}$ is given by the formula

$$\psi_i^{JJ'} = \begin{cases} 1 & \text{if } i \notin J' - J, \\ \varphi_i & \text{if } i \in J' - J. \end{cases}$$

In this formula, as in Remark 3.3.6, $\varphi_i : \mathbb{C}[x_i] \longrightarrow \mathbb{C}[x_i, x_i^{-1}]$ is the natural inclusion.

Exercise 3.5.4. Let the notation be as in Notation 3.5.3. The reader can check the following:

(i) The natural inclusion $R \longrightarrow S_J$ factors uniquely as

$$R \xrightarrow{\alpha_{f_J}} R[1/f_J] \xrightarrow{\varphi_J} S_J ;$$

furthermore, the ring homomorphism $\varphi_J : R[1/f_J] \longrightarrow S_J$ is an isomorphism.

(ii) The map τ of Notation 3.5.3(v) is an isomorphism $\tau : C_J \longrightarrow S_J$.

(iii) Suppose $J \subset J' \subset \{0, 1, \ldots, n\}$ are two subsets. The following two squares are commutative

$$\begin{array}{ccc} R[1/f_J] & \xrightarrow{\alpha_{f_{J'}}^{f_J}} & R[1/f_{J'}] \\ {\scriptstyle \varphi_J}\big\downarrow & & \big\downarrow{\scriptstyle \varphi_{J'}} \\ S_J & \xrightarrow[\text{inclusion}]{} & S_{J'} \end{array} \qquad \begin{array}{ccc} C_J & \xrightarrow{\psi^{JJ'}} & C_{J'} \\ {\scriptstyle \tau}\big\downarrow & & \big\downarrow{\scriptstyle \tau} \\ S_J & \xrightarrow[\text{inclusion}]{} & S_{J'} \end{array} .$$

Combining them, we obtain a single commutative square of ring homomorphisms

$$\begin{array}{ccc} C_J & \xrightarrow{\psi^{JJ'}} & C_{J'} \\ {\scriptstyle \{\varphi_J\}^{-1}\tau}\big\downarrow & & \big\downarrow{\scriptstyle \{\varphi_{J'}\}^{-1}\tau} \\ R[1/f_J] & \xrightarrow[\alpha_{f_{J'}}^{f_J}]{} & R[1/f_{J'}] \end{array} ,$$

where the vertical maps are isomorphisms.

Remark 3.5.5. Exercise 3.3.5, Remark 3.3.6, Example 3.5.1 and Exercise 3.5.2 are all special cases of Exercise 3.5.4. They are the special cases where $J = \{0, 1, \ldots, j\}$ and $J' = \{0, 1, \ldots, k\}$ are the subsets containing the first $(j+1)$ (respectively the first $(k+1)$) integers in $\{0, 1, \ldots, n\}$. The only difficulty in treating the more general case is the setting up of the notation; see Notation 3.5.3.

Remark 3.5.6. In Definition 3.4.6 we set

$$\Gamma(X_f, \mathcal{O}) \quad = \quad \Gamma(X_f, \widetilde{R}) \quad = \quad R[1/f] \ .$$

In the special case where $R = \mathbb{C}[x_0, x_1, \ldots, x_n]$ is the polynomial ring in $(n+1)$ variables, where $f = f_J$ as in Notation 3.5.3(i), and where $X_J = X_{f_J}$ as in Notation 3.5.3(ii), we obtain

$$\Gamma(X_J, \mathcal{O}) \quad = \quad R[1/f_J] \ .$$

In Definition 3.4.7 we declared that, for open sets $X_f \subset X_g \subset X$, the restriction map

$$\mathrm{res}^{X_g}_{X_f} : \Gamma(X_g, \mathcal{O}) \longrightarrow \Gamma(X_f, \mathcal{O})$$

is defined to be

$$\alpha^g_f : R[1/g] \longrightarrow R[1/f] \ .$$

If $J \subset J' \subset \{0, 1, \ldots, n\}$ are subsets, we have an inclusion of basic open sets

$$X_{J'} = X_{f_{J'}} \quad \subset \quad X_J = X_{f_J} \quad \subset \quad X \ ,$$

and the definition tells us that the restriction map

$$\mathrm{res}^{X_J}_{X_{J'}} : \Gamma(X_J, \mathcal{O}) \longrightarrow \Gamma(X_{J'}, \mathcal{O})$$

is nothing other than

$$\alpha^{f_J}_{f_{J'}} : R[1/f_J] \longrightarrow R[1/f_{J'}] \ .$$

The commutative squares of Exercise 3.5.4(iii) become

$$\begin{array}{ccc} C_J & \xrightarrow{\psi^{JJ'}} & C_{J'} \\ {\scriptstyle \{\varphi_J\}^{-1}T} \downarrow & & \downarrow {\scriptstyle \{\varphi_{J'}\}^{-1}T} \\ \Gamma(X_J, \mathcal{O}) & \xrightarrow[\mathrm{res}^{X_J}_{X_{J'}}]{} & \Gamma(X_{J'}, \mathcal{O}) \end{array} \qquad \begin{array}{ccc} S_J & \xrightarrow{\text{inclusion}} & S_{J'} \\ \downarrow {\scriptstyle \{\varphi_J\}^{-1}} & & \downarrow {\scriptstyle \{\varphi_{J'}\}^{-1}} \\ \Gamma(X_J, \mathcal{O}) & \xrightarrow[\mathrm{res}^{X_J}_{X_{J'}}]{} & \Gamma(X_{J'}, \mathcal{O}) \ , \end{array}$$

where the vertical maps are isomorphisms. We concede that this is not particularly profound; we have been discussing a baby example of the general theory. Still, at least in this simple case, we have understood the restriction maps.

3.6 Maps of ringed spaces $\left(\operatorname{Spec}(S), \widetilde{S}\right) \longrightarrow \left(\operatorname{Spec}(R), \widetilde{R}\right)$

Starting from a ring R we have learned how to torture it and produce a ringed space $\left(\operatorname{Spec}(R), \widetilde{R}\right)$. The details of the construction might be complicated, but we should never lose sight of the fact that $\left(\operatorname{Spec}(R), \widetilde{R}\right)$ contains no information that was not already in the ring R. No amount of torture can extract information that is not there. It seems only natural to wonder if a ring homomorphism $\varphi : R \longrightarrow S$ induces a map of ringed spaces, in one direction or the other, between $\left(\operatorname{Spec}(R), \widetilde{R}\right)$ and $\left(\operatorname{Spec}(S), \widetilde{S}\right)$. We have agreed that the ringed space $\left(\operatorname{Spec}(R), \widetilde{R}\right)$ contains no data which was not already in the ring R, and any map $\varphi : R \longrightarrow S$, preserving all the structure of a ring, must surely induce a map between $\left(\operatorname{Spec}(R), \widetilde{R}\right)$ and $\left(\operatorname{Spec}(S), \widetilde{S}\right)$. In this section we will produce such a map.

We begin with an easy fact from commutative algebra, that will permit us to define the map between the topogical spaces $\operatorname{Spec}(R)$ and $\operatorname{Spec}(S)$.

Lemma 3.6.1. *Let $\varphi : R \longrightarrow S$ be a ring homomorphism. Then, given a prime ideal $\mathfrak{p} \subset S$, the set $\varphi^{-1}\mathfrak{p}$ is a prime ideal in R.*

Proof. If $\mathfrak{p}$ is prime then $S/\mathfrak{p}$ is an integral domain. The composition of maps

$$R \xrightarrow{\quad \varphi \quad} S \xrightarrow{\qquad} S/\mathfrak{p}$$

is a map from the ring R to the integral domain $S/\mathfrak{p}$, and its kernel $\varphi^{-1}\mathfrak{p}$ is a prime ideal in R. □

Definition 3.6.2. *Given any ring homomorphism $\varphi : R \longrightarrow S$, then*

$$\operatorname{Spec}(\varphi) : \operatorname{Spec}(S) \longrightarrow \operatorname{Spec}(R)$$

is defined to be the map taking a prime ideal $\mathfrak{p} \subset S$ to the prime ideal $\varphi^{-1}\mathfrak{p} \subset R$.

Lemma 3.6.3. *Given a ring homomorphism $\varphi : R \longrightarrow S$, the map $\operatorname{Spec}(\varphi)$ is continuous.*

Proof. The closed sets of $\operatorname{Spec}(R)$ are all of the form $V(\mathfrak{a})$, where $\mathfrak{a} \subset R$ is an ideal. It therefore suffices to show $\{\operatorname{Spec}(\varphi)\}^{-1}V(\mathfrak{a})$ is closed in $\operatorname{Spec}(S)$, for each ideal $\mathfrak{a} \subset R$.

Now $\operatorname{Spec}(\varphi)$ is the map taking a prime ideal $\mathfrak{p} \subset S$ to $\varphi^{-1}\mathfrak{p} \subset R$. Therefore a prime ideal $\mathfrak{p} \subset S$ is in $\{\operatorname{Spec}(\varphi)\}^{-1}V(\mathfrak{a})$ if and only if

$\varphi^{-1}\mathfrak{p} \supset \mathfrak{a}$. This is equivalent to $\mathfrak{p} \supset \varphi(\mathfrak{a})$. Let $\mathfrak{b}$ be the ideal of S generated by $\varphi(\mathfrak{a})$; then

$$\{\varphi^{-1}\mathfrak{p} \supset \mathfrak{a}\} \qquad \Longleftrightarrow \qquad \{\mathfrak{p} \supset \mathfrak{b}\} ,$$

that is $\{\mathrm{Spec}(\varphi)\}^{-1}V(\mathfrak{a}) = V(\mathfrak{b})$. □

Remark 3.6.4. Let us also note what happens to the basic open subsets of Section 3.2 under the map $\mathrm{Spec}(\varphi)$. Put $X = \mathrm{Spec}(S)$ and $Y = \mathrm{Spec}(R)$; then we have seen above that $\varphi : R \longrightarrow S$ induces a continuous map

$$\mathrm{Spec}(\varphi) : X \longrightarrow Y .$$

What can we say about $\{\mathrm{Spec}(\varphi)\}^{-1}Y_f$, where $Y_f \subset Y$ is a basic open set as in Section 3.2?

By the proof of Lemma 3.6.3 we have that $\{\mathrm{Spec}(\varphi)\}^{-1}V(Rf) = V\big(S\varphi(f)\big)$. Taking complements, we conclude

$$\{\mathrm{Spec}(\varphi)\}^{-1}Y_f \quad = \quad X_{\varphi(f)} .$$

Now we will see how to extend the map $\mathrm{Spec}(\varphi) : X \longrightarrow Y$ to a map of ringed spaces.

Proposition 3.6.5. *Let $\varphi : R \longrightarrow S$ be a ring homomorphism. As in Remark 3.6.4 put $X = \mathrm{Spec}(S)$ and $Y = \mathrm{Spec}(R)$. The continuous map $\mathrm{Spec}(\varphi) : X \longrightarrow Y$, of Definition 3.6.2 and Lemma 3.6.3, may be extended to a morphism of ringed spaces*

$$(\mathrm{Spec}(\varphi), \widetilde{\varphi}) : (X, \widetilde{S}) \longrightarrow (Y, \widetilde{R}) .$$

The ring homomorphism

$$\widetilde{\varphi}_Y : \Gamma(Y, \widetilde{R}) \longrightarrow \Gamma\big(\{\mathrm{Spec}(\varphi)\}^{-1}Y, \widetilde{S}\big) \quad = \quad \Gamma(X, \widetilde{S})$$

agrees with $\varphi : R \longrightarrow S$. Furthermore, the map of ringed spaces we will define is the unique map

$$\big(\mathrm{Spec}(\varphi), \Phi^*\big) : (X, \widetilde{S}) \longrightarrow (Y, \widetilde{R}) ,$$

such that the underlying continuous map $X = \mathrm{Spec}(S) \longrightarrow \mathrm{Spec}(R) = Y$ is $\mathrm{Spec}(\varphi)$, and so that the map $\Phi^*_Y : R \longrightarrow S$ agrees with $\varphi : R \longrightarrow S$.

Remark 3.6.6. A little explanation of the proposition might be in order, before we launch into the proof. We are interested in morphisms of ringed spaces

$$(\mathrm{Spec}(\varphi), \Phi^*) : (X, \widetilde{S}) \longrightarrow (Y, \widetilde{R}) ,$$

for which the underlying continuous map of topological spaces is $\mathrm{Spec}(\varphi)$:

$X \longrightarrow Y$. Any such map will induce, for every open set $U \subset Y$, a ring homomorphism $\Phi_U^* : \Gamma(U, \widetilde{R}) \longrightarrow \Gamma\left(\{\mathrm{Spec}(\varphi)\}^{-1}U \ , \ \widetilde{S}\right)$. In the special case where $U = Y$ we have $\{\mathrm{Spec}(\varphi)\}^{-1}Y = X$, and hence, at the level of global sections (see Remark 2.2.6 for the jargon), we must be given a ring homomorphism

$$\Phi_Y^* : R \quad = \quad \Gamma(Y, \widetilde{R}) \longrightarrow \Gamma(X, \widetilde{S}) \quad = \quad S \ .$$

Proposition 3.6.5 asserts that there exists a unique morphism of ringed spaces $(\mathrm{Spec}(\varphi), \Phi^*)$, for which $\Phi_Y^* : R \longrightarrow S$ agrees with $\varphi : R \longrightarrow S$. We call this unique morphism $(\mathrm{Spec}(\varphi), \widetilde{\varphi})$.

Proof. We have a continuous map $\mathrm{Spec}(\varphi) : X \longrightarrow Y$. By Remark 3.6.4 the inverse image of Y_f is $X_{\varphi(f)}$. To have a map of ringed spaces $\big(\mathrm{Spec}(\varphi), \Phi^*\big)$ we must therefore define ring homomorphisms

$$\Phi_{Y_f}^* : \Gamma(Y_f, \widetilde{R}) \longrightarrow \Gamma(X_{\varphi(f)}, \widetilde{S}) \ .$$

That is, we need maps $\Phi_{Y_f}^* : R[1/f] \longrightarrow S[1/\varphi(f)]$. Assume that, in the special case where $f = 1$, the map $\Phi_Y^* : R \longrightarrow S$ agrees with $\varphi : R \longrightarrow S$. The definition of maps of ringed spaces (see Definition 2.2.12(iii)) gives a commutative square

$$\begin{array}{ccc} \Gamma(Y, \widetilde{R}) & \xrightarrow{\Phi_Y^*} & \Gamma(X, \widetilde{S}) \\ {\scriptstyle \mathrm{res}_{Y_f}^Y}\downarrow & & \downarrow{\scriptstyle \mathrm{res}_{X_{\varphi(f)}}^X} \\ \Gamma(Y_f, \widetilde{R}) & \xrightarrow[\Phi_{Y_f}^*]{} & \Gamma(X_{\varphi(f)}, \widetilde{S}) \end{array} \quad ,$$

and in our case we know three of the maps; they are

$$\begin{array}{ccc} R & \xrightarrow{\varphi} & S \\ {\scriptstyle \alpha_f}\downarrow & & \downarrow{\scriptstyle \alpha_{\varphi(f)}} \\ R[1/f] & & S[1/\varphi(f)] \end{array} \quad .$$

The map φ takes f to $\varphi(f)$, while the map $\alpha_{\varphi(f)}$ takes $\varphi(f)$ to an invertible element of $S[1/\varphi(f)]$. We conclude that the composite $\alpha_{\varphi(f)}\varphi$ must factor uniquely (see Proposition 3.3.2) through α_f. In other words there is a unique choice of a map $\Phi_{Y_f}^*$ rendering commutative the square

$$\begin{array}{ccc} R & \xrightarrow{\varphi} & S \\ {\scriptstyle \alpha_f}\downarrow & & \downarrow{\scriptstyle \alpha_{\varphi(f)}} \\ R[1/f] & \xrightarrow[\Phi_{Y_f}^*]{} & S[1/\varphi(f)] \end{array} \quad .$$

Define $\Phi^*_{Y_f}$ to be this unique map. Now observe that, if we have basic open sets $Y_f \subset Y_g \subset Y$, then there are two commutative diagrams

$$
\begin{array}{ccc}
R & \xrightarrow{\ \varphi\ } & S \\
{\scriptstyle \alpha_g}\downarrow & & \downarrow{\scriptstyle \alpha_{\varphi(g)}} \\
R[1/g] & & S[1/\varphi(g)] \\
{\scriptstyle \alpha^g_f}\downarrow & & \downarrow{\scriptstyle \alpha^{\varphi(g)}_{\varphi(f)}} \\
R[1/f] & \xrightarrow[\ \Phi^*_{Y_f}\ ]{} & S[1/\varphi(f)]
\end{array}
\qquad\qquad
\begin{array}{ccc}
R & \xrightarrow{\ \varphi\ } & S \\
{\scriptstyle \alpha_g}\downarrow & & \downarrow{\scriptstyle \alpha_{\varphi(g)}} \\
R[1/g] & \xrightarrow[\ \Phi^*_{Y_g}\ ]{} & S[1/\varphi(g)] \\
 & & \downarrow{\scriptstyle \alpha^{\varphi(g)}_{\varphi(f)}} \\
 & & S[1/\varphi(f)]
\end{array}
\quad .
$$

The commutativity of the diagram on the right is just by the definition of $\Phi^*_{Y_g}$. The commutativity of the diagram on the left comes from the definition of $\Phi^*_{Y_f}$, combined with Lemma 3.4.3 which asserts that $\alpha_f = \alpha^g_f \alpha_g$ and $\alpha_{\varphi(f)} = \alpha^{\varphi(g)}_{\varphi(f)} \alpha_{\varphi(g)}$. Taken together these commutative diagrams tell us that the three composite maps $R \longrightarrow S[1/\varphi(f)]$, given below, are equal to each other:

$$
\begin{array}{ccc}
R & & \\
{\scriptstyle \alpha_g}\downarrow & & \\
R[1/g] & \xrightarrow{\ \Phi^*_{Y_g}\ } & S[1/\varphi(g)] \\
{\scriptstyle \alpha^g_f}\downarrow & & \downarrow{\scriptstyle \alpha^{\varphi(g)}_{\varphi(f)}} \\
R[1/f] & \xrightarrow[\ \Phi^*_{Y_f}\ ]{} & S[1/\varphi(f)]
\end{array}
\qquad\qquad
\begin{array}{ccc}
R & \xrightarrow{\ \varphi\ } & S \\
 & & \downarrow{\scriptstyle \alpha_{\varphi(g)}} \\
 & & S[1/\varphi(g)] \\
 & & \downarrow{\scriptstyle \alpha^{\varphi(g)}_{\varphi(f)}} \\
 & & S[1/\varphi(f)]
\end{array}
\quad .
$$

The three equal maps all take $g \in R$ to an invertible element of $S[1/\varphi(f)]$. But now the diagram on the left gives us two factorizations of this map $R \longrightarrow S[1/\varphi(f)]$ through $\alpha_g : R \longrightarrow R[1/g]$; by the uniqueness part of Proposition 3.3.2 they are equal. We deduce that the square below is commutative

$$
\begin{array}{ccc}
R[1/g] & \xrightarrow{\ \Phi^*_{Y_g}\ } & S[1/\varphi(g)] \\
{\scriptstyle \alpha^g_f}\downarrow & & \downarrow{\scriptstyle \alpha^{\varphi(g)}_{\varphi(f)}} \\
R[1/f] & \xrightarrow[\ \Phi^*_{Y_f}\ ]{} & S[1/\varphi(f)]
\end{array}
\quad .
$$

This square is precisely

$$\begin{array}{ccc}\Gamma(Y_g,\widetilde{R}) & \xrightarrow{\Phi^*_{Y_g}} & \Gamma(X_{\varphi(g)},\widetilde{S}) \\ {\scriptstyle \mathrm{res}^{Y_g}_{Y_f}}\downarrow & & \downarrow{\scriptstyle \mathrm{res}^{X_{\varphi(g)}}_{X_{\varphi(f)}}} \\ \Gamma(Y_f,\widetilde{R}) & \xrightarrow[\Phi^*_{Y_f}]{} & \Gamma(X_{\varphi(f)},\widetilde{S})\end{array} \quad .$$

So far we have shown that the maps Φ^*_V can be defined, and are unique, when $V = Y_f$ is a basic open set. To extend to all open sets we proceed as in the proof of Lemma 3.4.10. Let V be an arbitrary open subset of Y, and let $U \subset X$ be its inverse image; that is, $U = \{\mathrm{Spec}(\varphi)\}^{-1}V$. Take the maximal cover of V by open sets Y_{f_i}; that is, let the index set I below be the set of all $f_i \in R$ with $Y_{f_i} \subset V$. We have a commutative diagram with exact rows

$$\begin{array}{ccccc} 0 \longrightarrow \Gamma(V,\widetilde{R}) & \longrightarrow & \prod_{i\in I}\Gamma(Y_{f_i},\widetilde{R}) & \longrightarrow & \prod_{i,j\in I}\Gamma(Y_{f_i}\cap Y_{f_j},\widetilde{R}) \\ & & \downarrow & & \downarrow \\ 0 \longrightarrow \Gamma(U,\widetilde{S}) & \longrightarrow & \prod_{i\in I}\Gamma(X_{\varphi(f_i)},\widetilde{S}) & \longrightarrow & \prod_{i,j\in I}\Gamma(X_{\varphi(f_i)}\cap X_{\varphi(f_j)},\widetilde{S}) \end{array}$$

and we define $\Phi^*_V : \Gamma(V,\widetilde{R}) \longrightarrow \Gamma(U,\widetilde{S})$ to be the unique map making the diagram commute. We leave it to the reader to check that, in the case where $V = Y_f$ is a basic open set, this agrees with the definition we already had. And the reader should also check that, for any open sets $V \subset V' \subset Y$, the following square commutes:

$$\begin{array}{ccc}\Gamma(V',\widetilde{R}) & \xrightarrow{\Phi^*_{V'}} & \Gamma\Big(\{\mathrm{Spec}(\varphi)\}^{-1}V',\widetilde{S}\Big) \\ {\scriptstyle \mathrm{res}^{V'}_{V}}\downarrow & & \downarrow{\scriptstyle \mathrm{res}^{\{\mathrm{Spec}(\varphi)\}^{-1}V'}_{\{\mathrm{Spec}(\varphi)\}^{-1}V}} \\ \Gamma(V,\widetilde{R}) & \xrightarrow[\Phi^*_V]{} & \Gamma\Big(\{\mathrm{Spec}(\varphi)\}^{-1}V,\widetilde{S}\Big)\end{array} \quad .$$

□

3.7 Some immediate consequences

In this section we will prove two fairly immediate consequences of Proposition 3.6.5. The first says that the maps $(\mathrm{Spec}(\varphi),\widetilde{\varphi})$ compose well. The second asserts that, in the special case of the ring homomorphism $\alpha_f : R \longrightarrow R[1/f]$, the map of ringed spaces $\Big(\mathrm{Spec}(\alpha_f),\widetilde{\alpha_f}\Big)$ identifies $\Big(\mathrm{Spec}(R[1/f]),\widetilde{R[1/f]}\Big)$ as an open subset (with the induced sheaf of rings) of $\left(\operatorname{Spec}(R), \widetilde{R}\right)$. In fact if we put $X = \operatorname{Spec}(R)$, and if X_f is the basic open subset of Section 3.2, then $\left(\operatorname{Spec}(R[1/f]), \widetilde{R[1/f]}\right)$ is identified with $(X_f, \widetilde{R}|_{X_f})$. In particular the open subset $X_f \subset X$, with its induced sheaf of rings $\widetilde{R}|_{X_f}$, is naturally identified as being itself of the form $\left(\operatorname{Spec}(S), \widetilde{S}\right)$.

Proposition 3.7.1. *Given a pair of ring homomorphisms $\varphi : R \longrightarrow S$ and $\psi : S \longrightarrow T$, we have maps of ringed spaces*

$$\left(\operatorname{Spec}(T), \widetilde{T}\right) \xrightarrow{(\operatorname{Spec}(\psi), \widetilde{\psi})} \left(\operatorname{Spec}(S), \widetilde{S}\right) \xrightarrow{(\operatorname{Spec}(\varphi), \widetilde{\varphi})} \left(\operatorname{Spec}(R), \widetilde{R}\right) .$$

We assert that the composite agrees with

$$\left(\operatorname{Spec}(T), \widetilde{T}\right) \xrightarrow{(\operatorname{Spec}(\psi\varphi), \widetilde{\psi\varphi})} \left(\operatorname{Spec}(R), \widetilde{R}\right) .$$

Proof. First we show that the map of topological spaces is the same. The map $\operatorname{Spec}(\psi)$ takes a prime ideal $\mathfrak{p} \subset T$ to $\psi^{-1}\mathfrak{p} \subset S$, while the map $\operatorname{Spec}(\varphi)$ takes $\psi^{-1}\mathfrak{p}$ to $\varphi^{-1}\psi^{-1}\mathfrak{p} = \{\psi\varphi\}^{-1}\mathfrak{p}$, which is precisely the image of $\mathfrak{p}$ by $\operatorname{Spec}(\psi\varphi)$.

Now on global sections we have maps

$$\Gamma\left(\operatorname{Spec}(R), \widetilde{R}\right) \xrightarrow{\widetilde{\varphi}_{\operatorname{Spec}(R)}} \Gamma\left(\operatorname{Spec}(S), \widetilde{S}\right) \xrightarrow{\widetilde{\psi}_{\operatorname{Spec}(S)}} \Gamma\left(\operatorname{Spec}(T), \widetilde{T}\right)$$

which, by Proposition 3.6.5, must agree with

$$R \xrightarrow{\varphi} S \xrightarrow{\psi} T ;$$

the composite is $\psi\varphi : R \longrightarrow T$, which, again by Proposition 3.6.5, is just

$$\widetilde{\psi\varphi}_{\operatorname{Spec}(R)} : \Gamma\left(\operatorname{Spec}(R), \widetilde{R}\right) \longrightarrow \Gamma\left(\operatorname{Spec}(T), \widetilde{T}\right) .$$

But the uniqueness statement of Proposition 3.6.5 tells us that the map of ringed spaces $\left(\operatorname{Spec}(T), \widetilde{T}\right) \longrightarrow \left(\operatorname{Spec}(R), \widetilde{R}\right)$, which is $\operatorname{Spec}(\psi\varphi)$ on underlying topological spaces and $\psi\varphi : R \longrightarrow T$ on global sections, is unique. □

Before we state the next proposition it is handy to make a definition.

Definition 3.7.2. *A morphism $(\Phi, \Phi^*) : (X, \mathcal{O}_X) \longrightarrow (Y, \mathcal{O}_Y)$ is called* *an* open immersion† *if the image* $\Phi X \subset Y$ *is an open subset of* Y *and if, in the induced factorization of* (Φ, Φ^*)

$$(X, \mathcal{O}_X) \xrightarrow{(\Psi,\Psi^*)} (\Phi X, \mathcal{O}|_{\Phi X}) \longrightarrow (Y, \mathcal{O}_Y)$$

the map of ringed spaces (Ψ, Ψ^*) *is an isomorphism.*

Put more concisely, an open immersion identifies $(X, \mathcal{O}_X)$ as homeomorphic to an open subset of Y, together with the sheaf of rings. Next we give an example, induced by the map $\alpha_f : R \longrightarrow R[1/f]$.

Proposition 3.7.3. *Let* R *be a ring, and put* $X = \mathrm{Spec}(R)$. *Pick any element* $f \in R$. *Proposition 3.6.5 tells us that the ring homomorphism* $\alpha_f : R \longrightarrow R[1/f]$ *induces a morphism of ringed spaces*

$$\left(\mathrm{Spec}(\alpha_f), \widetilde{\alpha_f}\right) : \left(\mathrm{Spec}(R[1/f]), \widetilde{R[1/f]}\right) \longrightarrow \left(\mathrm{Spec}(R), \widetilde{R}\right) .$$

We assert that this map is an open immersion. Furthermore, the image of the continuous map $\mathrm{Spec}(\alpha_f)$ *is precisely the open set* $X_f \subset X = \mathrm{Spec}(R)$.

Proof. The proof of Proposition 3.7.3 will occupy most of the rest of this section. Because of this in the rest of the section the notation is as in Proposition 3.7.3. We will break up the proof into a sequence of easy lemmas.

Lemma 3.7.4. *Let* $X = \mathrm{Spec}(R)$. *The continuous map of topological spaces*

$$\mathrm{Spec}(\alpha_f) : \mathrm{Spec}(R[1/f]) \longrightarrow \mathrm{Spec}(R)$$

is injective, and the image is precisely $X_f \subset X = \mathrm{Spec}(R)$.

Proof. Let us write $W = \mathrm{Spec}(R[1/f])$ and $X = \mathrm{Spec}(R)$, and we denote by $\mathrm{Spec}(\alpha_f) : W \longrightarrow X$ the map induced by the natural ring homomorphism $\alpha_f : R \longrightarrow R[1/f]$. By Remark 3.6.4 we know that the inverse image of $X_f \subset X$ is $W_{\alpha_f(f)} \subset W$. But $\alpha_f(f)$ is a unit in $R[1/f]$, and Remark 3.2.3 says that $W_{\alpha_f(f)} = W$. That is the inverse image of X_f is all of W. This proves that the image of the map $\mathrm{Spec}(\alpha_f)$ is contained in $X_f \subset X$.

Next we prove that the image contains X_f. Let $\mathfrak{p}$ be an element of X_f, that is a prime ideal not containing f. Then f is non-zero in the

† In algebraic geometry all immersions are assumed injective. This is not the usual convention of differential geometry, where an immersion is only locally injective. To a differential geometer the phrase "open immersion" suggests a local diffeomorphism. The algebraic geometer would refer to this as an étale map. In this book we follow the algebro-geometric conventions.

integral domain $R/\mathfrak{p}$, and invertible in its quotient field $k(\mathfrak{p})$. We have a ring homomorphism $R \longrightarrow k(\mathfrak{p})$ which takes f to an invertible element, and hence (see Proposition 3.3.2) must factor uniquely through

$$R \xrightarrow{\alpha_f} R[1/f] \xrightarrow{\varphi} k(\mathfrak{p}) .$$

The kernel of φ is a prime ideal $\mathfrak{q} \subset R[1/f]$, with $\alpha_f^{-1}\mathfrak{q} = \mathfrak{p}$.

Finally we have to prove the uniqueness of the prime $\mathfrak{q}$ with $\alpha_f^{-1}\mathfrak{q} = \mathfrak{p}$. For this it is irrelevant that $\mathfrak{q}$ and $\mathfrak{p}$ are prime ideals; every ideal of $\mathfrak{q} \subset R[1/f]$, prime or not, is generated by $\alpha_f\alpha_f^{-1}\mathfrak{q}$. After all any element of $R[1/f]$ is of the form

$$\frac{x}{f^n} \quad = \quad x \cdot \frac{1}{f^n} \quad .$$

If x/f^n lies in an ideal $\mathfrak{q}$, then $x = f^n \cdot (x/f^n)$ also lies in the ideal $\mathfrak{q}$, but it is an element of R. In other words x lies in the inverse image of $\mathfrak{q}$ under the map $\alpha_f : R \longrightarrow R[1/f]$; in symbols $x \in \alpha_f^{-1}\mathfrak{q}$. Therefore $x/f^n = (1/f^n) \cdot x$ is the product of $1/f^n \in R[1/f]$ and $x \in \alpha_f\alpha_f^{-1}\mathfrak{q}$. Thus $\mathfrak{q}$ is generated by $\alpha_f\alpha_f^{-1}\mathfrak{q}$. If we have two ideals $\mathfrak{q}, \mathfrak{q}'$, with $\alpha_f^{-1}\mathfrak{q} = \alpha_f^{-1}\mathfrak{q}'$, then $\alpha_f\alpha_f^{-1}\mathfrak{q} = \alpha_f\alpha_f^{-1}\mathfrak{q}'$, and the ideals they generate in $R[1/f]$ must also be equal; that is $\mathfrak{q} = \mathfrak{q}'$. □

Lemma 3.7.5. *Put* $X = \mathrm{Spec}(R)$. *The continuous map*

$$\mathrm{Spec}(\alpha_f) : \mathrm{Spec}(R[1/f]) \longrightarrow \mathrm{Spec}(R)$$

gives a homeomorphism of $\mathrm{Spec}(R[1/f])$ *onto its image* X_f.

Proof. By Lemma 3.7.4 we already know that the map is a continuous bijection onto its image X_f. It remains only to prove that any open set in $\mathrm{Spec}(R[1/f])$ is the inverse image of an open subset in $X_f \subset X$. Since Section 3.2 gives a basis for the topology of $W = \mathrm{Spec}(R[1/f])$, we only need to exhibit basic open sets W_h as inverse images of open subsets of X_f.

Therefore let h be any element of $R[1/f]$. Then $h = g/f^n$, with $g \in R$. Now Lemma 3.2.5 tells us that

$$W_{hf^{n+1}} = W_h \cap W_{f^{n+1}} \; .$$

But f is a unit in $R[1/f]$, and so $W_{f^{n+1}} = W$. The above tells us that $W_{hf^{n+1}} = W_h$. Noting that $hf^{n+1} = (g/f^n) \cdot f^{n+1} = fg$ we have $W_{fg} = W_h$. This expresses W_h as W_{fg} with f and g in R. In all of this we have committed the usual notational crime of letting f and g stand both for the elements of R and for their images by the map $\alpha_f : R \longrightarrow R[1/f]$.

Now we use Remark 3.6.4; we know that $\left\{\mathrm{Spec}(\alpha_f)\right\}^{-1} X_{fg}$ is W_{fg}, and so we have expressed the open set W_h as the inverse image of an open set $X_{fg} = X_f \cap X_g$ contained in X_f. □

At this point we are ready to complete the proof of Proposition 3.7.3. Let the notation be as in the proofs of Lemmas 3.7.4 and 3.7.5; that is, $W = \mathrm{Spec}(R[1/f])$, $X = \mathrm{Spec}(R)$, and $\mathrm{Spec}(\alpha_f) : W \longrightarrow X$ is the natural map giving a homeomorphism of W with the open set $X_f \subset X$. We need to show that the map of ringed spaces

$$\left(\mathrm{Spec}(\alpha_f), \widetilde{\alpha_f}\right) : \left(\mathrm{Spec}(R[1/f]), \widetilde{R[1/f]}\right) \longrightarrow \left(\mathrm{Spec}(R), \widetilde{R}\right)$$

identifies $\left(W, \widetilde{R[1/f]}\right)$ with $\left(X_f, \widetilde{R}|_{X_f}\right)$. Recall that the open subsets $X_f \cap X_g = X_{fg}$ of X_f form a basis, and $\left\{\mathrm{Spec}(\alpha_f)\right\}^{-1} X_{fg} = W_{fg}$. We need only check that, for each element $g \in R$, the ring homomorphism

$$\left\{\widetilde{\alpha_f}\right\}_{X_{fg}} : \Gamma(X_{fg}, \widetilde{R}) \longrightarrow \Gamma(W_{fg}, \widetilde{R[1/f]})$$

is an isomorphism. But we have a commutative square

$$\begin{array}{ccc}
\Gamma(X, \widetilde{R}) & \xrightarrow{\{\widetilde{\alpha_f}\}_X} & \Gamma(W, \widetilde{R[1/f]}) \\
{\scriptstyle \mathrm{res}^X_{X_{fg}}}\downarrow & & \downarrow{\scriptstyle \mathrm{res}^W_{W_{fg}}} \\
\Gamma(X_{fg}, \widetilde{R}) & \xrightarrow[\{\widetilde{\alpha_f}\}_{X_{fg}}]{} & \Gamma(W_{fg}, \widetilde{R[1/f]})
\end{array}$$

which comes to

$$\begin{array}{ccc}
R & \longrightarrow & R[1/f] \\
\downarrow & & \downarrow \\
R[1/fg] & \longrightarrow & \{R[1/f]\}[1/fg]
\end{array} ,$$

and we are reduced to the simple observation that the natural map of R–algebras $R[1/fg] \longrightarrow \{R[1/f]\}[1/fg]$ is an isomorphism. □

Remark 3.7.6. As in Proposition 3.7.3 we let $(X, \widetilde{R})$ be the ringed space $\left(\mathrm{Spec}(R), \widetilde{R}\right)$, and let $X_f \subset X$ be a basic open set. Proposition 3.7.3 proves, among other things, that the ringed space $(X_f, \widetilde{R}|_{X_f})$ is of the form $\left(\mathrm{Spec}(R'), \widetilde{R'}\right)$ for some suitable R'. As it happens the proposition also tells us that $R' = R[1/f]$, but the observation I want to make here is that, ignoring for a second the exact rings involved, we learn that any ringed space $(X, \mathcal{O}_X)$, isomorphic to a $\left(\mathrm{Spec}(R), \widetilde{R}\right)$, admits a basis of open sets each of which is also isomorphic, as a ringed space, to some $\left(\mathrm{Spec}(R'), \widetilde{R'}\right)$.

This exactly parallels what happens with $\mathcal{C}^k$–manifolds. The ringed space $(\mathbb{R}^n, \mathcal{O}_{\mathbb{R}^n})$ has a basis of open sets each of which is isomorphic to $(\mathbb{R}^n, \mathcal{O}_{\mathbb{R}^n})$. This means that, even if we shrink neighborhoods, manifolds remain built up of pieces of the form $(\mathbb{R}^n, \mathcal{O}_{\mathbb{R}^n})$. Based on Proposition 3.7.3 we would expect the analog to hold for ringed spaces whose building blocks are $\left(\mathrm{Spec}(R), \widetilde{R}\right)$.

3.8 A reminder of Hilbert's Nullstellensatz

In Section 3.9 we will want to use Hilbert's Nullstellensatz so, in this section, we give a brief reminder of the statement, as well as references for the proof. The Nullstellensatz is a theorem about finitely generated algebras over a field k. Recall therefore that a k–algebra R is a (commutative) ring R, together with a ring homomorphism $k \longrightarrow R$. The k–algebra R is *finitely generated* if there is a finite set $\{a_1, a_2, \ldots, a_n\}$ of elements of R which generate R; that is every element of R can be written as a polynomial in the $\{a_1, a_2, \ldots, a_n\}$. Another way of saying this is that, if $k[x_1, x_2, \ldots, x_n]$ is the polynomial ring, and the k–algebra homomorphism

$$k[x_1, x_2, \ldots, x_n] \xrightarrow{\ \psi\ } R$$

is defined by the formula $\psi(x_i) = a_i$, then ψ is surjective.

The Nullstellensatz is a theorem about finitely generated algebras R over a field k, which tells us something about the maximal ideals in R. One cheap way to produce maximal ideals is as kernels of k–algebra homomorphisms $\varphi : R \longrightarrow k$. The composite

$$k \longrightarrow R \xrightarrow{\ \varphi\ } k$$

is the identity by assumption; we assumed that the map $R \longrightarrow k$ is a k–algebra homomorphism. This means that φ must be a surjective homomorphism onto a field, and so its kernel is maximal. Among other things the Nullstellensatz gives conditions under which these are the only maximal ideals.

Reminder 3.8.1. Let k be an algebraically closed field. Let R be a finitely generated algebra over k. The Nullstellensatz tells us two things:

(i) All maximal ideals $\mathfrak{m} \subset R$ are kernels of k–algebra homomorphisms $R \longrightarrow k$.

(ii) If $\mathfrak{p}$ is any prime ideal in R then $\mathfrak{p}$ is the intersection of all the maximal ideals containing it.

Remark 3.8.2. Reminder 3.8.1(i) is usually called the "weak Nullstellensatz". The weak Nullstellensatz is somewhat more general than the statement we gave; it also tells us something, about maximal ideals in finitely generated k–algebras, over fields k which are not algebraically closed. The reader can find the more general statement, as well as the observation that the special case where k is an algebraically closed field is as stated above, in [1, Corollary 7.10].

Reminder 3.8.1(ii) usually goes by the name "strong Nullstellensatz". Once again, the strong Nullstellensatz is more general than the assertion we gave. Not only does it cover the case where k is not algebraically closed, the statement extends to ideals $\mathfrak{p}$ which are not prime but only *radical.* The reader can find the more general statement, as well as a proof, in [1, Exercise 14 of Chapter 7].

Remark 3.8.3. In all our applications k will be the field $\mathbb{C}$ of complex numbers. We know that $\mathbb{C}$ is algebraically closed, so the version of the Nullstellensatz given in Reminder 3.8.1 suffices for us. Many of the assertions generalize to other fields, but we do not care. GAGA is a statement about algebras over $\mathbb{C}$.

3.9 Ringed spaces over $\mathbb{C}$

In this section we define ringed spaces over $\mathbb{C}$ and prove that, if R and S are finitely generated algebras over $\mathbb{C}$, then there are no spurious maps

$$(\Phi, \Phi^*) : \left(\mathrm{Spec}(S), \widetilde{S}\right) \longrightarrow \left(\mathrm{Spec}(R), \widetilde{R}\right)$$

of ringed spaces over $\mathbb{C}$. Let us begin with the definition.

Definition 3.9.1. *A* ringed space over $\mathbb{C}$ *is a ringed space* $(X, \mathcal{O})$ *together with a ring homomorphism* $\rho_X : \mathbb{C} \longrightarrow \Gamma(X, \mathcal{O})$*. A* morphism of ringed spaces over $\mathbb{C}$ *is a map of ringed spaces*

$$(\phi, \phi^*) : (X, \mathcal{O}_X) \longrightarrow (Y, \mathcal{O}_Y)$$

so that the following triangle of ring homomorphisms commutes

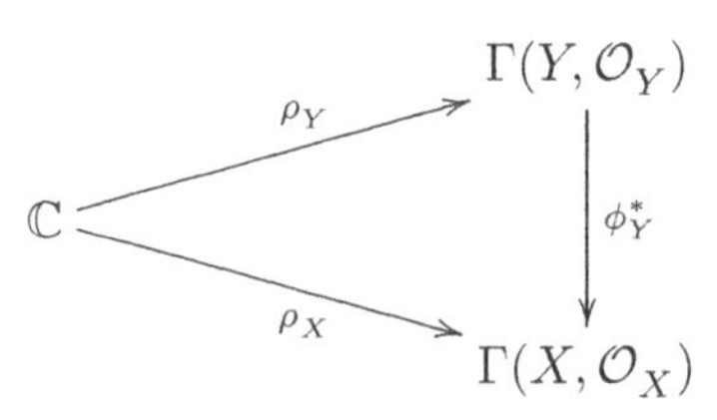

Remark 3.9.2. Note that, if $(X, \mathcal{O})$ is a ringed space over $\mathbb{C}$, then, for any open set $U \subset X$, the ring $\Gamma(U, \mathcal{O})$ is a $\mathbb{C}$–algebra. After all we have a ring homomorphism

$$\mathbb{C} \xrightarrow{\rho_X} \Gamma(X, \mathcal{O}) \xrightarrow{\mathrm{res}_U^X} \Gamma(U, \mathcal{O}) .$$

Obviously the ring homomorphisms $\{ \mathrm{res}_U^X \}\rho_X : \mathbb{C} \longrightarrow \Gamma(U, \mathcal{O})$ are compatible with the restriction maps. In other words $\mathcal{O}$ is not just a sheaf of rings on X, it is actually a sheaf of $\mathbb{C}$–algebras. And morphisms of ringed spaces over $\mathbb{C}$ are precisely the maps respecting the structure.

Remark 3.9.3. Let R be a $\mathbb{C}$–algebra; that is, assume we are given a ring homomorphism $\rho : \mathbb{C} \longrightarrow R$. Then the ringed space $\left(\mathrm{Spec}(R), \widetilde{R}\right)$ is naturally a ringed space over $\mathbb{C}$. This is obvious since we were given a ring homomorphism

$$\rho : \mathbb{C} \longrightarrow R \quad = \quad \Gamma\left(\mathrm{Spec}(R), \widetilde{R}\right) .$$

Furthermore, if $\varphi : R \longrightarrow S$ is a homomorphism of $\mathbb{C}$–algebras, then the map of ringed spaces

$$(\mathrm{Spec}(\varphi), \widetilde{\varphi}) : \left(\mathrm{Spec}(S), \widetilde{S}\right) \longrightarrow \left(\mathrm{Spec}(R), \widetilde{R}\right)$$

is a morphism of ringed spaces over $\mathbb{C}$. After all Proposition 3.6.5 guarantees that the map of global sections

$$\Gamma\left(\mathrm{Spec}(R), \widetilde{R}\right) \longrightarrow \Gamma\left(\mathrm{Spec}(S), \widetilde{S}\right)$$

is just the given $\mathbb{C}$–algebra homomorphism $\varphi : R \longrightarrow S$.

The main theorem in this section asserts

Theorem 3.9.4. *Let R and S be finitely generated algebras over $\mathbb{C}$. Then every morphism of ringed spaces over $\mathbb{C}$*

$$(\Phi, \Phi^*) : \left(\mathrm{Spec}(S), \widetilde{S}\right) \longrightarrow \left(\mathrm{Spec}(R), \widetilde{R}\right)$$

is equal to $(\mathrm{Spec}(\varphi), \widetilde{\varphi})$*, for some $\mathbb{C}$–algebra homomorphism $\varphi : R \longrightarrow S$.*

Proof. Put $X = \mathrm{Spec}(S)$ and $Y = \mathrm{Spec}(R)$. Suppose we have any morphism of ringed spaces over $\mathbb{C}$

$$(\Phi, \Phi^*) : (X, \widetilde{S}) \longrightarrow (Y, \widetilde{R}) .$$

We need to prove that it is of the form $(\mathrm{Spec}(\varphi), \widetilde{\varphi})$. There is only one choice for φ; Proposition 3.6.5 tells us that, given a map of ringed spaces $(\mathrm{Spec}(\varphi), \widetilde{\varphi})$, we can recover φ as the map of global sections

$$R \quad = \quad \Gamma(Y, \widetilde{R}) \xrightarrow{\Phi_Y^*} \Gamma(X, \widetilde{S}) \quad = \quad S .$$

That is φ must be the map $\Phi_Y^* : R \longrightarrow S$. Our problem is to prove that, with $\varphi = \Phi_Y^* : R \longrightarrow S$, the map $(\text{Spec}(\varphi), \widetilde{\varphi})$ agrees with (Φ, Φ^*).

Now we remember the uniqueness part of the statement of Proposition 3.6.5. If we could show that the continuous map $\Phi : X \longrightarrow Y$ agrees with $\text{Spec}(\varphi) : X \longrightarrow Y$, then we would be done. We would have a morphism of ringed spaces $\big(\text{Spec}(\varphi), \Phi^*\big)$ which, on global sections, induces the map $\Phi_Y^* = \varphi : R \longrightarrow S$. Proposition 3.6.5 would allow us to conclude that (Φ, Φ^*) must equal $(\text{Spec}(\varphi), \widetilde{\varphi})$.

Hence it remains only to prove that $\Phi : X \longrightarrow Y$ agrees with $\text{Spec}(\varphi)$. The rest of this section is devoted to the proof. We begin with a lemma that does not depend on the algebras R and S being finitely generated over $\mathbb{C}$.

Lemma 3.9.5. *Let R and S be rings, and*

$$(\Phi, \Phi^*) : \Big(\text{Spec}(S), \widetilde{S}\Big) \longrightarrow \Big(\text{Spec}(R), \widetilde{R}\Big)$$

any morphism of ringed spaces. As above, let φ be the map on global sections

$$R \quad = \quad \Gamma\big(Y, \widetilde{R}\big) \xrightarrow{\Phi_Y^*} \Gamma\big(X, \widetilde{S}\big) \quad = \quad S\,.$$

We assert that, for any $\mathfrak{q} \in X$ (that is for any $\mathfrak{q}$ a prime ideal of S), the point $\Phi(\mathfrak{q}) \in Y$ is a prime ideal of R with

$$\varphi^{-1}\mathfrak{q} \subset \Phi(\mathfrak{q})\,.$$

Proof. We are given a prime ideal $\mathfrak{q}$ of S. The ring $S/\mathfrak{q}$ is an integral domain, and its quotient field $k(\mathfrak{q})$ is a field. We have a ring homomorphism $\psi : S \longrightarrow k(\mathfrak{q})$ with kernel $\mathfrak{q}$. By Proposition 3.6.5 it induces a map of ringed spaces

$$\Big(\text{Spec}(\psi), \widetilde{\psi}\Big) : \Big(\text{Spec}(k(\mathfrak{q})), \widetilde{k(\mathfrak{q})}\Big) \longrightarrow \Big(\text{Spec}(S), \widetilde{S}\Big)\,.$$

We will prove the lemma by studying the composite morphism of ringed spaces

$$\Big(\text{Spec}(k(\mathfrak{q})), \widetilde{k(\mathfrak{q})}\Big) \xrightarrow{(\text{Spec}(\psi), \widetilde{\psi})} \Big(\text{Spec}(S), \widetilde{S}\Big) \xrightarrow{(\Phi, \Phi^*)} \Big(\text{Spec}(R), \widetilde{R}\Big)\,.$$

On the level of topological spaces we have continuous maps

$$\text{Spec}(k(\mathfrak{q})) \xrightarrow{\text{Spec}(\psi)} \text{Spec}(S) \xrightarrow{\Phi} \text{Spec}(R)\,.$$

Now $\text{Spec}(k(\mathfrak{q}))$ is the 1–point space; the only prime ideal in the field $k(\mathfrak{q})$ is $\{0\}$. The map $\text{Spec}(\psi)$ takes the unique point of $\text{Spec}(k(\mathfrak{q}))$ to $\psi^{-1}\{0\} = \mathfrak{q}$. Therefore the image of the unique point in $\text{Spec}(k(\mathfrak{q}))$ by the composite $\Phi \circ \big\{\text{Spec}(\psi)\big\}$ is $\Phi(\mathfrak{q}) \in Y$.

Let $f \in R - \Phi(\mathfrak{q})$; that is, f is an element of the ring R not in the prime ideal $\Phi(\mathfrak{q})$. Then $\Phi(\mathfrak{q}) \in Y_f$, where Y_f is the basic open set of Section 3.2. The inverse image of Y_f under the composite $\Phi \circ \{\operatorname{Spec}(\psi)\}$ contains the unique point in $\operatorname{Spec}(k(\mathfrak{q}))$; the inverse image must be all of $\operatorname{Spec}(k(\mathfrak{q}))$. We deduce a commutative diagram of ring homomorphisms

$$\begin{array}{ccccc}
\Gamma(Y, \widetilde{R}) & \xrightarrow{\Phi_Y^*} & \Gamma(X, \widetilde{S}) & \xrightarrow{\widetilde{\psi}_X} & \Gamma(\operatorname{Spec}(k(\mathfrak{q})), \widetilde{k(\mathfrak{q})}) \\
{\scriptstyle \operatorname{res}^Y_{Y_f}} \downarrow & & \downarrow {\scriptstyle \operatorname{res}^X_{\Phi^{-1}Y_f}} & & \downarrow {\scriptstyle \text{identity}} \\
\Gamma(Y_f, \widetilde{R}) & \xrightarrow[\Phi_{Y_f}^*]{} & \Gamma(\Phi^{-1}Y_f, \widetilde{S}) & \xrightarrow[\widetilde{\psi}_{\Phi^{-1}Y_f}]{} & \Gamma(\operatorname{Spec}(k(\mathfrak{q})), \widetilde{k(\mathfrak{q})})
\end{array} .$$

Forgetting the middle term at the bottom, we have a commutative square

$$\begin{array}{ccc}
R & \xrightarrow{\varphi} & S \\
{\scriptstyle \alpha_f} \downarrow & & \downarrow {\scriptstyle \psi} \\
R[1/f] & \longrightarrow & k(\mathfrak{q})
\end{array} .$$

From the composite

$$\begin{array}{ccc}
R & & \\
{\scriptstyle \alpha_f} \downarrow & & \\
R[1/f] & \longrightarrow & k(\mathfrak{q})
\end{array}$$

we see that the image of $f \in R$ must be non-zero, and the fact that the map is equal to

$$\begin{array}{ccc}
R & \xrightarrow{\varphi} & S \\
 & & \downarrow {\scriptstyle \psi} \\
 & & k(\mathfrak{q})
\end{array}$$

shows that $\varphi(f)$ does not lie in $\mathfrak{q}$, which is the kernel of $\psi : S \longrightarrow k(\mathfrak{q})$. In conclusion we have proved that $f \notin \Phi(\mathfrak{q})$ implies $f \notin \varphi^{-1}\mathfrak{q}$. □

It now remains to see how to deduce Theorem 3.9.4 from Lemma 3.9.5. We need to prove that, in the special case where R and S are finitely generated algebras over $\mathbb{C}$, the map $\Phi : X \longrightarrow Y$ is forced to agree with $\operatorname{Spec}(\varphi)$. The idea is that it is a very easy consequence of Hilbert's Nullstellensatz.

Let $\mathfrak{m}$ be a maximal ideal of S. Because S is a finitely generated algebra over $\mathbb{C}$ the weak form of the Nullstellensatz (see Reminder 3.8.1(i)) tells us that $\mathfrak{m}$ is the kernel of a $\mathbb{C}$–algebra homomorphism $\psi : S \longrightarrow \mathbb{C}$.

The inverse image $\varphi^{-1}\mathfrak{m}$ is the kernel of the composite

$$R \xrightarrow{\ \varphi\ } S \xrightarrow{\ \psi\ } \mathbb{C} ,$$

which must be onto, the longer composite $\mathbb{C} \longrightarrow R \xrightarrow{\varphi} S \xrightarrow{\psi} \mathbb{C}$ being the identity. It follows that $\varphi^{-1}\mathfrak{m}$ is the kernel of a surjective homomorphism to a field, hence a maximal ideal. The inclusion $\varphi^{-1}\mathfrak{m} \subset \Phi(\mathfrak{m})$ must be an equality.

Remark 3.1.10 tells us that a prime ideal $\mathfrak{p}$ is contained in another $\mathfrak{q}$ if and only if $\mathfrak{q}$ is in the closure $\overline{\mathfrak{p}}$ of $\mathfrak{p}$. The map Φ being continuous means that $\mathfrak{q} \in \overline{\mathfrak{p}}$ must imply $\Phi(\mathfrak{q}) \in \overline{\Phi(\mathfrak{p})}$. In other words if $\mathfrak{p} \subset \mathfrak{q}$ then $\Phi(\mathfrak{p}) \subset \Phi(\mathfrak{q})$. In the special case where $\mathfrak{q} = \mathfrak{m}$ is maximal, we conclude that if $\mathfrak{p} \subset \mathfrak{m}$ then

$$\Phi(\mathfrak{p}) \subset \Phi(\mathfrak{m}) = \varphi^{-1}\mathfrak{m} \quad .$$

Therefore

$$\Phi(\mathfrak{p}) \quad \subset \quad \bigcap_{\substack{\mathfrak{m} \supset \mathfrak{p} \\ \mathfrak{m} \text{ a maximal ideal} \\ \text{of the ring } S}} \quad \varphi^{-1}\mathfrak{m} \quad .$$

But the strong form of the Nullstellensatz (see Reminder 3.8.1(ii)) tells us that every prime ideal $\mathfrak{p} \subset S$ is the intersection of the maximal ideals containing it. That is

$$\mathfrak{p} \quad = \quad \bigcap_{\substack{\mathfrak{m} \supset \mathfrak{p} \\ \mathfrak{m} \text{ a maximal ideal} \\ \text{of the ring } S}} \quad \mathfrak{m} \quad .$$

Inverse images of sets respect intersections, hence

$$\varphi^{-1}\mathfrak{p} \quad = \quad \bigcap_{\substack{\mathfrak{m} \supset \mathfrak{p} \\ \mathfrak{m} \text{ a maximal ideal} \\ \text{of the ring } S}} \quad \varphi^{-1}\mathfrak{m} \quad ,$$

and we conclude that $\Phi(\mathfrak{p}) \subset \varphi^{-1}\mathfrak{p}$. But Lemma 3.9.5 gives us the reverse inclusion $\varphi^{-1}\mathfrak{p} \subset \Phi(\mathfrak{p})$, and the proof is complete. □

3.10 Schemes of finite type over $\mathbb{C}$

Finally we are ready to define schemes of finite type over $\mathbb{C}$.

Definition 3.10.1. *A* scheme locally of finite type over $\mathbb{C}$ *is a ringed space over* $\mathbb{C}$ *which is locally isomorphic, as a ringed space over* $\mathbb{C}$*, to* $\left(\operatorname{Spec}(R_i), \widetilde{R_i}\right)$*, with* R_i *finitely generated algebras over* $\mathbb{C}$*.*

Remark 3.10.2. Perhaps once again it is best to make this a little more explicit. A scheme locally of finite type over $\mathbb{C}$ is first of all a ringed space $(X, \mathcal{O})$. By virtue of being a ringed space over $\mathbb{C}$ it comes with a ring homomorphism

$$\rho_X : \mathbb{C} \longrightarrow \Gamma(X, \mathcal{O}) .$$

And we also need the existence of the following:

(i) An open cover $X = \cup_{i \in I} U_i$.
(ii) For each i an isomorphism of ringed spaces over $\mathbb{C}$ of $(U_i, \mathcal{O}|_{U_i})$ with $\left(\operatorname{Spec}(R_i), \widetilde{R_i}\right)$, with each R_i a finitely generated algebra over $\mathbb{C}$.

Definition 3.10.3. *A* scheme of finite type over $\mathbb{C}$ *is a scheme locally of finite type over* $\mathbb{C}$ *as above, where the open cover in Remark 3.10.2(i) may be taken to be finite.*

Definition 3.10.4. *A scheme of finite type over* $\mathbb{C}$ *is called* affine *if the open cover in Remark 3.10.2(i) may be taken to consist of precisely one open set. That is if* $(X, \mathcal{O})$ *is isomorphic, as a ringed space over* $\mathbb{C}$*, to* $\left(\operatorname{Spec}(R), \widetilde{R}\right)$*, for some finitely generated* $\mathbb{C}$*–algebra* R*.*

Remark 3.10.5. The open cover of Remark 3.10.2(i) is usually called an *affine open cover* for X. I suppose in analogy with manifolds we might be tempted to call it an atlas, but tradition is against us. The reason for the tradition is that we are covering X by affine open sets, as in Definition 3.10.4.

Lemma 3.10.6. *Let* R *be a finitely generated algebra over* $\mathbb{C}$*. Put* $X = \operatorname{Spec}(R)$*. Then the ringed space* $\left(X_f, \widetilde{R}|_{X_f}\right)$ *is an affine scheme of finite type over* $\mathbb{C}$*.*

Proof. Proposition 3.7.3 gave us an isomorphism

$$\left(\operatorname{Spec}(R[1/f]), \widetilde{R[1/f]}\right) \longrightarrow \left(X_f, \widetilde{R}|_{X_f}\right) ,$$

and this is a map of ringed spaces over $\mathbb{C}$. Furthermore, since R is finitely generated as an algebra over $\mathbb{C}$, so is $R[1/f]$. □

Proposition 3.10.7. *Let $(X, \mathcal{O})$ be a scheme locally of finite type over $\mathbb{C}$. Let $U \subset X$ be any open subset. Then $(U, \mathcal{O}|_U)$ is also a scheme locally of finite type over $\mathbb{C}$.*

Proof. Because X is locally of finite type over $\mathbb{C}$ it has an open cover $X = \cup_{i \in I} U_i$, with each $(U_i, \mathcal{O}|_{U_i})$ isomorphic to $\left(\operatorname{Spec}(R_i), \widetilde{R_i}\right)$ for some finitely generated $\mathbb{C}$–algebra R_i. Then

$$U = \bigcup_{i \in I} \{U \cap U_i\} .$$

Each $U \cap U_i$ is open in $U_i \cong \operatorname{Spec}(R_i)$. But any open set in $\operatorname{Spec}(R_i)$ is a union of basic open sets; hence $U \cap U_i$ is a union of open sets of the form $\{\operatorname{Spec}(R_i)\}_f \subset \operatorname{Spec}(R_i)$, and by Lemma 3.10.6 all these open subsets of $U \cap U_i$ have the property that they are isomorphic, as ringed spaces over $\mathbb{C}$, to $\left(\operatorname{Spec}(S), \widetilde{S}\right)$ for some suitable finitely generated $\mathbb{C}$–algebras S. □

Remark 3.10.8. Proposition 3.10.7 provides us with some examples of schemes locally of finite type over $\mathbb{C}$. Now we return to more general considerations. What can we say about the transition functions? If $X = \cup_{i \in I} U_i$ is the open cover of Remark 3.10.2(i), how does $U_i \cap U_j$ embed in U_i and U_j? The answer is that $U_i \cap U_j$ can be covered by open sets which are basic open subsets (see Section 3.2) in both $U_i = \operatorname{Spec}(R_i)$ and $U_j = \operatorname{Spec}(R_j)$, and the transition map of ringed spaces is natural. The following proposition formulates a slight generalization of this, somewhat more precisely.

Proposition 3.10.9. *Let $(Z, \mathcal{O})$ be a ringed space over $\mathbb{C}$, and let U and V be open subsets of Z. Suppose we have isomorphisms of ringed spaces over $\mathbb{C}$*

$$(\Phi, \Phi^*) : \left(\operatorname{Spec}(S), \widetilde{S}\right) \longrightarrow (U, \mathcal{O}|_U) ,$$
$$(\Psi, \Psi^*) : \left(\operatorname{Spec}(R), \widetilde{R}\right) \longrightarrow (V, \mathcal{O}|_V) .$$

Put $X = \operatorname{Spec}(S)$ and $Y = \operatorname{Spec}(R)$.

Let $W \subset U \cap V$ be any open subset. We assert that the space W has an open cover

$$W = \bigcup_{i \in I} W_i$$

so that, for each $i \in I$, there exists an $f_i \in S$ and a $g_i \in R$ with $\Phi^{-1} W_i = X_{f_i}$ and $\Psi^{-1} W_i = Y_{g_i}$.

Remark 3.10.10. This means that, for each W_i in the cover, we have elements $g \in R$ and $f \in S$ and isomorphisms of ringed spaces

$$\left(X_f, \widetilde{S}|_{X_f}\right) \xrightarrow{(\Phi,\Phi^*)} (W_i, \mathcal{O}|_{W_i}) \xleftarrow{(\Psi,\Psi^*)} \left(Y_g, \widetilde{R}|_{Y_g}\right) .$$

The composite isomorphism $\left(X_f, \widetilde{S}\right) \longrightarrow \left(Y_g, \widetilde{R}\right)$ is, by Proposition 3.7.3, an isomorphism

$$\left(\mathrm{Spec}(S[1/f]), \widetilde{S[1/f]}\right) \longrightarrow \left(\mathrm{Spec}(R[1/g]), \widetilde{R[1/g]}\right) .$$

By Theorem 3.9.4 this map must be of the form $(\mathrm{Spec}(\varphi), \widetilde{\varphi})$, where $\varphi : R[1/g] \longrightarrow S[1/f]$ is an isomorphism of $\mathbb{C}$–algebras.

Proof. We want an open cover for W; it suffices to show that every point p in W has an open neighborhood $W_p \subset W$ of the required form. Choose therefore a point $p \in W$. We are given a homeomorphism $\Phi :$ $\mathrm{Spec}(S) \longrightarrow U$. The inverse image of $W \subset U$ is an open subset of $X = \mathrm{Spec}(S)$ containing $\Phi^{-1}(p)$. We can choose an $f \in S$ with $\Phi^{-1}(p) \in$ $X_f \subset \Phi^{-1}\{W\}$. By Lemma 3.10.6 we know that the scheme $\left(X_f, \widetilde{S}|_{X_f}\right)$ of finite type over $\mathbb{C}$ is affine and, by the proof, we know more specifically that it is $\left(\mathrm{Spec}(S[1/f]), \widetilde{S[1/f]}\right)$. This becomes identified with an open ringed subspace of $\left(W, \mathcal{O}|_W\right)$ which itself is open in $\left(\mathrm{Spec}(R), \widetilde{R}\right)$. We deduce an open immersion of ringed spaces

$$(\Theta, \Theta^*) : \left(\mathrm{Spec}(S[1/f]), \widetilde{S[1/f]}\right) \longrightarrow \left(\mathrm{Spec}(R), \widetilde{R}\right) .$$

Theorem 3.9.4 says that all such morphisms are of the form $\left(\mathrm{Spec}(\theta), \widetilde{\theta}\right)$ for some $\theta : R \longrightarrow S[1/f]$. The map $\mathrm{Spec}(\theta)$ must embed $X_f =$ $\mathrm{Spec}(S[1/f])$ as an open subset of $Y = \mathrm{Spec}(R)$. Choose an element $g \in R$ so that the image of $\mathrm{Spec}(\theta)$ contains Y_g, and so that $\Psi^{-1}(p) \in Y_g$. We have a diagram of ring homomorphisms

$$\begin{array}{ccc} R & \xrightarrow{\theta} & S[1/f] \\ {\scriptstyle \alpha_g}\downarrow & & \downarrow{\scriptstyle \alpha_{\theta(g)}} \\ R[1/g] & & \{S[1/f]\}[1/\theta(g)] \end{array} ,$$

and under the composite

$$\begin{array}{ccc} R & \xrightarrow{\theta} & S[1/f] \\ & & \downarrow{\scriptstyle \alpha_{\theta(g)}} \\ & & \{S[1/f]\}[1/\theta(g)] \end{array}$$

$g \in R$ maps to an invertible element. We can therefore uniquely complete to a commutative square of ring homomorphisms

$$\begin{array}{ccc} R & \xrightarrow{\theta} & S[1/f] \\ {\scriptstyle \alpha_g}\downarrow & & \downarrow{\scriptstyle \alpha_{\theta(g)}} \\ R[1/g] & \xrightarrow[\varphi]{} & \{S[1/f]\}[1/\theta(g)] \end{array} .$$

Now $\theta(g)$, being an element of $S[1/f]$, must be of the form $\theta(g) = f'/f^n$. Hence the ring $\{S[1/f]\}[1/\theta(g)]$ simplifies to $S[1/ff']$, and the commutative square becomes the slightly more palatable

$$\begin{array}{ccc} R & \xrightarrow{\theta} & S[1/f] \\ {\scriptstyle \alpha_g}\downarrow & & \downarrow{\scriptstyle \alpha_{\theta(g)}} \\ R[1/g] & \xrightarrow[\varphi]{} & S[1/ff'] \end{array} .$$

We deduce a commutative diagram of ringed spaces over $\mathbb{C}$

$$\begin{array}{ccc} \left(\mathrm{Spec}(S[1/ff']), \widetilde{S[1/ff']}\right) & \xrightarrow{(\mathrm{Spec}(\varphi),\widetilde{\varphi})} & \left(\mathrm{Spec}(R[1/g]), \widetilde{R[1/g]}\right) \\ {\scriptstyle \left(\mathrm{Spec}(\alpha_{\theta(g)}),\widetilde{\alpha_{\theta(g)}}\right)}\downarrow & & \downarrow{\scriptstyle \left(\mathrm{Spec}(\alpha_g),\widetilde{\alpha_g}\right)} \\ \left(\mathrm{Spec}(S[1/f]), \widetilde{S[1/f]}\right) & \xrightarrow[(\mathrm{Spec}(\theta),\widetilde{\theta})]{} & \left(\mathrm{Spec}(R), \widetilde{R}\right) \end{array} .$$

The vertical maps are open immersions by Proposition 3.7.3; all maps induced by an $\alpha_h : T \longrightarrow T[1/h]$ are open immersions. The map at the bottom was an open immersion by construction. Hence so is the top map in the square; that is $\varphi : R[1/g] \longrightarrow S[1/ff']$ induces an open immersion $\left(\mathrm{Spec}(S[1/ff']), \widetilde{S[1/ff']}\right) \longrightarrow \left(\mathrm{Spec}(R[1/g]), \widetilde{R[1/g]}\right)$.

Forgetting the sheaves of rings and remembering only the topological spaces, we have a diagram of open inclusions

$$\begin{array}{ccc} \{X_f\}_{\theta(g)} & \longrightarrow & Y_g \\ \downarrow & & \downarrow \\ X_f & \xrightarrow[\mathrm{Spec}(\theta)]{} & Y \end{array} .$$

The open subset $Y_g \subset Y$ was chosen to be contained in the image of $\mathrm{Spec}(\theta)$, and by Remark 3.6.4 $\{X_f\}_{\theta(g)} \subset X_f$ is precisely the inverse image of $Y_g \subset Y$. We conclude that the map $\mathrm{Spec}(\varphi) : \{X_f\}_{\theta(g)} \longrightarrow Y_g$ is a homeomorphism. That is, the open immersion

$$\left(\mathrm{Spec}(S[1/ff']), \widetilde{S[1/ff']}\right) \xrightarrow{(\mathrm{Spec}(\varphi),\widetilde{\varphi})} \left(\mathrm{Spec}(R[1/g]), \widetilde{R[1/g]}\right)$$

is an isomorphism of ringed spaces. $\square$

Remark 3.10.11. We have already spent much time belaboring the parallel with manifolds, so it cannot hurt much more to continue. Proposition 3.10.9 is exactly what we would expect for manifolds if, as we did, we allowed only the ringed space $\left(\mathbb{R}^n, \mathcal{O}_{\mathbb{R}^n}\right)$ as a building block. In general, if X is a manifold and $X = \cup_{i\in I} U_i$ is an atlas, we can insist that $\left(U_i, \mathcal{O}|_{U_i}\right)$ should be isomorphic to $\left(\mathbb{R}^n, \mathcal{O}_{\mathbb{R}^n}\right)$, but do not usually expect that $U_i \cap U_j$ should be of this form. All we usually know is that $U_i \cap U_j$ can be covered by open subsets homeomorphic to $\mathbb{R}^n$. This is exactly the situation of Remark 3.10.8; we do not know that $U_i \cap U_j$ is affine, but we can cover $U_i \cap U_j$ by affines. This is already in Remark 3.7.6. What we learn here is that the affine covering $U_i \cap U_j = \cup_{\lambda\in\Lambda} V_\lambda$ can be chosen so that the open immersions of each $(V_\lambda, \mathcal{O}|_{V_\lambda})$ into $(U_i, \mathcal{O}|_{U_i})$ and into $(U_j, \mathcal{O}|_{U_j})$ are especially nice, embedding each open affine V_λ as a basic open subset in both U_i and U_j.

Definition 3.10.12. *A* morphism of schemes, *either locally or globally of finite type over $\mathbb{C}$, is just any morphism between them as ringed spaces over $\mathbb{C}$. See Definition 3.9.1 for a precise formulation of what is a morphism of ringed spaces over $\mathbb{C}$.*

Remark 3.10.13. Suppose $(\Phi, \Phi^*) : (X, \mathcal{O}_X) \longrightarrow (Y, \mathcal{O}_Y)$ is a morphism of schemes locally of finite type over $\mathbb{C}$. I assert that, locally in X and Y, the map is just

$$(\mathrm{Spec}(\varphi), \widetilde{\varphi}) : \left(\mathrm{Spec}(S), \widetilde{S}\right) \longrightarrow \left(\mathrm{Spec}(R), \widetilde{R}\right) ,$$

with $\varphi : R \longrightarrow S$ a homomorphism of algebras finitely generated over $\mathbb{C}$.

Let us prove the assertion. Assume x is a point in X, and let $\Phi(x) \in Y$ be its image. Choose an open subset $V \subset Y$, so that $\Phi(x) \in V$ and $(V, \{\mathcal{O}_Y\}|_V) \cong \left(\mathrm{Spec}(R), \widetilde{R}\right)$. The open set $\Phi^{-1}V \subset X$ contains x. Proposition 3.10.7 tells us that $(\Phi^{-1}V \ , \ \{\mathcal{O}_X\}|_{\Phi^{-1}V})$ is a scheme locally of finite type over $\mathbb{C}$; we may choose an open subset $U \subset \Phi^{-1}V$, containing x, and so that $(U, \{\mathcal{O}_X\}|_U) \cong \left(\mathrm{Spec}(S), \widetilde{S}\right)$. We deduce a commutative square, in which the vertical maps are open immersions

$$\begin{array}{ccc} (U, \{\mathcal{O}_X\}|_U) & \longrightarrow & (V, \{\mathcal{O}_Y\}|_V) \\ \downarrow & & \downarrow \\ (X, \mathcal{O}_X) & \xrightarrow[(\Phi,\Phi^*)]{} & (Y, \mathcal{O}_Y) \end{array} ,$$

and the construction above tells us that

(i) $x \in U$.

(ii) The top row, in the commutative square above, is isomorphic to a map $\left(\operatorname{Spec}(S), \widetilde{S}\right) \longrightarrow \left(\operatorname{Spec}(R), \widetilde{R}\right)$.

By Theorem 3.9.4 all such maps are of the form $(\operatorname{Spec}(\varphi), \widetilde{\varphi})$.

4
The complex topology

We have now defined schemes locally of finite type over $\mathbb{C}$. Let $(X, \mathcal{O})$ be such a scheme. The scheme $(X, \mathcal{O})$ is a ringed space, and in particular X is a topological space. The topology is a little strange, as we saw in Remark 3.1.10: the space X is almost never Hausdorff, even worse, not every point in X is closed. It is natural enough to look at the subset of closed points of X. We can give it the subspace topology, in which case we denote it by $\mathrm{Max}(X)$. But the set $\mathrm{Max}(X)$ turns out to have another topology, called the complex topology. We denote this topological space X^{an}. As sets of points $\mathrm{Max}(X)$ and X^{an} agree; only the topologies differ.

4.1 Synopsis of the main results

We now briefly summarize the main results of this chapter. The notation is as in the paragraph above. This chapter will prove:

4.1.1. Let $(X, \mathcal{O})$ be a scheme locally of finite type over $\mathbb{C}$. Then the natural map $\lambda_X : X^{\mathrm{an}} \longrightarrow X$ is continuous. This means the following: when we forget the topology X^{an} is simply the set of all closed points in X, and there is an obvious inclusion map $\lambda_X : X^{\mathrm{an}} \longrightarrow X$. We are asserting that, if we give X its Zariski topology and X^{an} its complex topology, then the map λ_X is continuous.

4.1.2. Suppose $(\Phi, \Phi^*) : (X, \mathcal{O}_X) \longrightarrow (Y, \mathcal{O}_Y)$ is a morphism of schemes locally of finite type over $\mathbb{C}$. Then there exists a unique continuous map $\Phi^{\mathrm{an}} : X^{\mathrm{an}} \longrightarrow Y^{\mathrm{an}}$ making the following square commute

$$\begin{array}{ccc} X^{\mathrm{an}} & \xrightarrow{\Phi^{\mathrm{an}}} & Y^{\mathrm{an}} \\ {\scriptstyle \lambda_X}\downarrow & & \downarrow{\scriptstyle \lambda_Y} \\ X & \xrightarrow[\Phi]{} & Y \end{array} .$$

Note that on the level of sets, if we forget all about the topology, the vertical maps λ_X and λ_Y are simply the inclusions of the closed points. So, at the level of maps of sets, what is being asserted is that the function $\Phi : X \longrightarrow Y$ takes closed points in X to closed points in Y; the uniqueness of Φ^{an} is clear. The topological claim is that the restriction of the map Φ to the subsets of closed points, that is the map $\Phi^{\mathrm{an}} : X^{\mathrm{an}} \longrightarrow Y^{\mathrm{an}}$, is continuous in the complex topology.

4.1.3. Let $(\Phi, \Phi^*) : (X, \mathcal{O}_X) \longrightarrow (Y, \mathcal{O}_Y)$ be an open immersion of schemes locally of finite type over $\mathbb{C}$. Then the map $\Phi^{\mathrm{an}} : X^{\mathrm{an}} \longrightarrow Y^{\mathrm{an}}$ is an embedding of X^{an} as the open subset $Y^{\mathrm{an}} \cap \Phi X \subset Y^{\mathrm{an}}$. Note that this makes sense: The set $\Phi X \subset Y$ is open in Y since the morphism (Φ, Φ^*) is an open immersion. And, as the inclusion $\lambda_Y : Y^{\mathrm{an}} \longrightarrow Y$ is continuous, the set $Y^{\mathrm{an}} \cap \Phi X$ is open in Y^{an}. We assert that this is the image of the map $\Phi^{\mathrm{an}} : X^{\mathrm{an}} \longrightarrow Y^{\mathrm{an}}$, and that the map takes X^{an} homeomorphically onto its image.

The way the construction works is local; we will begin by assuming $(X, \mathcal{O}) = \left(\mathrm{Spec}(R), \widetilde{R}\right)$, and define the complex topology on the set $\left\{ \mathrm{Spec}(R) \right\}^{\mathrm{an}}$. Since every scheme $(X, \mathcal{O})$ is locally $\left(\mathrm{Spec}(R), \widetilde{R}\right)$, we can then glue the local data to define the complex topology on X^{an}. We have tried to explain all the details in a clear and transparent way, breaking up the argument into a string of very easy lemmas. The intention is to provide the interested reader with as much detail as she wants. The reader willing to believe that the construction works can skip ahead to Chapter 5.

4.2 The subspace $\mathrm{Max}(X) \subset X$

Given any toplogical space X we now define, in glorious generality, a subspace $\mathrm{Max}(X)$.

Definition 4.2.1. *Let X be a topological space. The subspace* $\mathrm{Max}(X) \subset X$ *is simply the set of all closed points in X, with the subspace topology.*

Example 4.2.2. Let R be a ring, and $X = \mathrm{Spec}(R)$. The points of X are the prime ideals in R. Definition 4.2.1 tells us that the points in $\mathrm{Max}(X)$ are the closed points of X, which, by Remark 3.1.10, are precisely the maximal ideals of the ring R. In the special case where $X = \mathrm{Spec}(R)$ we will drop one of the pairs of brackets in the notation $\mathrm{Max}(X) = \mathrm{Max}(\mathrm{Spec}(R))$, and refer to the space as $\mathrm{MaxSpec}(R)$. This is short for the *maximal spectrum* of R.

Lemma 4.2.3. *Let R be a finitely generated $\mathbb{C}$–algebra. Then every maximal ideal $\mathfrak{m} \subset R$ is the kernel of a unique $\mathbb{C}$–algebra homomorphism $\varphi : R \longrightarrow \mathbb{C}$.*

Proof. The weak Nullstellensatz, which you can find in Remark 3.8.1(i), tells us that all maximal ideals in R are kernels of $\mathbb{C}$–algebra homomorphisms $\varphi : R \longrightarrow \mathbb{C}$. The problem is to prove uniqueness. Therefore let $\varphi, \varphi' : R \longrightarrow \mathbb{C}$ be two $\mathbb{C}$–algebra homomorphisms with the same kernel $\mathfrak{m} = \ker(\varphi) = \ker(\varphi')$. Let $\rho : \mathbb{C} \longrightarrow R$ be the ring homomorphism giving the ring R its structure as a $\mathbb{C}$–algebra. Then we have, for all $m \in \mathfrak{m}$ and all $\lambda \in \mathbb{C}$,

$$\varphi(m) = \varphi'(m) = 0, \qquad\qquad \varphi\rho(\lambda) = \varphi'\rho(\lambda) = \lambda.$$

But any element $r \in R$ is equal to $\big(r - \rho\varphi(r)\big) + \rho\varphi(r)$, and $r - \rho\varphi(r)$ lies in $\ker(\varphi) = \mathfrak{m}$. Thus every element of R can be written as $m + \rho(\lambda)$ for $m \in \mathfrak{m}$ and $\lambda \in \mathbb{C}$. We conclude that $\varphi = \varphi'$. □

We want to rephrase Lemma 4.2.3 slightly.

Proposition 4.2.4. *Let R be a finitely generated $\mathbb{C}$–algebra. Then there is a natual bijection*

$$\left\{ \begin{array}{c} \text{Closed points } x \\ \text{in Spec}(R) \end{array} \right\} \underset{\Delta'}{\overset{\Delta}{\rightleftarrows}} \left\{ \begin{array}{c} \mathbb{C}\text{–algebra} \\ \text{homomorphisms} \\ \varphi : R \longrightarrow \mathbb{C} \end{array} \right\} .$$

The map Δ takes the closed point $x \in \mathrm{Spec}(R)$, that is the maximal ideal $x \subset R$, to the unique φ whose kernel is x. The map Δ' takes $\varphi : R \longrightarrow \mathbb{C}$ to the image of $\mathrm{Spec}(\varphi) : \mathrm{Spec}(\mathbb{C}) \longrightarrow \mathrm{Spec}(R)$.

Proof. Lemma 4.2.3 established that Δ is well-defined; given a closed point $x \in \mathrm{Spec}(R)$ there is a unique φ to which we can send it. The recipe for the map Δ' tells us that it should take $\varphi : R \longrightarrow \mathbb{C}$ to the image of

$$\mathrm{Spec}(\varphi) : \mathrm{Spec}(\mathbb{C}) \longrightarrow \mathrm{Spec}(R) .$$

We need to show that this image is a closed point in $\mathrm{Spec}(R)$. There is a unique point in $\mathrm{Spec}(\mathbb{C})$, namely the maximal ideal $\{0\} \subset \mathbb{C}$. The map $\mathrm{Spec}(\varphi)$ takes it to $\varphi^{-1}\{0\}$. Since $\varphi : R \longrightarrow \mathbb{C}$ must be surjective its kernel $\varphi^{-1}\{0\}$ is maximal. It is a closed point in $\mathrm{Spec}(R)$, and the map Δ' is well-defined.

It remains to show that the composites $\Delta\Delta'$ and $\Delta'\Delta$ are both identities. This is clear from the above computation of Δ'; the map Δ' takes

φ to its kernel $\mathfrak{m}$, and the map Δ takes $\mathfrak{m}$ to the unique φ having kernel $\mathfrak{m}$. □

Remark 4.2.5. Proposition 4.2.4 gives a bijection between points $x \in \mathrm{MaxSpec}(R)$ and $\mathbb{C}$–algebra homomorphisms $\varphi : R \longrightarrow \mathbb{C}$. Theorem 3.9.4 tells us that there is a bijection between $\mathbb{C}$–algebra homomorphisms $\varphi : R \longrightarrow \mathbb{C}$ and maps of ringed spaces over $\mathbb{C}$

$$(\Phi, \Phi^*) : \left(\mathrm{Spec}(\mathbb{C}), \widetilde{\mathbb{C}}\right) \longrightarrow \left(\mathrm{Spec}(R), \widetilde{R}\right) .$$

Combining Proposition 4.2.4 with Theorem 3.9.4 gives a bijection between closed points $x \in \mathrm{Spec}(R)$ and maps of ringed spaces (Φ, Φ^*) as above. The next lemma tells us that the bijection, between closed points $x \in \mathrm{Max}(X)$ and maps of ringed spaces, generalizes to arbitrary schemes locally of finite type over $\mathbb{C}$.

Lemma 4.2.6. *Let $(X, \mathcal{O})$ be a scheme locally of finite type over $\mathbb{C}$. Any morphism of ringed spaces over $\mathbb{C}$*

$$(\Phi, \Phi^*) : \left(\mathrm{Spec}(\mathbb{C}), \widetilde{\mathbb{C}}\right) \longrightarrow (X, \mathcal{O})$$

has the property that the image, of the unique point in $\mathrm{Spec}(\mathbb{C})$ *by the continuous map $\Phi : \mathrm{Spec}(\mathbb{C}) \longrightarrow X$, is a closed point of X, and any closed point in X is the image of a Φ for a unique map of ringed spaces (Φ, Φ^*) as above.*

Proof. Let us first prove the existence of a morphism (Φ, Φ^*). Suppose therefore that we are given a closed point $x \in X$. Choose an open subset $U \subset X$ containing x, with $(U, \mathcal{O}|_U)$ isomorphic, as a ringed space over $\mathbb{C}$, to $\left(\mathrm{Spec}(R), \widetilde{R}\right)$, with R a finitely generated $\mathbb{C}$–algebra. The point x, being closed in all of X, must be closed in the subspace U. By Proposition 4.2.4 there exists a $\mathbb{C}$–algebra homomorphism $\varphi : R \longrightarrow \mathbb{C}$, so that $\mathrm{Spec}(\varphi)$ takes the unique point of $\mathrm{Spec}(\mathbb{C})$ to $x \in \mathrm{Spec}(R)$. The map (Φ, Φ^*) can be taken to be the composite

$$\left(\mathrm{Spec}(\mathbb{C}), \widetilde{\mathbb{C}}\right) \xrightarrow{(\mathrm{Spec}(\varphi), \widetilde{\varphi})} \left(\mathrm{Spec}(R), \widetilde{R}\right) \xrightarrow{(i, i^*)} (X, \mathcal{O}) ,$$

where (i, i^*) is the open immersion of $(U, \mathcal{O}|_U) \cong \left(\mathrm{Spec}(R), \widetilde{R}\right)$.

Next we suppose we are given the map

$$(\Phi, \Phi^*) : \left(\mathrm{Spec}(\mathbb{C}), \widetilde{\mathbb{C}}\right) \longrightarrow (X, \mathcal{O}) .$$

We must prove that the image of Φ is a closed point of X, and we must also prove the uniqueness of (Φ, Φ^*) given the image of Φ. In any case we know that Φ takes the unique point of $\mathrm{Spec}(\mathbb{C})$ to a point $x \in X$.

Given any open subset $V \subset X$ containing $\Phi\{\mathrm{Spec}(\mathbb{C})\} = x$, the map of ringed spaces (Φ, Φ^*) must factor as

$$\left(\mathrm{Spec}(\mathbb{C}), \widetilde{\mathbb{C}}\right) \longrightarrow (V, \mathcal{O}|_V) \xrightarrow{(j,j^*)} (X, \mathcal{O}) ,$$

where (j, j^*) is the open immersion of $(V, \mathcal{O}|_V)$ into $(X, \mathcal{O})$. If $(V, \mathcal{O}|_V)$ happens to be affine, that is isomorphic to $\left(\mathrm{Spec}(S), \widetilde{S}\right)$ for some finitely generated algebra S over $\mathbb{C}$, then the map

$$\left(\mathrm{Spec}(\mathbb{C}), \widetilde{\mathbb{C}}\right) \longrightarrow (V, \mathcal{O}|_V) \quad = \quad \left(\mathrm{Spec}(S), \widetilde{S}\right)$$

must, by Theorem 3.9.4, be induced by a $\mathbb{C}$–algebra homomorphism $\psi : S \longrightarrow \mathbb{C}$. Proposition 4.2.4, more specifically the fact that the map Δ' of the proposition is well-defined, tells us that the image of the unique point in $\mathrm{Spec}(\mathbb{C})$, by $\mathrm{Spec}(\psi) : \mathrm{Spec}(\mathbb{C}) \longrightarrow \mathrm{Spec}(S)$, is closed in $\mathrm{Spec}(S) = V$. This means that the point $x \in X$ must be closed in $V = \mathrm{Spec}(S)$. Hence $V - \{x\}$ is open in V which is open in X, making $V - \{x\}$ open in X.

We have proved that, for any open affine V containing x, the set $V - \{x\}$ is open. Open affines V not containing x satisfy $V - \{x\} = V$, which is most certainly open. Now X has an open cover $X = \cup_{i \in I} U_i$ with each U_i affine. This makes

$$X - \{x\} \quad = \quad \bigcup_{i \in I} \{U_i - \{x\}\}$$

with each $U_i - \{x\}$ open; we conclude that $X - \{x\}$ is open, or equivalently that x is a closed point in X.

As for the map (Φ, Φ^*), we have shown that it factors

$$\left(\mathrm{Spec}(\mathbb{C}), \widetilde{\mathbb{C}}\right) \xrightarrow{(\mathrm{Spec}(\psi), \widetilde{\psi})} \left(\mathrm{Spec}(S), \widetilde{S}\right) \xrightarrow{(j,j^*)} (X, \mathcal{O}) ,$$

where (j, j^*) is the open immersion of $(V, \mathcal{O}|_V) \cong \left(\mathrm{Spec}(S), \widetilde{S}\right)$ into $(X, \mathcal{O})$, and ψ is the unique homomorphism of Lemma 4.2.3. Thus the map (Φ, Φ^*) is unique. □

An immediate consequence is

Lemma 4.2.7. *Let $(\Psi, \Psi^*) : (X, \mathcal{O}_X) \longrightarrow (Y, \mathcal{O}_Y)$ be a morphism of schemes locally of finite type over $\mathbb{C}$. Then the continuous map $\Psi : X \longrightarrow Y$ takes the subspace* $\mathrm{Max}(X) \subset X$ *to the subspace* $\mathrm{Max}(Y) \subset Y$.

Proof. We need to prove that Ψ takes closed points in X to closed points in Y. Let x be a closed point in X. By Lemma 4.2.6 there is a morphism of ringed spaces over $\mathbb{C}$

$$(\Phi, \Phi^*) : \left(\mathrm{Spec}(\mathbb{C}), \widetilde{\mathbb{C}}\right) \longrightarrow (X, \mathcal{O}_X)$$

so that $\Phi\{\mathrm{Spec}(\mathbb{C})\} = x$. But then the composite morphism

$$\left(\mathrm{Spec}(\mathbb{C}), \widetilde{\mathbb{C}}\right) \xrightarrow{(\Phi,\Phi^*)} (X, \mathcal{O}_X) \xrightarrow{(\Psi,\Psi^*)} (Y, \mathcal{O}_Y)$$

satisfies $\Psi\Phi\{\mathrm{Spec}(\mathbb{C})\} = \Psi(x)$ and, using Lemma 4.2.6 again, we conlude that $\Psi(x)$ is closed in Y. □

Definition 4.2.8. *Let $(\Psi, \Psi^*) : (X, \mathcal{O}_X) \longrightarrow (Y, \mathcal{O}_Y)$ be a morphism of schemes locally of finite type over $\mathbb{C}$. The induced map from the subspace $\mathrm{Max}(X) \subset X$ to the subspace $\mathrm{Max}(Y) \subset Y$ will be called*

$$\mathrm{Max}(\Psi) : \mathrm{Max}(X) \longrightarrow \mathrm{Max}(Y) .$$

Corollary 4.2.9. *Let $(Y, \mathcal{O})$ be a scheme locally of finite type over $\mathbb{C}$ and let $X \subset Y$ be an open subset. Then the inclusion map $\Psi : X \longrightarrow Y$ embeds $\mathrm{Max}(X)$ homeomorphically onto the open subset $\Psi X \cap \mathrm{Max}(Y)$.*

Proof. Proposition 3.10.7 tells us that $(X, \mathcal{O}|_X)$ is also a scheme locally of finite type over $\mathbb{C}$. Therefore Lemma 4.2.7 applies to the natural open immersion $(\Psi, \Psi^*) : (X, \mathcal{O}|_X) \longrightarrow (Y, \mathcal{O})$, and we conclude that $\Psi : X \longrightarrow Y$ carries $\mathrm{Max}(X)$ to $\mathrm{Max}(Y)$. But the map $\Psi : X \longrightarrow Y$ is an embedding, and hence the restriction to the subspace $\mathrm{Max}(X)$ must be an embedding; that is the map is injective and $\mathrm{Max}(X) \subset \mathrm{Max}(Y)$ has the subspace topology. We need to show that the image is all of the open set $\Psi X \cap \mathrm{Max}(Y)$. Note that ΨX is open in Y by hypothesis, and hence $\Psi X \cap \mathrm{Max}(Y)$ is open in $\mathrm{Max}(Y)$.

Take any point $y \in \Psi X \cap \mathrm{Max}(Y)$. Then y is a closed point in Y and y lies in the subspace $\Psi X \subset Y$. Hence y is closed in ΨX. But the map Ψ is a homeomorphism of X with its image, and hence $y = \Psi(x)$ for a closed point $x \in X$. □

Remark 4.2.10. Let $(X, \mathcal{O})$ be a scheme locally of finite type over $\mathbb{C}$. Given any open cover $X = \cup_{i \in I} U_i$, we have a string of equalities, the last of which comes from Corollary 4.2.9:

$$\begin{aligned} \mathrm{Max}(X) &= \left\{\bigcup_{i\in I} U_i\right\} \cap \mathrm{Max}(X) = \bigcup_{i\in I} \{U_i \cap \mathrm{Max}(X)\} \\ &= \bigcup_{i\in I} \mathrm{Max}(U_i) . \end{aligned}$$

In other words, the sets $\mathrm{Max}(U_i) \subset \mathrm{Max}(X)$ provide an open cover for $\mathrm{Max}(X)$.

Remark 4.2.11. In the special case where the open cover $X = \cup_{i \in I} U_i$ is affine, we have that

$$\mathrm{Max}(U_i) \quad \cong \quad \mathrm{MaxSpec}(R_i) \ ;$$

in other words we have a cover of the space $\mathrm{Max}(X)$ by open subsets homeomorphic to $\mathrm{MaxSpec}(R_i)$. Since any scheme $(X, \mathcal{O})$, locally of finite type over $\mathbb{C}$, has an affine cover, it follows that $\mathrm{Max}(X)$ can always be covered by spaces $\mathrm{MaxSpec}(R_i)$.

4.3 The correspondence between maximal ideals and $\varphi : R \longrightarrow \mathbb{C}$

In the light of Remark 4.2.11 it seems natural to study more closely the spaces $\mathrm{MaxSpec}(R)$. After all they are the building blocks of $\mathrm{Max}(X)$, for every scheme $(X, \mathcal{O})$ locally of finite type over $\mathbb{C}$. Proposition 4.2.4 tells us that the points of $\mathrm{MaxSpec}(R)$ are in bijection with $\mathbb{C}$–algebra homomorphisms $\varphi : R \longrightarrow \mathbb{C}$. In this section we want to understand the bijection better. The space $\mathrm{MaxSpec}(R)$ is a topological space, and it is only natural to ask what is the topology on the set of $\mathbb{C}$–algebra homomorphisms $\varphi : R \longrightarrow \mathbb{C}$. Any $\mathbb{C}$–algebra homomorphism $\theta : R \longrightarrow S$ induces a map $\mathrm{MaxSpec}(S) \longrightarrow \mathrm{MaxSpec}(R)$; we want to understand what this map does to a homomorphism $\psi : S \longrightarrow \mathbb{C}$. If $\theta : R \longrightarrow S$ happens to be surjective we can identify the image of the map $\mathrm{MaxSpec}(\theta)$. All of this will become useful in Section 4.4, where we actually compute $\mathrm{MaxSpec}(R)$.

Our first step will be to understand the image, of the open sets in $\mathrm{MaxSpec}(R)$, under the identification of $\mathrm{MaxSpec}(R)$ with the set of $\mathbb{C}$–algebra homomorphisms $\varphi : R \longrightarrow \mathbb{C}$.

Lemma 4.3.1. *By Proposition 4.2.4 there is a natural identification*

$$\mathrm{MaxSpec}(R) \quad \cong \quad \left\{ \begin{array}{c} \mathbb{C}\text{–algebra homomorphisms} \\ \varphi : R \longrightarrow \mathbb{C} \end{array} \right\} .$$

Under this identification the basic open sets $X_f \cap \mathrm{Max}(X)$ are mapped to the sets

$$\mathrm{Hom}(R, \mathbb{C})_f \quad = \quad \left\{ \begin{array}{c} \mathbb{C}\text{–algebra homomorphisms} \\ \varphi : R \longrightarrow \mathbb{C} \\ \text{with } \varphi(f) \neq 0 \end{array} \right\} .$$

Proof. Remark 3.2.2 tells us that a prime ideal $\mathfrak{p} \in \operatorname{Spec}(R)$ lies in X_f if and only if $f \notin \mathfrak{p}$. In the case of a maximal ideal $\mathfrak{m} = \ker(\varphi)$, where $\varphi : R \longrightarrow \mathbb{C}$ is a homomorphism of $\mathbb{C}$–algebras, this says that $\mathfrak{m} \in X_f$ is equivalent to $\varphi(f) \neq 0$. □

The next lemma tells us how the correspondence of Proposition 4.2.4 behaves under ring homomorphisms.

Lemma 4.3.2. *For any finitely generated $\mathbb{C}$–algebras R and S, Proposition 4.2.4 identified*

$$\operatorname{MaxSpec}(R) \cong \left\{ \begin{array}{c} \mathbb{C}\text{–}maps \\ \varphi : R \longrightarrow \mathbb{C} \end{array} \right\}, \qquad \operatorname{MaxSpec}(S) \cong \left\{ \begin{array}{c} \mathbb{C}\text{–}maps \\ \psi : S \longrightarrow \mathbb{C} \end{array} \right\}.$$

Let $\theta : R \longrightarrow S$ be a homomorphism of $\mathbb{C}$–algebras. Then the map

$$\operatorname{MaxSpec}(\theta) : \operatorname{MaxSpec}(S) \longrightarrow \operatorname{MaxSpec}(R) ,$$

of Definitions 3.6.2 and 4.2.8, is the map taking $\psi : S \longrightarrow \mathbb{C}$ to $\psi\theta : R \longrightarrow \mathbb{C}$.

Proof. The identification of Proposition 4.2.4 is such that $\psi : S \longrightarrow \mathbb{C}$ is sent to the maximal ideal $\ker(\psi) \subset S$. The map $\operatorname{Spec}(\theta) : \operatorname{Spec}(S) \longrightarrow \operatorname{Spec}(R)$ takes the maximal ideal $\ker(\psi)$ to

$$\theta^{-1} \ker(\psi) = \ker(\psi\theta) ,$$

and this is the maximal ideal corresponding, under the identification of Proposition 4.2.4, with $\psi\theta : R \longrightarrow \mathbb{C}$. □

Now we would like to specialize Lemma 4.3.2 to the case where the ring homomorphism $\theta : R \longrightarrow S$ is surjective. Before dealing with the map on maximal spectra $\operatorname{MaxSpec}(S) \longrightarrow \operatorname{MaxSpec}(R)$, let us observe what happens on ordinary spectra.

Lemma 4.3.3. *Let $\theta : R \longrightarrow S$ be a surjective map of finitely generated algebras over $\mathbb{C}$. Then the map $\operatorname{Spec}(\theta) : \operatorname{Spec}(S) \longrightarrow \operatorname{Spec}(R)$ is injective, and $\operatorname{Spec}(S)$ maps homeomorphically to its image.*

Proof. The map $\operatorname{Spec}(\theta)$ takes a prime ideal $\mathfrak{p} \subset S$ to $\theta^{-1}\mathfrak{p} \subset R$. Since θ is surjective, $\theta^{-1}\mathfrak{p} = \theta^{-1}\mathfrak{p}'$ clearly implies $\mathfrak{p} = \mathfrak{p}'$; hence the injectivity of $\operatorname{Spec}(\theta)$.

We need to show that any open set in $X = \operatorname{Spec}(S)$ is the inverse image of an open set in $Y = \operatorname{Spec}(R)$. It is enough to prove this for basic open sets X_f. Therefore let f be an arbitrary element in S. Since $\theta : R \longrightarrow S$ is surjective there exists an element $f' \in R$ with $\theta(f') = f$. By Remark 3.6.4 the inverse image of $Y_{f'} \subset Y$ is $X_{\theta(f')} = X_f \subset X$. □

Lemma 4.3.4. *Let $\theta : R \longrightarrow S$ be a surjective map of finitely generated algebras over $\mathbb{C}$. Let I be the kernel of θ; that is R/I maps isomorphically to S. Then the map* $\text{MaxSpec}(\theta)$ *embeds* $\text{MaxSpec}(S)$, *homeomorphically, into a subspace of* $\text{MaxSpec}(R)$. *The image of* $\text{MaxSpec}(\theta)$ *is identified with all $\varphi : R \longrightarrow \mathbb{C}$ such that $\varphi(I) = 0$.*

Proof. We have a commutative square of maps of topological spaces

$$\begin{array}{ccc} \text{MaxSpec}(S) & \longrightarrow & \text{MaxSpec}(R) \\ {\scriptstyle i_1}\downarrow & & \downarrow{\scriptstyle i_2} \\ \text{Spec}(S) & \xrightarrow[\text{Spec}(\theta)]{} & \text{Spec}(R) \end{array} .$$

By Lemma 4.3.3 the map $\text{Spec}(\theta)$ is an embedding. The maps i_1 and i_2 are embeddings because the topology of $\text{Max}(X)$ is defined, for every X, to be the subspace topology via its inclusion in X. It follows that all spaces in the diagram are subspaces of $\text{Spec}(R)$, and in particular the map $\text{MaxSpec}(S) \longrightarrow \text{MaxSpec}(R)$ is an embedding. It only remains to identify the image of the map.

Lemma 4.3.2 tells us that the map $\text{MaxSpec}(\theta)$ takes $\psi : S \longrightarrow \mathbb{C}$ to $\psi\theta : R \longrightarrow \mathbb{C}$. A homomorphism $\varphi : R \longrightarrow \mathbb{C}$ lies in the image of $\text{MaxSpec}(\theta)$ precisely if it factors through θ, which happens exactly when φ annihilates the kernel I of θ. □

4.4 The special case of the polynomial ring

In Section 4.2 we developed the general theory of the subspaces $\text{Max}(X)$, where $(X, \mathcal{O})$ is a scheme locally of finite type over $\mathbb{C}$. In this section we will be more lowbrow and look at the concrete case of $\left(\text{Spec}(R), \widetilde{R}\right)$, where $R = \mathbb{C}[x_1, x_2, \ldots, x_n]$ is the polynomial ring in n variables. Recall that, by Proposition 4.2.4, the points of $\text{MaxSpec}(R)$ correspond bijectively to homomorphisms $\varphi : R \longrightarrow \mathbb{C}$. Understanding the points of $\text{MaxSpec}(R)$ is equivalent to classifying the $\mathbb{C}$–algebra homomorphisms $\varphi : R \longrightarrow \mathbb{C}$. In Section 4.3 we developed the bijective correspondence of Proposition 4.2.4 further. In the current section we plan to make use of all this theory to explicitly compute the spaces, the topology and the maps.

Lemma 4.4.1. *Let $R = \mathbb{C}[x_1, x_2, \ldots, x_n]$ be the polynomial ring in n variables. We assert that there is a bijective correspondence between $\mathbb{C}$–algebra homomorphisms $\varphi : R \longrightarrow \mathbb{C}$ and points in $\mathbb{C}^n$. Given a point $a = (a_1, a_2, \ldots, a_n) \in \mathbb{C}^n$, we can define the homomorphism φ_a by the* *formula*

$$\varphi_a(f) = f(a_1, a_2, \ldots, a_n) \ .$$

Given $\varphi : R \longrightarrow \mathbb{C}$ *we can determine a point* $(a_1, a_2, \ldots, a_n) \in \mathbb{C}^n$ *by*

$$\varphi(x_1) = a_1, \quad \varphi(x_2) = a_2, \quad \ldots, \quad \varphi(x_n) = a_n \ ,$$

and we assert that $\varphi = \varphi_a$.

Proof. One direction is clear: given a point $a = (a_1, a_2, \ldots, a_n) \in \mathbb{C}^n$ the map $\varphi_a : R \longrightarrow \mathbb{C}$, sending $f \in R$ to its value $f(a) \in \mathbb{C}$, is clearly a homomorphism of $\mathbb{C}$–algebras. We need to know that every φ is of this form.

Therefore let $\varphi : R \longrightarrow \mathbb{C}$ be a homomorphism, and put $\varphi(x_1) = a_1$, $\varphi(x_2) = a_2, \ldots, \varphi(x_n) = a_n$. Every element of $R = \mathbb{C}[x_1, x_2, \ldots, x_n]$ can be written

$$f(x_1, x_2, \ldots, x_n) = \sum \lambda_{i_1, i_2, \ldots, i_n} x_1^{i_1} x_2^{i_2} \cdots x_n^{i_n} \ .$$

As φ is a $\mathbb{C}$–algebra homomorphism we deduce that

$$\varphi(f) = \sum \lambda_{i_1, i_2, \ldots, i_n} a_1^{i_1} a_2^{i_2} \cdots a_n^{i_n} \ ,$$

that is $\varphi(f) = f(a_1, a_2, \ldots, a_n)$. □

Remark 4.4.2. Let R be any finitely generated $\mathbb{C}$–algebra, and let $X = \mathrm{Spec}(R)$. Lemma 4.3.1 tells us how the basic open sets X_f intersect $\mathrm{Max}(X)$. For any $f \in R$ the intersection $X_f \cap \mathrm{Max}(X)$ was

$$\mathrm{Hom}(R, \mathbb{C})_f \quad = \quad \left\{ \begin{array}{c} \mathbb{C}\text{–algebra homomorphisms} \\ \varphi : R \longrightarrow \mathbb{C} \\ \text{with } \varphi(f) \neq 0 \end{array} \right\} \ ,$$

where Proposition 4.2.4 identifies the points of $\mathrm{Max}(X)$ with homomorphisms $\varphi : R \longrightarrow \mathbb{C}$. In the special case where R is the ring $R = \mathbb{C}[x_1, x_2, \ldots, x_n]$, that is the polynomial ring in n variables, Lemma 4.4.1 gave an identification of the sets $\mathbb{C}^n$ and $\mathrm{Max}(X) = \mathrm{MaxSpec}(R)$. The next lemma combines the two, telling us what are the Zariski open subsets of $\mathbb{C}^n$ under its identification with $\mathrm{MaxSpec}(R)$.

Lemma 4.4.3. *Let* $R = \mathbb{C}[x_1, x_2, \ldots, x_n]$ *be the polynomial ring in* n *variables, and put* $X = \mathrm{Spec}(R)$. *Under the identification of* $\mathbb{C}^n$ *with* $\mathrm{Max}(X) = \mathrm{MaxSpec}(R)$ *of Lemma 4.4.1, the open subsets* $X_f \cap \mathrm{Max}(X)$ *identify to the sets* $\mathbb{C}^n_f$ *given by*

$$\mathbb{C}^n_f \quad = \quad \{x \in \mathbb{C}^n \mid f(x) \neq 0\} \ .$$

Proof. Lemma 4.4.1 identifies a point $a \in \mathbb{C}^n$ with the homomorphism $\varphi_a : R \longrightarrow \mathbb{C}$, given by the formula $\varphi_a(f) = f(a)$. Lemma 4.3.1 tells us that the open set $X_f \cap \mathrm{Max}(X)$ is the set of all $\varphi : R \longrightarrow \mathbb{C}$ with $\varphi(f) \neq 0$. But this means that, under the identification of $X_f \cap \mathrm{Max}(X)$ with a subset of $\mathbb{C}^n$, we have the set of all $a \in \mathbb{C}^n$ such that

$$f(a) = \varphi_a(f) \neq 0\,.$$

□

Notation 4.4.4. Next we deal with homomorphisms between polynomial rings. Let R and S be two polynomial rings, that is $R = \mathbb{C}[x_1, x_2, \ldots, x_n]$ and $S = \mathbb{C}[y_1, y_2, \ldots, y_m]$. Any $\mathbb{C}$–algebra homomorphism $\theta : R \longrightarrow S$ determines, by Lemma 4.2.7, a map

$$\mathrm{MaxSpec}(\theta) : \mathrm{MaxSpec}(S) \longrightarrow \mathrm{MaxSpec}(R)\ .$$

With the identifications of Lemma 4.4.1 we may rewrite this as a function $\mathrm{MaxSpec}(\theta) : \mathbb{C}^m \longrightarrow \mathbb{C}^n$. We want to understand this map.

Now the ring homomorphism $\theta : R \longrightarrow S$ takes elements of R to elements of $S = \mathbb{C}[y_1, y_2, \ldots, y_m]$, that is to polynomials in $y_1, y_2, \ldots, y_m$. Put

$$\theta(x_1) = f_1(y_1, y_2, \ldots, y_m)\,, \quad \ldots \quad , \theta(x_n) = f_n(y_1, y_2, \ldots, y_m)\,.$$

We prove

Lemma 4.4.5. *Let the notation be as in Notation 4.4.4. The map*

$$\mathrm{MaxSpec}(\theta) : \mathbb{C}^m \longrightarrow \mathbb{C}^n$$

takes a point $a = (a_1, a_2, \ldots, a_m) \in \mathbb{C}^m$ to the point

$$\big(f_1(a), f_2(a), \ldots, f_n(a)\big) \in \mathbb{C}^n\,.$$

Remark 4.4.6. In other words: the map $\mathrm{MaxSpec}(\theta)$ is just a polynomial function from $\mathbb{C}^m$ to $\mathbb{C}^n$.

Proof. We need to see what the function $\mathrm{MaxSpec}(\theta)$ does to a point $a = (a_1, a_2, \ldots, a_m) \in \mathbb{C}^m$. Let $\varphi_a : S \longrightarrow \mathbb{C}$ be the homomorphism corresponding to $a \in \mathbb{C}^n$ by Lemma 4.4.1; that is, $\varphi_a(f) = f(a)$. Lemma 4.3.2 tells us that the map $\mathrm{MaxSpec}(\theta)$ takes φ_a to $\varphi_a\theta : R \longrightarrow \mathbb{C}$. The corresponding point in $\mathbb{C}^n$ is, by Lemma 4.4.1 again, the point with coordinates

$$\big(\varphi_a\theta(x_1), \varphi_a\theta(x_2), \ldots, \varphi_a\theta(x_n)\big) \in \mathbb{C}^n\,.$$

In Notation 4.4.4 we wrote $\theta(x_i) = f_i(y_1, y_2, \ldots, y_m)$, which makes

$$\varphi_a\theta(x_i) = f_i(a_1, a_2, \ldots, a_m)\,.$$

In other words the point $a = (a_1, a_2, \dots, a_m) \in \mathbb{C}^m$ goes to

$$\bigl(f_1(a), f_2(a), \dots, f_n(a)\bigr) \in \mathbb{C}^n .$$

□

Lemma 4.4.7. *Let the notation be as in Notation 4.4.4. Let $P \in R$ be any polynomial; that is*

$$P \quad \in \quad R \quad = \quad \mathbb{C}[x_1, x_2, \dots, x_n]$$

is a polynomial in n variables. We know that $\theta(P) \in S$ is a polynomial

$$\theta(P) \quad \in \quad S \quad = \quad \mathbb{C}[y_1, y_2, \dots, y_m] .$$

We assert that the value of $\theta(P)$, at the point $a = (a_1, a_2, \dots, a_m) \in \mathbb{C}^m$, is equal to the value of P at the image of $a \in \mathbb{C}^m$ under the map $\mathrm{MaxSpec}(\theta) : \mathbb{C}^m \longrightarrow \mathbb{C}^n$.

Proof. Since θ is a ring homomorphism, we have

$$\theta\bigl(P(x_1, x_2, \dots, x_n)\bigr) \quad = \quad P\bigl(\theta(x_1), \theta(x_2), \dots, \theta(x_n)\bigr) .$$

Using the identity $\theta(x_i) = f_i(y_1, \dots, y_m)$ this becomes

$$\{\theta(P)\}(y_1, y_2, \dots, y_m) \quad = \quad P\bigl(f_1(y_1, \dots, y_m), \dots, f_n(y_1, \dots, y_m)\bigr) .$$

The result now follows by evaluating both sides at $a = (a_1, a_2, \dots, a_m) \in \mathbb{C}^m$ and recalling Lemma 4.4.5. □

Remark 4.4.8. Let $R = \mathbb{C}[x_1, x_2, \dots, x_n]$ and $S = \mathbb{C}[y_1, y_2, \dots, y_m]$. Let us be given a map of ringed spaces over $\mathbb{C}$

$$(\Theta, \Theta^*) : \left(\mathrm{Spec}(S), \widetilde{S}\right) \longrightarrow \left(\mathrm{Spec}(R), \widetilde{R}\right) ,$$

which by Theorem 3.9.4 must be of the form $\left(\mathrm{Spec}(\theta), \widetilde{\theta}\right)$, for some $\mathbb{C}$–algebra homomorphism $\theta : R \longrightarrow S$. From Lemma 4.4.5 we learn that, on the sets of closed points, the continuous map $\Theta = \mathrm{Spec}(\theta)$ induces a polynomial map

$$\Theta^{\mathrm{an}} : \mathbb{C}^m \longrightarrow \mathbb{C}^n .$$

Lemma 4.4.7 teaches us that the map $\theta = \Theta^*_{\mathrm{Spec}(R)}$ takes an element P of $R = \mathbb{C}[x_1, x_2, \dots, x_n]$, which can be viewed as a polynomial function on $\mathbb{C}^n$, to the composite $P \circ \mathrm{MaxSpec}(\theta)$. This is reminiscent of what happens with manifolds; see Theorem 2.3.2.

4.5 The complex topology on MaxSpec(R)

In the case where $R = \mathbb{C}[x_1, x_2, \dots, x_n]$ is a polynomial ring we have now understood MaxSpec(R); as a set it is naturally identified with $\mathbb{C}^n$, and the open sets $\mathbb{C}^n_f$ of Lemma 4.4.3 form a basis for the Zariski topology. We remind the reader of the definition of $\mathbb{C}^n_f$: take a polynomial $f \in \mathbb{C}[x_1, x_2, \dots, x_n]$, and set

$$\mathbb{C}^n_f \quad = \quad \{x \in \mathbb{C}^n \mid f(x) \neq 0\} \ .$$

The Zariski topology is very coarse. The open sets are all unions of $\mathbb{C}^n_f$'s, and if $f \neq 0$ the set $\mathbb{C}^n_f$ has a complement of Lebesgue measure zero. Now $\mathbb{C}^n$ has another topology, the usual one. What we want to do in this section is define the usual topology on every MaxSpec(S), where S is any finitely generated $\mathbb{C}$–algebra.

Therefore we let S be a finitely generated algebra over $\mathbb{C}$. If we choose generators $\{a_1, a_2, \dots, a_n\} \subset S$ they induce a surjective homomorphism from the polynomial ring

$$\theta : \mathbb{C}[x_1, x_2, \dots, x_n] \longrightarrow S \ .$$

The map is the one sending the indeterminate $x_i \in \mathbb{C}[x_1, x_2, \dots, x_n]$ to the generator $\theta(x_i) = a_i \in S$. Put $R = \mathbb{C}[x_1, x_2, \dots, x_n]$. Definition 4.2.8 gives us a map

$$\mathrm{MaxSpec}(\theta) : \mathrm{MaxSpec}(S) \longrightarrow \mathrm{MaxSpec}(R) \ ,$$

and Lemma 4.4.1 tells us that MaxSpec(R) $= \mathbb{C}^n$. Lemma 4.3.4 guarantees that MaxSpec(θ) is injective; not only is it injective, we also know that the Zariski topology on MaxSpec(S) is the subspace Zariski topology via its embedding in MaxSpec(R). But now we care about the complex topology. It seems natural to make the following definition:

Definition 4.5.1. *Let S be a finitely generated algebra over $\mathbb{C}$. Choose generators $\{a_1, a_2, \dots, a_n\} \subset S$. Put $R = \mathbb{C}[x_1, x_2, \dots, x_n]$ and let $\theta : R \longrightarrow S$ be the natural surjection, as above. The* complex topology on MaxSpec(S) *is its topology as a subspace of $\mathbb{C}^n$, via the injective map*

$$\mathrm{MaxSpec}(\theta) : \mathrm{MaxSpec}(S) \longrightarrow \mathrm{MaxSpec}(R) \quad \cong \quad \mathbb{C}^n \ .$$

The symbol $\{\,\mathrm{Spec}(S)\}^{\mathrm{an}}$ will denote the set MaxSpec(S)*, with its complex topology.*

Remark 4.5.2. Perhaps we should refine the last sentence of Definition 4.5.1. For now we do not even know the topology is independent of the choice of generators, and know nothing about the way various natural maps behave with respect to the topology. For this section and the next, many of the diagrams we will draw will be such that we know something about the continuity of some of the maps in the complex topology, and use this to prove statements about the others. When we wish to refer to MaxSpec(S) as a set, or as a topological space with its Zariski topology, we will call it MaxSpec(S). Only when we are very sure of the continuity will we call it $\{\,\mathrm{Spec}(S)\}^{\mathrm{an}}$. As we progress we will use the term MaxSpec(S) less and less.

The first thing we need to prove is

Lemma 4.5.3. *The complex topology on* MaxSpec(S) *is independent of the choice of generators.*

Proof. Let $\{a_1, a_2, \ldots, a_n\}$ and $\{b_1, b_2, \ldots, b_m\}$ be two sets of generators. Let $R = \mathbb{C}[x_1, x_2, \ldots, x_n]$ and $R' = \mathbb{C}[y_1, y_2, \ldots, y_m]$ be the polynomial rings. We have two surjective homomorphisms

$$\theta : R \longrightarrow S, \qquad \theta' : R' \longrightarrow S,$$

where $\theta(x_i) = a_i$ and $\theta'(y_i) = b_i$. Each of the two maps MaxSpec(θ) and MaxSpec(θ') embeds the set MaxSpec(S) onto a subspace of, respectively, $\mathbb{C}^n$ and $\mathbb{C}^m$; we need to show that the two subspace topologies on MaxSpec(S) agree. From the symmetry it is enough to show that one topology is finer than the other. We will prove that the identity on MaxSpec(S), viewed as a map from the subspace of $\mathbb{C}^m$ to the subspace of $\mathbb{C}^n$, is continuous.

We know that $b_1, b_2, \ldots, b_m$ generate S, while $a_1, a_2, \ldots, a_n$ are elements of S. It follows that we can write the a_j's as polynomials in the b_i; there exist identities in S

$$a_j = P_j(b_1, b_2, \ldots, b_m)$$

with P_j polynomials with coefficients in $\mathbb{C}$. Now we define a homomorphism $\varphi : R \longrightarrow R'$ by setting

$$\varphi(x_j) = P_j(y_1, y_2, \ldots, y_m)\ .$$

We deduce a commutative triangle of ring homomorphisms

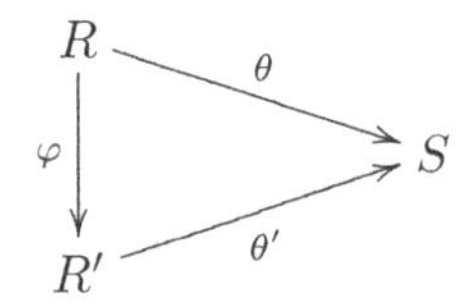

Taking the induced maps on maximal spectra we deduce a commutative

triangle

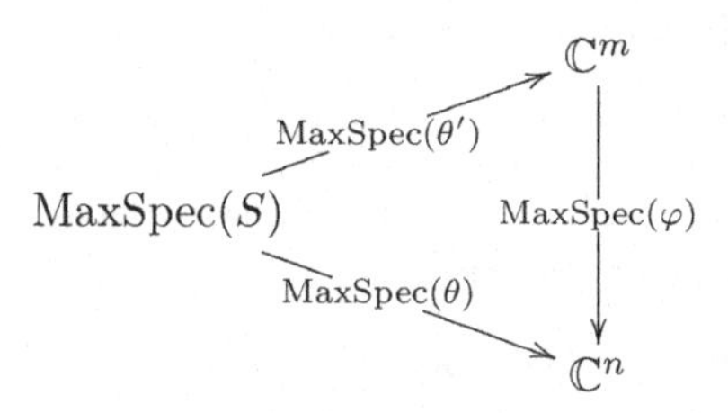

By Remark 4.4.6 we know that the map $\mathrm{MaxSpec}(\varphi) : \mathbb{C}^m \longrightarrow \mathbb{C}^n$ is a polynomial map; it is continuous in the usual, complex topology on $\mathbb{C}^m$ and $\mathbb{C}^n$. It follows that it induces a continuous map from $\mathrm{MaxSpec}(S) \subset \mathbb{C}^m$ to $\mathrm{MaxSpec}(S) \subset \mathbb{C}^n$. This means that the identity map from MaxSpec(S) to itself is continuous, when viewed as a map from one subspace topology to the other. □

Lemma 4.5.4. *Let S be a finitely generated $\mathbb{C}$–algebra. The natural inclusion*

$$\mathrm{MaxSpec}(S) \longrightarrow \mathrm{Spec}(S)$$

is a continuous map of topological spaces, if we give Spec(S) *the Zariski topology and* MaxSpec(S) *its complex topology.*

Proof. Let $\{a_1, a_2, \ldots, a_n\}$ be a set of generators for S. Suppose $R = \mathbb{C}[x_1, x_2, \ldots, x_n]$ is the polynomial ring, and let $\theta : R \longrightarrow S$ be the natural surjection. We have a commutative square

$$\begin{array}{ccc} \mathrm{MaxSpec}(S) & \longrightarrow & \mathrm{Spec}(S) \\ {\scriptstyle i_1}\downarrow & & \downarrow{\scriptstyle i_2} \\ \mathrm{MaxSpec}(R) & \longrightarrow & \mathrm{Spec}(R) \end{array} .$$

The vertical maps are embeddings of subspaces. For i_1 this is the definition, of the complex topology of MaxSpec(S) via the embedding into $\mathrm{MaxSpec}(R) = \mathbb{C}^n$; see Definition 4.5.1. For i_2, Lemma 4.3.3 proved that Spec(S) embeds in Spec(R). The commutativity means that it suffices to prove that the inclusion $\mathrm{MaxSpec}(R) \longrightarrow \mathrm{Spec}(R)$ is continuous; the restriction to the subspaces would then have to be continuous.

But $\mathrm{MaxSpec}(R) \cong \mathbb{C}^n$, together with its subspace topology via the inclusion into Spec(R), was identified in Lemma 4.4.3. A basis for the Zariski topology on $\mathbb{C}^n$ is given by the subsets $\mathbb{C}^n_f = f^{-1}\big(\mathbb{C} - \{0\}\big)$. And these are all open sets in the complex topology. □

Notation 4.5.5. As already indicated, in Definition 4.5.1 and Remark 4.5.2, we will refer to MaxSpec(S) as $\big\{\mathrm{Spec}(S)\big\}^{\mathrm{an}}$ when we want to understand it as having the complex topology. We will feel free to refer to the map of Lemma 4.5.4 as

$$\lambda_S : \left\{ \text{Spec}(S) \right\}^{\text{an}} \longrightarrow \text{Spec}(S) .$$

Lemma 4.5.6. *Let $\varphi : S \longrightarrow S'$ be a homomorphism of finitely generated $\mathbb{C}$–algebras. The natural map*

$$\text{MaxSpec}(\varphi) : \text{MaxSpec}(S') \longrightarrow \text{MaxSpec}(S)$$

is continuous in the complex topology; we will in future feel free to write it as $\left\{ \text{Spec}(\varphi) \right\}^{\text{an}} : \left\{ \text{Spec}(S') \right\}^{\text{an}} \longrightarrow \left\{ \text{Spec}(S) \right\}^{\text{an}}$. Furthermore, the following square commutes

$$\begin{array}{ccc} \left\{ \text{Spec}(S') \right\}^{\text{an}} & \xrightarrow{\left\{ \text{Spec}(\varphi) \right\}^{\text{an}}} & \left\{ \text{Spec}(S) \right\}^{\text{an}} \\ {\scriptstyle \lambda_{S'}} \Big\downarrow & & \Big\downarrow {\scriptstyle \lambda_S} \\ \text{Spec}(S') & \xrightarrow[\text{Spec}(\varphi)]{} & \text{Spec}(S) \end{array} .$$

Proof. The comutativity of the square is a statement that the composites are equal as maps of sets; this is clear, since the map $\text{MaxSpec}(\varphi)$ was defined as the restriction of $\text{Spec}(\varphi)$ to the subsets of closed points. We need to prove the continuity of $\text{MaxSpec}(\varphi)$ in the complex topology.

Let $\{a_1, a_2, \ldots, a_n\} \subset S$ and $\{b_1, b_2, \ldots, b_m\} \subset S'$ be generators of S and S' respectively. Let $R = \mathbb{C}[x_1, x_2, \ldots, x_n]$ and $R' = \mathbb{C}[y_1, y_2, \ldots, y_m]$ be the polynomial rings. We have two surjective homomorphisms

$$\theta : R \longrightarrow S, \qquad \theta' : R' \longrightarrow S',$$

where $\theta(x_i) = a_i$ and $\theta'(y_i) = b_i$.

We know that $b_1, b_2, \ldots, b_m$ generate S', while $\varphi(a_1), \varphi(a_2), \ldots, \varphi(a_n)$ are elements of S'. It follows that we can write the $\varphi(a_j)$'s as polynomials in the b_i; there exist identities in S'

$$\varphi(a_j) = P_j(b_1, b_2, \ldots, b_m) ,$$

where P_j are polynomials with coefficients in $\mathbb{C}$. Now we define a homomorphism $\psi : R \longrightarrow R'$ by setting

$$\psi(x_j) = P_j(y_1, y_2, \ldots, y_m) .$$

We deduce a commutative square of ring homomorphisms

$$\begin{array}{ccc} R & \xrightarrow{\psi} & R' \\ {\scriptstyle \theta} \Big\downarrow & & \Big\downarrow {\scriptstyle \theta'} \\ S & \xrightarrow[\varphi]{} & S' \end{array} .$$

This gives us a commutative square

$$\begin{array}{ccccccc} & & \mathrm{MaxSpec}(S') & \xrightarrow{\mathrm{MaxSpec}(\varphi)} & \mathrm{MaxSpec}(S) & & \\ & & {\scriptstyle \mathrm{MaxSpec}(\theta')}\big\downarrow & & \big\downarrow{\scriptstyle \mathrm{MaxSpec}(\theta)} & & \\ \mathbb{C}^m & \cong & \mathrm{MaxSpec}(R') & \xrightarrow[\mathrm{MaxSpec}(\psi)]{} & \mathrm{MaxSpec}(R) & \cong & \mathbb{C}^n \end{array} .$$

In the complex topology the vertical maps are inclusions of subspaces by definition; Definition 4.5.1 gives the complex topology on MaxSpec(S) using its embedding in $\mathbb{C}^n$. The map $\mathrm{MaxSpec}(\psi) : \mathbb{C}^m \longrightarrow \mathbb{C}^n$ is a polynomial map by Remark 4.4.6, and hence continuous in the complex topology. It follows that the map of subspaces MaxSpec(φ) is also continuous in the complex topology. □

Corollary 4.5.7. *Suppose $\varphi : S \longrightarrow S'$ is an isomorphism of finitely generated $\mathbb{C}$–algebras. Then the induced map*

$$\left\{ \mathrm{Spec}(\varphi) \right\}^{\mathrm{an}} : \left\{ \mathrm{Spec}(S') \right\}^{\mathrm{an}} \longrightarrow \left\{ \mathrm{Spec}(S) \right\}^{\mathrm{an}}$$

is a homeomorphism.

Proof. Let $\varphi^{-1} : S' \longrightarrow S$ be the inverse of φ. Then the continuous map

$$\left\{ \mathrm{Spec}(\varphi^{-1}) \right\}^{\mathrm{an}} : \left\{ \mathrm{Spec}(S) \right\}^{\mathrm{an}} \longrightarrow \left\{ \mathrm{Spec}(S') \right\}^{\mathrm{an}}$$

is clearly a 2–sided inverse for $\left\{ \mathrm{Spec}(\varphi) \right\}^{\mathrm{an}}$. □

Lemma 4.5.8. *Suppose $\varphi : S \longrightarrow S'$ is a surjective homomorphism of finitely generated $\mathbb{C}$–algebras. Then the induced map of Lemma 4.5.6*

$$\left\{ \mathrm{Spec}(\varphi) \right\}^{\mathrm{an}} : \left\{ \mathrm{Spec}(S') \right\}^{\mathrm{an}} \longrightarrow \left\{ \mathrm{Spec}(S) \right\}^{\mathrm{an}}$$

is an embedding, mapping $\left\{ \mathrm{Spec}(S') \right\}^{\mathrm{an}}$ homeomorphically to its image in $\left\{ \mathrm{Spec}(S) \right\}^{\mathrm{an}}$.

Proof. Let us choose generators $a_1, a_2, \ldots, a_n$ for the ring S, and now let $R = \mathbb{C}[x_1, x_2, \ldots, x_n]$ be the polynomial ring. Define $\theta : R \longrightarrow S$ to be the surjection with $\theta(x_i) = a_i$. Then R surjects also to S' via the composite

$$R \xrightarrow{\ \theta\ } S \xrightarrow{\ \varphi\ } S' .$$

This means that the complex topology, on each of $\left\{ \mathrm{Spec}(S') \right\}^{\mathrm{an}}$ and $\left\{ \mathrm{Spec}(S) \right\}^{\mathrm{an}}$, can be defined as the subspace topology via the embedding into $\left\{ \mathrm{Spec}(R) \right\}^{\mathrm{an}} = \mathbb{C}^n$. We have maps

$$\left\{ \mathrm{Spec}(S') \right\}^{\mathrm{an}} \xrightarrow{\left\{ \mathrm{Spec}(\varphi) \right\}^{\mathrm{an}}} \left\{ \mathrm{Spec}(S) \right\}^{\mathrm{an}} \xrightarrow{\left\{ \mathrm{Spec}(\theta) \right\}^{\mathrm{an}}} \left\{ \mathrm{Spec}(R) \right\}^{\mathrm{an}} ,$$

which must make $\{\,\mathrm{Spec}(\varphi)\}^{\mathrm{an}}$ the inclusion of one subspace of the ambient $\{\,\mathrm{Spec}(R)\}^{\mathrm{an}} = \mathbb{C}^n$ into a larger subspace. □

Lemma 4.5.9. *Let $R = \mathbb{C}[x_1, x_2, \ldots, x_n]$ be the polynomial ring, and let f be an element of R. By Lemma 4.5.6 the ring homomorphism $\alpha_f : R \longrightarrow R[1/f]$ induces a continuous map*

$$\{\,\mathrm{Spec}(\alpha_f)\}^{\mathrm{an}} : \{\,\mathrm{Spec}(R[1/f])\}^{\mathrm{an}} \longrightarrow \{\,\mathrm{Spec}(R)\}^{\mathrm{an}} .$$

We assert that this map embeds $\{\,\mathrm{Spec}(R[1/f])\}^{\mathrm{an}}$ as a subspace of $\{\,\mathrm{Spec}(R)\}^{\mathrm{an}}$.

Proof. The case where $f = 0$ is obvious, so assume $f \neq 0$. The ring R is the polynomial ring $\mathbb{C}[x_1, x_2, \ldots, x_n]$, and the ring $R[1/f]$ has generators $\{x_1, x_2, \ldots, x_n, 1/f\}$. Consider therefore the polynomial ring $R' = \mathbb{C}[x_1, x_2, \ldots, x_n, x_{n+1}]$, let $\varphi : R \longrightarrow R'$ be the inclusion, and let $\theta : R' \longrightarrow R[1/f]$ be the map

$$\theta(x_i) = \begin{cases} x_i & \text{if } 1 \leq i \leq n, \\ \frac{1}{f(x_1, x_2, \ldots, x_n)} & \text{if } i = n+1. \end{cases}$$

The composite $\theta\varphi$ agrees with $\alpha_f : R \longrightarrow R[1/f]$. We deduce a factorization of $\{\,\mathrm{Spec}(\alpha_f)\}^{\mathrm{an}}$ as

$$\{\,\mathrm{Spec}(R[1/f])\}^{\mathrm{an}} \xrightarrow{\{\,\mathrm{Spec}(\theta)\}^{\mathrm{an}}} \{\,\mathrm{Spec}(R')\}^{\mathrm{an}} \xrightarrow{\{\,\mathrm{Spec}(\varphi)\}^{\mathrm{an}}} \{\,\mathrm{Spec}(R)\}^{\mathrm{an}} ,$$

and now we propose to compute these maps.

We have identified $\{\,\mathrm{Spec}(R')\}^{\mathrm{an}} \cong \mathbb{C}^{n+1}$ and $\{\,\mathrm{Spec}(R)\}^{\mathrm{an}} \cong \mathbb{C}^n$, and Lemma 4.4.5 tells us further that the map $\{\,\mathrm{Spec}(\varphi)\}^{\mathrm{an}}$ is just the projection $\mathbb{C}^{n+1} \longrightarrow \mathbb{C}^n$ taking

$$(a_1, a_2, \ldots, a_n, a_{n+1}) \in \mathbb{C}^{n+1} \qquad \text{to} \qquad (a_1, a_2, \ldots, a_n) \in \mathbb{C}^n .$$

Because the map $\theta : R' \longrightarrow R[1/f]$ is surjective, the induced map

$$\{\,\mathrm{Spec}(\theta)\}^{\mathrm{an}} : \{\,\mathrm{Spec}(R[1/f])\}^{\mathrm{an}} \longrightarrow \{\,\mathrm{Spec}(R')\}^{\mathrm{an}}$$

is an embedding of a subspace. To show that the composite embeds $\{\,\mathrm{Spec}(R[1/f])\}^{\mathrm{an}}$ it suffices to establish that the image of $\{\,\mathrm{Spec}(\theta)\}^{\mathrm{an}}$ is contained in some subspace of $\Gamma \subset \mathbb{C}^{n+1}$, on which the projection map to $\mathbb{C}^n$ is an embedding. Next we will explicitly give a Γ that works.

The Γ we wish to consider will be the set $\Gamma \subset \mathbb{C}^{n+1}$ of all points $(a_1, a_2, \ldots, a_n, a_{n+1}) \in \mathbb{C}^{n+1}$ satisfying the equation

$$a_{n+1} = \frac{1}{f(a_1, a_2, \ldots, a_n)} .$$

We will first show that the map from $\Gamma \subset \mathbb{C}^{n+1}$ to $\mathbb{C}^n$ is an embedding of a subspace, and then that the image of $\{\operatorname{Spec}(\theta)\}^{\text{an}}$ is contained in Γ.

In this paragraph we show that the map from $\Gamma \subset \mathbb{C}^{n+1}$ to $\mathbb{C}^n$ is an embedding of a subspace. To see this note that there is a continuous function

$$\tfrac{1}{f} : \mathbb{C}^n_f \longrightarrow \mathbb{C} ,$$

where $\mathbb{C}^n_f \subset \mathbb{C}^n$ is the subset of all $a \in \mathbb{C}^n$ with $f(a) \neq 0$. The graph of this function is the subset

$$\Gamma \quad \subset \quad \mathbb{C}^n_f \times \mathbb{C} \quad \subset \quad \mathbb{C}^n \times \mathbb{C} \quad = \quad \mathbb{C}^{n+1} .$$

Now consider the commutative diagram

$$\begin{array}{ccccc} \Gamma & \subset & \mathbb{C}^n_f \times \mathbb{C} & \xrightarrow{\ i\ } & \mathbb{C}^n \times \mathbb{C} \\ & & {\scriptstyle \pi}\big\downarrow & & \big\downarrow{\scriptstyle \{\operatorname{Spec}(\varphi)\}^{\text{an}}} \\ & & \mathbb{C}^n_f & \xrightarrow[j]{} & \mathbb{C}^n \end{array} .$$

The composite

$$\begin{array}{ccc} \Gamma & \subset & \mathbb{C}^n_f \times \mathbb{C} \\ & & {\scriptstyle \pi}\big\downarrow \\ & & \mathbb{C}^n_f \end{array}$$

is a homeomorphism; it has a continuous inverse taking $a \in \mathbb{C}^n_f$ to $\big(a, 1/f(a)\big) \in \Gamma \subset \mathbb{C}^n_f \times \mathbb{C}$. This makes the composite

$$\begin{array}{cccc} \Gamma & \subset & \mathbb{C}^n_f \times \mathbb{C} & \\ & & {\scriptstyle \pi}\big\downarrow & \\ & & \mathbb{C}^n_f & \xrightarrow[j]{} \mathbb{C}^n \end{array}$$

a homeomorphism from Γ to the open subset $\mathbb{C}^n_f \subset \mathbb{C}^n$. But the commutativity says that this is equal to the composite

$$\begin{array}{ccccc} \Gamma & \subset & \mathbb{C}^n_f \times \mathbb{C} & \xrightarrow{\ i\ } & \mathbb{C}^n \times \mathbb{C} \\ & & & & \big\downarrow{\scriptstyle \{\operatorname{Spec}(\varphi)\}^{\text{an}}} \\ & & & & \mathbb{C}^n \end{array}$$

which is the restriction to $\Gamma \subset \mathbb{C}^n \times \mathbb{C} = \mathbb{C}^{n+1}$ of the projection $\{\operatorname{Spec}(\varphi)\}^{\text{an}} : \mathbb{C}^{n+1} \longrightarrow \mathbb{C}^n$.

It remains to show that the image of the map

$$\{\operatorname{Spec}(\theta)\}^{\mathrm{an}} : \{\operatorname{Spec}(R[1/f])\}^{\mathrm{an}} \longrightarrow \mathbb{C}^{n+1}$$

is contained in Γ. Let us therefore be given a point in $\{\operatorname{Spec}(R[1/f])\}^{\mathrm{an}}$, which by Proposition 4.2.4 we identify with a homomorphism of $\mathbb{C}$–algebras $\varphi : R[1/f] \longrightarrow \mathbb{C}$. By Lemma 4.3.4 its image in $\{\operatorname{Spec}(R')\}^{\mathrm{an}}$ identifies with $\varphi\theta : R' \longrightarrow \mathbb{C}$. The identification of $\{\operatorname{Spec}(R')\}^{\mathrm{an}}$ with $\mathbb{C}^{n+1}$ takes $\varphi\theta$ to the vector

$$\big(\varphi\theta(x_1), \varphi\theta(x_2), \ldots, \varphi\theta(x_n), \varphi\theta(x_{n+1})\big) \in \mathbb{C}^{n+1},$$

and we need to show that this vector lies in Γ. But for $0 \le i \le n$ we have $\theta(x_i) = x_i$, while $\theta(x_{n+1}) = 1/f(x_1, x_2, \ldots, x_n)$. The vector above becomes

$$\left(\varphi(x_1), \varphi(x_2), \ldots, \varphi(x_n), \varphi\left(\frac{1}{f(x_1, x_2, \ldots, x_n)}\right)\right) \in \mathbb{C}^{n+1}.$$

As φ is a homomorphism this point must lie on Γ. □

Proposition 4.5.10. *Let S be a finitely generated algebra over $\mathbb{C}$, and let f be an element of S. The natural map*

$$\{\operatorname{Spec}(\alpha_f)\}^{\mathrm{an}} : \{\operatorname{Spec}(S[1/f])\}^{\mathrm{an}} \longrightarrow \{\operatorname{Spec}(S)\}^{\mathrm{an}}$$

embeds $\{\operatorname{Spec}(S[1/f])\}^{\mathrm{an}}$ as the open subset $\{\operatorname{Spec}(S)\}_f \cap \{\operatorname{Spec}(S)\}^{\mathrm{an}}$ of the space $\{\operatorname{Spec}(S)\}^{\mathrm{an}}$.

Proof. Choose generators $a_1, a_2, \ldots, a_n$ for the ring S, suppose $R = \mathbb{C}[x_1, x_2, \ldots, x_n]$ is the polynomial ring, and let $\theta : R \longrightarrow S$ be the surjection with $\theta(x_i) = a_i$. Pick an element $g \in R$ with $\theta(g) = f$. We have a diagram of rings

$$\begin{array}{ccc} R & \xrightarrow{\alpha_g} & R[1/g] \\ {\scriptstyle\theta}\downarrow & & \\ S & \xrightarrow[\alpha_f]{} & S[1/f] \end{array} .$$

The composite

$$\begin{array}{ccc} R & & \\ {\scriptstyle\theta}\downarrow & & \\ S & \xrightarrow[\alpha_f]{} & S[1/f] \end{array}$$

takes g to the invertible element $f/1 \in S[1/f]$, and hence must factor through α_g. We deduce a commutative square

$$\begin{array}{ccc} R & \xrightarrow{\alpha_g} & R[1/g] \\ {\scriptstyle\theta}\downarrow & & \downarrow{\scriptstyle\theta'} \\ S & \xrightarrow[\alpha_f]{} & S[1/f] \end{array} .$$

Furthermore, the homomorphism $\theta' : R[1/g] \longrightarrow S[1/f]$ is surjective. Every element of $S[1/f]$ can be written as s/f^n with $s \in S$. Since $\theta : R \longrightarrow S$ is surjective we can write s as $\theta(r)$, and this makes $s/f^n = \theta(r)/\theta(g)^n = \theta'(r/g^n)$. There is a commutative square of continuous maps of topological spaces

$$\begin{array}{ccc} \left\{\operatorname{Spec}(S[1/f])\right\}^{\text{an}} & \xrightarrow{\left\{\operatorname{Spec}(\alpha_f)\right\}^{\text{an}}} & \left\{\operatorname{Spec}(S)\right\}^{\text{an}} \\ {\scriptstyle\left\{\operatorname{Spec}(\theta')\right\}^{\text{an}}}\downarrow & & \downarrow{\scriptstyle\left\{\operatorname{Spec}(\theta)\right\}^{\text{an}}} \\ \left\{\operatorname{Spec}(R[1/g])\right\}^{\text{an}} & \xrightarrow[\left\{\operatorname{Spec}(\alpha_g)\right\}^{\text{an}}]{} & \left\{\operatorname{Spec}(R)\right\}^{\text{an}} \cong \mathbb{C}^n \end{array} .$$

Both the vertical maps induce embeddings of subspaces because the ring homomorphisms are surjective; see Lemma 4.5.8. The bottom map is an embedding of a subspace by Lemma 4.5.9. Hence all the spaces in the diagram are identified as subspaces of $\left\{\operatorname{Spec}(R)\right\}^{\text{an}} \cong \mathbb{C}^n$, and the top map must be an embedding.

It remains only to identify the image in $\left\{\operatorname{Spec}(S)\right\}^{\text{an}}$ and show that it is open. Put $X = \operatorname{Spec}(S)$. Lemma 3.7.5 tells us that the map

$$\operatorname{Spec}(\alpha_f) : \operatorname{Spec}(S[1/f]) \longrightarrow \operatorname{Spec}(S) = X$$

is an embedding whose image is precisely the open set $X_f \subset X$. Corollary 4.2.9 says that the image of the map $\left\{\operatorname{Spec}(S[1/f])\right\}^{\text{an}} \longrightarrow \operatorname{Max}(X)$ is precisely $X_f \cap \operatorname{Max}(X)$. But Lemma 4.5.4 says that the inclusion $\operatorname{Max}(X) \longrightarrow X$ is continuous, where X has the Zariski topology and $\operatorname{Max}(X)$ the complex topology. Hence the intersection of the open set $X_f \subset X$ with $\operatorname{Max}(X)$ is open in the complex topology. □

4.6 The complex topology on schemes

Until now we have defined the complex topology for spaces $\operatorname{MaxSpec}(R)$, where R is a finitely generated algebra over $\mathbb{C}$. Now we want to show that $\operatorname{Max}(X)$ has a complex topology, for any scheme $(X, \mathcal{O})$ locally of finite type over $\mathbb{C}$. The idea is the following. We have an open cover $X = \cup_{i\in I} U_i$, together with isomorphisms

$$(U_i, \mathcal{O}|_{U_i}) \quad \cong \quad \left(\operatorname{Spec}(R_i), \widetilde{R_i}\right) ,$$

for some finitely generated $\mathbb{C}$–algebras R_i. By Remark 4.2.10

$$\operatorname{Max}(X) \quad = \quad \bigcup_{i\in I} \operatorname{Max}(U_i) \quad \cong \quad \bigcup_{i\in I} \operatorname{MaxSpec}(R_i)$$

is a cover of $\operatorname{Max}(X)$. Remark 4.2.10 also proves that it is an open cover in the Zariski topology, but this does not help us much with the complex topology. What we want to do is define the complex topology on $\operatorname{Max}(X)$ by declaring the above to be an open cover in the complex topology, by sets homeomorphic to $\{\operatorname{Spec}(R_i)\}^{\text{an}}$. The complex topology of $\{\operatorname{Spec}(R_i)\}^{\text{an}}$ already exists; what we need to check is that the $\{\operatorname{Spec}(R_i)\}^{\text{an}}$ glue well. Modulo technicalities, the key lemma is that $\operatorname{Max}(U_i)\cap\operatorname{Max}(U_j)$ is open in the complex topologies of each of $\operatorname{Max}(U_i)$ and $\operatorname{Max}(U_j)$, and that the two subspace topologies on it agree.

Lemma 4.6.1. *Let $(Z, \mathcal{O})$ be a scheme locally of finite type over $\mathbb{C}$, and let U and V be open subsets of Z. Suppose we have isomorphisms of ringed spaces over $\mathbb{C}$*

$$(\Phi, \Phi^*) : \left(\operatorname{Spec}(S), \widetilde{S}\right) \longrightarrow (U, \mathcal{O}|_U) ,$$
$$(\Psi, \Psi^*) : \left(\operatorname{Spec}(R), \widetilde{R}\right) \longrightarrow (V, \mathcal{O}|_V) .$$

Put $X = \operatorname{Spec}(S)$ and $Y = \operatorname{Spec}(R)$. We assert that $\operatorname{Max}(U)\cap\operatorname{Max}(V)$ is open in each of $\{\operatorname{Spec}(R)\}^{\text{an}}$ and $\{\operatorname{Spec}(S)\}^{\text{an}}$, and that the two subspace topologies, induced on $\operatorname{Max}(U)\cap\operatorname{Max}(V)$ by the embeddings

$$\operatorname{Max}(U)\cap\operatorname{Max}(V) \subset \{\operatorname{Spec}(R)\}^{\text{an}}$$
$$\operatorname{Max}(U)\cap\operatorname{Max}(V) \subset \{\operatorname{Spec}(S)\}^{\text{an}} ,$$

agree with each other.

Proof. Let us first prove that $\operatorname{Max}(U)\cap\operatorname{Max}(V)$ is open in the complex topology of $\operatorname{Max}(U) = \{\operatorname{Spec}(S)\}^{\text{an}}$; the fact that it is open in $\{\operatorname{Spec}(R)\}^{\text{an}}$ follows by interchanging U and V. The ringed space $(U, \mathcal{O}|_U)$ is locally of finite type over $\mathbb{C}$ (in fact it is even affine), and $U\cap V$ is an open subset. By Corollary 4.2.9

$$\operatorname{Max}(U\cap V) \quad = \quad \{U\cap V\}\cap\operatorname{Max}(U) .$$

By symmetry

$$\operatorname{Max}(U\cap V) \quad = \quad \{U\cap V\}\cap\operatorname{Max}(V) ,$$

and, combining the two equalities, we have

$$\operatorname{Max}(U\cap V) \quad = \quad \{U\cap V\}\cap\operatorname{Max}(U)\cap\operatorname{Max}(V) \quad = \quad \operatorname{Max}(U)\cap\operatorname{Max}(V).$$

The last equality is because $U \cap V$ contains $\operatorname{Max}(U) \cap \operatorname{Max}(V)$.

We have so far shown that $\operatorname{Max}(U) \cap \operatorname{Max}(V) = \operatorname{Max}(U \cap V)$, and we want to show $\operatorname{Max}(U) \cap \operatorname{Max}(V)$ open in the complex topology of $\operatorname{Max}(U) = \left\{ \operatorname{Spec}(S) \right\}^{\text{an}}$. Since the sets are equal we will prove that $\operatorname{Max}(U \cap V)$ is open in $\operatorname{Max}(U) = \left\{ \operatorname{Spec}(S) \right\}^{\text{an}}$. By the above we know that $\operatorname{Max}(U \cap V) = \{U \cap V\} \cap \operatorname{Max}(U)$, which expresses $\operatorname{Max}(U \cap V)$ as the intersection of a Zariski open set $U \cap V$ in X with $\operatorname{Max}(U) = \left\{ \operatorname{Spec}(S) \right\}^{\text{an}}$. Lemma 4.5.4 tells us that the map

$$\lambda_S : \left\{ \operatorname{Spec}(S) \right\}^{\text{an}} \longrightarrow \operatorname{Spec}(S)$$

is continuous; hence the inverse image of the open set $U \cap V \subset \operatorname{Spec}(S)$ is open in $\left\{ \operatorname{Spec}(S) \right\}^{\text{an}}$.

Next we want to show that the two complex topologies on $\operatorname{Max}(U) \cap \operatorname{Max}(V)$, the one it inherits as a subspace of $\left\{ \operatorname{Spec}(S) \right\}^{\text{an}}$ and the one it inherits as a subspace of $\left\{ \operatorname{Spec}(R) \right\}^{\text{an}}$, agree with each other. To do this we recall Proposition 3.10.9. The proposition tells us that the space $U \cap V$ has an open cover

$$U \cap V \quad = \quad \bigcup_{i \in I} W_i$$

so that, for each $i \in I$, there exists an $f_i \in S$ and a $g_i \in R$ with $\Phi^{-1} W_i = X_{f_i}$ and $\Psi^{-1} W_i = Y_{g_i}$. The idea of the proof is to use this open cover.

The W_i provide an open cover of $U \cap V$; Remark 4.2.10 says that

$$\operatorname{Max}(U \cap V) \quad = \quad \bigcup_{i \in I} \operatorname{Max}(W_i)\ .$$

We also know that $W_i \subset U$ is open, and in Corollary 4.2.9 we learned that $\operatorname{Max}(W_i) = W_i \cap \operatorname{Max}(U)$; therefore $\operatorname{Max}(W_i)$ is Zariski open in $\operatorname{Max}(U)$. The continuity of the inclusion of $\operatorname{Max}(U) = \left\{ \operatorname{Spec}(S) \right\}^{\text{an}}$ into $U = \operatorname{Spec}(S)$ tells us that $\operatorname{Max}(W_i) \subset \operatorname{Max}(U)$ is open in the complex topology as well. Symmetry tells us that $\operatorname{Max}(W_i) \subset \operatorname{Max}(V)$ is also open in the complex topology on $\operatorname{Max}(V)$. The expression

$$\operatorname{Max}(U \cap V) \quad = \quad \bigcup_{i \in I} \operatorname{Max}(W_i)$$

therefore gives an open cover of $\operatorname{Max}(U \cap V)$, in either the complex topology of $\operatorname{Max}(U) = \left\{ \operatorname{Spec}(S) \right\}^{\text{an}}$ or in the complex topology of

$\text{Max}(V) = \{\text{Spec}(R)\}^{\text{an}}$. For any subset $D \subset \text{Max}(U \cap V)$ we have an equality

$$D = \bigcup_{i \in I} D \cap \text{Max}(W_i) ,$$

and D will be open in $\text{Max}(U)$ (respectively $\text{Max}(V)$) if and only if all the $D \cap \text{Max}(W_i)$ are. We are therefore reduced to showing that, for each i, the two complex topologies on $\text{Max}(W_i)$ agree; we must show that a subset $D \subset \text{Max}(W_i)$ is open in $\text{Max}(U) = \{\text{Spec}(S)\}^{\text{an}}$ if and only if it is open in in $\text{Max}(V) = \{\text{Spec}(R)\}^{\text{an}}$. Fix the index i and write $W = W_i$. As in the statement of Proposition 3.10.9 put $X = \text{Spec}(S)$ and $Y = \text{Spec}(R)$. Then $W \cong X_f \cong Y_g$, and we are reduced to showing that the subspace complex topologies on $X_f \cap \text{Max}(X)$ and $Y_g \cap \text{Max}(Y)$ agree.

Now Proposition 4.5.10 tells us that the ring homomorphism $\alpha_f : S \longrightarrow S[1/f]$ induces an open embedding in the complex topology

$$\{\text{Spec}(\alpha_f)\}^{\text{an}} : \{\text{Spec}(S[1/f])\}^{\text{an}} \longrightarrow \{\text{Spec}(S)\}^{\text{an}}$$

whose image is precisely $X_f \cap \text{Max}(X)$. Similarly $Y_g \cap \text{Max}(Y)$ is identified, with its complex subspace topology, as the image of the open embedding

$$\{\text{Spec}(\alpha_g)\}^{\text{an}} : \{\text{Spec}(R[1/g])\}^{\text{an}} \longrightarrow \{\text{Spec}(R)\}^{\text{an}} .$$

We are therefore reduced to showing that the identification of the set $\text{MaxSpec}(S[1/f])$ with the set $\text{MaxSpec}(R[1/g])$ respects the complex topology. But the identification comes from the isomorphism of ringed spaces over $\mathbb{C}$

$$\left(\text{Spec}(S[1/f]), \widetilde{S[1/f]}\right) \cong (W, \mathcal{O}|_W) \cong \left(\text{Spec}(R[1/g]), \widetilde{R[1/g]}\right) .$$

This gives an isomorphism of ringed spaces over $\mathbb{C}$

$$(\Phi, \Phi^*) : \left(\text{Spec}(S[1/f]), \widetilde{S[1/f]}\right) \longrightarrow \left(\text{Spec}(R[1/g]), \widetilde{R[1/g]}\right) ,$$

which, by Theorem 3.9.4, must be of the form $(\text{Spec}(\varphi), \widetilde{\varphi})$ for some isomorphism φ. The induced map

$$\{\text{Spec}(\varphi)\}^{\text{an}} : \{\text{Spec}(S[1/f])\}^{\text{an}} \longrightarrow \{\text{Spec}(R[1/g])\}^{\text{an}}$$

is a homeomorphism by Corollary 4.5.7. □

Remark 4.6.2. In the special case, where $U = V$ in the statement of

Lemma 4.6.1, we deduce that, if R and S are finitely generated algebras over $\mathbb{C}$ and we have isomorphisms

$$\left(\operatorname{Spec}(R), \widetilde{R}\right) \xrightarrow{(\Phi,\Phi^*)} (U, \mathcal{O}|_U) \xleftarrow{(\Psi,\Psi^*)} \left(\operatorname{Spec}(S), \widetilde{S}\right) ,$$

then the natural map $\operatorname{MaxSpec}(R) \longrightarrow \operatorname{MaxSpec}(S)$ is a homeomorphism in the complex topology. Of course this is also an easy consequence of Theorem 3.9.4 and Corollary 4.5.7.

Modulo technicalities Lemma 4.6.1 is the key, and it is time to use it in the global construction of the complex topology.

Reminder 4.6.3. We will define the complex topology of a scheme as the weak topology with respect to certain inclusions. We should first remind the reader of the weak topology. Let X be a set, and let $\{\varphi_i : U_i \longrightarrow X\}$ be a set of functions from topological spaces U_i to the set X. The weak topology of X, with respect to these maps φ_i, is the finest topology under which all the functions φ_i are continuous. That is, a subset $V \subset X$ is open in X if and only if all the $\varphi_i^{-1}V \subset U_i$ are open.

Definition 4.6.4. *Let $(X, \mathcal{O})$ be a scheme locally of finite type over $\mathbb{C}$. Let I be the set of all open immersions of ringed spaces over $\mathbb{C}$*

$$(\Phi_i, \Phi_i^*) : \left(\operatorname{Spec}(R_i), \widetilde{R_i}\right) \longrightarrow (X, \mathcal{O}) ,$$

with R_i a finitely generated $\mathbb{C}$–algebra. By Lemma 4.2.7 we have, for each $i \in I$, a map

$$\operatorname{Max}(\Phi_i) : \left\{\operatorname{Spec}(R_i)\right\}^{\mathrm{an}} \longrightarrow \operatorname{Max}(X) .$$

The complex topology on $\operatorname{Max}(X)$ is the weak topology with respect to the maps $\operatorname{Max}(\Phi_i)$.

Notation 4.6.5. Let $(X, \mathcal{O})$ be a scheme locally of finite type over $\mathbb{C}$. The set $\operatorname{Max}(X)$, endowed with its complex topology, will be denoted X^{an}.

Lemma 4.6.6. *Let $(X, \mathcal{O})$ be a scheme locally of finite type over $\mathbb{C}$. Assume we are given an open immersion of ringed spaces over $\mathbb{C}$*

$$\Psi : \left(\operatorname{Spec}(R), \widetilde{R}\right) \longrightarrow (X, \mathcal{O}) .$$

Then the map $\operatorname{Max}(\Psi) : \left\{\operatorname{Spec}(R)\right\}^{\mathrm{an}} \longrightarrow X^{\mathrm{an}}$ is, in the complex topology, a homeomorphism of $\left\{\operatorname{Spec}(R)\right\}^{\mathrm{an}}$ onto its image in X^{an}, and the image in X^{an} is open.

Proof. By Corollary 4.2.9 the map $\mathrm{Max}(\Psi)$ is injective. It is continuous since $\Psi : \left(\mathrm{Spec}(R), \widetilde{R}\right) \longrightarrow X$ is one of the elements of the set I of Definition 4.6.4; the complex topology on X^{an} is the weak topology with respect to several maps into the set $\mathrm{Max}(X) = X^{\mathrm{an}}$, among them the map $\mathrm{Max}(\Psi) : \left\{\mathrm{Spec}(R)\right\}^{\mathrm{an}} \longrightarrow \mathrm{Max}(X)$. The continuity is guaranteed. We need to prove that the image is open, and that $\mathrm{Max}(\Psi)$ maps $\left\{\mathrm{Spec}(R)\right\}^{\mathrm{an}}$ homeomorphically to its image. Concretely this reduces to showing that the image of any open set $W \subset \left\{\mathrm{Spec}(R)\right\}^{\mathrm{an}}$ is open in $\mathrm{Max}(X)$.

Therefore fix an open set $W \subset \left\{\mathrm{Spec}(R)\right\}^{\mathrm{an}}$. Then ΨW is a subset of $\mathrm{Max}(X)$ and we must show that, for all $i \in I$, the inverse image of ΨW via the map

$$\mathrm{Max}(\Psi_i) : \left\{\mathrm{Spec}(R_i)\right\}^{\mathrm{an}} \longrightarrow \mathrm{Max}(X)$$

is open in $\left\{\mathrm{Spec}(R_i)\right\}^{\mathrm{an}}$. Let us fix i. That is we are given some open immersion

$$\Phi : \left(\mathrm{Spec}(S), \widetilde{S}\right) \longrightarrow (X, \mathcal{O}) ;$$

it induces a map

$$\mathrm{Max}(\Phi) : \left\{\mathrm{Spec}(S)\right\}^{\mathrm{an}} \longrightarrow \mathrm{Max}(X) ,$$

and we want to study the inverse image of ΨW by the map. As in Lemma 4.6.1 let U and V be, respectively, the images of the two embeddings

$$\Phi : \mathrm{Spec}(S) \longrightarrow X , \qquad \Psi : \mathrm{Spec}(R) \longrightarrow X .$$

What we are given is that $\Psi W \subset \mathrm{Max}(V)$ is open in the complex topology that comes from the bijection of sets $\mathrm{Max}(V) \cong \mathrm{MaxSpec}(R)$. What is required is to show that $\mathrm{Max}(U) \cap \Psi W$ is open in the topology that comes from the bijection $\mathrm{Max}(U) \cong \mathrm{MaxSpec}(S)$.

But by Lemma 4.6.1 we know that $\mathrm{Max}(U) \cap \mathrm{Max}(V)$ is open as a subset of $\mathrm{Max}(V) \cong \mathrm{MaxSpec}(R)$. Since $\Psi W \subset \mathrm{Max}(V)$ is open by hypothesis, the intersection $\Psi W \cap \mathrm{Max}(U) \cap \mathrm{Max}(V) = \Psi W \cap \mathrm{Max}(U)$ is also open in $\mathrm{Max}(V) \cong \mathrm{MaxSpec}(R)$. But by Lemma 4.6.1 the topological space $\mathrm{Max}(U) \cap \mathrm{Max}(V)$ has the same complex topology, whether viewed as an open subspace of $\mathrm{Max}(U)$ or of $\mathrm{Max}(V)$. Hence the subset $\Psi W \cap \mathrm{Max}(U) \subset \mathrm{Max}(U) \cap \mathrm{Max}(V)$ is open in $\mathrm{Max}(U)$. □

Lemma 4.6.7. *Let* $(X, \mathcal{O})$ *be a scheme locally of finite type over* $\mathbb{C}$. *The inclusion* $\mathrm{Max}(X) \subset X$ *is a continuous map, when* $\mathrm{Max}(X)$ *is given the complex topology and* X *its Zariski topology.*

Proof. The problem is local in X and $\text{Max}(X)$. Choose any point in $x \in \text{Max}(X)$, that is a closed point $x \in X$. Let $U \subset X$ be an open affine containing x. That is, U is open and there is an isomorphism

$$(\Phi, \Phi^*) : \left(\text{Spec}(R), \widetilde{R}\right) \longrightarrow (U, \mathcal{O}|_U) .$$

We have a commutative square

$$\begin{array}{ccc} \text{MaxSpec}(R) & \xrightarrow{\lambda_R} & \text{Spec}(R) \\ {\scriptstyle i_1}\downarrow & & \downarrow{\scriptstyle i_2} \\ \text{Max}(X) & \longrightarrow & X \end{array} .$$

The map i_2 is an open embedding by hypothesis. The map i_1 is an open embedding in the complex topology by Lemma 4.6.6. The map λ_R is continuous by Lemma 4.5.4. We conclude that the map $\text{Max}(X) \longrightarrow X$ is continuous in a neighborhood of $x \in \text{Max}(X)$. But $x \in \text{Max}(X)$ was arbitrary. □

Lemma 4.6.8. *Let $(\Phi, \Phi^*) : (X, \mathcal{O}_X) \longrightarrow (Y, \mathcal{O}_Y)$ be a morphism of schemes locally of finite type over $\mathbb{C}$. Then the induced map*

$$\text{Max}(\Phi) : \text{Max}(X) \longrightarrow \text{Max}(Y)$$

is continuous in the complex topology.

Proof. Pick a point $x \in \text{Max}(X)$ and we will prove continuity at x. By Remark 3.10.13 it is possible to choose a commutative diagram

$$\begin{array}{ccc} \left(\text{Spec}(S), \widetilde{S}\right) & \xrightarrow{(\text{Spec}(\varphi), \widetilde{\varphi})} & \left(\text{Spec}(R), \widetilde{R}\right) \\ {\scriptstyle (i_1, i_1^*)}\downarrow & & \downarrow{\scriptstyle (i_2, i_2^*)} \\ (X, \mathcal{O}_X) & \xrightarrow[(\Phi, \Phi^*)]{} & (Y, \mathcal{O}_Y) \end{array} ,$$

where the vertical maps are open immersions, and where x belongs to $\text{Spec}(S) \subset X$. But then, taking closed points everywhere, we get a commutative square of maps of sets

$$\begin{array}{ccc} \text{MaxSpec}(S) & \xrightarrow{\text{MaxSpec}(\varphi)} & \text{MaxSpec}(R) \\ {\scriptstyle \text{Max}(i_1)}\downarrow & & \downarrow{\scriptstyle \text{Max}(i_2)} \\ \text{Max}(X) & \xrightarrow[\text{Max}(\Phi)]{} & \text{Max}(Y) \end{array} .$$

By Lemma 4.6.6 the vertical maps are open immersions in the complex topology. The top horizontal map is continuous in the complex topology by Lemma 4.5.6. We conclude that the bottom map is continuous at $x \in \text{Max}(X)$. □

Lemma 4.6.9. *Let $(\Phi, \Phi^*) : (X, \mathcal{O}_X) \longrightarrow (Y, \mathcal{O}_Y)$ be an open immersion of schemes of finite type over $\mathbb{C}$. Then the induced map*

$$\text{Max}(\Phi) : \text{Max}(X) \longrightarrow \text{Max}(Y)$$

embeds $\text{Max}(X)$ onto the open subset $X \cap \text{Max}(Y)$ in the complex topology.

Proof. The fact that the map is injective and the image is the set $X \cap \text{Max}(Y)$ may be found in Corollary 4.2.9. The fact that $X \cap \text{Max}(Y)$ is open in the complex topology of $\text{Max}(Y)$ comes from the continuity of the inclusion $\text{Max}(Y) \longrightarrow Y$, proved in Lemma 4.6.7. It only remains to prove that the map is an embedding in the complex topology. But this is a local question. Any point $x \in \text{Max}(X)$ is in the image of some open immersion

$$(\Psi, \Psi^*) : \left(\text{Spec}(R), \widetilde{R}\right) \longrightarrow (X, \mathcal{O}_X) .$$

The composite

$$\left(\text{Spec}(R), \widetilde{R}\right) \xrightarrow{(\Psi,\Psi^*)} (X, \mathcal{O}_X) \xrightarrow{(\Phi,\Phi^*)} (Y, \mathcal{O}_Y) .$$

is the composite of two open immersions, hence an open immersion. It follows from Lemma 4.6.6 that both of the maps

$$\text{Max}(\Psi) \ : \text{MaxSpec}(R) \longrightarrow \text{Max}(X)$$

$$\text{Max}(\Phi\Psi) : \text{MaxSpec}(R) \longrightarrow \text{Max}(Y)$$

are open immersions in the complex topology. Therefore the map $\text{Max}(\Phi) : \text{Max}(X) \longrightarrow \text{Max}(Y)$ is an embedding near $x \in \text{Max}(X)$. □

Summary 4.6.10. What we have proved in this section, so far, is the following. Given a scheme $(X, \mathcal{O}_X)$, locally of finite type over $\mathbb{C}$, it is possible to attach a complex topology to the set $\text{Max}(X)$. Let us call this topological space X^{an}. Given a morphism of schemes locally of finite type over $\mathbb{C}$

$$(\Phi, \Phi^*) : (X, \mathcal{O}_X) \longrightarrow (Y, \mathcal{O}_Y) ,$$

the map $\Phi : X \longrightarrow Y$ induces a map from the subset $\text{Max}(X) \subset X$ to the subset $\text{Max}(Y) \subset Y$ of closed points, and this map is continuous in the complex topology. We will denote it

$$\Phi^{\text{an}} : X^{\text{an}} \longrightarrow Y^{\text{an}} .$$

If (Φ, Φ^*) is an open immersion then the map Φ^{an} is an open embedding with image $X \cap Y^{\text{an}} \subset Y^{\text{an}}$. And finally, the inclusion $\text{Max}(X) \longrightarrow X$ is continuous, if we endow X with the Zariski topology and $\text{Max}(X)$ with the complex topology. We will denote this map

$$\lambda_X : X^{\text{an}} \longrightarrow X.$$

Next we give a couple of easy consequences.

Corollary 4.6.11. *Given two composable morphisms of schemes locally of finite type over $\mathbb{C}$*

$$(X, \mathcal{O}_X) \xrightarrow{(\Phi,\Phi^*)} (Y, \mathcal{O}_Y) \xrightarrow{(\Psi,\Psi^*)} (Z, \mathcal{O}_Z)$$

the composite

$$X^{\text{an}} \xrightarrow{\Phi^{\text{an}}} Y^{\text{an}} \xrightarrow{\Psi^{\text{an}}} Z^{\text{an}}$$

agrees with $\{\Psi\Phi\}^{\text{an}} : X^{\text{an}} \longrightarrow Z^{\text{an}}$.

Proof. This is just a statement that the maps agree at the level of sets. Since all three are subsets of closed points in, respectively, X, Y and Z, we are just saying that the restriction to $\text{Max}(X) \subset X$ of the composite $\Psi\Phi$ is the composite of the restrictions. □

Corollary 4.6.12. *Let $(\Phi, \Phi^*) : (X, \mathcal{O}_X) \longrightarrow (Y, \mathcal{O}_Y)$ be a map of schemes locally of finite type over $\mathbb{C}$. Then the following square of continuous maps commutes*

$$\begin{array}{ccc} X^{\text{an}} & \xrightarrow{\Phi^{\text{an}}} & Y^{\text{an}} \\ {\scriptstyle \lambda_X}\downarrow & & \downarrow{\scriptstyle \lambda_Y} \\ X & \xrightarrow[\Phi]{} & Y \end{array}$$

Proof. Once again we know the continuity of the maps; the only question is whether the square commutes, on the level of maps of sets. The fact that it commutes is the definition of $\Phi^{\text{an}} = \text{Max}(\Phi)$. The map was defined as the restriction of Φ to the subset $\text{Max}(X) \subset X$. □

5

The analytification of a scheme

From Chapter 3 we know what it means for a ringed space $(X, \mathcal{O})$ to be a scheme locally of finite type over $\mathbb{C}$. In Chapter 4 we learned about the complex topology on the subset $X^{\mathrm{an}} \subset X$ of closed points. Now we want to turn X^{an} into a ringed space. We want to define the sheaf of rings of analytic functions on X^{an}. We will produce a sheaf of rings $\mathcal{O}^{\mathrm{an}}$ on X^{an}, in such a way that the ringed space $(X^{\mathrm{an}}, \mathcal{O}^{\mathrm{an}})$ will be an *analytic space.*

Remark 5.0.1. We highly recommend that the reader learn some of the theory of analytic functions in several complex variables. It is a beautiful theory. In this book we will not develop it nearly as much as we have been explaining the algebraic theory. We will try to say enough for our assertions to be comprehensible, referring the reader elsewhere for proofs. In particular we will not explain in any detail the properties of analytic spaces. For us they are ringed spaces, and we explicitly produce some examples.

Let us provide a brief summary of the main results in this chapter.

5.1 Synopsis of the main results

5.1.1. Let $(X, \mathcal{O})$ be a scheme of finite type over $\mathbb{C}$. There is a sheaf of rings $\mathcal{O}^{\mathrm{an}}$ on X^{an}, and a map of ringed spaces over $\mathbb{C}$

$$(\lambda_X, \lambda_X^*) : (X^{\mathrm{an}}, \mathcal{O}^{\mathrm{an}}) \longrightarrow (X, \mathcal{O}) .$$

The continuous map $\lambda_X : X^{\mathrm{an}} \longrightarrow X$ is the inclusion of the closed points of X into X, as in 4.1.1. The ringed space $(X^{\mathrm{an}}, \mathcal{O}^{\mathrm{an}})$ is an analytic space, and is called the *analytification* of the scheme $(X, \mathcal{O})$.

5.1.2. Let $(\Phi, \Phi^*) : (X, \mathcal{O}_X) \longrightarrow (Y, \mathcal{O}_Y)$ be a morphism of schemes locally of finite type over $\mathbb{C}$. Then there is a morphism of analytic spaces

$$\left(\Phi^{\mathrm{an}}, \{\Phi^*\}^{\mathrm{an}}\right) : (X^{\mathrm{an}}, \mathcal{O}_X^{\mathrm{an}}) \longrightarrow (Y^{\mathrm{an}}, \mathcal{O}_Y^{\mathrm{an}})$$

which renders the following square commutative

$$\begin{array}{ccc} (X^{\mathrm{an}}, \mathcal{O}_X^{\mathrm{an}}) & \xrightarrow{\left(\Phi^{\mathrm{an}}, \{\Phi^*\}^{\mathrm{an}}\right)} & (Y^{\mathrm{an}}, \mathcal{O}_Y^{\mathrm{an}}) \\ {\scriptstyle (\lambda_X, \lambda_X^*)}\downarrow & & \downarrow{\scriptstyle (\lambda_Y, \lambda_Y^*)} \\ (X, \mathcal{O}_X) & \xrightarrow[(\Phi, \Phi^*)]{} & (Y, \mathcal{O}_Y) \end{array} .$$

The continuous map $\Phi^{\mathrm{an}} : X^{\mathrm{an}} \longrightarrow Y^{\mathrm{an}}$ is the restriction to the subset X^{an}, the set of closed points in X, of the map $\Phi : X \longrightarrow Y$; see 4.1.2. The morphism of ringed spaces $\left(\Phi^{\mathrm{an}}, \{\Phi^*\}^{\mathrm{an}}\right)$ is called the analytification of the morphism of schemes (Φ, Φ^*).

5.1.3. Analytification respects composition: given two composable morphisms, of schemes of finite type over $\mathbb{C}$

$$(X, \mathcal{O}_X) \xrightarrow{(\Phi, \Phi^*)} (Y, \mathcal{O}_Y) \xrightarrow{(\Psi, \Psi^*)} (Z, \mathcal{O}_Z) ,$$

we deduce two composable morphisms

$$(X^{\mathrm{an}}, \mathcal{O}_X^{\mathrm{an}}) \xrightarrow{\left(\Phi^{\mathrm{an}}, \{\Phi^*\}^{\mathrm{an}}\right)} (Y^{\mathrm{an}}, \mathcal{O}_Y^{\mathrm{an}}) \xrightarrow{\left(\Psi^{\mathrm{an}}, \{\Psi^*\}^{\mathrm{an}}\right)} (Z^{\mathrm{an}}, \mathcal{O}_Z^{\mathrm{an}}) .$$

We assert that $\left\{(\Psi, \Psi^*)(\Phi, \Phi^*)\right\}^{\mathrm{an}} = \left(\Psi^{\mathrm{an}}, \{\Psi^*\}^{\mathrm{an}}\right)\left(\Phi^{\mathrm{an}}, \{\Phi^*\}^{\mathrm{an}}\right)$. Also, the analytification of the identity map $(1,1) : (X, \mathcal{O}) \longrightarrow (X, \mathcal{O})$ is the identity on $(X^{\mathrm{an}}, \mathcal{O}^{\mathrm{an}})$.

5.1.4. Let $(\Phi, \Phi^*) : (X, \mathcal{O}_X) \longrightarrow (Y, \mathcal{O}_Y)$ be an open immersion of schemes locally of finite type over $\mathbb{C}$. Then the morphism of analytic spaces

$$\left(\Phi^{\mathrm{an}}, \{\Phi^*\}^{\mathrm{an}}\right) : (X^{\mathrm{an}}, \mathcal{O}_X^{\mathrm{an}}) \longrightarrow (Y^{\mathrm{an}}, \mathcal{O}_Y^{\mathrm{an}})$$

is also an open immersion. It identifies $(X^{\mathrm{an}}, \mathcal{O}_X^{\mathrm{an}})$, as a ringed space, with the open subspace $\left(X \cap Y^{\mathrm{an}}, \{\mathcal{O}_Y^{\mathrm{an}}\}|_{X \cap Y^{\mathrm{an}}}\right)$.

Remark 5.1.5. In more category theoretic language this says the following. We have a functor taking the category of schemes of finite type over $\mathbb{C}$ to analytic spaces. The objects $(X, \mathcal{O})$ go to $(X^{\mathrm{an}}, \mathcal{O}^{\mathrm{an}})$, and the morphisms (Φ, Φ^*) go to the morphisms $\left(\Phi^{\mathrm{an}}, \{\Phi^*\}^{\mathrm{an}}\right)$. Call this functor A. Now both the category of schemes of finite type over $\mathbb{C}$ and the category of analytic spaces include into the category of all ringed spaces over $\mathbb{C}$. Let the inclusion be I. There is also a natural transformation

$\lambda : IA \Longrightarrow I$. And finally the functor A takes open immersions to open immersions.

Modulo the fact that we do not include the proofs of most of the analytic statements we make, we have tried to explain the constructions in some detail. As in Chapter 4 the construction is local. When $(X, \mathcal{O}) = \left(\mathrm{Spec}(S), \widetilde{S}\right)$ it is relatively clear what to do, at least after choosing coordinates. It needs to be checked that the object constructed is independent of the choice of coordinates. And then, in passing from the special case $(X, \mathcal{O}) = \left(\mathrm{Spec}(S), \widetilde{S}\right)$ to the general case, we need to understand how to glue the local data. The reader willing to believe that the local data glue can skip much of this chapter. The statements above are what we will need, not their proofs.

5.2 The Hilbert Basis Theorem

The time has come when we need to remember one more fact from commutative algebra, namely the Hilbert Basis Theorem. Again we do not state the most general version, but restrict ourselves to one which we will use.

Theorem 5.2.1. *Let k be a field, and let R be a finitely generated k–algebra. Then any ideal $I \subset R$ is finitely generated. That is, there exist elements $i_1, i_2, \ldots, i_n$ in I, so that every element $i \in I$ can be written*

$$i = a_1 i_1 + a_2 i_2 + \ldots + a_n i_n ,$$

for some $a_1, a_2, \ldots, a_n$ in R.

Proof. The proof may be found in [1, Chapter 7]. More precisely: the statement may be found in [1, Corollary 7.7], in the part that starts "In particular...". The theorem, as stated and proved in [1, Chapter 7], is more general than Theorem 5.2.1. □

In this book we only care about the field $k = \mathbb{C}$. As I have said before much of algebraic geometry deals with other fields, and many of the statements we make generalize quite easily. We leave all generalizations to more advanced books.

In the rest of the section we observe some immediate corollaries of the Hilbert Basis Theorem.

Corollary 5.2.2. *Let R be a finitely generated algebra over $\mathbb{C}$. Every open subset in $X = \mathrm{Spec}(R)$ can be covered by finitely many basic open subsets X_{f_i}.*

Proof. Equivalently we can show that any closed subset $V(I)$ is the intersection of finitely many closed sets $V(Rf) = \{X - X_f\}$. Therefore let I be an ideal in R. By Theorem 5.2.1 the ideal I is finitely generated; it has generators $f_1, f_2, \ldots, f_n$. By Lemma 3.1.9 we know that

$$V(I) \quad = \quad \bigcap_{i=1}^{n} V(Rf_i) \ ,$$

and this expresses $V(I)$ as a finite intersection of $V(Rf_i)$. □

Corollary 5.2.3. *Let $(X, \mathcal{O})$ be a scheme of finite type over $\mathbb{C}$. Then any open subset of X is quasicompact†.*

Proof. Let V be an open subset of X. Definition 3.10.3 tells us that, since X is of finite type, it has a finite open cover $X = \cup_{i=1}^n U_i$ by spaces U_i homeomorphic to $\text{Spec}(R)$. It is enough to show that any open cover of $V \cap U_i$ has a finite subcover, and we are reduced to the case where $X = U_i$ is affine. Suppose therefore that V is an open subset of $X = \text{Spec}(R)$. By Corollary 5.2.2 we know that V has a finite open cover by basic open sets X_{f_i}. It suffices to show that each X_{f_i} is quasicompact. Therefore let $X = \text{Spec}(R)$ be affine, let $X_f \subset X$ be a basic open set, and we will show that any open cover of X_f has a finite subcover.

The sets X_g form a basis for the topology of X. Any open cover of X_f can be refined to an open cover by basic open sets X_{g_α}. But Lemma 3.3.8 tells us that any cover of X_f by basic open sets X_{g_α} has a finite subcover. □

Corollary 5.2.4. *Suppose that $(X, \mathcal{O})$ is a scheme of finite type over $\mathbb{C}$ and $V \subset X$ is any open subset. Then $(V, \mathcal{O}|_V)$ is also a scheme of finite type over $\mathbb{C}$.*

Proof. Proposition 3.10.7 says that $(V, \mathcal{O}|_V)$ is a scheme locally of finite type over $\mathbb{C}$. That is, the space V has an open cover $V = \cup_{i \in I} U_i$, with $(U_i, \mathcal{O}|_{U_i}) \cong \left(\text{Spec}(R_i), \widetilde{R_i}\right)$ and R_i finitely generated over $\mathbb{C}$. But Corollary 5.2.3 proves that V is quasicompact, hence there is a finite subcover. We conclude that $(V, \mathcal{O}|_V)$ is a scheme of finite type over $\mathbb{C}$. □

† Quasicompact means that every open cover has a finite subcover, but with no assumption that the space is Hausdorff. The Zariski topology is practically never Hausdorff, and the "quasi" part of quasicompact is meant as a warning to the reader.

5.3 The sheaf of analytic functions on an affine scheme

Let $(X, \mathcal{O})$ be a scheme locally of finite type over $\mathbb{C}$. We are aiming to show that there is a natural way to define a sheaf of rings $\mathcal{O}^{\text{an}}$ on the topological space X^{an}. In this section and in Section 6.1 we will treat the case where $(X, \mathcal{O}) \cong \left(\text{Spec}(S), \widetilde{S}\right)$ is affine. In the rest of the chapter we will worry about gluing the local data to define a global sheaf on X^{an}.

Remark 5.3.1. In this book I make an attempt to develop the algebraic theory in a reasonably self-contained way. There is no similar effort to be complete in the treatment of the theory of analytic functions of several complex variables. It is a beautiful theory, it is recommended that the reader learn some of it, but in this book we feel free to state analytic results with no real indication of the proof. For references we appeal to Gunning and Rossi's excellent book [3]. There are many other fine books on the subject. Among this rich literature there are two obvious choices, two books which develop the analytic theory in a way that closely parallels what we have done for schemes. The two books in question are Gunning and Rossi [3] and Grauert and Remmert [2]. Both are lovely books, highly recommended. I chose the older [3] because I deemed it more elementary, easier for my intended audience to read.

After the disclaimer of Remark 5.3.1, it is time to return to the construction of a sheaf of rings on $\{\text{Spec}(S)\}^{\text{an}}$. To begin with we identify the set of points in $\{\text{Spec}(S)\}^{\text{an}}$.

Notation 5.3.2. Let S be a finitely generated $\mathbb{C}$–algebra, suppose $a_1, a_2, \ldots, a_n$ are generators for the ring S, let $R = \mathbb{C}[x_1, x_2, \ldots, x_n]$ be the polynomial ring, and let $\theta : R \longrightarrow S$ be the surjection with $\theta(x_i) = a_i$. Let $I \subset R$ be the kernel of θ. By Theorem 5.2.1 the ideal I is finitely generated. Choose generators $f_1, f_2, \ldots, f_m$ for the ideal I. This notation will be fixed throughout the section. All the constructions of this section depend on a choice of generators $a_1, a_2, \ldots, a_n$ for the ring S. Later in the chapter we will worry about making them more coordinate-free, so that gluing the data, on schemes locally of finite type over $\mathbb{C}$, proceeds more smoothly.

Lemma 5.3.3. *Let the notation be as in Notation 5.3.2. The image of the embedding*

$$\{\text{Spec}(\theta)\}^{\text{an}} : \{\text{Spec}(S)\}^{\text{an}} \longrightarrow \{\text{Spec}(R)\}^{\text{an}} = \mathbb{C}^n$$

is the set

$$V(I) \quad = \quad \{x \in \mathbb{C}^n \mid f_i(x) = 0 \quad \forall 1 \le i \le m\} \quad .$$

Proof. The fact that $\left\{\operatorname{Spec}(\theta)\right\}^{\mathrm{an}}$ is an embedding is by definition of the complex topology on $\left\{\operatorname{Spec}(S)\right\}^{\mathrm{an}}$ as the subspace topology; see Definition 4.5.1. We only need to identify the set-theoretic image of the map. Lemma 4.3.4 helps. It tells us that the image is precisely the set of $\mathbb{C}$–algebra homomorphisms $\varphi : R \longrightarrow \mathbb{C}$ such that $\varphi(I) = 0$. The identification of Lemma 4.4.1, where points $a \in \mathbb{C}^n$ are identified with homomorphisms $\varphi : R \longrightarrow \mathbb{C}$, takes $a \in \mathbb{C}^n$ to the homomorphism φ_a evaluating $f \in R$ at $a \in \mathbb{C}^n$. The point $a \in \mathbb{C}^n$ will lie in the image of $\left\{\operatorname{Spec}(\theta)\right\}^{\mathrm{an}}$ iff $f(a) = \varphi_a(f)$ vanishes for all $f \in I$, or equivalently iff $f_i(a)$ vanishes for all the generators $f_1, f_2, \ldots, f_m$ of the ideal I. □

In other words the complex space $\left\{\operatorname{Spec}(S)\right\}^{\mathrm{an}}$ is given as the zero-set of finitely many polynomials. But, more generally, the zero-set of finitely many holomorphic functions has a well-known sheaf of rings on it. We remind the reader.

Remark 5.3.4. First we will describe a sheaf of rings $\mathcal{O}^{\mathrm{an}}$ on all of $\mathbb{C}^n$, and then we will show that this sheaf really lives on $\left\{\operatorname{Spec}(S)\right\}^{\mathrm{an}} \subset \mathbb{C}^n$. The way we will describe the sheaf on $\mathbb{C}^n$ is by giving the rings $\Gamma(U, \mathcal{O}^{\mathrm{an}})$ for U belonging to a suitable basis of the topology of $\mathbb{C}^n$. We therefore begin with a description of the open sets $\Delta(g; w; r)$ which form our preferred basis for the complex topology of $\mathbb{C}^n$.

Definition 5.3.5. *A* generalized polydisc *in* $\mathbb{C}^n$ *is a set* $\Delta(g; w; r) \subset \mathbb{C}^n$ *of the form*

$$\begin{aligned} \Delta(g; w; r) \quad &= \quad \Delta(g_1, g_2, \ldots, g_\ell; w_1, w_2, \ldots, w_\ell; r_1, r_2, \ldots, r_\ell) \\ &= \quad \left\{x \in \mathbb{C}^n \;\middle|\; |g_i(x) - w_i| < r_i \quad \forall 1 \le i \le \ell\right\} . \end{aligned}$$

Here g_i *are elements of* $R = \mathbb{C}[x_1, x_2, \ldots, x_n]$, *that is polynomials in* $x_1, x_2, \ldots, x_n$. *The point* $w = (w_1, w_2, \ldots, w_\ell) \in \mathbb{C}^\ell$ *is the generalized center of the polydisc, and*

$$r = (r_1, r_2, \ldots, r_\ell) \in \mathbb{R}^\ell, \quad r_i > 0$$

is the generalized polyradius.

Example 5.3.6. Observe that $\mathbb{C}^n$ is always a generalized polydisc in $\mathbb{C}^n$. It is the polydisc $\Delta(g; w; r)$, where $\ell = 1$ and

(i) $g : \mathbb{C}^n \longrightarrow \mathbb{C}$ is the zero polynomial.
(ii) The polycenter $w \in \mathbb{C}$ is $w = 0$.

(iii) The polyradius is $r = 1$.

In other words, $\mathbb{C}^n$ is the inverse image of the unit disc in $\mathbb{C}$ by the zero map $g : \mathbb{C}^n \longrightarrow \mathbb{C}$.

Remark 5.3.7. The usual polydiscs (see [3, page 1]) are the special generalized polydiscs where $n = \ell$ and, for any $x = (x_1, x_2, \ldots, x_n)$, we have

$$g_1(x) = x_1,\ g_2(x) = x_2,\ \ldots,\ g_n(x) = x_n\ .$$

A usual polydisc is denoted $\Delta(w; r)$. The polydisc $\Delta(w; r) \subset \mathbb{C}^\ell$ would be the set of all $y = (y_1, y_2, \ldots, y_\ell) \in \mathbb{C}^\ell$ such that

$$|y_i - w_i| < r_i \qquad \forall 1 \le i \le \ell\,.$$

The generalized polydisc $\Delta(g; w; r)$ is the inverse image of a usual polydisc $\Delta(w; r) \subset \mathbb{C}^\ell$ by the polynomial map $g : \mathbb{C}^n \longrightarrow \mathbb{C}^\ell$, sending $x \in \mathbb{C}^n$ to

$$\big(g_1(x), g_2(x), \ldots, g_\ell(x)\big)\,.$$

In this book when we use the term "polydisc", with no adjective, we will mean a generalized polydisc as in Definition 5.3.5. On the rare occasions when we want to specifically refer to the usual polydiscs, we will be explicit and call them "usual polydiscs".

Aside for the Experts 5.3.8. The graph of the polynomial function $g : \mathbb{C}^n \longrightarrow \mathbb{C}^\ell$ is a closed analytic submanifold of $\Gamma \subset \mathbb{C}^n \times \mathbb{C}^\ell$. The intersection of Γ with $\mathbb{C}^n \times \Delta(w; r)$ is biholomorphic to the generalized polydisc $\Delta(g; w; r)$. This makes $\Delta(g; w; r)$ a closed, analytic submanifold of the Stein domain $\mathbb{C}^n \times \Delta(w; r)$. Hence $\Delta(g; w; r)$ is a Stein manifold. Also we know that $\mathbb{C}^n \times \Delta(w; r)$ is holomorphically convex in $\mathbb{C}^n \times \mathbb{C}^\ell$, and hence its intersection with the graph Γ of $g : \mathbb{C}^n \longrightarrow \mathbb{C}^\ell$ is holomorphically convex in Γ. Via the biholomorphic map from Γ to $\mathbb{C}^n$ we conclude that the generalized polydisc $\Delta(g; w; r)$ is holomorphically convex in $\mathbb{C}^n$.

Lemma 5.3.9. *The generalized polydiscs form a basis for the complex topology of $\mathbb{C}^n$.*

Proof. It is clear that every open subset of $\mathbb{C}^n$ is a union of generalized polydiscs; the usual polydiscs will do. We need to show that the intersection of two generalized polydiscs is a generalized polydisc. Let us therefore be given two generalized polydiscs $\Delta(g; w; r)$ and $\Delta(g'; w'; r')$. Then there are two polynomial maps

$$g : \mathbb{C}^n \longrightarrow \mathbb{C}^\ell \qquad \text{and} \qquad g' : \mathbb{C}^n \longrightarrow \mathbb{C}^{\ell'}\,,$$

and two ordinary polydiscs $\Delta(w;r) \subset \mathbb{C}^\ell$ and $\Delta(w';r') \subset \mathbb{C}^{\ell'}$ so that

$$\Delta(g;w;r) = g^{-1}\Delta(w;r) \qquad \text{and} \qquad \Delta(g';w';r') = \{g'\}^{-1}\Delta(w';r').$$

The map $(g,g') : \mathbb{C}^n \longrightarrow \mathbb{C}^\ell \times \mathbb{C}^{\ell'}$ is a polynomial map, and the inverse image of $\Delta(w;r) \times \Delta(w';r')$ is $\Delta(g;w;r) \cap \Delta(g';w';r')$. □

Remark 5.3.10. One obvious sheaf of rings on $\mathbb{C}^n$ is the sheaf of holomorphic functions. One denotes it $\mathcal{O}^{\mathrm{an}}_{\mathbb{C}^n}$, and the definition is

$$\Gamma\big(U, \mathcal{O}^{\mathrm{an}}_{\mathbb{C}^n}\big) \quad = \quad \{f : U \longrightarrow \mathbb{C} \mid f \text{ is holomorphic}\} \quad .$$

In words (rather than symbols) this says that the sections of $\mathcal{O}^{\mathrm{an}}_{\mathbb{C}^n}$, on an open set $U \subset \mathbb{C}^n$, are the holomorphic functions on U.

The sheaf $\mathcal{O}^{\mathrm{an}}_{\mathbb{C}^n}$ is a sheaf of rings on $\mathbb{C}^n$. The polynomials can be viewed as holomorphic functions on all of $\mathbb{C}^n$, which means that we have an inclusion ring homomorphism

$$R \quad = \quad \mathbb{C}[x_1, x_2, \dots, x_n] \longrightarrow \Gamma\big(\mathbb{C}^n, \mathcal{O}^{\mathrm{an}}_{\mathbb{C}^n}\big) \ .$$

If $U \subset \mathbb{C}^n$ is any open subset we can compose to get a ring homomorphism

$$R \longrightarrow \Gamma\big(\mathbb{C}^n, \mathcal{O}^{\mathrm{an}}_{\mathbb{C}^n}\big) \xrightarrow{\mathrm{res}^{\mathbb{C}^n}_U} \Gamma\big(U, \mathcal{O}^{\mathrm{an}}_{\mathbb{C}^n}\big) \ .$$

The image of the ideal $I \subset R$ generates an ideal in $\Gamma\big(U, \mathcal{O}^{\mathrm{an}}_{\mathbb{C}^n}\big)$, which we will denote $I \cdot \Gamma\big(U, \mathcal{O}^{\mathrm{an}}_{\mathbb{C}^n}\big)$. We define, for every open set $U \subset \mathbb{C}^n$, a ring homomorphism

$$\Gamma\big(U, \mathcal{O}^{\mathrm{an}}_{\mathbb{C}^n}\big) \longrightarrow \frac{\Gamma\big(U, \mathcal{O}^{\mathrm{an}}_{\mathbb{C}^n}\big)}{I \cdot \Gamma\big(U, \mathcal{O}^{\mathrm{an}}_{\mathbb{C}^n}\big)} \ .$$

If $U \subset V \subset \mathbb{C}^n$ are open sets, there is an obvious restriction map

$$\mathrm{res}^V_U : \frac{\Gamma\big(V, \mathcal{O}^{\mathrm{an}}_{\mathbb{C}^n}\big)}{I \cdot \Gamma\big(V, \mathcal{O}^{\mathrm{an}}_{\mathbb{C}^n}\big)} \longrightarrow \frac{\Gamma\big(U, \mathcal{O}^{\mathrm{an}}_{\mathbb{C}^n}\big)}{I \cdot \Gamma\big(U, \mathcal{O}^{\mathrm{an}}_{\mathbb{C}^n}\big)} \ .$$

We have defined a presheaf of rings $\frac{\Gamma\big(U, \mathcal{O}^{\mathrm{an}}_{\mathbb{C}^n}\big)}{I \cdot \Gamma\big(U, \mathcal{O}^{\mathrm{an}}_{\mathbb{C}^n}\big)}$ on $\mathbb{C}^n$. It does not in general satisfy the sheaf axiom. However, in the special case where $U = \Delta(g;w;r)$ is a polydisc we can define

$$\Gamma\big(\Delta(g;w;r), \mathcal{O}^{\mathrm{an}}\big) \quad = \quad \frac{\Gamma\big(\Delta(g;w;r), \mathcal{O}^{\mathrm{an}}_{\mathbb{C}^n}\big)}{I \cdot \Gamma\big(\Delta(g;w;r), \mathcal{O}^{\mathrm{an}}_{\mathbb{C}^n}\big)} \ ,$$

and it is a theorem, in the theory of several complex variables, that this does extend to a sheaf. We state it below.

Theorem 5.3.11. *There is a sheaf of rings $\mathcal{O}^{\rm an}$ on $\mathbb{C}^n$ whose value, on the polydisc $\Delta(g;w;r)$, is the $\Gamma\big(\Delta(g;w;r),\mathcal{O}^{\rm an}\big)$ of Remark 5.3.10, and where, for any pair of polydiscs $\Delta(g;w;r)\subset\Delta(g';w';r')$, the restriction map*

$$\mathrm{res}^{\Delta(g';w';r')}_{\Delta(g;w;r)} \quad : \quad \Gamma\big(\Delta(g';w';r'),\mathcal{O}^{\rm an}\big) \longrightarrow \Gamma\big(\Delta(g;w;r),\mathcal{O}^{\rm an}\big)$$

is as in Remark 5.3.10. Furthermore there is a map of sheaves on $\mathbb{C}^n$

$$\mathcal{O}^{\rm an}_{\mathbb{C}^n} \longrightarrow \mathcal{O}^{\rm an}$$

which, on polydiscs $\Delta(g;w;r)$, induces the natural quotient map

$$\Gamma\big(\Delta(g;w;r),\mathcal{O}^{\rm an}_{\mathbb{C}^n}\big) \longrightarrow \frac{\Gamma\big(\Delta(g;w;r),\mathcal{O}^{\rm an}_{\mathbb{C}^n}\big)}{I\cdot\Gamma\big(\Delta(g;w;r),\mathcal{O}^{\rm an}_{\mathbb{C}^n}\big)} = \Gamma\big(\Delta(g;w;r),\mathcal{O}^{\rm an}\big)\ .$$

Remark 5.3.12. By Remark 3.4.9 we know that, if the sheaf $\mathcal{O}^{\rm an}$ exists, then it is unique up to canonical isomorphism. In principle it might be possible to check the existence using Lemma 3.4.10; that is, checking the exactness of certain sequences involving covers of basic open sets by basic open sets. I am not aware of any such proof.

Proof. The usual way to prove Theorem 5.3.11 is to define a sheaf $\mathcal{O}^{\rm an}$ as the cokernel in an exact sequence of sheaves on $\mathbb{C}^n$

$$\{\mathcal{O}^{\rm an}_{\mathbb{C}^n}\}^m \longrightarrow \mathcal{O}^{\rm an}_{\mathbb{C}^n} \longrightarrow \mathcal{O}^{\rm an} \longrightarrow 0\ .$$

The map $\{\mathcal{O}^{\rm an}_{\mathbb{C}^n}\}^m \longrightarrow \mathcal{O}^{\rm an}_{\mathbb{C}^n}$ is given by the matrix $(f_1,f_2,\ldots,f_m)$. That is, it takes the m-tuple of holomorphic functions $(a_1,a_2,\ldots,a_m)$ on the open set $U\subset\mathbb{C}^n$ to the holomorphic function

$$a_1f_1+a_2f_2+\cdots+a_mf_m\ .$$

Being the quotient of a map of coherent analytic sheaves the sheaf $\mathcal{O}^{\rm an}$ is coherent; see [3, page 136, Definition 2]. Because generalized polydiscs in $\mathbb{C}^n$ are Stein manifolds (see Aside 5.3.8) Cartan's Theorem B tells us that taking sections over a generalized polydisc is exact on coherent sheaves; see [3, page 243, Theorem 14]. Put more succinctly: we prove the existence of a sheaf $\mathcal{O}^{\rm an}$ some other way and then establish that, for any polydisc $\Delta(g;w;r)\subset\mathbb{C}^n$, the sequence

$$\Gamma\big(\Delta(g;w;r),\{\mathcal{O}^{\rm an}_{\mathbb{C}^n}\}^m\big)\longrightarrow\Gamma\big(\Delta(g;w;r),\mathcal{O}^{\rm an}_{\mathbb{C}^n}\big)\longrightarrow\Gamma\big(\Delta(g;w;r),\mathcal{O}^{\rm an}\big)\longrightarrow 0$$

is exact, meaning that $\Gamma\big(\Delta(g;w;r),\mathcal{O}^{\rm an}\big)$ has to be the quotient of $\Gamma\big(\Delta(g;w;r),\mathcal{O}^{\rm an}_{\mathbb{C}^n}\big)$ by the ideal generated by $f_1,f_2,\ldots,f_m$. □

The next string of lemmas will prove that the sheaf $\mathcal{O}^{\rm an}$ is supported on $\big\{\mathrm{Spec}(S)\big\}^{\rm an}\subset\mathbb{C}^n$.

Lemma 5.3.13. *Every point* $p \in \mathbb{C}^n - \{\operatorname{Spec}(S)\}^{\mathrm{an}}$ *has a neighborhood* $U \subset \mathbb{C}^n - \{\operatorname{Spec}(S)\}^{\mathrm{an}}$ *so that, on any polydisc* $\Delta(g;w;r)$ *contained in* U,

$$\Gamma\big(\Delta(g;w;r),\mathcal{O}^{\mathrm{an}}\big) \quad = \quad 0\,.$$

Proof. In Lemma 5.3.3 we computed what it means for a point $p \in \mathbb{C}^n$ to lie in the subset $\{\operatorname{Spec}(S)\}^{\mathrm{an}}$. From Lemma 5.3.3, if $p \in \mathbb{C}^n$ does not lie in $\{\operatorname{Spec}(S)\}^{\mathrm{an}}$ then one of the polynomials $f_1, f_2, \ldots, f_m$ is non-zero at p. Assume without loss that $f_1(p) \neq 0$. But then f_1 is invertible in some neighborhood $p \in U \subset \mathbb{C}^n - \{\operatorname{Spec}(S)\}^{\mathrm{an}}$. On the set U the multiplication by f_1, viewed as a map $\mathcal{O}^{\mathrm{an}}_{\mathbb{C}^n} \longrightarrow \mathcal{O}^{\mathrm{an}}_{\mathbb{C}^n}$, is invertible. It follows that, for any polydisc $\Delta(g;w;r) \subset U$, the ideal $I \cdot \Gamma\big(\Delta(g;w;r),\mathcal{O}^{\mathrm{an}}_{\mathbb{C}^n}\big)$ is all of the ring $\Gamma\big(\Delta(g;w;r),\mathcal{O}^{\mathrm{an}}_{\mathbb{C}^n}\big)$. Hence the quotient vanishes, that is

$$\Gamma\big(\Delta(g;w;r),\mathcal{O}^{\mathrm{an}}\big) \quad = \quad 0\,.$$

□

Lemma 5.3.14. *The restriction of the sheaf* $\mathcal{O}^{\mathrm{an}}$ *to the open set* $\mathbb{C}^n - \{\operatorname{Spec}(S)\}^{\mathrm{an}}$ *vanishes. In other words,* $\Gamma(V,\mathcal{O}^{\mathrm{an}}) = 0$ *for all open sets* $V \subset \mathbb{C}^n - \{\operatorname{Spec}(S)\}^{\mathrm{an}}$.

Proof. Let V be any open set contained in $\mathbb{C}^n - \{\operatorname{Spec}(S)\}^{\mathrm{an}}$, and let σ be a section in $\Gamma(V,\mathcal{O}^{\mathrm{an}})$. We need to prove that σ vanishes. But every point $p \in V$ has a neighborhood U as in Lemma 5.3.13. The set $U \cap V$ is a neighborhood of p, and hence contains a polydisc $\Delta(g;w;r)$ containing p. Because

$$\Delta(g;w;r) \quad \subset \quad U \cap V \quad \subset \quad U$$

Lemma 5.3.13 tells us that

$$\Gamma\big(\Delta(g;w;r),\mathcal{O}^{\mathrm{an}}\big) \quad = \quad 0\,.$$

On the other hand $\Delta(g;w;r) \subset V$, and we deduce that the map

$$\operatorname{res}^V_{\Delta(g;w;r)} \quad : \quad \Gamma\big(V,\mathcal{O}^{\mathrm{an}}\big) \longrightarrow \Gamma\big(\Delta(g;w;r),\mathcal{O}^{\mathrm{an}}\big)$$

takes $\sigma \in \Gamma(V,\mathcal{O}^{\mathrm{an}})$ to zero. Since the point $p \in V$ was arbitrary we can choose the polydiscs $\Delta(g;w;r)$ to cover V. The sheaf axioms, more particularly Definition 2.2.7(i), guarantee that $\sigma = 0$. □

Lemma 5.3.15. *Let* A *be the set* $A = \mathbb{C}^n - \{\operatorname{Spec}(S)\}^{\mathrm{an}}$. *Let* $V \subset \mathbb{C}^n$ *be an arbitrary open set. The restriction map*

$$\operatorname{res}^{A\cup V}_V : \Gamma(A \cup V,\mathcal{O}^{\mathrm{an}}) \longrightarrow \Gamma(V,\mathcal{O}^{\mathrm{an}})$$

is an isomorphism.

Proof. We need to prove the map injective and surjective. Let us prove injectivity first. Suppose $\sigma \in \Gamma(A \cup V, \mathcal{O}^{\mathrm{an}})$ lies in the kernel. The open set $A \cup V$ can be covered by two open sets A and V. The restriction of σ to V vanishes by hypothesis. The restriction of σ to A is a section in $\Gamma(A, \mathcal{O}^{\mathrm{an}})$ and must vanish because, by Lemma 5.3.14, $\Gamma(A, \mathcal{O}^{\mathrm{an}}) = 0$. Definition 2.2.7(i) now tells us that $\sigma = 0$.

Next we prove surjectivity. Choose a section $\sigma \in \Gamma(V, \mathcal{O}^{\mathrm{an}})$. Consider the pair of sections

$$\sigma \in \Gamma(V, \mathcal{O}^{\mathrm{an}}), \qquad 0 \in \Gamma(A, \mathcal{O}^{\mathrm{an}}).$$

The intersection $A \cap V$ is a subset of $A = \mathbb{C}^n - \{\operatorname{Spec}(S)\}^{\mathrm{an}}$, and Lemma 5.3.14 tells us that

$$\Gamma(A \cap V, \mathcal{O}^{\mathrm{an}}) = 0.$$

The restrictions of $\sigma \in \Gamma(V, \mathcal{O}^{\mathrm{an}})$ and $0 \in \Gamma(A, \mathcal{O}^{\mathrm{an}})$ must agree in the trivial ring $\Gamma(A \cap V, \mathcal{O}^{\mathrm{an}})$. Definition 2.2.7(ii) therefore tells us that they are the restrictions of some $\sigma' \in \Gamma(A \cup V, \mathcal{O}^{\mathrm{an}})$. □

Remark 5.3.16. The previous string of lemmas was intended to show that $\Gamma(V, \mathcal{O}^{\mathrm{an}})$ depends only on the intersection of V with $\{\operatorname{Spec}(S)\}^{\mathrm{an}}$. Suppose V and V' have the same intersection with $\{\operatorname{Spec}(S)\}^{\mathrm{an}}$. Put $A = \mathbb{C}^n - \{\operatorname{Spec}(S)\}^{\mathrm{an}}$ as in Lemma 5.3.15. Then $A \cup V = A \cup V'$, and Lemma 5.3.15 gives isomorphisms

$$\Gamma\big(V, \mathcal{O}^{\mathrm{an}}\big) \xleftarrow{\operatorname{res}_V^{A\cup V}} \Gamma\big(A \cup V, \mathcal{O}^{\mathrm{an}}\big) \xrightarrow{\operatorname{res}_{V'}^{A\cup V}} \Gamma\big(V', \mathcal{O}^{\mathrm{an}}\big).$$

Construction 5.3.17. We construct a sheaf of rings on $\{\operatorname{Spec}(S)\}^{\mathrm{an}} \subset \mathbb{C}^n$. We will abuse the notation by calling it by the same name as the sheaf on $\mathbb{C}^n$; we will use the symbol $\mathcal{O}^{\mathrm{an}}$ for both. As the reader will see this makes some sense; the sheaves carry basically the same information, even though they are defined on different spaces.

Take any open set $U \subset \{\operatorname{Spec}(S)\}^{\mathrm{an}}$. If $A = \mathbb{C}^n - \{\operatorname{Spec}(S)\}^{\mathrm{an}}$ as in Lemma 5.3.15, then $A \cup U$ is an open subset of $\mathbb{C}^n$. We define $\Gamma\big(U, \mathcal{O}^{\mathrm{an}}\big)$ by the formula

$$\Gamma\big(U, \mathcal{O}^{\mathrm{an}}\big) = \Gamma\big(A \cup U, \mathcal{O}^{\mathrm{an}}\big).$$

For open sets $U \subset V \subset \{\operatorname{Spec}(S)\}^{\mathrm{an}}$ we define res_U^V to be $\operatorname{res}_{A\cup U}^{A\cup V}$.

We leave it to the reader to verify the sheaf axioms. Note that, for the empty set $\emptyset \subset \{\operatorname{Spec}(S)\}^{\mathrm{an}}$, we need Lemma 5.3.15 to prove $\Gamma(\emptyset, \mathcal{O}^{\mathrm{an}}) = 0$.

The sheaf on the subset $\{\operatorname{Spec}(S)\}^{\mathrm{an}} \subset \mathbb{C}^n$ carries all the information contained in the sheaf on the larger set $\mathbb{C}^n$. For any open subset $V \subset \mathbb{C}^n$ Lemma 5.3.15 tells us that the natural map

$$\operatorname{res}_V^{A\cup V} : \Gamma(A \cup V, \mathcal{O}^{\mathrm{an}}) \longrightarrow \Gamma(V, \mathcal{O}^{\mathrm{an}})$$

is an isomorphism. But the ring on the left is nothing other than

$$\Gamma\big(V \cap \{\operatorname{Spec}(S)\}^{\mathrm{an}}, \mathcal{O}^{\mathrm{an}}\big) \quad .$$

That is, the value of the sheaf $\mathcal{O}^{\mathrm{an}}$ on the open set $V \subset \mathbb{C}^n$ is canonically determined by the value of the sheaf, which by abuse of notation we have been calling by the same name, on the open set $V \cap \{\operatorname{Spec}(S)\}^{\mathrm{an}} \subset \{\operatorname{Spec}(S)\}^{\mathrm{an}}$.

Remark 5.3.18. We have defined a sheaf of rings on $\{\operatorname{Spec}(S)\}^{\mathrm{an}}$. The sheaf depends on choices we made in Notation 5.3.2. We chose a set of elements $\{a_1, a_2, \dots, a_n\} \subset S$ which generate S as an algebra over $\mathbb{C}$, and we chose a set of elements $\{f_1, f_2, \dots, f_m\} \subset I$ which generate the ideal I over S. The choice of the generators for the ideal I was not used in the constructions. But it is far less clear that what we have done is independent of the choice of generators for the ring S. We deal with this in the rest of the chapter.

5.4 A reminder about Fréchet spaces

We need to briefly remind the reader of Fréchet spaces. Let us first introduce the weaker notion of a pre-Fréchet space.

Definition 5.4.1. *Let V be a vector space over $\mathbb{C}$. For V to be a* pre-Fréchet space *it needs*

(i) *A translation-invariant pseudometric†; that is, we need a function $f : V \longrightarrow \mathbb{R}^+$ so that $d(v, w) = f(v - w)$ is a pseudometric on V.*
(ii) *The ball of radius r is convex; that is for any pair $v, w \in V$ satisfying $f(v), f(w) < r$, and for any pair of complex numbers λ and μ with $|\lambda| + |\mu| \le 1$, we have $f(\lambda v + \mu w) < r$.*
(iii) *The multiplication map $\mathbb{C} \times V \longrightarrow V$ is continuous.*

A pre-Fréchet ring *will be a pre-Fréchet space which is also a $\mathbb{C}$–algebra, and where the multipication is continuous.*

† A pseudometric is a distance function $d : V \times V \longrightarrow \mathbb{R}^+$ satisfying the triangle inequality, but where we permit $d(x, y) = 0$ for $x \neq y$.

Remark 5.4.2. Any continuous linear map $\varphi : V \longrightarrow W$ of pre-Fréchet spaces is uniformly continuous.

Definition 5.4.3. *Two functions $f, \tilde{f} : V \longrightarrow \mathbb{R}^+$ as in Definition 5.4.1 will be called* equivalent *if the induced topologies on V are the same.*

Remark 5.4.4. In other words the pseudometrics f and $\tilde{f}$ are equivalent if the identity map $1 : V \longrightarrow V$ induces continuous maps, in both directions, between the metric spaces (V, f) and $(V, \tilde{f})$. By Remark 5.4.2 the identity is then automatically uniformly continuous. Cauchy sequences with respect to the pseudometric induced by f are the same as Cauchy sequences with respect to the pseudometric induced by $\tilde{f}$. The completions with respect to the two pseudometrics are homeomorphic.

This means that the completion of a pre-Fréchet space depends only on the topology. Once we know the topology is pre-Fréchet a translation-invariant metric exists, and is unique up to equivalence. And the completion depends only on the equivalence class of the pseudometric.

Now for the definition of a Fréchet space:

Definition 5.4.5. *Let V be a vector space over $\mathbb{C}$. For V to be a* Fréchet space *it must be a pre-Fréchet space, the pseudometric must be a metric, and V must be complete with respect to this metric. A* Fréchet ring *is a Fréchet space which is also a $\mathbb{C}$–algebra, and where the multiplication map is continuous.*

Remark 5.4.6. It is more customary to give the topology in terms of a sequence of seminorms†. That is, in the literature you will usually see a pre-Fréchet space defined as a topological vector space, with a sequence of seminorms $p_\ell : V \longrightarrow \mathbb{R}^+$, so that the sets

$$B(m, \varepsilon) = \left\{ v \in V \;\middle|\; \max_{1 \leq \ell \leq m} \left(p_\ell(v)\right) < \varepsilon \right\}$$

give a neighborhood basis at the point $0 \in V$. Next we will sketch how to see that the customary definition is equivalent to the one we gave in Definition 5.4.1.

Suppose we are given a function f as in Definition 5.4.1; we need to produce a sequence of seminorms p_ℓ. The recipe is that, for every integer $\ell > 0$, the function $p_\ell : V \longrightarrow \mathbb{R}^+$ is given by the formula

$$p_\ell(v) = \inf\{r > 0 \mid f(v/\ell r) < 1/\ell\} .$$

The reader can check that p_ℓ is a seminorm for each ℓ.

† A seminorm $p : V \longrightarrow \mathbb{R}^+$ is the analogue of a norm, but we allow the possibility that $p(v) = 0$.

Next assume we are given a sequence of seminorms p_ℓ; we want to construct a function $\tilde{f}$. The recipe is to use the formula

$$\tilde{f}(v) = \max\left(\frac{1 - e^{-p_\ell(v)}}{\ell}\right) .$$

We know that $p_\ell(v) \geq 0$, from which we deduce

$$0 \quad \leq \quad 1 - e^{-p_\ell(v)} \quad < \quad 1 .$$

Hence $\frac{1-e^{-p_\ell(v)}}{\ell}$ is a sequence of positive numbers tending to zero, and a maximum must exist. We leave it to the reader to check that if we start with f, and produce first the sequence $\{p_\ell\}$ and then $\tilde{f}$ by the formulas given above, then we obtain an equivalent pseudometric $\tilde{f}$.

Exercise 5.4.7. Let V be a vector space over $\mathbb{C}$, and let $p_\ell : V \longrightarrow \mathbb{R}^+$ be a sequence of seminorms. Let $\widetilde{f}$ be the pseudometric we cooked up, out of the sequence $\{p_\ell\}$, in Remark 5.4.6. Let $\{v_1, v_2, v_3, \ldots\}$ be a sequence of elements of V. The reader should check the following:

(i) The sequence $\{v_1, v_2, v_3, \ldots\}$ is Cauchy, with respect to the pseudometric $\widetilde{f}$, if and only if it is Cauchy with respect to each seminorm p_ℓ.

(ii) The sequence $\{v_1, v_2, v_3, \ldots\}$ converges to zero, with respect to the pseudometric $\widetilde{f}$, if and only if it converges to zero with respect to each seminorm p_ℓ.

Example 5.4.8. Let $U \subset \mathbb{C}^n$ be an open set, and let V be the vector space of all holomorphic functions $f : U \longrightarrow \mathbb{C}$. I assert that V is naturally a Fréchet space. By Remark 5.4.6 it suffices to produce a sequence of seminorms, and prove that the space is complete with respect to the metric induced by the function $\tilde{f}$ of Remark 5.4.6. Let us begin by producing the seminorms.

For each integer $\ell > 0$ we want a seminorm p_ℓ. We first define the subset $U_\ell \subset U$ to be the set of points $z = (z_1, z_2, \ldots, z_n)$ satisfying:

$$U_\ell \quad = \quad \left\{ z \in U \;\middle|\; \begin{array}{cc} |z_i| \leq \ell & \text{for } 1 \leq i \leq n \\ \|z - y\| \geq 1/\ell & \text{for all } y \notin U \end{array} \right\} .$$

In words rather than symbols: U_ℓ is the set of points, in the closure of the usual polydisc of polyradius $(\ell, \ell, \ldots, \ell)$, of distance at least $1/\ell$ from the complement of U. Because U_ℓ is contained in the polydisc of polyradius $(\ell, \ell, \ldots, \ell)$ it is bounded, and it is clearly closed. Hence it is compact. For each ℓ we have that U_ℓ is contained in the interior of the set $U_{\ell+1}$. And it is also clear that U is the increasing union of the countably many compact subsets U_ℓ.

Now define the seminorm $p_\ell : V \longrightarrow \mathbb{R}$ by the formula

$$p_\ell(f) = \sup_{z \in U_\ell} |f(z)| \ .$$

By the compactness of U_ℓ and the continuity of f the number $p_\ell(f)$ is well defined, and it is easy to check that this formula gives a seminorm.

Using Exercise 5.4.7, the reader can verify that a Cauchy sequence in V is a sequence of of functions $\{f_i\}$ which converge uniformly on the compact sets U_ℓ. The fact that V is a Fréchet space, that is that the seminorms, taken together, give a complete metric space, reduces to the assertion that if a sequence of holomorphic functions converges to a limit, uniformly on compact sets, then the limit is holomorphic.

Note that the Fréchet space V is in fact a Fréchet ring.

Notation 5.4.9. Let $U \subset \mathbb{C}^n$ be an open set. In Example 5.4.8 we learned that the vector space of all holomorphic functions on U has a natural Fréchet topology. We will denote this vector space, together with its Fréchet topology, by the symbol $\Gamma(U, \mathcal{O}^{\mathrm{an}}_{\mathbb{C}^n})$.

In this context there is a useful little lemma, whose proof we leave to the reader:

Exercise 5.4.10. Let U be an open set in $\mathbb{C}^n$ and let U' be an open set in $\mathbb{C}^{n'}$. Suppose $\varphi : U \longrightarrow U'$ is a holomorphic map. Then there is an induced map

$$\varphi^* : \Gamma(U', \mathcal{O}^{\mathrm{an}}_{\mathbb{C}^{n'}}) \longrightarrow \Gamma(U, \mathcal{O}^{\mathrm{an}}_{\mathbb{C}^n}) \ ,$$

taking $\psi \in \Gamma(U', \mathcal{O}^{\mathrm{an}}_{\mathbb{C}^{n'}})$ to $\varphi^*(\psi) = \psi\varphi$. That is, φ^* takes a function $\psi : U' \longrightarrow \mathbb{C}$ to the composite

$$U \xrightarrow{\ \varphi\ } U' \xrightarrow{\ \psi\ } \mathbb{C} \ .$$

We assert that φ^* is continuous in the Fréchet topologies of $\Gamma(U, \mathcal{O}^{\mathrm{an}}_{\mathbb{C}^n})$, $\Gamma(U', \mathcal{O}^{\mathrm{an}}_{\mathbb{C}^{n'}})$.

For a little while let us leave the examples from complex analysis, and return to the general theory of Fréchet spaces. For us, the most important theorem about Fréchet spaces will be the Open Mapping Theorem:

Theorem 5.4.11. *Let $\varphi : V \longrightarrow W$ be a continuous linear map between Fréchet spaces. If φ is onto then the image of any open set in V is open in W.*

Proof. The reader may find a proof in [3, page 292, Theorem 6]. □

Corollary 5.4.12. *Any continuous linear bijection $\varphi : V \longrightarrow W$ of Fréchet spaces is a homeomorphism.*

In Example 5.4.8 and Notation 5.4.9 we saw that, if $U \subset \mathbb{C}^n$ is an open set, then there is a simple way to give a Fréchet topology to $\Gamma(U, \mathcal{O}^{\text{an}}_{\mathbb{C}^n})$. The next theorem asserts that this topology can be defined in great generality, not just for open subsets of $\mathbb{C}^n$. The fact that this topology exists, not only for open subsets $U \subset \mathbb{C}^n$ but also for more complicated U, requires some work to prove; we will simply state the theorem and cite some references. We will not state the theorems in the greatest possible generality; all we need are the versions that work for closed analytic subspaces of open subsets in $\mathbb{C}^n$. Before we state the theorems, about the Fréchet structure on the rings $\Gamma(U^{\text{an}}, \mathcal{O}^{\text{an}})$, let us therefore remind the reader of some facts about subsets $U \subset \mathbb{R}^n$.

Reminder 5.4.13. The topological space $\mathbb{R}^n$ has a countable basis $\{B_m; m \in \mathbb{N}\}$ of open sets. When $n = 1$ we can take the open intervals (a, b) with a, b rational. For larger n we take products of such intervals. It follows that every subspace $X \subset \mathbb{R}^n$ has a countable basis of open sets; take a basis for $\mathbb{R}^n$ and intersect with X. Any open cover $X = \cup U_i$ has a countable subcover: for each B_m, in the basis of open sets, choose a U_i containing it (if such a U_i exists). This gives a countable subcover. The countability statements above make analysis on subspaces of $\mathbb{R}^n$ more pleasant than analysis in general. We will state the next theorems for subspaces of $\mathbb{C}^n$. There are generalizations, but we do not care.

The key facts from complex analysis which we want to use are encapsulated in the following two theorems.

Theorem 5.4.14. *Let X be a complex analytic subspace of $\mathbb{C}^n$. Then, for any open subset $U \subset X$, the ring $\Gamma(U, \mathcal{O}^{\text{an}})$ is a Fréchet ring. Furthermore, if $U \subset V$ are open subsets of X, then the restriction ring homomorphism*

$$\text{res}^V_U : \Gamma(V, \mathcal{O}^{\text{an}}) \longrightarrow \Gamma(U, \mathcal{O}^{\text{an}})$$

is continuous.

Proof. If the space X is reduced this may be found in [3, page 158, Theorem 5]. The general case follows from [3, page 239, Theorem 8]. □

Theorem 5.4.15. *Let $\Delta(w;r) \subset \mathbb{C}^n$ be a usual polydisc (not yet a generalized one). Then the $\mathbb{C}$–algebra homomorphism*

$$\mathbb{C}[x_1, x_2, \ldots, x_n] \longrightarrow \Gamma\big(\Delta(w;r), \mathcal{O}^{\mathrm{an}}\big)$$

has a dense image in the Fréchet ring $\Gamma\big(\Delta(w;r), \mathcal{O}^{\mathrm{an}}\big)$. That is, the polynomials are dense in the holomorphic functions on the polydisc.

Proof. See [3, page 37, Theorem 2]. □

5.5 The ring of analytic functions as a completion

In Section 5.3 we defined, starting with a finitely generated $\mathbb{C}$–algebra S and a set of generators $\{a_1, a_2, \ldots, a_n\} \subset S$, a sheaf of rings $\mathcal{O}^{\mathrm{an}}$ on the space $\big\{\operatorname{Spec}(S)\big\}^{\mathrm{an}}$. This means that for every open set $V \subset \big\{\operatorname{Spec}(S)\big\}^{\mathrm{an}}$ we defined a ring $\Gamma(V, \mathcal{O}^{\mathrm{an}})$. In this section we will show that there is a natural ring homomorphism $\mu_V : S \longrightarrow \Gamma(V, \mathcal{O}^{\mathrm{an}})$, and that if $V = \Delta(g; w; r) \cap \big\{\operatorname{Spec}(S)\big\}^{\mathrm{an}}$ is a polydisc then the image of the map μ_V is dense in the Fréchet topology of $\Gamma(V, \mathcal{O}^{\mathrm{an}})$. This means that we can view $\Gamma(V, \mathcal{O}^{\mathrm{an}})$ as the completion of S with respect to a pre-Fréchet topology. In future sections we will show that this pre-Fréchet topology on S is independent of the choice of generators.

We therefore return to the situation of Section 5.3. In particular let the conventions be as in Notation 5.3.2: R is again the polynomial ring $R = \mathbb{C}[x_1, x_2, \ldots, x_n]$, we have a surjective homomorphism $\theta : R \longrightarrow S$, and we constructed a sheaf of rings $\mathcal{O}^{\mathrm{an}}$ on

$$\big\{\operatorname{Spec}(S)\big\}^{\mathrm{an}} \quad \subset \quad \big\{\operatorname{Spec}(R)\big\}^{\mathrm{an}} \quad \cong \quad \mathbb{C}^n .$$

The way the construction worked was that we first defined a sheaf of rings $\mathcal{O}^{\mathrm{an}}$ on $\mathbb{C}^n$, then showed that this sheaf vanishes on every open set contained in $A = \mathbb{C}^n - \big\{\operatorname{Spec}(S)\big\}^{\mathrm{an}}$, and used this to conclude that there is a reasonable sheaf on $\big\{\operatorname{Spec}(S)\big\}^{\mathrm{an}}$ which contains all the data. For an open set $V \subset \big\{\operatorname{Spec}(S)\big\}^{\mathrm{an}}$ we defined the ring of sections over V as the ring of sections of the sheaf on $\mathbb{C}^n$ over the open set $A \cup V \subset \mathbb{C}^n$. In symbols

$$\Gamma(V, \mathcal{O}^{\mathrm{an}}) \quad = \quad \Gamma(A \cup V, \mathcal{O}^{\mathrm{an}}) .$$

Now we make the observation

Observation 5.5.1. Let U be any open set in $\mathbb{C}^n$. The continuous map of Fréchet rings

$$\operatorname{res}_U^{A \cup U} : \Gamma(A \cup U, \mathcal{O}^{\mathrm{an}}) \longrightarrow \Gamma(U, \mathcal{O}^{\mathrm{an}})$$

is a homeomorphism.

Proof. The map is continuous by Theorem 5.4.14, while Lemma 5.3.15 informs us that it is bijective. Corollary 5.4.12 then forces it to be a homeomorphism. □

Remark 5.5.2. The significance of Observation 5.5.1 is that we can use any open subset $U \subset \mathbb{C}^n$, with $U \cap \mathrm{Spec}(S) = V$, to compute $\Gamma(V, \mathcal{O}^{\mathrm{an}})$. Remark 5.3.16 told us that the rings $\Gamma(U, \mathcal{O}^{\mathrm{an}})$ all canonically agree if $U \cap \mathrm{Spec}(S) = V$. Now we know that they even agree as Fréchet spaces.

Next we note

Lemma 5.5.3. *There is a natural ring homomorphism, defined in the proof of this lemma,*

$$\widetilde{\mu} : R \longrightarrow \Gamma\left(\left\{\mathrm{Spec}(S)\right\}^{\mathrm{an}}, \mathcal{O}^{\mathrm{an}}\right) .$$

Proof. We define $\widetilde{\mu}$ as the composite

$$R \longrightarrow \Gamma(\mathbb{C}^n, \mathcal{O}^{\mathrm{an}}_{\mathbb{C}^n}) \longrightarrow \Gamma\left(\left\{\mathrm{Spec}(S)\right\}^{\mathrm{an}}, \mathcal{O}^{\mathrm{an}}\right) .$$

The map $R \longrightarrow \Gamma\left(\mathbb{C}^n, \mathcal{O}^{\mathrm{an}}_{\mathbb{C}^n}\right)$ is the map taking a polynomial on $\mathbb{C}^n$ to itself, viewed as a holomorphic function. The map $\Gamma(\mathbb{C}^n, \mathcal{O}^{\mathrm{an}}_{\mathbb{C}^n}) \longrightarrow \Gamma\left(\left\{\mathrm{Spec}(S)\right\}^{\mathrm{an}}, \mathcal{O}^{\mathrm{an}}\right)$ is the ring homomorphism induced by the map of sheaves $\mathcal{O}^{\mathrm{an}}_{\mathbb{C}^n} \longrightarrow \mathcal{O}^{\mathrm{an}}$ of Theorem 5.3.11. □

Lemma 5.5.4. *The map $\widetilde{\mu} : R \longrightarrow \Gamma\left(\left\{\mathrm{Spec}(S)\right\}^{\mathrm{an}}, \mathcal{O}^{\mathrm{an}}\right)$ of Lemma 5.5.3 factors uniquely through*

$$R \xrightarrow{\ \theta\ } S \xrightarrow{\ \mu\ } \Gamma\left(\left\{\mathrm{Spec}(S)\right\}^{\mathrm{an}}, \mathcal{O}^{\mathrm{an}}\right) .$$

Proof. In Construction 5.3.17 we defined the sheaf $\mathcal{O}^{\mathrm{an}}$, on the space $\left\{\mathrm{Spec}(S)\right\}^{\mathrm{an}}$, by the formula

$$\Gamma(V, \mathcal{O}^{\mathrm{an}}) \quad = \quad \Gamma(A \cup V, \mathcal{O}^{\mathrm{an}}) .$$

In the case where $V = \left\{\mathrm{Spec}(S)\right\}^{\mathrm{an}}$, this gives

$$\Gamma\left(\left\{\mathrm{Spec}(S)\right\}^{\mathrm{an}}, \mathcal{O}^{\mathrm{an}}\right) \quad = \quad \Gamma(\mathbb{C}^n, \mathcal{O}^{\mathrm{an}}) .$$

Example 5.3.6 tells us that $\mathbb{C}^n$ is a generalized polydisc in $\mathbb{C}^n$. For this polydisc, the formula of Theorem 5.3.11 tells us that $\widetilde{\mu}$ is, very explicitly, the composite

$$R \longrightarrow \Gamma(\mathbb{C}^n, \mathcal{O}^{\mathrm{an}}_{\mathbb{C}^n}) \longrightarrow \frac{\Gamma(\mathbb{C}^n, \mathcal{O}^{\mathrm{an}}_{\mathbb{C}^n})}{I \cdot \Gamma(\mathbb{C}^n, \mathcal{O}^{\mathrm{an}}_{\mathbb{C}^n})} \quad = \quad \Gamma(\mathbb{C}^n, \mathcal{O}^{\mathrm{an}}) .$$

And it is very clear that the ideal $I \subset R$ maps to zero, and hence the map factors, uniquely, through θ. □

Definition 5.5.5. *For any open set $V \subset \{\operatorname{Spec}(S)\}^{\mathrm{an}}$ define the map $\mu_V : S \longrightarrow \Gamma(V, \mathcal{O}^{\mathrm{an}})$ as the composite*

$$S \xrightarrow{\ \mu\ } \Gamma(\{\operatorname{Spec}(S)\}^{\mathrm{an}}, \mathcal{O}^{\mathrm{an}}) \xrightarrow{\ \mathrm{res}_V^{\{\operatorname{Spec}(S)\}^{\mathrm{an}}}\ } \Gamma(V, \mathcal{O}^{\mathrm{an}}) .$$

Define $\widetilde{\mu}_V : R \longrightarrow \Gamma(V, \mathcal{O}^{\mathrm{an}})$ to be the composite

$$R \xrightarrow{\ \theta\ } S \xrightarrow{\ \mu_V\ } \Gamma(V, \mathcal{O}^{\mathrm{an}}) .$$

Remark 5.5.6. If $V \subset V' \subset \mathbb{C}^n$ are open sets, then the following triangle commutes:

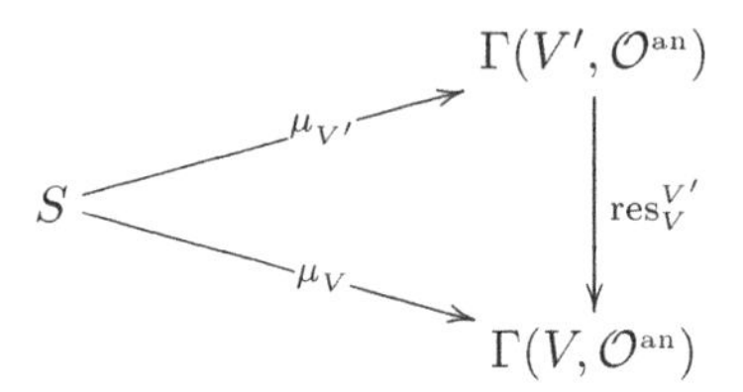

The point of the exercise is

Proposition 5.5.7. *Let $V = \{\operatorname{Spec}(S)\}^{\mathrm{an}} \cap \Delta(g; w; r)$. The natural map $\mu_V : S \longrightarrow \Gamma(V, \mathcal{O}^{\mathrm{an}})$ of Definition 5.5.5 has a dense image, where $\Gamma(V, \mathcal{O}^{\mathrm{an}})$ is given its Fréchet topology of Theorem 5.4.14.*

Proof. It clearly suffices to prove that the image of the composite

$$R \xrightarrow{\ \theta\ } S \xrightarrow{\ \mu_V\ } \Gamma(V, \mathcal{O}^{\mathrm{an}})$$

is dense in $\Gamma(V, \mathcal{O}^{\mathrm{an}})$. Now we have a commutative square of embeddings

$$\begin{array}{ccc} \Delta(g; w; r) & \longrightarrow & \mathbb{C}^n \times \Delta(w; r) \\ \downarrow & & \downarrow \\ \mathbb{C}^n & \xrightarrow[(1,g)]{} & \mathbb{C}^n \times \mathbb{C}^\ell \end{array} ,$$

where the map $(1, g) : \mathbb{C}^n \longrightarrow \mathbb{C}^n \times \mathbb{C}^\ell$ is the embedding of $\mathbb{C}^n$ as the graph of the polynomial map $g : \mathbb{C}^n \longrightarrow \mathbb{C}^\ell$. This gives a commutative diagram of ring homomorphisms

$$\begin{array}{ccccc} \Gamma(\mathbb{C}^n \times \mathbb{C}^\ell, \mathcal{O}^{\mathrm{an}}_{\mathbb{C}^n \times \mathbb{C}^\ell}) & \longrightarrow & \Gamma(\mathbb{C}^n, \mathcal{O}^{\mathrm{an}}_{\mathbb{C}^n}) & & \\ \downarrow & & \downarrow & & \\ \Gamma\big(\mathbb{C}^n \times \Delta(w;r), \mathcal{O}^{\mathrm{an}}_{\mathbb{C}^{n+\ell}}\big) & \longrightarrow & \Gamma\big(\Delta(g;w;r), \mathcal{O}^{\mathrm{an}}_{\mathbb{C}^n}\big) & \xrightarrow[\beta]{} & \Gamma\big(\Delta(g;w;r), \mathcal{O}^{\mathrm{an}}\big) \end{array} .$$

In this diagram the map $\beta : \Gamma\big(\Delta(g;w;r), \mathcal{O}^{\mathrm{an}}_{\mathbb{C}^n}\big) \longrightarrow \Gamma\big(\Delta(g;w;r), \mathcal{O}^{\mathrm{an}}\big)$ is the natural quotient map

$$\Gamma\big(\Delta(g;w;r), \mathcal{O}^{\mathrm{an}}_{\mathbb{C}^n}\big) \longrightarrow \frac{\Gamma\big(\Delta(g;w;r), \mathcal{O}^{\mathrm{an}}_{\mathbb{C}^n}\big)}{I \cdot \Gamma\big(\Delta(g;w;r), \mathcal{O}^{\mathrm{an}}_{\mathbb{C}^n}\big)} = \Gamma\big(\Delta(g;w;r), \mathcal{O}^{\mathrm{an}}\big)$$

of Theorem 5.3.11. Now polynomials on $\mathbb{C}^{n+\ell}$ (respectively on $\mathbb{C}^n$) include into the holomorphic functions, and we therefore have a commutative diagram

$$\begin{array}{ccccc} \mathbb{C}[x_1,\ldots,x_n,y_1,\ldots,y_\ell] & \longrightarrow & \mathbb{C}[x_1,\ldots,x_n] & & \\ \downarrow{\scriptstyle\varphi} & & \downarrow & & \\ \Gamma\big(\mathbb{C}^n \times \Delta(w;r), \mathcal{O}^{\mathrm{an}}_{\mathbb{C}^{n+\ell}}\big) & \xrightarrow[\alpha]{} & \Gamma\big(\Delta(g;w;r), \mathcal{O}^{\mathrm{an}}_{\mathbb{C}^n}\big) & \xrightarrow[\beta]{} & \Gamma\big(\Delta(g;w;r), \mathcal{O}^{\mathrm{an}}\big)\,. \end{array}$$

The maps α and β are surjective; β by definition (see above), and α by [3, page 243, Theorem 14]. Furthermore both α and β are continuous; this follows in a straightforward way from the separability of the spaces and [3, page 237, Theorem 7(2)]. The map φ has a dense image by Theorem 5.4.15. From the commutativity we deduce that the composite

$$\begin{array}{ccc} \mathbb{C}[x_1,\ldots,x_n] & & \\ \downarrow & & \\ \Gamma\big(\Delta(g;w;r), \mathcal{O}^{\mathrm{an}}_{\mathbb{C}^n}\big) & \xrightarrow[\beta]{} & \Gamma\big(\Delta(g;w;r), \mathcal{O}^{\mathrm{an}}\big) \end{array}$$

has a dense image. But this composite factors as

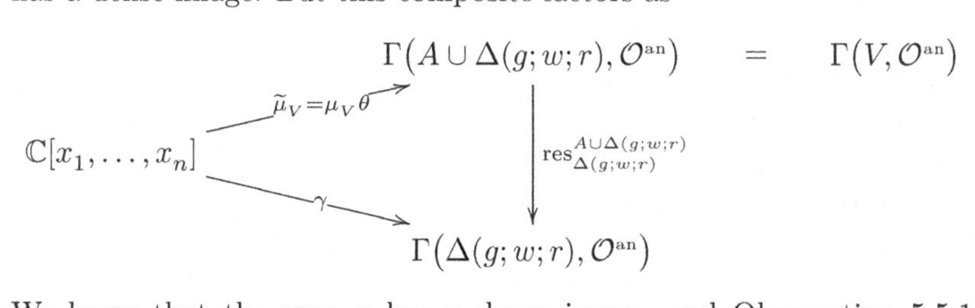

We know that the map γ has a dense image, and Observation 5.5.1 tells us that $\mathrm{res}^{A\cup\Delta(g;w;r)}_{\Delta(g;w;r)}$ is a homeomorphism. Hence the image of $\widetilde{\mu}_V = \mu_V\theta$ is dense. □

Corollary 5.5.8. *The map* $\mu : S \longrightarrow \Gamma\big(\{\,\mathrm{Spec}(S)\}^{\mathrm{an}}, \mathcal{O}^{\mathrm{an}}\big)$ *has a dense image.*

Proof. This is the special case of Proposition 5.5.7, in which we let $V = \{\,\mathrm{Spec}(S)\}^{\mathrm{an}}$. Note that $\{\,\mathrm{Spec}(S)\}^{\mathrm{an}} = \{\,\mathrm{Spec}(S)\}^{\mathrm{an}} \cap \mathbb{C}^n$, and Example 5.3.6 assures us that $\mathbb{C}^n$ is a polydisc in $\mathbb{C}^n$. The open set $V = \{\,\mathrm{Spec}(S)\}^{\mathrm{an}}$ satisfies the hypothesis of Proposition 5.5.7. □

5.6 Allowing the ring and the generators to vary

In Sections 5.3 and 5.5 we fixed a finitely generated $\mathbb{C}$–algebra S and a set of generators $\{a_1, a_2, \dots, a_n\} \subset S$. Starting from this we defined a sheaf of rings $\mathcal{O}^{\text{an}}$ on $\{\operatorname{Spec}(S)\}^{\text{an}}$, and for every open set $V \subset \{\operatorname{Spec}(S)\}^{\text{an}}$ we defined a natural ring homomorphism $\mu_V : S \longrightarrow \Gamma(V, \mathcal{O}^{\text{an}})$, whose image is sometimes dense in the Fréchet topology. We care mostly about the case where S is dense in $\Gamma(V, \mathcal{O}^{\text{an}})$, since $\Gamma(V, \mathcal{O}^{\text{an}})$ is then the completion of S with respect to the pre-Fréchet topology it inherits from the map μ_V. So the first step is to define the subsets $V \subset \{\operatorname{Spec}(S)\}^{\text{an}}$ we are willing to permit, and then discuss the topology on S defined by such V's. In this section we will show that this topology is independent of the choice of generators for S, and then continue to study the way it changes as we vary V and as we vary S. First we need two definitions.

Definition 5.6.1. *Let S be a finitely generated algebra over $\mathbb{C}$. If we choose generators $\{a_1, a_2, \dots, a_n\} \subset S$ then we have an embedding $\{\operatorname{Spec}(S)\}^{\text{an}} \subset \mathbb{C}^n$. Using this we define*

5.6.1.1. *A* polydisc *in $\{\operatorname{Spec}(S)\}^{\text{an}}$ is the intersection of $\{\operatorname{Spec}(S)\}^{\text{an}}$ with a polydisc $\Delta(g; w; r) \subset \mathbb{C}^n$.*

5.6.1.2. *Let V be a polydisc in $\{\operatorname{Spec}(S)\}^{\text{an}}$. The pre-Fréchet ring $[S, V]$ is the $\mathbb{C}$–algebra S, endowed with the inverse image of the Fréchet topology of $\Gamma(V, \mathcal{O}^{\text{an}})$ by the ring homomorphism*

$$\mu_V : S \longrightarrow \Gamma(V, \mathcal{O}^{\text{an}}) \ .$$

Example 5.6.2. In Definition 5.6.1.1, note that $\{\operatorname{Spec}(S)\}^{\text{an}}$ is always a polydisc in $\{\operatorname{Spec}(S)\}^{\text{an}}$. By Example 5.3.6 we know that, for a suitable choice of $(g; w; r)$, we have $\Delta(g; w; r) = \mathbb{C}^n$. For this particular $(g; w; r)$, the intersection of $\mathbb{C}^n = \Delta(g; w; r)$ with $\{\operatorname{Spec}(S)\}^{\text{an}}$ is all of $\{\operatorname{Spec}(S)\}^{\text{an}}$.

Remark 5.6.3. Definition 5.6.1.2 means very concretely the following. By Definition 5.5.5 we have a ring homomorphism $\mu_V : S \longrightarrow \Gamma(V, \mathcal{O}^{\text{an}})$. Because $\Gamma(V, \mathcal{O}^{\text{an}})$ is a Fréchet ring there is a metric, given by a function $f : \Gamma(V, \mathcal{O}^{\text{an}}) \longrightarrow \mathbb{R}^+$ as in Definition 5.4.1(i). Then $f\mu_V : S \longrightarrow \mathbb{R}^+$ also satisfies the axioms of Definition 5.4.1. It gives a topology on S. The open sets in S are precisely the sets of the form $\mu_V^{-1}U$, with U an open subset of $\Gamma(V, \mathcal{O}^{\text{an}})$. Endowing S with this topology we obtain a pre-Fréchet ring denoted $[S, V]$.

By Remark 5.4.2 all $\mathbb{C}$–linear continuous maps $[S, V] \longrightarrow [S', V']$ are uniformly continuous and extend to the completions.

Proposition 5.6.4. *Let S be a finitely generated $\mathbb{C}$–algebra. We assert*

(i) *The polydiscs in $\{\operatorname{Spec}(S)\}^{\mathrm{an}}$ are independent of the choice of generators.*

(ii) *Let $V \subset \{\operatorname{Spec}(S)\}^{\mathrm{an}}$ be a polydisc. The pre-Fréchet topology on $[S, V]$ is independent of the choice of generators.*

Proof. (cf. Lemma 4.5.3): Let $\{a_1, a_2, \ldots, a_n\}$ and $\{b_1, b_2, \ldots, b_m\}$ be two sets of generators. By symmetry it suffices to prove that if $V \subset \{\operatorname{Spec}(S)\}^{\mathrm{an}}$ is a polydisc, with respect to the generators $\{a_1, a_2, \ldots, a_n\}$, then it is also a polydisc with respect to the generators $\{b_1, b_2, \ldots, b_m\}$, and that the identity map $1 : S \longrightarrow S$ is continuous if we give one S the topology of the generators $\{a_1, a_2, \ldots, a_n\}$ and the other S the topology of the generators $\{b_1, b_2, \ldots, b_m\}$. This is what we will do.

Let $R = \mathbb{C}[x_1, x_2, \ldots, x_n]$ and $R' = \mathbb{C}[y_1, y_2, \ldots, y_m]$ be the polynomial rings. We have two surjective homomorphisms

$$\theta : R \longrightarrow S\,, \qquad \theta' : R' \longrightarrow S\,,$$

where $\theta(x_i) = a_i$ and $\theta'(y_i) = b_i$. We know that $b_1, b_2, \ldots, b_m$ generate S, while $a_1, a_2, \ldots, a_n$ are elements of S. It follows that we can write the a_j as polynomials in the b_i; there exist identities in S

$$a_j = P_j(b_1, b_2, \ldots, b_m)\,,$$

with P_j polynomials with coefficients in $\mathbb{C}$. Now we define a homomorphism $\varphi : R \longrightarrow R'$ by setting

$$\varphi(x_j) = P_j(y_1, y_2, \ldots, y_m)\,.$$

We deduce a commutative triangle of ring homomorphisms

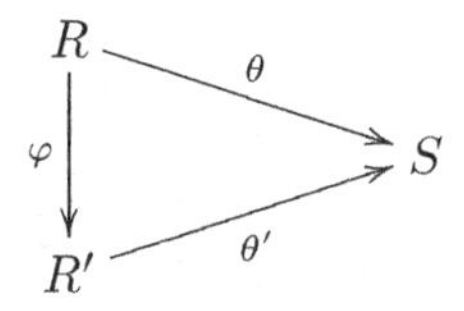

5.6.4.1. *Aside, which we will need later:* Let $I \subset R$ be the kernel of $\theta : R \longrightarrow S$, and let $I' \subset R'$ be the kernel of $\theta' : R' \longrightarrow S$. The commutativity of the diagram above ensures that $\varphi(I) \subset I'$.

Take the commutative triangle above Aside 5.6.4.1. If we apply to it the construction $\{\operatorname{Spec}(-)\}^{\mathrm{an}}$, we deduce a commutative triangle of spaces

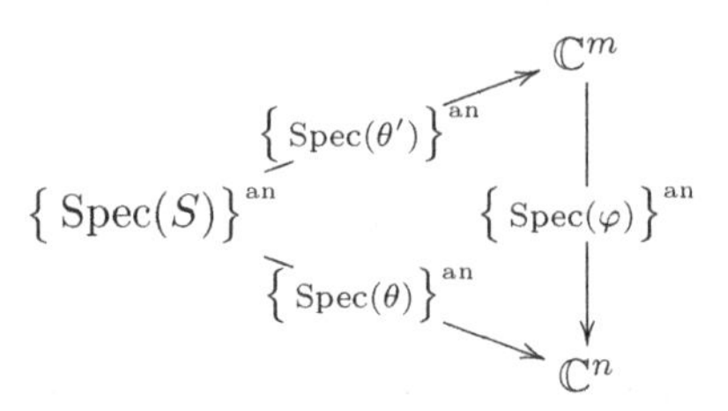

This already allows us to prove Proposition 5.6.4(i). Any generalized polydisc $\Delta(g;w;r) \subset \mathbb{C}^n$ is $g^{-1}\Delta(w;r)$, where $g : \mathbb{C}^n \longrightarrow \mathbb{C}^\ell$ is a polynomial map and $\Delta(w;r)$ is a usual polydisc in $\mathbb{C}^\ell$; see Remark 5.3.7. This means that the inverse image, by the polynomial map $\{\operatorname{Spec}(\varphi)\}^{\mathrm{an}} : \mathbb{C}^m \longrightarrow \mathbb{C}^n$, will be a polydisc in $\mathbb{C}^m$. The intersections with $\{\operatorname{Spec}(S)\}^{\mathrm{an}}$ agree, from the commutativity of the triangle above. Thus we have shown that, if V can be expressed as $V = \Delta(g;w;r) \cap \{\operatorname{Spec}(S)\}^{\mathrm{an}}$ with $\Delta(g;w;r)$ a polydisc in $\mathbb{C}^n$, then it can also be expressed as $V = \Delta(g';w;r) \cap \{\operatorname{Spec}(S)\}^{\mathrm{an}}$ with $\Delta(g';w;r)$ a polydisc in $\mathbb{C}^m$. By symmetry we deduce Proposition 5.6.4(i).

Now we know that the sets $V = \Delta(g;w;r) \cap \{\operatorname{Spec}(S)\}^{\mathrm{an}}$ are independent of the choice of generators. Let V be a polydisc in $\{\operatorname{Spec}(S)\}^{\mathrm{an}}$, which by the above can be expressed as $V = \Delta(g;w;r) \cap \{\operatorname{Spec}(S)\}^{\mathrm{an}}$ with $\Delta(g;w;r) \subset \mathbb{C}^n$. To compactify the notation put $U = \Delta(g;w;r)$, and let $U' = \Delta(g';w;r)$ be the inverse image of U by the continuous map $\{\operatorname{Spec}(\varphi)\}^{\mathrm{an}} : \mathbb{C}^m \longrightarrow \mathbb{C}^n$.

Next let $\xi : \Gamma(U, \mathcal{O}^{\mathrm{an}}_{\mathbb{C}^n}) \longrightarrow \Gamma(U', \mathcal{O}^{\mathrm{an}}_{\mathbb{C}^m})$ be the map taking a holomorphic function $\psi : U \longrightarrow \mathbb{C}$ to the composite

$$U' \xrightarrow{\{\operatorname{Spec}(\varphi)\}^{\mathrm{an}}} U \xrightarrow{\ \psi\ } \mathbb{C}.$$

We have a commutative diagram of ring homomorphisms

$$\begin{array}{ccccccc}
R & \longrightarrow & \Gamma(U, \mathcal{O}^{\mathrm{an}}_{\mathbb{C}^n}) & \longrightarrow & \dfrac{\Gamma(U, \mathcal{O}^{\mathrm{an}}_{\mathbb{C}^n})}{I \cdot \Gamma(U, \mathcal{O}^{\mathrm{an}}_{\mathbb{C}^n})} & = & \Gamma(V, \mathcal{O}^{\mathrm{an}}) \\
{\scriptstyle\varphi}\downarrow & & \downarrow{\scriptstyle\xi} & & \downarrow{\scriptstyle\gamma} & & \\
R' & \longrightarrow & \Gamma(U', \mathcal{O}^{\mathrm{an}}_{\mathbb{C}^m}) & \longrightarrow & \dfrac{\Gamma(U', \mathcal{O}^{\mathrm{an}}_{\mathbb{C}^m})}{I' \cdot \Gamma(U', \mathcal{O}^{\mathrm{an}}_{\mathbb{C}^m})} & = & \Gamma(V, \mathcal{O}^{\mathrm{an}})
\end{array} .$$

The commutativity of the square on the left, that is

$$\begin{array}{ccc}
R & \longrightarrow & \Gamma(U, \mathcal{O}^{\mathrm{an}}_{\mathbb{C}^n}) \\
{\scriptstyle\varphi}\downarrow & & \downarrow{\scriptstyle\xi} \\
R' & \longrightarrow & \Gamma(U', \mathcal{O}^{\mathrm{an}}_{\mathbb{C}^m})
\end{array}$$

is by Lemma 4.4.7; that lemma showed us that the map ξ, which takes P to its composite with $\{\operatorname{Spec}(\varphi)\}^{\text{an}}$, agrees on polynomials with the map φ, which takes P to $\varphi(P)$. By 5.6.4.1 the map φ takes the ideal $I \subset R$ to $I' \subset R'$, allowing us to extend the commutative diagram to include the map γ.

Caution 5.6.5. Note that until now we have not allowed the choice of generators to vary. The result is that there is now a notational confusion. Given a set of generators for S, and an open subset $V \subset \{\operatorname{Spec}(S)\}^{\text{an}}$, we constructed a ring $\Gamma(V, \mathcal{O}^{\text{an}})$. Now we have two sets of generators, hence two rings. As the notation suggests the rings $\Gamma(V, \mathcal{O}^{\text{an}})$ depend not on $U \subset \mathbb{C}^n$ or $U' \subset \mathbb{C}^m$ but only on the intersection $V = U \cap \{\operatorname{Spec}(S)\}^{\text{an}} = U' \cap \{\operatorname{Spec}(S)\}^{\text{an}}$; see Remark 5.3.16 and Observation 5.5.1. But the sheaf $\mathcal{O}^{\text{an}}$ at least seemed to depend on a choice of generators. Until now the choice was fixed, so there was no problem. Now, at least for the duration of this proof, the $\Gamma(V, \mathcal{O}^{\text{an}})$ on the top row must be viewed as different from the $\Gamma(V, \mathcal{O}^{\text{an}})$ on the bottom row of the commutative diagram above.

In any case the maps

$$\begin{array}{ccccc}
\Gamma(U, \mathcal{O}^{\text{an}}_{\mathbb{C}^n}) & \longrightarrow & \dfrac{\Gamma(U, \mathcal{O}^{\text{an}}_{\mathbb{C}^n})}{I \cdot \Gamma(U, \mathcal{O}^{\text{an}}_{\mathbb{C}^n})} & = & \Gamma(V, \mathcal{O}^{\text{an}}) \\
\Big\downarrow \xi & & & & \\
\Gamma(U', \mathcal{O}^{\text{an}}_{\mathbb{C}^m}) & \longrightarrow & \dfrac{\Gamma(U', \mathcal{O}^{\text{an}}_{\mathbb{C}^m})}{I' \cdot \Gamma(U', \mathcal{O}^{\text{an}}_{\mathbb{C}^m})} & = & \Gamma(V, \mathcal{O}^{\text{an}})
\end{array}$$

are continuous in the Fréchet topology; for the map ξ this is by Exercise 5.4.10, while for the horizontal maps it is by the proof of Theorem 5.4.14. We remind the reader that the proof of Theorem 5.4.14 amounts to the construction of the Fréchet topology on $\Gamma(V, \mathcal{O}^{\text{an}})$, and the proof is not included in this book. References were given in Theorem 5.4.14.

The top horizontal map is continuous by the above and is obviously surjective, hence open by Theorem 5.4.11. Therefore the map

$$\gamma \quad : \quad \frac{\Gamma(U, \mathcal{O}^{\text{an}}_{\mathbb{C}^n})}{I \cdot \Gamma(U, \mathcal{O}^{\text{an}}_{\mathbb{C}^n})} \longrightarrow \frac{\Gamma(U', \mathcal{O}^{\text{an}}_{\mathbb{C}^m})}{I' \cdot \Gamma(U', \mathcal{O}^{\text{an}}_{\mathbb{C}^m})}$$

must be continuous. The composites

$$\begin{array}{ccccccc}
R & \longrightarrow & \Gamma(U, \mathcal{O}^{\text{an}}_{\mathbb{C}^n}) & \longrightarrow & \dfrac{\Gamma(U, \mathcal{O}^{\text{an}}_{\mathbb{C}^n})}{I \cdot \Gamma(U, \mathcal{O}^{\text{an}}_{\mathbb{C}^n})} & = & \Gamma(V, \mathcal{O}^{\text{an}}) \\
R' & \longrightarrow & \Gamma(U', \mathcal{O}^{\text{an}}_{\mathbb{C}^m}) & \longrightarrow & \dfrac{\Gamma(U', \mathcal{O}^{\text{an}}_{\mathbb{C}^m})}{I' \cdot \Gamma(U', \mathcal{O}^{\text{an}}_{\mathbb{C}^m})} & = & \Gamma(V, \mathcal{O}^{\text{an}})
\end{array}$$

are identified with the maps $\widetilde{\mu}_V : R \longrightarrow \Gamma(V, \mathcal{O}^{\text{an}})$ and $\widetilde{\mu}_V : R' \longrightarrow \Gamma(V, \mathcal{O}^{\text{an}})$ of Definition 5.5.5, where once again our notation is in dangerous territory for the reasons explained in Caution 5.6.5. Definition 5.5.5 allows us to factor these uniquely through θ and θ'; we deduce a commutative diagram

$$\begin{array}{ccccccc} R & \xrightarrow{\theta} & S & \xrightarrow{\mu_V} & \dfrac{\Gamma(U, \mathcal{O}^{\text{an}}_{\mathbb{C}^n})}{I \cdot \Gamma(U, \mathcal{O}^{\text{an}}_{\mathbb{C}^n})} & = & \Gamma(V, \mathcal{O}^{\text{an}}) \\ {\scriptstyle\varphi}\downarrow & & {\scriptstyle 1}\downarrow & & \downarrow{\scriptstyle\gamma} & & \\ R' & \xrightarrow[\theta']{} & S & \xrightarrow[\mu_V]{} & \dfrac{\Gamma(U', \mathcal{O}^{\text{an}}_{\mathbb{C}^m})}{I' \cdot \Gamma(U', \mathcal{O}^{\text{an}}_{\mathbb{C}^m})} & = & \Gamma(V, \mathcal{O}^{\text{an}}) \end{array} .$$

This means that the μ_V that comes from the embedding into $\mathbb{C}^m$ factors through the μ_V that comes from the embedding into $\mathbb{C}^n$, with γ a continuous map. Hence the topology induced on S by $V \subset \mathbb{C}^n$ is finer than the topology induced by $V \subset \mathbb{C}^m$. By symmetry the topologies must agree. □

A very easy fact is

Lemma 5.6.6. *Let $V \subset V' \subset \left\{ \text{Spec}(S) \right\}^{\text{an}}$ be polydiscs. Then the identity map from S to itself is a continuous map $[S, V'] \longrightarrow [S, V]$.*

Proof. By Remark 5.5.6 we have a commutative triangle

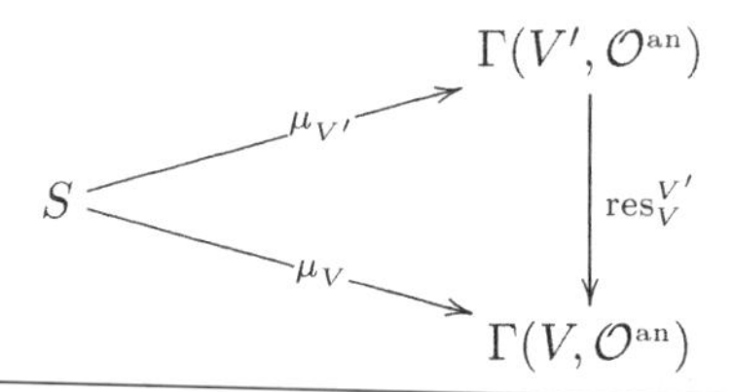

The restriction map $\text{res}_V^{V'}$ is continuous by Theorem 5.4.14, and hence the topology induced on S by the map $\mu_{V'}$ is finer than the topology induced by the map μ_V. □

Next we see what happens with ring homomorphisms.

Lemma 5.6.7. *Let $\varphi : S \longrightarrow S'$ be a homomorphism of finitely generated $\mathbb{C}$–algebras. Suppose we are given a polydisc $V \subset \left\{ \text{Spec}(S) \right\}^{\text{an}}$. Then we assert:*

(i) *The inverse image of V by the continuous map*

$$\left\{ \text{Spec}(\varphi) \right\}^{\text{an}} : \left\{ \text{Spec}(S') \right\}^{\text{an}} \longrightarrow \left\{ \text{Spec}(S) \right\}^{\text{an}}$$

is a polydisc in $\left\{ \text{Spec}(S') \right\}^{\text{an}}$.

(ii) *Let the inverse image of V by the map $\left\{ \operatorname{Spec}(\varphi)\right\}^{\mathrm{an}}$ be denoted V'. We assert that the ring homomorphism $\varphi : S \longrightarrow S'$ is continuous when viewed as a map of pre-Fréchet rings $\varphi : [S,V] \longrightarrow [S',V']$.*

Proof. (cf. proof of Lemma 4.5.6): Suppose we have two sets of generators $\{a_1, a_2, \ldots, a_n\} \subset S$ and $\{b_1, b_2, \ldots, b_m\} \subset S'$ for the $\mathbb{C}$–algebras S and S' respectively. Let $R = \mathbb{C}[x_1, x_2, \ldots, x_n]$ and $R' = \mathbb{C}[y_1, y_2, \ldots, y_m]$ be the polynomial rings. We have two surjective homomorphisms

$$\theta : R \longrightarrow S\,, \qquad \theta' : R' \longrightarrow S'\,,$$

where $\theta(x_i) = a_i$ and $\theta'(y_i) = b_i$.

We know that $b_1, b_2, \ldots, b_m$ generate S', while $\varphi(a_1), \varphi(a_2), \ldots, \varphi(a_n)$ are elements of S'. It follows that we can write the $\varphi(a_j)$ as polynomials in the b_i; there exist identities in S'

$$\varphi(a_j) = P_j(b_1, b_2, \ldots, b_m)\;,$$

with P_j polynomials with coefficients in $\mathbb{C}$. Now we define a homomorphism $\psi : R \longrightarrow R'$ by setting

$$\psi(x_j) = P_j(y_1, y_2, \ldots, y_m)\;.$$

We deduce a commutative square of ring homomorphisms

$$\begin{array}{ccc} R & \xrightarrow{\ \psi\ } & R' \\ {\scriptstyle\theta}\big\downarrow & & \big\downarrow{\scriptstyle\theta'} \\ S & \xrightarrow[\ \varphi\ ]{} & S' \end{array} \quad .$$

Let I be the kernel of the $\mathbb{C}$–algebra homomorphism $\theta : R \longrightarrow S$, and let I' be the kernel of the $\mathbb{C}$–algebra homomorphism $\theta' : R' \longrightarrow S'$. The commutative square of rings gives us a commutative square of topological spaces

$$\begin{array}{ccc} \left\{ \operatorname{Spec}(S')\right\}^{\mathrm{an}} & \xrightarrow{\ \left\{ \operatorname{Spec}(\theta')\right\}^{\mathrm{an}}\ } & \mathbb{C}^m \\ {\scriptstyle\left\{ \operatorname{Spec}(\varphi)\right\}^{\mathrm{an}}}\big\downarrow & & \big\downarrow{\scriptstyle\left\{ \operatorname{Spec}(\psi)\right\}^{\mathrm{an}}} \\ \left\{ \operatorname{Spec}(S)\right\}^{\mathrm{an}} & \xrightarrow[\ \left\{ \operatorname{Spec}(\theta)\right\}^{\mathrm{an}}\ ]{} & \mathbb{C}^n \end{array}$$

and we propose to prove the lemma by studying this square.

The assertion of Lemma 5.6.7(i) is already immediate. A polydisc in $V \subset \left\{ \operatorname{Spec}(S)\right\}^{\mathrm{an}}$ is a set of the form $g^{-1}\Delta(w;r) \cap \left\{ \operatorname{Spec}(S)\right\}^{\mathrm{an}}$, where $g : \mathbb{C}^n \longrightarrow \mathbb{C}^\ell$ is a polynomial map and $\Delta(w;r)$ is a usual polydisc in $\mathbb{C}^\ell$.

The inverse image of $g^{-1}\Delta(w;r)$ by the map $\mathbb{C}^m \longrightarrow \mathbb{C}^n$ is a polydisc in $\mathbb{C}^m$, and the intersection with $\{\operatorname{Spec}(S)\}^{\mathrm{an}}$ is the inverse image of V.

It remains to prove Lemma 5.6.7(ii). Let V be some polydisc in $\{\operatorname{Spec}(S)\}^{\mathrm{an}}$; that is $V = \Delta(g;w;r) \cap \{\operatorname{Spec}(S)\}^{\mathrm{an}}$, for a polydisc $U = \Delta(g;w;r)$ in $\mathbb{C}^n$. By the above the inverse image of $U = \Delta(g;w;r)$, by the polynomial map $\{\operatorname{Spec}(\psi)\}^{\mathrm{an}} : \mathbb{C}^m \longrightarrow \mathbb{C}^n$, is also a polydisc $U' = \Delta(g';w;r) \subset \mathbb{C}^m$.

As in the proof of Proposition 5.6.4, we have a commutative diagram of ring homomorphisms

$$\begin{array}{ccccccc} R & \longrightarrow & \Gamma(U, \mathcal{O}^{\mathrm{an}}_{\mathbb{C}^n}) & \longrightarrow & \dfrac{\Gamma(U, \mathcal{O}^{\mathrm{an}}_{\mathbb{C}^n})}{I \cdot \Gamma(U, \mathcal{O}^{\mathrm{an}}_{\mathbb{C}^n})} & = & \Gamma(V, \mathcal{O}^{\mathrm{an}}) \\ \psi \downarrow & & \downarrow \xi & & \downarrow \gamma & & \\ R' & \longrightarrow & \Gamma(U', \mathcal{O}^{\mathrm{an}}_{\mathbb{C}^m}) & \longrightarrow & \dfrac{\Gamma(U', \mathcal{O}^{\mathrm{an}}_{\mathbb{C}^m})}{I' \cdot \Gamma(U', \mathcal{O}^{\mathrm{an}}_{\mathbb{C}^m})} & = & \Gamma(V', \mathcal{O}^{\mathrm{an}}) \end{array} .$$

Also as in the proof of Proposition 5.6.4, the maps

$$\begin{array}{ccccc} \Gamma(U, \mathcal{O}^{\mathrm{an}}_{\mathbb{C}^n}) & \longrightarrow & \dfrac{\Gamma(U, \mathcal{O}^{\mathrm{an}}_{\mathbb{C}^n})}{I \cdot \Gamma(U, \mathcal{O}^{\mathrm{an}}_{\mathbb{C}^n})} & = & \Gamma(V, \mathcal{O}^{\mathrm{an}}) \\ \downarrow \xi & & & & \\ \Gamma(U', \mathcal{O}^{\mathrm{an}}_{\mathbb{C}^m}) & \longrightarrow & \dfrac{\Gamma(U', \mathcal{O}^{\mathrm{an}}_{\mathbb{C}^m})}{I' \cdot \Gamma(U', \mathcal{O}^{\mathrm{an}}_{\mathbb{C}^m})} & = & \Gamma(V', \mathcal{O}^{\mathrm{an}}) \end{array}$$

are continuous in the Fréchet topology. The top horizontal map is surjective, hence open by Theorem 5.4.11. Therefore the map

$$\gamma \quad : \quad \frac{\Gamma(U, \mathcal{O}^{\mathrm{an}}_{\mathbb{C}^n})}{I \cdot \Gamma(U, \mathcal{O}^{\mathrm{an}}_{\mathbb{C}^n})} \longrightarrow \frac{\Gamma(U', \mathcal{O}^{\mathrm{an}}_{\mathbb{C}^m})}{I' \cdot \Gamma(U', \mathcal{O}^{\mathrm{an}}_{\mathbb{C}^m})}$$

must be continuous. The composites

$$R \longrightarrow \Gamma(U, \mathcal{O}^{\mathrm{an}}_{\mathbb{C}^n}) \longrightarrow \frac{\Gamma(U, \mathcal{O}^{\mathrm{an}}_{\mathbb{C}^n})}{I \cdot \Gamma(U, \mathcal{O}^{\mathrm{an}}_{\mathbb{C}^n})} = \Gamma(V, \mathcal{O}^{\mathrm{an}})$$

$$R' \longrightarrow \Gamma(U', \mathcal{O}^{\mathrm{an}}_{\mathbb{C}^m}) \longrightarrow \frac{\Gamma(U', \mathcal{O}^{\mathrm{an}}_{\mathbb{C}^m})}{I' \cdot \Gamma(U', \mathcal{O}^{\mathrm{an}}_{\mathbb{C}^m})} = \Gamma(V', \mathcal{O}^{\mathrm{an}})$$

are identified with the maps $\widetilde{\mu}_V : R \longrightarrow \Gamma(V, \mathcal{O}^{\mathrm{an}})$ and $\widetilde{\mu}_{V'} : R' \longrightarrow \Gamma(V', \mathcal{O}^{\mathrm{an}})$ of Definition 5.5.5, and they factor uniquely through θ and θ'; we deduce a commutative diagram

$$\begin{array}{ccccccc} R & \xrightarrow{\theta} & S & \xrightarrow{\mu_V} & \dfrac{\Gamma(U, \mathcal{O}^{\mathrm{an}}_{\mathbb{C}^n})}{I \cdot \Gamma(U, \mathcal{O}^{\mathrm{an}}_{\mathbb{C}^n})} & = & \Gamma(V, \mathcal{O}^{\mathrm{an}}) \\ \psi \downarrow & & \varphi \downarrow & & \downarrow \gamma & & \\ R' & \xrightarrow[\theta']{} & S' & \xrightarrow[\mu_{V'}]{} & \dfrac{\Gamma(U', \mathcal{O}^{\mathrm{an}}_{\mathbb{C}^m})}{I' \cdot \Gamma(U', \mathcal{O}^{\mathrm{an}}_{\mathbb{C}^m})} & = & \Gamma(V', \mathcal{O}^{\mathrm{an}}) \end{array} .$$

Take an open set in S', with respect to its $[S',V']$ topology. Such a set must be of the form $\mu_{V'}^{-1}W$ with W an open set in $\Gamma(V',\mathcal{O}^{\mathrm{an}})$. The inverse image via the map $\varphi : S \longrightarrow S'$ is

$$\varphi^{-1}\mu_{V'}^{-1}W = \mu_V^{-1}\gamma^{-1}W \ .$$

But $\gamma^{-1}W$ is open in $\Gamma(V,\mathcal{O}^{\mathrm{an}})$ by the continuity of γ and hence, by the definition of the $[S,V]$ topology on S, the set $\mu_V^{-1}\gamma^{-1}W$ must be open in S. □

Remark 5.6.8. Lemma 5.6.7 tells us about general $\mathbb{C}$–algebra homomorphisms $\varphi : S \longrightarrow S'$. A very useful special case is the ring homomorphism $\alpha_f : S \longrightarrow S[1/f]$. In this case we can be much more specific.

Lemma 5.6.9. *Let S be a a finitely generated $\mathbb{C}$–algebra, and let $f \in S$. Put $(X,\mathcal{O}) = \Big(\mathrm{Spec}(S), \widetilde{S}\Big)$, and therefore $X^{\mathrm{an}} = \big\{\mathrm{Spec}(S)\big\}^{\mathrm{an}}$. Let $V \subset X^{\mathrm{an}}$ be a polydisc of $X^{\mathrm{an}} = \big\{\mathrm{Spec}(S)\big\}^{\mathrm{an}}$, and assume V is entirely contained in the open set $X_f \cap X^{\mathrm{an}}$.*

Let $\alpha_f : S \longrightarrow S[1/f]$ be the canonical map. With V as above, Lemma 5.6.7 gives a continuous map of pre-Fréchet rings $[S,V] \longrightarrow \big[S[1/f],V'\big]$. This map induces an isomorphism on the completions.

Remark 5.6.10. With $(X,\mathcal{O}) = \Big(\mathrm{Spec}(S), \widetilde{S}\Big)$ as above we learn, from Proposition 4.5.10, that the map

$$\big\{\mathrm{Spec}(\alpha_f)\big\}^{\mathrm{an}} : \big\{\mathrm{Spec}(S[1/f])\big\}^{\mathrm{an}} \longrightarrow \big\{\mathrm{Spec}(S)\big\}^{\mathrm{an}}$$

is a homeomorphism of $\big\{\mathrm{Spec}(S[1/f])\big\}^{\mathrm{an}}$ with its image $X_f \cap X^{\mathrm{an}} \subset X^{\mathrm{an}}$. Therefore V', which in Lemma 5.6.7 is the inverse image of $V \subset X^{\mathrm{an}}$ by the map $\big\{\mathrm{Spec}(\alpha_f)\big\}^{\mathrm{an}}$, is just $V \cap X_f$. Since we are assuming $V \subset X_f$ we conclude $V' = V$.

Proof. Choose generators $a_1, a_2, \ldots, a_n$ for the ring S, suppose $R = \mathbb{C}[x_1, x_2, \ldots, x_n]$ is the polynomial ring, and let $\theta : R \longrightarrow S$ be the homomorphism with $\theta(x_i) = a_i$. The ring homomorphism θ is surjective, so we may choose an element $\tilde{f} \in R$ with $\theta(\tilde{f}) = f$. Now we let $R' = \mathbb{C}[x_1, x_2, \ldots, x_n, x_{n+1}]$ be the polynomial ring in $(n+1)$ variables, let $\varphi : R \longrightarrow R'$ be the inclusion, and let $\theta' : R' \longrightarrow S[1/f]$ be the homomorphism

$$\theta'(x_i) = \begin{cases} a_i & \text{if } 1 \le i \le n, \\ 1/f & \text{if } i = n+1. \end{cases}$$

Let I be the kernel of $\theta : R \longrightarrow S$, and let I' be the kernel of $\theta' : R' \longrightarrow$

$S[1/f]$. We have a commutative square of ring homomorphisms

$$\begin{array}{ccc} R & \xrightarrow{\varphi} & R' \\ {\scriptstyle\theta}\downarrow & & \downarrow{\scriptstyle\theta'} \\ S & \xrightarrow[\alpha_f]{} & S[1/f] \end{array}$$

with θ and θ' surjective. Applying $\{\operatorname{Spec}(-)\}^{\mathrm{an}}$ to the diagram we have

$$\begin{array}{ccc} \{\operatorname{Spec}(S[1/f])\}^{\mathrm{an}} & \xrightarrow{\{\operatorname{Spec}(\theta')\}^{\mathrm{an}}} & \mathbb{C}^{n+1} \\ {\scriptstyle \{\operatorname{Spec}(\alpha_f)\}^{\mathrm{an}}}\downarrow & & \downarrow{\scriptstyle \{\operatorname{Spec}(\varphi)\}^{\mathrm{an}}} \\ \{\operatorname{Spec}(S)\}^{\mathrm{an}} & \xrightarrow[\{\operatorname{Spec}(\theta)\}^{\mathrm{an}}]{} & \mathbb{C}^{n} \end{array} .$$

In this diagram the map $\{\operatorname{Spec}(\varphi)\}^{\mathrm{an}}$ is particularly simple; it is the projection taking a point

$$(b_1, b_2, \ldots, b_n, b_{n+1}) \in \mathbb{C}^{n+1} \quad \text{to} \quad (b_1, b_2, \ldots, b_n) \in \mathbb{C}^n ;$$

see Lemma 4.4.5. Let $U = \Delta(g; w; r)$ be an arbitrary polydisc in $\mathbb{C}^n$, and let U' be its inverse image in $\mathbb{C}^{n+1}$. Let $V = U \cap \{\operatorname{Spec}(S)\}^{\mathrm{an}}$, and let $V' = U' \cap \{\operatorname{Spec}(S[1/f])\}^{\mathrm{an}}$. In the proof of Lemma 5.6.7 we produced a commutative diagram

$$\begin{array}{ccccccc} R & \xrightarrow{\theta} & S & \xrightarrow{\mu_V} & \dfrac{\Gamma(U, \mathcal{O}^{\mathrm{an}}_{\mathbb{C}^n})}{I \cdot \Gamma(U, \mathcal{O}^{\mathrm{an}}_{\mathbb{C}^n})} & = & \Gamma(V, \mathcal{O}^{\mathrm{an}}) \\ {\scriptstyle\varphi}\downarrow & & {\scriptstyle\alpha_f}\downarrow & & \downarrow{\scriptstyle\gamma} & & \\ R' & \xrightarrow[\theta']{} & S[1/f] & \xrightarrow[\mu_{V'}]{} & \dfrac{\Gamma(U', \mathcal{O}^{\mathrm{an}}_{\mathbb{C}^m})}{I' \cdot \Gamma(U', \mathcal{O}^{\mathrm{an}}_{\mathbb{C}^m})} & = & \Gamma(V', \mathcal{O}^{\mathrm{an}}) \end{array}$$

and the main step in the proof is

5.6.10.1. *If the polydisc $U \subset \mathbb{C}^n$ is contained in the open set $\mathbb{C}^n_{\tilde{f}}$ then the map γ above is a homeomorphism.*

Reminder 5.6.11. We remind the reader that the subset $\mathbb{C}^n_{\tilde{f}} \subset \mathbb{C}^n$ is the set of all points $a \in \mathbb{C}^n$ with $\tilde{f}(a) \neq 0$. The element $\tilde{f} \in R$ was chosen so that $\theta(\tilde{f}) = f$; see the first paragraph of the proof.

Now we prove 5.6.10.1. Suppose $\tilde{f}$ does not vanish on the polydisc U.

The commutative diagram above came from a diagram

$$\begin{array}{ccccccc} R & \longrightarrow & \Gamma(U,\mathcal{O}^{\mathrm{an}}_{\mathbb{C}^n}) & \longrightarrow & \dfrac{\Gamma(U,\mathcal{O}^{\mathrm{an}}_{\mathbb{C}^n})}{I\cdot\Gamma(U,\mathcal{O}^{\mathrm{an}}_{\mathbb{C}^n})} & = & \Gamma(V,\mathcal{O}^{\mathrm{an}}) \\ {\scriptstyle\varphi}\downarrow & & \downarrow{\scriptstyle\xi} & & \downarrow{\scriptstyle\gamma} & & \\ R' & \longrightarrow & \Gamma(U',\mathcal{O}^{\mathrm{an}}_{\mathbb{C}^m}) & \longrightarrow & \dfrac{\Gamma(U',\mathcal{O}^{\mathrm{an}}_{\mathbb{C}^m})}{I'\cdot\Gamma(U',\mathcal{O}^{\mathrm{an}}_{\mathbb{C}^m})} & = & \Gamma(V',\mathcal{O}^{\mathrm{an}}) \end{array} .$$

and we have to compute the relation between the ideal $I\cdot\Gamma(U,\mathcal{O}^{\mathrm{an}}_{\mathbb{C}^n})$ and the ideal $I'\cdot\Gamma(U',\mathcal{O}^{\mathrm{an}}_{\mathbb{C}^m})$. With such computations it is never clear how much detail to provide. Let us give some hints.

Our hypothesis, that $\tilde{f}$ is non-zero on U, means that $1/\tilde{f}$ is a holomorphic function on U. We can extend the diagram to

$$\begin{array}{ccccccc} R & \longrightarrow & R[1/\tilde{f}] & \longrightarrow & \Gamma(U,\mathcal{O}^{\mathrm{an}}_{\mathbb{C}^n}) & \longrightarrow & \dfrac{\Gamma(U,\mathcal{O}^{\mathrm{an}}_{\mathbb{C}^n})}{I\cdot\Gamma(U,\mathcal{O}^{\mathrm{an}}_{\mathbb{C}^n})} \\ {\scriptstyle\varphi}\downarrow & & {\scriptstyle\varphi[1/\tilde{f}]}\downarrow & & \downarrow{\scriptstyle\xi} & & \downarrow{\scriptstyle\gamma} \\ R' & \longrightarrow & R'[1/\tilde{f}] & \longrightarrow & \Gamma(U',\mathcal{O}^{\mathrm{an}}_{\mathbb{C}^m}) & \longrightarrow & \dfrac{\Gamma(U',\mathcal{O}^{\mathrm{an}}_{\mathbb{C}^m})}{I'\cdot\Gamma(U',\mathcal{O}^{\mathrm{an}}_{\mathbb{C}^m})} \end{array}$$

We have ideals

$$I\subset R,\quad I\cdot R[1/\tilde{f}]\subset R[1/\tilde{f}],\quad I'\subset R'\quad\text{and}\quad I'\cdot R'[1/\tilde{f}]\subset R'[1/\tilde{f}].$$

The proof is based on understanding the relation among these ideals. To compare them it is helpful to note that the map $\varphi[1/\tilde{f}] : R[1/\tilde{f}] \longrightarrow R'[1/\tilde{f}]$ has a splitting, $\psi : R'[1/\tilde{f}] \longrightarrow R[1/\tilde{f}]$, with $\psi(x_{n+1}) = 1/\tilde{f}$. That is we have a commutative triangle

$$\begin{array}{ccc} R[1/\tilde{f}] & \searrow^{1} & \\ {\scriptstyle\varphi[1/\tilde{f}]}\downarrow & & R[1/\tilde{f}] \\ R'[1/\tilde{f}] & \nearrow_{\psi} & \end{array} .$$

We extend this to a larger commutative diagram

$$\begin{array}{ccccccc} R & \xrightarrow{\alpha_{\tilde{f}}} & R[1/\tilde{f}] & \searrow^{1} & & & \\ {\scriptstyle\varphi}\downarrow & & {\scriptstyle\varphi[1/\tilde{f}]}\downarrow & & R[1/\tilde{f}] & \xrightarrow{\delta} & S[1/f] \\ R' & \xrightarrow[\alpha_{\tilde{f}}]{} & R'[1/\tilde{f}] & \nearrow_{\psi} & & & \end{array}$$

and the kernels we wish to compute are all kernels of various composites in this diagram, especially composites ending in $S[1/f]$. The idea is to use the commutativity for our computations.

The ideal $I \subset R$ is the kernel of $\theta : R \longrightarrow S$; that is we have an exact sequence

$$0 \longrightarrow I \longrightarrow R \xrightarrow{\theta} S \longrightarrow 0\,.$$

By [1, Proposition 3.3] the ideal $I \cdot R[1/\tilde{f}] \subset R[1/\tilde{f}]$ is the kernel of the homomorphism $\delta : R[1/\tilde{f}] \longrightarrow S[1/f]$. Similarly the ideal $I' \subset R'$ is the kernel of $\theta' : R' \longrightarrow S[1/f]$, and hence the ideal $I' \cdot R'[1/\tilde{f}] \subset R'[1/\tilde{f}]$ must be the kernel of $\delta\psi : R'[1/\tilde{f}] \longrightarrow S[1/f]$. Now the kernel of ψ is easy; it is generated by the single element $x_{n+1} - (1/\tilde{f}) \in R'[1/\tilde{f}]$. The commutativity tells us that the kernel of $\delta\psi$ is generated by the kernel of ψ and the kernel of $\delta = \delta \circ \psi \circ \{\varphi[1/f]\}$. Hence $I' \cdot R'[1/\tilde{f}]$ is generated by the ideal $I \cdot R'[1/\tilde{f}]$ and the single element $x_{n+1} - (1/\tilde{f})$.

Now we turn this into a statement about rings of holomorphic functions. The map γ is the homomorphism

$$\frac{\Gamma(U, \mathcal{O}^{\text{an}}_{\mathbb{C}^n})}{I \cdot \Gamma(U, \mathcal{O}^{\text{an}}_{\mathbb{C}^n})} \xrightarrow{\gamma} \frac{\Gamma(U', \mathcal{O}^{\text{an}}_{\mathbb{C}^m})}{I' \cdot \Gamma(U', \mathcal{O}^{\text{an}}_{\mathbb{C}^m})}\,,$$

and we already noted that the element $\tilde{f}$ is invertible both in $\Gamma(U, \mathcal{O}^{\text{an}}_{\mathbb{C}^n})$ and in $\Gamma(U', \mathcal{O}^{\text{an}}_{\mathbb{C}^m})$. That is,

$$R[1/\tilde{f}]{\cdot}\Gamma(U, \mathcal{O}^{\text{an}}_{\mathbb{C}^n}) = \Gamma(U, \mathcal{O}^{\text{an}}_{\mathbb{C}^n})\,, \qquad R'[1/\tilde{f}]{\cdot}\Gamma(U', \mathcal{O}^{\text{an}}_{\mathbb{C}^m}) = \Gamma(U', \mathcal{O}^{\text{an}}_{\mathbb{C}^m})\,.$$

The homomorphism γ can be rewritten

$$\frac{\Gamma(U, \mathcal{O}^{\text{an}}_{\mathbb{C}^n})}{\{I \cdot R[1/\tilde{f}]\} \cdot \Gamma(U, \mathcal{O}^{\text{an}}_{\mathbb{C}^n})} \xrightarrow{\gamma} \frac{\Gamma(U', \mathcal{O}^{\text{an}}_{\mathbb{C}^m})}{\{I' \cdot R'[1/\tilde{f}]\} \cdot \Gamma(U', \mathcal{O}^{\text{an}}_{\mathbb{C}^m})}\,.$$

The computation above tells us that $I' \cdot R'[1/\tilde{f}]$ is generated by $I \cdot R[1/\tilde{f}]$ and the single element $x_{n+1} - (1/\tilde{f})$. Now recall that the polydisc U', being the inverse image of U by the projection, is just $U \times \mathbb{C}$. The ring homomorphism

$$\Gamma(U, \mathcal{O}^{\text{an}}_{\mathbb{C}^n}) \longrightarrow \Gamma(U \times \mathbb{C}, \mathcal{O}^{\text{an}}_{\mathbb{C}^m}) \longrightarrow \frac{\Gamma(U \times \mathbb{C}, \mathcal{O}^{\text{an}}_{\mathbb{C}^m})}{\{x_{n+1} - (1/\tilde{f})\} \cdot \Gamma(U \times \mathbb{C}, \mathcal{O}^{\text{an}}_{\mathbb{C}^m})}$$

is clearly an isomorphism, and dividing further by the ideal I we deduce that γ is bijective. It is continuous by the proof of Lemma 5.6.7, and hence a homeomorphism by Corollary 5.4.12.

So far we know that, if U is a polydisc contained in $\mathbb{C}^n_{\tilde{f}}$, then, in the commutative square

$$\begin{array}{ccccc} S & \xrightarrow{\mu_V} & \dfrac{\Gamma(U,\mathcal{O}^{\mathrm{an}}_{\mathbb{C}^n})}{I\cdot\Gamma(U,\mathcal{O}^{\mathrm{an}}_{\mathbb{C}^n})} & = & \Gamma(V,\mathcal{O}^{\mathrm{an}}) \\ {\scriptstyle\alpha_f}\Big\downarrow & & \Big\downarrow{\scriptstyle\gamma} & & \\ S[1/f] & \xrightarrow[\mu_{V'}]{} & \dfrac{\Gamma(U',\mathcal{O}^{\mathrm{an}}_{\mathbb{C}^m})}{I'\cdot\Gamma(U',\mathcal{O}^{\mathrm{an}}_{\mathbb{C}^m})} & = & \Gamma(V',\mathcal{O}^{\mathrm{an}}) \end{array} ,$$

the map γ is a homeomorphism. This proves 5.6.10.1; now we deduce the lemma.

Recall the notation $X = \mathrm{Spec}(S)$. In the statement of the lemma we were allowed to take any polydisc $V \subset X_f \cap X^{\mathrm{an}}$. Being a polydisc in X^{an} the set V can be written $V = \Delta(g;w;r) \cap X^{\mathrm{an}}$, with $\Delta(g;w;r)$ a polydisc in $\mathbb{C}^n$. But we are not guaranteed that $\Delta(g;w;r)$ will be a subset of $\mathbb{C}^n_{\tilde{f}}$, and hence cannot immediately apply what we proved above.

However, we do know that $V = \Delta(g;w;r) \cap X^{\mathrm{an}}$ is a subset of $\mathbb{C}^n_{\tilde{f}}$. The reason for this is the following. We have a map of ringed spaces

$$(X,\mathcal{O}_X) = \left(\mathrm{Spec}(S),\widetilde{S}\right) \xrightarrow{(\mathrm{Spec}(\theta),\widetilde{\theta})} \left(\mathrm{Spec}(R),\widetilde{R}\right) = (Y,\mathcal{O}_Y) .$$

This induces an inclusion

$$\left\{\mathrm{Spec}(\theta)\right\}^{\mathrm{an}} : X^{\mathrm{an}} \longrightarrow Y^{\mathrm{an}} = \mathbb{C}^n .$$

Lemma 4.4.3 identified $\mathbb{C}^n_{\tilde{f}} \subset \mathbb{C}^n$ with $Y_{\tilde{f}} \cap Y^{\mathrm{an}}$. The inverse image by the inclusion $\left\{\mathrm{Spec}(\theta)\right\}^{\mathrm{an}}$ is $X_f \cap X^{\mathrm{an}}$ (see Remark 3.6.4), which contains V by hypothesis. Thus V is contained in $\mathbb{C}^n_{\tilde{f}}$.

The polydiscs form a basis for the topology of $\mathbb{C}^n$, and therefore the open subset $\mathbb{C}^n_{\tilde{f}} \subset \mathbb{C}^n$ can be covered by polydiscs $\Delta(g'_i;w'_i;r'_i)$. We can cover V by sets of the form

$$X^{\mathrm{an}} \cap \Delta(g;w;r) \cap \Delta(g'_i;w'_i;r'_i)$$

where all the $\Delta(g;w;r) \cap \Delta(g'_i;w'_i;r'_i)$ are contained in $\mathbb{C}^n_{\tilde{f}} \subset \mathbb{C}^n$. Assume therefore that $V = \cup V_i$, where each V_i is the intersection of X^{an} with a polydisc contained in $\mathbb{C}^n_{\tilde{f}}$. Let V'_i be the inverse image of V_i under the map $\{\operatorname{Spec}(\alpha_f)\}^{\text{an}}$. We have a commutative diagram with exact rows

$$\begin{array}{ccccccc}
0 \longrightarrow & \Gamma(V,\mathcal{O}^{\text{an}}) & \xrightarrow{\alpha} & \prod_{i\in I}\Gamma(V_i,\mathcal{O}^{\text{an}}) & \xrightarrow{\beta} & \prod_{i,j\in I}\Gamma(V_i\cap V_j,\mathcal{O}^{\text{an}}) \\
& \downarrow{\scriptstyle\gamma} & & \downarrow{\scriptstyle\prod\gamma_i} & & \downarrow{\scriptstyle\prod\gamma_{ii}} \\
0 \longrightarrow & \Gamma(V',\mathcal{O}^{\text{an}}) & \xrightarrow[\alpha']{} & \prod_{i\in I}\Gamma(V_i',\mathcal{O}^{\text{an}}) & \xrightarrow[\beta']{} & \prod_{i,j\in I}\Gamma(V_i'\cap V_j',\mathcal{O}^{\text{an}})
\end{array} .$$

By 5.6.10.1 we know that $\prod\gamma_i$ and $\prod\gamma_{ij}$ are bijections. Hence so is γ. In the proof of Lemma 5.6.7 we learnt that γ is continuous, and the above shows it is bijective; γ is a bijective, continuous map of Fréchet spaces. Corollary 5.4.12 informs us that it must be a homeomorphism.

We have learnt that, in the commutative square

$$\begin{array}{ccc}
S & \xrightarrow{\mu_V} & \Gamma(V,\mathcal{O}^{\text{an}}) \\
{\scriptstyle\alpha_f}\downarrow & & \downarrow{\scriptstyle\gamma} \\
S[1/f] & \xrightarrow[\mu_{V'}]{} & \Gamma(V',\mathcal{O}^{\text{an}})
\end{array} ,$$

the map γ is a homeomorphism. By Proposition 5.5.7 we know that the map μ_V embeds S as a dense subset of $\Gamma(V,\mathcal{O}^{\text{an}})$, and trivially the image of $S[1/f]$, which is bigger, must be dense in the homeomorphic space $\Gamma(V',\mathcal{O}^{\text{an}})$. Hence the completions of $[S,V]$ and $\big[S[1/f],V'\big]$ agree. □

5.7 Affine schemes, done without coordinates

Remark 5.7.1. Let S be a finitely generated $\mathbb{C}$–algebra. Write

$$(X,\mathcal{O}_X) = \left(\operatorname{Spec}(S),\widetilde{S}\right) .$$

For every polydisc $V \subset X^{\text{an}} = \{\operatorname{Spec}(S)\}^{\text{an}}$ we produced, in Definition 5.6.1, a pre-Fréchet topology $[S,V]$ on S. This topology is independent of choices of generators by Proposition 5.6.4. We can define

(i) $\Gamma(V,\mathcal{O}_X^{\text{an}})$ is the completion of the Fréchet ring $[S,V]$.

(ii) If $V \subset V'$ are polydiscs, the map

$$\operatorname{res}_V^{V'} : \Gamma(V',\mathcal{O}_X^{\text{an}}) \longrightarrow \Gamma(V,\mathcal{O}_X^{\text{an}})$$

is the completion of the identity map from S to itself, viewed as a continuous homomophism $1 : [S,V'] \longrightarrow [S,V]$. The continuity is guaranteed by Lemma 5.6.6.

This defines $\Gamma(V, \mathcal{O}_X^{\mathrm{an}})$ for all polydiscs V, and defines restriction maps. We know that there is a sheaf extension to all open sets $U \subset X^{\mathrm{an}}$. The reason we know this is by construction. If we choose a set of generators for S, we can use the embedding of $\{\operatorname{Spec}(S)\}^{\mathrm{an}}$ into $\mathbb{C}^n$ to define a sheaf of rings $\mathcal{O}^{\mathrm{an}}$ on $\{\operatorname{Spec}(S)\}^{\mathrm{an}}$. As we explained in Caution 5.6.5 the notation is slightly criminal, since it does not make the dependence on the choice of generators clear. Anyway, the topological space $[S, V]$ was defined to be S, with the topology it inherits from the $\mathbb{C}$–algebra homomorphism

$$\mu_V : S \longrightarrow \Gamma(V, \mathcal{O}^{\mathrm{an}})$$

of Definition 5.5.5. Proposition 5.5.7 tells us that the image of S is dense. That is the completion of $[S, V]$ is precisely the map

$$\mu_V : S \longrightarrow \Gamma(V, \mathcal{O}^{\mathrm{an}}) .$$

The extension exists because we have exhibited one, namely $\mathcal{O}^{\mathrm{an}}$.

But the extension is canonically unique if we know the restriction of the sheaf to a basis of the open sets. Because the construction of $\mathcal{O}_X^{\mathrm{an}}$, made in (i) and (ii) above, made no mention of generators we conclude that the sheaf $\mathcal{O}^{\mathrm{an}}$, which is canonically isomorphic to the sheaf $\mathcal{O}_X^{\mathrm{an}}$, is a sheaf of S–algebras on the topological space $X^{\mathrm{an}} = \{\operatorname{Spec}(S)\}^{\mathrm{an}}$, independent of the choice of generators. We will denote the ringed space $(X^{\mathrm{an}}, \mathcal{O}_X^{\mathrm{an}})$ by the symbol $\left(\{\operatorname{Spec}(S)\}^{\mathrm{an}}, \widetilde{S}^{\mathrm{an}}\right)$.

Lemma 5.7.2. *Let $\varphi : S \longrightarrow S'$ be a ring homomorphism. Put*

$$(X^{\mathrm{an}}, \mathcal{O}_X^{\mathrm{an}}) = \left(\{\operatorname{Spec}(S')\}^{\mathrm{an}}, \widetilde{S'}^{\mathrm{an}}\right), \quad (Y^{\mathrm{an}}, \mathcal{O}_Y^{\mathrm{an}}) = \left(\{\operatorname{Spec}(S)\}^{\mathrm{an}}, \widetilde{S}^{\mathrm{an}}\right).$$

The continuous map of topological spaces $\{\operatorname{Spec}(\varphi)\}^{\mathrm{an}} : X^{\mathrm{an}} \longrightarrow Y^{\mathrm{an}}$, which we constructed in Lemma 4.5.6, can be extended to a natural map of ringed spaces

$$\left(\{\operatorname{Spec}(\varphi)\}^{\mathrm{an}}, \widetilde{\varphi}^{\mathrm{an}}\right) : (X^{\mathrm{an}}, \mathcal{O}_X^{\mathrm{an}}) \longrightarrow (Y^{\mathrm{an}}, \mathcal{O}_Y^{\mathrm{an}}) .$$

Furthermore, the following square of ring homomorphisms commutes

$$\begin{array}{ccc} S & \xrightarrow{\ \varphi\ } & S' \\ {\scriptstyle\mu}\big\downarrow & & \big\downarrow{\scriptstyle\mu'} \\ \Gamma(Y^{\mathrm{an}}, \mathcal{O}_Y^{\mathrm{an}}) & \xrightarrow[\widetilde{\varphi}^{\mathrm{an}}_{Y^{\mathrm{an}}}]{} & \Gamma(X^{\mathrm{an}}, \mathcal{O}_X^{\mathrm{an}}) \end{array} .$$

Proof. Let V be a polydisc in $\{\operatorname{Spec}(S)\}^{\mathrm{an}}$, and let V' be its inverse image in $\{\operatorname{Spec}(S')\}^{\mathrm{an}}$. Lemma 5.6.7 tells us that V' is a polydisc, and that the map $\varphi : [S, V] \longrightarrow [S', V']$ is a continuous map of Fréchet rings. Hence we deduce a continuous map of completions

$$\widetilde{\varphi}_V^{\text{an}} : \Gamma(V, \widetilde{S}^{\text{an}}) \longrightarrow \Gamma(V', \widetilde{S'}^{\text{an}}) .$$

Furthermore given two polydiscs $V \subset W \subset \{\operatorname{Spec}(S)\}^{\text{an}}$, if the inverse images by the map $\{\operatorname{Spec}(\varphi)\}^{\text{an}}$ are denoted $V' \subset W' \subset \{\operatorname{Spec}(S')\}^{\text{an}}$, we have a commutative square of ring homomorphisms of the completions

$$\begin{array}{ccc} \Gamma(W, \widetilde{S}^{\text{an}}) & \xrightarrow{\widetilde{\varphi}_W^{\text{an}}} & \Gamma\big(W', \widetilde{S'}^{\text{an}}\big) \\ {\scriptstyle \text{res}_V^W}\downarrow & & \downarrow{\scriptstyle \text{res}_{V'}^{W'}} \\ \Gamma(V, \widetilde{S}^{\text{an}}) & \xrightarrow{\widetilde{\varphi}_V^{\text{an}}} & \Gamma\big(V', \widetilde{S'}^{\text{an}}\big) \end{array} .$$

In other words on the polydiscs V we define the maps $\widetilde{\varphi}_V^{\text{an}}$ as completions, and the maps are compatible with restrictions. The polydiscs form a basis for the topology of $Y^{\text{an}} = \{\operatorname{Spec}(S)\}^{\text{an}}$, and there is a unique way to extend $\widetilde{\varphi}_V^{\text{an}}$ to all open $V \subset Y^{\text{an}}$. See the last paragraph of the proof of Proposition 3.6.5.

Now we come to the commutativity of the square

$$\begin{array}{ccc} S & \xrightarrow{\varphi} & S' \\ {\scriptstyle \mu}\downarrow & & \downarrow{\scriptstyle \mu'} \\ \Gamma(Y^{\text{an}}, \widetilde{S}^{\text{an}}) & \xrightarrow[\widetilde{\varphi}_{Y^{\text{an}}}^{\text{an}}]{} & \Gamma\big(X^{\text{an}}, \widetilde{S'}^{\text{an}}\big) \end{array} .$$

Example 5.6.2 tells us that $Y^{\text{an}} = \{\operatorname{Spec}(S)\}^{\text{an}}$ is a polydisc. Therefore, with $V = Y^{\text{an}}$, we may complete $[S, Y^{\text{an}}]$ and $[S', X^{\text{an}}]$ to obtain the rings $\Gamma(Y^{\text{an}}, \widetilde{S}^{\text{an}})$ and $\Gamma\big(X^{\text{an}}, \widetilde{S'}^{\text{an}}\big)$ respectively. The maps μ and μ' are the homomorphisms from S and S' to their completions, and the square obviously commutes. □

Lemma 5.7.3. *Suppose we have composable homomorphisms of finitely generated* $\mathbb{C}$*–algebras*

$$S \xrightarrow{\varphi} S' \xrightarrow{\varphi'} S''.$$

Put

$$(X^{\text{an}}, \mathcal{O}_X^{\text{an}}) = \Big(\{\operatorname{Spec}(S'')\}^{\text{an}}, \widetilde{S''}^{\text{an}}\Big), \quad (Y^{\text{an}}, \mathcal{O}_Y^{\text{an}}) = \Big(\{\operatorname{Spec}(S')\}^{\text{an}}, \widetilde{S'}^{\text{an}}\Big),$$
$$\text{and} \qquad (Z^{\text{an}}, \mathcal{O}_Z^{\text{an}}) = \Big(\{\operatorname{Spec}(S)\}^{\text{an}}, \widetilde{S}^{\text{an}}\Big).$$

We have composable maps of ringed spaces

$$(X^{\text{an}}, \mathcal{O}_X^{\text{an}}) \xrightarrow{\left(\{\operatorname{Spec}(\varphi')\}^{\text{an}}, \widetilde{\varphi'}^{\text{an}}\right)} (Y^{\text{an}}, \mathcal{O}_Y^{\text{an}}) \xrightarrow{\left(\{\operatorname{Spec}(\varphi)\}^{\text{an}}, \widetilde{\varphi}^{\text{an}}\right)} (Z^{\text{an}}, \mathcal{O}_Z^{\text{an}}) .$$

We assert that the composite map of ringed spaces agrees with the map of ringed spaces $\left(\left\{\operatorname{Spec}(\varphi'\varphi)\right\}^{\mathrm{an}}, \widetilde{\varphi'\varphi}^{\mathrm{an}}\right)$.

Proof. The assertion that the underlying continuous maps of topological spaces agree may be found in Corollary 4.6.11. We need to check that the ring homomorphisms agree. It is enough to check agreement on polydiscs in $Z^{\mathrm{an}} = \left\{\operatorname{Spec}(S)\right\}^{\mathrm{an}}$. Let V be a polydisc in Z^{an}. The inverse image by the map $\left\{\operatorname{Spec}(\varphi)\right\}^{\mathrm{an}}$ is a polydisc $V' \subset Y^{\mathrm{an}}$, and its inverse image by $\left\{\operatorname{Spec}(\varphi')\right\}^{\mathrm{an}}$ is a polydisc $V'' \subset X^{\mathrm{an}}$. The maps

$$\Gamma(V, \mathcal{O}_Z^{\mathrm{an}}) \xrightarrow{\widetilde{\varphi'}^{\mathrm{an}}_V} \Gamma(V', \mathcal{O}_Y^{\mathrm{an}}) \xrightarrow{\widetilde{\varphi}^{\mathrm{an}}_{V'}} \Gamma(V'', \mathcal{O}_X^{\mathrm{an}})$$

are just the completions of the maps

$$[S, V] \xrightarrow{\quad \varphi \quad} [S', V'] \xrightarrow{\quad \varphi' \quad} [S'', V''] \,,$$

and the composite of the completions clearly agrees with the completion of the composite. □

Lemma 5.7.4. *If* $\alpha_f : S \longrightarrow S[1/f]$ *is the localization map, then the map of ringed spaces*

$$\left(\left\{\operatorname{Spec}(S[1/f])\right\}^{\mathrm{an}}, \widetilde{S[1/f]}^{\mathrm{an}}\right) \xrightarrow{\left(\left\{\operatorname{Spec}(\alpha_f)\right\}^{\mathrm{an}}, \widetilde{\alpha_f}^{\mathrm{an}}\right)} \left(\left\{\operatorname{Spec}(S)\right\}^{\mathrm{an}}, \widetilde{S}^{\mathrm{an}}\right)$$

is an open immersion.

Proof. We already know, from Proposition 4.5.10, that the continuous map of topological spaces embeds $\left\{\operatorname{Spec}(S[1/f])\right\}^{\mathrm{an}}$ as an open subset of $\left\{\operatorname{Spec}(S)\right\}^{\mathrm{an}}$. We only need to check that the sheaves of rings agree. In other words we need to check that, for any open subset $V \subset \left\{\operatorname{Spec}(S[1/f])\right\}^{\mathrm{an}} \subset \left\{\operatorname{Spec}(S)\right\}^{\mathrm{an}}$, the map

$$\widetilde{\alpha}_f^{\mathrm{an}} : \Gamma\big(V, \widetilde{S}^{\mathrm{an}}\big) \longrightarrow \Gamma\big(V, \widetilde{S[1/f]}^{\mathrm{an}}\big)$$

is an isomorphism. It suffices to check this on a basis for the topology. Now the polydiscs in $\left\{\operatorname{Spec}(S)\right\}^{\mathrm{an}}$ form a basis for the topology of $\left\{\operatorname{Spec}(S)\right\}^{\mathrm{an}}$, and so the ones which happen to be contained in the open subset $\left\{\operatorname{Spec}(S[1/f])\right\}^{\mathrm{an}}$ form a basis for the topology of $\left\{\operatorname{Spec}(S[1/f])\right\}^{\mathrm{an}}$. Lemma 5.6.9 and Remark 5.6.10 tell us that, for such polydiscs V, the map $\alpha_f : [S, V] \longrightarrow \big[S[1/f], V\big]$ gives an isomorphism on the completions. But the map of completions is precisely $\widetilde{\alpha}_f^{\mathrm{an}}$. □

Remark 5.7.5. Let S be a finitely generated algebra over $\mathbb{C}$, and put

$$(X, \mathcal{O}_X) = \left(\operatorname{Spec}(S), \widetilde{S}\right) , \qquad (X^{\mathrm{an}}, \mathcal{O}_X^{\mathrm{an}}) = \left(\left\{\operatorname{Spec}(S)\right\}^{\mathrm{an}}, \widetilde{S}^{\mathrm{an}}\right) .$$

Let f be an element of S. By Lemma 5.7.2 we have a commutative square

$$\begin{array}{ccc} S & \xrightarrow{\alpha_f} & S[1/f] \\ {\scriptstyle\mu}\big\downarrow & & \big\downarrow{\scriptstyle\mu'} \\ \Gamma(X^{\mathrm{an}}, \widetilde{S}^{\mathrm{an}}) & \xrightarrow[\{\widetilde{\alpha}_f^{\mathrm{an}}\}_{X^{\mathrm{an}}}]{} & \Gamma(X_f^{\mathrm{an}}, \widetilde{S[1/f]}^{\mathrm{an}}) \end{array} .$$

For any open subset $V \subset X_f^{\mathrm{an}} = X_f \cap X^{\mathrm{an}}$ we have a further commutative square

$$\begin{array}{ccc} \Gamma(X^{\mathrm{an}}, \widetilde{S}^{\mathrm{an}}) & \xrightarrow{\{\widetilde{\alpha}_f^{\mathrm{an}}\}_{X^{\mathrm{an}}}} & \Gamma(X_f^{\mathrm{an}}, \widetilde{S[1/f]}^{\mathrm{an}}) \\ {\scriptstyle \mathrm{res}_V^{X^{\mathrm{an}}}}\big\downarrow & & \big\downarrow{\scriptstyle \mathrm{res}_V^{X_f^{\mathrm{an}}}} \\ \Gamma(V, \widetilde{S}^{\mathrm{an}}) & \xrightarrow[\{\widetilde{\alpha}_f^{\mathrm{an}}\}_V]{} & \Gamma(V, \widetilde{S[1/f]}^{\mathrm{an}}) \end{array} .$$

and the map $\{\widetilde{\alpha}_f^{\mathrm{an}}\}_V$ is an isomorphism by Lemma 5.7.4. We conclude that the map $\mu_V : S \longrightarrow \Gamma(V, \widetilde{S}^{\mathrm{an}})$ of Definition 5.5.5, which may be written as the composite

$$S \xrightarrow{\mu} \Gamma(X^{\mathrm{an}}, \widetilde{S}^{\mathrm{an}}) \xrightarrow{\mathrm{res}_V^{X^{\mathrm{an}}}} \Gamma(V, \widetilde{S}^{\mathrm{an}}) ,$$

can be factored through $\alpha_f : S \longrightarrow S[1/f]$. This means that f maps to an invertible element in $\Gamma(V, \widetilde{S}^{\mathrm{an}})$, and by the universal property of localization (see Proposition 3.3.2) the factorization is unique.

Remark 5.7.6. In Remark 5.7.5, consider the special case where $V \subset X_f^{\mathrm{an}}$ is a polydisc. Remark 5.7.5 produced for us some (unique) ring homomorphism $S[1/f] \longrightarrow \Gamma(V, \widetilde{S}^{\mathrm{an}})$. We assert that the image of $S[1/f]$ is dense, and that the topology induced on $S[1/f]$ agrees with $[S[1/f], V]$.

To see this, consider the commutative diagram

$$\begin{array}{ccc} S & \xrightarrow{\alpha_f} & S[1/f] \\ {\scriptstyle\mu}\big\downarrow & & \big\downarrow{\scriptstyle\mu'} \\ \Gamma(X^{\mathrm{an}}, \widetilde{S}^{\mathrm{an}}) & \xrightarrow[\{\widetilde{\alpha}_f^{\mathrm{an}}\}_{X^{\mathrm{an}}}]{} & \Gamma(X_f^{\mathrm{an}}, \widetilde{S[1/f]}^{\mathrm{an}}) \\ {\scriptstyle \mathrm{res}_V^{X^{\mathrm{an}}}}\big\downarrow & & \big\downarrow{\scriptstyle \mathrm{res}_V^{X_f^{\mathrm{an}}}} \\ \Gamma(V, \widetilde{S}^{\mathrm{an}}) & \xrightarrow[\{\widetilde{\alpha}_f^{\mathrm{an}}\}_V]{} & \Gamma(V, \widetilde{S[1/f]}^{\mathrm{an}}) \end{array} .$$

The map $\{\widetilde{\alpha}_f^{\mathrm{an}}\}_V$ is a continuous, bijective homomorphism of Fréchet spaces, hence a homeomorphism. By definition the composite $\mu'_V =$

$\mathrm{res}_V^{X_f^{\mathrm{an}}} \mu'$ is the homomorphism from $S[1/f]$ to its completion, with respect to the topology $\big[S[1/f], V\big]$. The composite

$$S[1/f] \xrightarrow{\ \mu'_V\ } \Gamma(V, \widetilde{S[1/f]}^{\mathrm{an}}) \xrightarrow{\ \{\widetilde{\alpha}_f^{\mathrm{an}}\}_V^{-1}\ } \Gamma(V, \widetilde{S}^{\mathrm{an}})$$

may be viewed, up to the homeomorphism $\{\widetilde{\alpha}_f^{\mathrm{an}}\}_V$, as the map to the same completion.

We will use this remark in the next section, where the case that will interest us most is where $V = X_f^{\mathrm{an}} = \big\{\operatorname{Spec}(S[1/f])\big\}^{\mathrm{an}}$ is the large polydisc of Example 5.6.2.

Caution 5.7.7. The reader should be cautioned at this point. An open subset $V \subset X_f^{\mathrm{an}} \subset X^{\mathrm{an}}$ may well be a polydisc when considered as a subset of X_f^{an}, without being a polydisc in X^{an}. For example: $X_f^{\mathrm{an}} \subset X_f^{\mathrm{an}}$ is clearly a polydisc in X_f^{an}, while it is almost never a polydisc in X^{an}. There are several way to see this, for example one can show that S is not dense in $\Gamma(X_f^{\mathrm{an}}, \mathcal{O}_X^{\mathrm{an}})$. We will see one explicit example of this phenomenon, with the seminorms and the completion of S worked out in detail, in Remark 5.8.29. But, using a little more complex analysis than we know, it is not difficult to show that this is what happens "most of the time".

Lemma 5.7.8. *Let S be a finitely generated $\mathbb{C}$–algebra. Put*

$$(X, \mathcal{O}_X) = \Big(\operatorname{Spec}(S), \widetilde{S}\Big)\,, \qquad (X^{\mathrm{an}}, \mathcal{O}_X^{\mathrm{an}}) = \Big(\big\{\operatorname{Spec}(S)\big\}^{\mathrm{an}}, \widetilde{S}^{\mathrm{an}}\Big)\,.$$

There is a unique map of ringed spaces

$$(\lambda_X, \lambda_X^*) : (X^{\mathrm{an}}, \mathcal{O}_X^{\mathrm{an}}) \longrightarrow (X, \mathcal{O}_X)$$

which on global sections gives the homomorphism

$$\mu : S \quad = \quad \Gamma(X, \mathcal{O}_X) \longrightarrow \Gamma(X^{\mathrm{an}}, \mathcal{O}_X^{\mathrm{an}})\,.$$

Proof. The continuous map $\lambda_X : X^{\mathrm{an}} \longrightarrow X$ was given to us in Notation 4.5.5. The assertion is about the existence and uniqueness of the map of sheaves of rings. Let X_f be a basic open subset of X. If there were a map of ringed spaces as above, the following square would have to commute

$$\begin{array}{ccccc}
S & = & \Gamma(X, \mathcal{O}_X) & \xrightarrow{\ \{\lambda_X^*\}_X\ } & \Gamma(X^{\mathrm{an}}, \mathcal{O}_X^{\mathrm{an}}) \\
{\scriptstyle \alpha_f}\big\downarrow & & {\scriptstyle \mathrm{res}_{X_f}^X}\big\downarrow & & \big\downarrow{\scriptstyle \mathrm{res}_{X_f^{\mathrm{an}}}^{X^{\mathrm{an}}}} \\
S[1/f] & = & \Gamma(X_f, \mathcal{O}_X) & \xrightarrow[\ \{\lambda_X^*\}_{X_f}\ ]{} & \Gamma(X_f^{\mathrm{an}}, \mathcal{O}_X^{\mathrm{an}})
\end{array}\quad.$$

In this diagram we are given the morphisms

$$\begin{array}{ccccc} S & = & \Gamma(X, \mathcal{O}_X) & \xrightarrow{\{\lambda_X^*\}_X} & \Gamma(X^{\mathrm{an}}, \mathcal{O}_X^{\mathrm{an}}) \\ {\scriptstyle \alpha_f}\downarrow & & {\scriptstyle \mathrm{res}^X_{X_f}}\downarrow & & \downarrow{\scriptstyle \mathrm{res}^{X^{\mathrm{an}}}_{X_f^{\mathrm{an}}}} \\ S[1/f] & = & \Gamma(X_f, \mathcal{O}_X) & & \Gamma(X_f^{\mathrm{an}}, \mathcal{O}_X^{\mathrm{an}}) \end{array} ,$$

the only question is the existence and uniqueness of $\{\lambda_X^*\}_{X_f}$. But the map $\{\lambda_X^*\}_X$ is assumed to be $\mu : S \longrightarrow \Gamma(X^{\mathrm{an}}, \mathcal{O}_X^{\mathrm{an}})$ and, by Remark 5.7.5, there is a unique factorization of the composite

$$\begin{array}{ccccc} S & = & \Gamma(X, \mathcal{O}_X) & \xrightarrow{\{\lambda_X^*\}_X} & \Gamma(X^{\mathrm{an}}, \mathcal{O}_X^{\mathrm{an}}) \\ & & & & \downarrow{\scriptstyle \mathrm{res}^{X^{\mathrm{an}}}_{X_f^{\mathrm{an}}}} \\ & & & & \Gamma(X_f^{\mathrm{an}}, \mathcal{O}_X^{\mathrm{an}}) \end{array}$$

through α_f. This proves the uniqueness and tells us how we must define the map $\{\lambda_X^*\}_{X_f}$. It remains to show that this unique possible definition works; that is, that it delivers a map of ringed spaces.

Now suppose we are given two basic open subsets $X_f \subset X_g \subset X$. We have a diagram

$$\begin{array}{ccccc} S & = & \Gamma(X, \mathcal{O}_X) & \xrightarrow{\{\lambda_X^*\}_X} & \Gamma(X^{\mathrm{an}}, \mathcal{O}_X^{\mathrm{an}}) \\ {\scriptstyle \alpha_g}\downarrow & & {\scriptstyle \mathrm{res}^X_{X_g}}\downarrow & & \downarrow{\scriptstyle \mathrm{res}^{X^{\mathrm{an}}}_{X_g^{\mathrm{an}}}} \\ S[1/g] & = & \Gamma(X_g, \mathcal{O}_X) & \xrightarrow[\{\lambda_X^*\}_{X_g}]{} & \Gamma(X_g^{\mathrm{an}}, \mathcal{O}_X^{\mathrm{an}}) \\ {\scriptstyle \alpha_f^g}\downarrow & & {\scriptstyle \mathrm{res}^{X_g}_{X_f}}\downarrow & & \downarrow{\scriptstyle \mathrm{res}^{X_g^{\mathrm{an}}}_{X_f^{\mathrm{an}}}} \\ S[1/f] & = & \Gamma(X_f, \mathcal{O}_X) & \xrightarrow[\{\lambda_X^*\}_{X_f}]{} & \Gamma(X_f^{\mathrm{an}}, \mathcal{O}_X^{\mathrm{an}}) \end{array} .$$

As in the proof of Proposition 3.6.5 we know the commutativity of two of the three squares in the diagram, and deduce that the three maps from top left to bottom right are equal. Hence, in the diagram

$$\begin{array}{ccccc} S & & & & \\ {\scriptstyle \alpha_g}\downarrow & & & & \\ S[1/g] & = & \Gamma(X_g, \mathcal{O}_X) & \xrightarrow{\{\lambda_X^*\}_{X_g}} & \Gamma(X_g^{\mathrm{an}}, \mathcal{O}_X^{\mathrm{an}}) \\ {\scriptstyle \alpha_f^g}\downarrow & & {\scriptstyle \mathrm{res}^{X_g}_{X_f}}\downarrow & & \downarrow{\scriptstyle \mathrm{res}^{X_g^{\mathrm{an}}}_{X_f^{\mathrm{an}}}} \\ S[1/f] & = & \Gamma(X_f, \mathcal{O}_X) & \xrightarrow[\{\lambda_X^*\}_{X_f}]{} & \Gamma(X_f^{\mathrm{an}}, \mathcal{O}_X^{\mathrm{an}}) \end{array} ,$$

we know that the composites from top left to bottom right agree. The two composites give two factorizations of a single ring homomorphism $S \longrightarrow \Gamma(X_f^{\text{an}}, \mathcal{O}_X^{\text{an}})$ through $\alpha_g : S \longrightarrow S[1/g]$; by the universal property of Propostition 3.3.2 they are equal. This means the square

$$\begin{array}{ccccc} S[1/g] & = & \Gamma(X_g, \mathcal{O}_X) & \xrightarrow{\{\lambda_X^*\}_{X_g}} & \Gamma(X_g^{\text{an}}, \mathcal{O}_X^{\text{an}}) \\ \alpha_f^g \big\downarrow & & \text{res}_{X_f}^{X_g} \big\downarrow & & \big\downarrow \text{res}_{X_f^{\text{an}}}^{X_g^{\text{an}}} \\ S[1/f] & = & \Gamma(X_f, \mathcal{O}_X) & \xrightarrow[\{\lambda_X^*\}_{X_f}]{} & \Gamma(X_f^{\text{an}}, \mathcal{O}_X^{\text{an}}) \end{array}$$

commutes. On the basic open sets in X we have defined the maps $\{\lambda_X^*\}_{X_f}$, and these are compatible with restrictions. As in the end of the proof of Proposition 3.6.5 there is a unique extension to all other open sets, giving a map of ringed spaces. □

Lemma 5.7.9. *Let $\varphi : S \longrightarrow S'$ be a homomorphism of finitely generated $\mathbb{C}$–algebras. Put*

$$(X, \mathcal{O}_X) = \left(\text{Spec}(S'), \widetilde{S'}\right), \qquad (X^{\text{an}}, \mathcal{O}_X^{\text{an}}) = \left(\left\{\text{Spec}(S')\right\}^{\text{an}}, \widetilde{S'}^{\text{an}}\right),$$
$$(Y, \mathcal{O}_Y) = \left(\text{Spec}(S), \widetilde{S}\right), \qquad (Y^{\text{an}}, \mathcal{O}_Y^{\text{an}}) = \left(\left\{\text{Spec}(S)\right\}^{\text{an}}, \widetilde{S}^{\text{an}}\right).$$

The following square of morphisms of ringed spaces commutes

$$\begin{array}{ccc} (X^{\text{an}}, \mathcal{O}_X^{\text{an}}) & \xrightarrow{\left(\left\{\text{Spec}(\varphi)\right\}^{\text{an}}, \widetilde{\varphi}^{\text{an}}\right)} & (Y^{\text{an}}, \mathcal{O}_Y^{\text{an}}) \\ (\lambda_X, \lambda_X^*) \big\downarrow & & \big\downarrow (\lambda_Y, \lambda_Y^*) \\ (X, \mathcal{O}_X) & \xrightarrow[(\text{Spec}(\varphi), \widetilde{\varphi})]{} & (Y, \mathcal{O}_Y) \end{array} .$$

Proof. The two composites give two morphisms of ringed spaces

$$(X^{\text{an}}, \mathcal{O}_X^{\text{an}}) \rightrightarrows (Y, \mathcal{O}_Y),$$

and we need to show them equal. The topological statement, that the two continuous maps $X^{\text{an}} \longrightarrow Y$ agree with each other, was proved already in Lemma 4.2.7. Given an open set $V \subset Y$ its inverse image, by either of the two (equal) composites, is an open subset $W \subset X^{\text{an}}$. We have two ring homomorphisms

$$\Gamma(V, \mathcal{O}_Y) \rightrightarrows \Gamma(W, \mathcal{O}_X^{\text{an}}),$$

and we need to show them equal. As always, it suffices to do so for V belonging to a basis of the open sets in Y; we will consider the basic open sets $Y_f \subset Y$.

Lemma 5.7.2 provides us with a commutative square of ring homomorphisms

$$\begin{array}{ccc} S & \xrightarrow{\varphi} & S' \\ \mu\Big\downarrow & & \Big\downarrow\mu' \\ \Gamma(Y^{\mathrm{an}}, \mathcal{O}_Y^{\mathrm{an}}) & \xrightarrow[\widetilde{\varphi}^{\mathrm{an}}_{Y^{\mathrm{an}}}]{} & \Gamma(X^{\mathrm{an}}, \mathcal{O}_X^{\mathrm{an}}) \end{array} .$$

Let f be an element of the ring S. Now $\left(\{\operatorname{Spec}(\varphi)\}^{\mathrm{an}}, \widetilde{\varphi}^{\mathrm{an}}\right)$ is a morphism of ringed spaces, and $Y_f^{\mathrm{an}} = Y_f \cap Y^{\mathrm{an}}$ is an open subset of the space Y^{an}. We deduce a second commutative square of ring homomorphisms

$$\begin{array}{ccc} \Gamma(Y^{\mathrm{an}}, \mathcal{O}_Y^{\mathrm{an}}) & \xrightarrow{\widetilde{\varphi}^{\mathrm{an}}_{Y^{\mathrm{an}}}} & \Gamma(X^{\mathrm{an}}, \mathcal{O}_X^{\mathrm{an}}) \\ \mathrm{res}^{Y^{\mathrm{an}}}_{Y_f^{\mathrm{an}}}\Big\downarrow & & \Big\downarrow\mathrm{res}^{X^{\mathrm{an}}}_{X^{\mathrm{an}}_{\varphi(f)}} \\ \Gamma(Y_f^{\mathrm{an}}, \mathcal{O}_Y^{\mathrm{an}}) & \xrightarrow[\widetilde{\varphi}^{\mathrm{an}}_{Y_f^{\mathrm{an}}}]{} & \Gamma(X^{\mathrm{an}}_{\varphi(f)}, \mathcal{O}_X^{\mathrm{an}}) \end{array} .$$

Combining the two produces a commutative square

$$\begin{array}{ccc} S & \xrightarrow{\varphi} & S' \\ \mu_{Y_f^{\mathrm{an}}}\Big\downarrow & & \Big\downarrow\mu'_{X^{\mathrm{an}}_{\varphi(f)}} \\ \Gamma(Y_f^{\mathrm{an}}, \mathcal{O}_Y^{\mathrm{an}}) & \xrightarrow[\widetilde{\varphi}^{\mathrm{an}}_{Y_f^{\mathrm{an}}}]{} & \Gamma(X^{\mathrm{an}}_{\varphi(f)}, \mathcal{O}_X^{\mathrm{an}}) \end{array} .$$

The construction of λ_X^* and λ_Y^*, given in the proof of Lemma 5.7.8, was to factor each of the vertical maps in this commutative square as follows:

$$\begin{array}{ccccccc} & & S & \xrightarrow{\varphi} & S' & & \\ & & \alpha_f\Big\downarrow & & \Big\downarrow\alpha_{\varphi(f)} & & \\ \Gamma(Y_f, \mathcal{O}_Y) & = & S[1/f] & & S'[1/\varphi(f)] & = & \Gamma(X_{\varphi(f)}, \mathcal{O}_X) \\ & & \{\lambda_Y^*\}_{Y_f}\Big\downarrow & & \Big\downarrow\{\lambda_X^*\}_{X_{\varphi(f)}} & & \\ & & \Gamma(Y_f^{\mathrm{an}}, \mathcal{O}_Y^{\mathrm{an}}) & \xrightarrow[\widetilde{\varphi}^{\mathrm{an}}_{Y_f^{\mathrm{an}}}]{} & \Gamma(X^{\mathrm{an}}_{\varphi(f)}, \mathcal{O}_X^{\mathrm{an}}) & & \end{array} .$$

Of course the map $\widetilde{\varphi}_{Y_f} : \Gamma(Y_f, \mathcal{O}_Y) \longrightarrow \Gamma(X_{\varphi(f)}, \mathcal{O}_X)$, which comes from the morphism of ringed spaces $(\operatorname{Spec}(\varphi), \widetilde{\varphi})$, renders commutative the square

$$\begin{array}{ccccccc}
\Gamma(Y,\mathcal{O}_Y) & = & S & \xrightarrow{\varphi=\widetilde{\varphi}_Y} & S' & = & \Gamma(X,\mathcal{O}_X) \\
 & & {\scriptstyle\alpha_f}\downarrow & & \downarrow{\scriptstyle\alpha_{\varphi(f)}} & & \\
\Gamma(Y_f,\mathcal{O}_Y) & = & S[1/f] & \xrightarrow[\widetilde{\varphi}_{Y_f}]{} & S'[1/\varphi(f)] & = & \Gamma(X_{\varphi(f)},\mathcal{O}_X)
\end{array} .$$

Once again we find ourselves with a diagram

$$\begin{array}{ccccccc}
 & & S & \xrightarrow{\varphi=\widetilde{\varphi}_Y} & S' & & \\
 & & {\scriptstyle\alpha_f}\downarrow & & \downarrow{\scriptstyle\alpha_{\varphi(f)}} & & \\
\Gamma(Y_f,\mathcal{O}_Y) & = & S[1/f] & \xrightarrow[\widetilde{\varphi}_{Y_f}]{} & S'[1/\varphi(f)] & = & \Gamma(X_{\varphi(f)},\mathcal{O}_X) \\
 & & {\scriptstyle\{\lambda_Y^*\}_{Y_f}}\downarrow & & \downarrow{\scriptstyle\{\lambda_X^*\}_{X_{\varphi(f)}}} & & \\
 & & \Gamma(Y_f^{\mathrm{an}},\mathcal{O}_Y^{\mathrm{an}}) & \xrightarrow[\widetilde{\varphi}^{\mathrm{an}}_{Y_f^{\mathrm{an}}}]{} & \Gamma(X_{\varphi(f)}^{\mathrm{an}},\mathcal{O}_X^{\mathrm{an}}) & &
\end{array}$$

in which we know that two of the squares are commutative, and wish to show the commutativity of the third. The three maps from top left to bottom right agree, and hence the composites

$$\begin{array}{ccccccc}
 & & S & & & & \\
 & & {\scriptstyle\alpha_f}\downarrow & & & & \\
\Gamma(Y_f,\mathcal{O}_Y) & = & S[1/f] & \xrightarrow{\widetilde{\varphi}_{Y_f}} & S'[1/\varphi(f)] & = & \Gamma(X_{\varphi(f)},\mathcal{O}_X) \\
 & & {\scriptstyle\{\lambda_Y^*\}_{Y_f}}\downarrow & & \downarrow{\scriptstyle\{\lambda_X^*\}_{X_{\varphi(f)}}} & & \\
 & & \Gamma(Y_f^{\mathrm{an}},\mathcal{O}_Y^{\mathrm{an}}) & \xrightarrow[\widetilde{\varphi}^{\mathrm{an}}_{Y_f^{\mathrm{an}}}]{} & \Gamma(X_{\varphi(f)}^{\mathrm{an}},\mathcal{O}_X^{\mathrm{an}}) & &
\end{array}$$

are equal. We have two factorizations of the same ring homomorphism through $\alpha_f : S \longrightarrow S[1/f]$. The factorizations must agree. The square

$$\begin{array}{ccc}
\Gamma(Y_f,\mathcal{O}_Y) & \xrightarrow{\widetilde{\varphi}_{Y_f}} & \Gamma(X_{\varphi(f)},\mathcal{O}_X) \\
{\scriptstyle\{\lambda_Y^*\}_{Y_f}}\downarrow & & \downarrow{\scriptstyle\{\lambda_X^*\}_{X_{\varphi(f)}}} \\
\Gamma(Y_f^{\mathrm{an}},\mathcal{O}_Y^{\mathrm{an}}) & \xrightarrow[\widetilde{\varphi}^{\mathrm{an}}_{Y_f^{\mathrm{an}}}]{} & \Gamma(X_{\varphi(f)}^{\mathrm{an}},\mathcal{O}_X^{\mathrm{an}})
\end{array}$$

therefore commutes. We have shown, as required, that the two composites

$$\Gamma(Y_f,\mathcal{O}_Y) \rightrightarrows \Gamma(X_{\varphi(f)}^{\mathrm{an}},\mathcal{O}_X^{\mathrm{an}})$$

are equal. □

Summary 5.7.10. At this point we have proved all the assertions of Section 5.1 in the case where $(X, \mathcal{O}_X)$ is an affine scheme $\left(\operatorname{Spec}(S), \widetilde{S}\right)$. We have constructed the ringed space $\left(\left\{\operatorname{Spec}(S)\right\}^{\text{an}}, \widetilde{S}^{\text{an}}\right)$, we have shown that morphisms of affine schemes over $\mathbb{C}$, which correspond to ring homomorphisms $\varphi : S \longrightarrow S'$, induce maps of analytic spaces, and we have produced the map

$$(\lambda_X, \lambda_X^*) : (X, \mathcal{O}_X) \longrightarrow (X^{\text{an}}, \mathcal{O}_X^{\text{an}})$$

and shown that it is natural in $(X, \mathcal{O}_X)$. We even showed that the ring homomorphism $\alpha_f : S \longrightarrow S[1/f]$ induces an open immersion of analytic spaces.

5.8 In the world of elementary examples

We now know that, if S is a finitely generated algebra over $\mathbb{C}$, then there is a way to canonically construct an analytic space $\left(\left\{\operatorname{Spec}(S)\right\}^{\text{an}}, \widetilde{S}^{\text{an}}\right)$. Furthermore, this space has all the wonderful properties we listed in Summary 5.7.10. In particular, we have a canonically defined sheaf of rings $\mathcal{O}^{\text{an}} = \widetilde{S}^{\text{an}}$ over $\left\{\operatorname{Spec}(S)\right\}^{\text{an}}$. For polydiscs $V \subset \left\{\operatorname{Spec}(S)\right\}^{\text{an}}$, the highbrow description is that $\Gamma(V, \mathcal{O}^{\text{an}})$ is the completion of the ring $S = [S, V]$ with respect to its pre-Fréchet topology.

If we actually want to compute what this sheaf is, in concrete examples, it is helpful to go back to the more pedestrian description in terms of generators. The theory tells us that

(i) The polydiscs $V \subset \left\{\operatorname{Spec}(S)\right\}^{\text{an}}$ are independent of the choice of generators (Proposition 5.6.4(i)).
(ii) If V is a polydisc in $\left\{\operatorname{Spec}(S)\right\}^{\text{an}}$, then the ring $\Gamma(V, \mathcal{O}^{\text{an}})$ is independent of the choice of generators (Proposition 5.6.4(ii)).

This is good to know; it means we are free to choose, in the computations, our very favorite generators.

Notation 5.8.1. In the remainder of this section S will be the polynomial ring $S = \mathbb{C}[x_1, x_2, \cdots, x_n]$. The ringed space $\left(\operatorname{Spec}(S), \widetilde{S}\right)$ will be denoted $(X, \mathcal{O})$, and its analytification $\left(\left\{\operatorname{Spec}(S)\right\}^{\text{an}}, \widetilde{S}^{\text{an}}\right)$ will be written $(X^{\text{an}}, \mathcal{O}^{\text{an}})$. We will frequently appeal to the fact that X^{an} may be identified with $\mathbb{C}^n$. Thus, we will use interchangeably the notation $(X^{\text{an}}, \mathcal{O}^{\text{an}}) = (\mathbb{C}^n, \mathcal{O}^{\text{an}})$.

Lemma 5.8.2. *We adopt the conventions of Notation 5.8.1. Then* $\mathcal{O}^{\mathrm{an}} = \widetilde{S}^{\mathrm{an}}$ *is the sheaf of holomorphic functions on* $\big\{\operatorname{Spec}(S)\big\}^{\mathrm{an}} = \mathbb{C}^n$.

Proof. We are free to choose any generators we wish for the ring S; we choose the set of generators $\{x_1, x_2, \cdots, x_n\} \subset S = \mathbb{C}[x_1, x_2, \cdots, x_n]$. With this choice of generators, we now go through the lowbrow construction of the sheaf $\mathcal{O}^{\mathrm{an}}$, as in Section 5.3.

Notation 5.3.2 gives us a ring homomorphism $\theta : R \longrightarrow S$, which in our case is the identity

$$1 : \mathbb{C}[x_1, x_2, \cdots, x_n] \longrightarrow \mathbb{C}[x_1, x_2, \cdots, x_n] \,.$$

This makes the ideal I, the kernel of θ, into the zero ideal $I = \{0\}$. Next, in Theorem 5.3.11, we produced a sheaf $\mathcal{O}^{\mathrm{an}}$, defined on all of $\mathbb{C}^n$, and a map of sheaves $\mathcal{O}^{\mathrm{an}}_{\mathbb{C}^n} \longrightarrow \mathcal{O}^{\mathrm{an}}$. On the polydisc $\Delta(g; w; r)$ the formula for this map is

$$\Gamma\big(\Delta(g; w; r), \mathcal{O}^{\mathrm{an}}_{\mathbb{C}^n}\big) \longrightarrow \frac{\Gamma\big(\Delta(g; w; r), \mathcal{O}^{\mathrm{an}}_{\mathbb{C}^n}\big)}{I \cdot \Gamma\big(\Delta(g; w; r), \mathcal{O}^{\mathrm{an}}_{\mathbb{C}^n}\big)} = \Gamma\big(\Delta(g; w; r), \mathcal{O}^{\mathrm{an}}\big) \,.$$

Since $I = 0$, this simplifies to an isomorphism

$$\Gamma\big(\Delta(g; w; r), \mathcal{O}^{\mathrm{an}}_{\mathbb{C}^n}\big) \quad = \quad \Gamma\big(\Delta(g; w; r), \mathcal{O}^{\mathrm{an}}\big) \,.$$

This map is an isomorphism on all polydiscs $\Delta(g; w; r)$, and the polydiscs form a basis for the topology by Lemma 5.3.9. Hence the map is an isomorphism of sheaves on $\mathbb{C}^n$. The sheaf $\mathcal{O}^{\mathrm{an}}$ agrees with the sheaf $\mathcal{O}^{\mathrm{an}}_{\mathbb{C}^n}$, of holomorphic functions on $\mathbb{C}^n$.

The next part of the recipe, in Section 5.3, was to turn $\mathcal{O}^{\mathrm{an}}$ into a sheaf supported on $\big\{\operatorname{Spec}(S)\big\}^{\mathrm{an}} \subset \mathbb{C}^n$. In our case we have $\big\{\operatorname{Spec}(S)\big\}^{\mathrm{an}} = \mathbb{C}^n$, making the complement $A = \mathbb{C}^n - \big\{\operatorname{Spec}(S)\big\}^{\mathrm{an}}$ empty. Construction 5.3.17 defined, for any open set $U \subset \big\{\operatorname{Spec}(S)\big\}^{\mathrm{an}}$,

$$\Gamma(U, \mathcal{O}^{\mathrm{an}}) \quad = \quad \Gamma(A \cup U, \mathcal{O}^{\mathrm{an}}) \,,$$

where on the left we understand by $\mathcal{O}^{\mathrm{an}}$ the sheaf of rings on $\big\{\operatorname{Spec}(S)\big\}^{\mathrm{an}}$, while on the right $\mathcal{O}^{\mathrm{an}}$ is the sheaf of rings on the (possibly) larger $\mathbb{C}^n$. Since A is empty the sheaves agree on all open sets U. The sheaf $\mathcal{O}^{\mathrm{an}}$, on $\big\{\operatorname{Spec}(S)\big\}^{\mathrm{an}}$, is simply the sheaf of holomorphic functions on $\mathbb{C}^n$. $\square$

Lemma 5.8.3. *We keep the conventions of Notation 5.8.1. The ring homomorphism* $\mu : S \longrightarrow \Gamma(X^{\mathrm{an}}, \mathcal{O}^{\mathrm{an}})$, *of Lemma 5.5.4, is the inclusion of the polynomials into the holomorphic functions on* $\mathbb{C}^n$.

Proof. Once again this is a matter of tracing through the definitions, on this occasion the ones at the beginning of Section 5.5. As in the proof of Lemma 5.8.2, the generators for the ring $S = \mathbb{C}[x_1, x_2, \cdots, x_n]$ will be $\{x_1, x_2, \cdots, x_n\} \subset S$, and the induced map $\theta : R \longrightarrow S$ is then the identity. In Lemma 5.5.3 the map $\widetilde{\mu} : R \longrightarrow \Gamma(X^{\mathrm{an}}, \mathcal{O}^{\mathrm{an}})$ was defined to be the inclusion of the polynomials into the holomorphic functions. In Lemma 5.5.4, the map $\mu : S \longrightarrow \Gamma(X^{\mathrm{an}}, \mathcal{O}^{\mathrm{an}})$ is defined to be the unique homomorphism, giving a factorization of $\widetilde{\mu} : R \longrightarrow \Gamma(X^{\mathrm{an}}, \mathcal{O}^{\mathrm{an}})$ as the composite

$$R \xrightarrow{\quad\theta\quad} S \xrightarrow{\quad\mu\quad} \Gamma(X^{\mathrm{an}}, \mathcal{O}^{\mathrm{an}}) \ .$$

In our case θ is the identity, making μ the inclusion of the polynomials in the holomorphic functions. □

Remark 5.8.4. Let f be an element in $S = \mathbb{C}[x_1, x_2, \cdots, x_n]$. Remark 5.7.5, in the special case where $V = X_f^{\mathrm{an}}$, gives us a commutative square

$$\begin{array}{ccc} S & \xrightarrow{\qquad\alpha_f\qquad} & S[1/f] \\ {\scriptstyle\mu}\Big\downarrow & & \Big\downarrow{\scriptstyle\nu} \\ \Gamma(X^{\mathrm{an}}, \mathcal{O}^{\mathrm{an}}) & \xrightarrow[\mathrm{res}^{X^{\mathrm{an}}}_{X_f^{\mathrm{an}}}]{} & \Gamma(X_f^{\mathrm{an}}, \mathcal{O}^{\mathrm{an}}) \end{array} \ .$$

Remark 5.7.6 tells us that the image of ν is dense, and that the topology induced on $S[1/f]$ is $\big[S[1/f], X_f^{\mathrm{an}}\big]$. Now X^{an} was identified with $\mathbb{C}^n$ in Lemma 4.4.1, and under this identification $X_f^{\mathrm{an}} = X_f \cap X^{\mathrm{an}}$ becomes $\mathbb{C}^n_f \subset \mathbb{C}^n$, as we showed in Lemma 4.4.3. Lemma 5.8.2 tells us that $\mathcal{O}^{\mathrm{an}}$ is the sheaf of holomorphic functions on $\mathbb{C}^n$.

All of this means:

Lemma 5.8.5. *Pick an element $f \in S$. Then $\Gamma(X_f^{\mathrm{an}}, \mathcal{O}^{\mathrm{an}}) = \Gamma(\mathbb{C}^n_f, \mathcal{O}^{\mathrm{an}})$ is the ring of holomorphic functions on the open set $\mathbb{C}^n_f \subset \mathbb{C}^n$. With these identifications, the map $\nu : S[1/f] \longrightarrow \Gamma(\mathbb{C}^n_f, \mathcal{O}^{\mathrm{an}})$ is the homomorphism from $S[1/f]$ to its completion with respect to the topology $\big[S[1/f], X_f^{\mathrm{an}}\big] = \big[S[1/f], \{\,\mathrm{Spec}(S[1/f])\}^{\mathrm{an}}\big]$.* □

Notation 5.8.6. The f's which we wish to study are the ones of Notation 3.5.3. We remind the reader: given a subset $J \subset \{1, 2, \ldots n\}$, we let $f_J = \prod_{j \in J} x_j$. We will now agree on more notation, valid for the rest of this section:

(i) We let S_J be the Laurent polynomial ring of Notation 3.5.3(ii). In Exercise 3.5.4(i) we constructed an isomorphism $\varphi_J : S[1/f_J] \longrightarrow S_J$.

(ii) We let $X_J \subset X$ be the open set $X_J = X_{f_J}$.

(iii) X_J^{an} will be the open set $X_J \cap X^{\mathrm{an}} \subset X^{\mathrm{an}}$. It is identified with the open subset $\mathbb{C}^n_{f_J} \subset \mathbb{C}^n$.

(iv) The map $\nu : S[1/f_J] \longrightarrow \Gamma(\mathbb{C}^n_{f_J}, \mathcal{O}^{\mathrm{an}})$, of Remark 5.8.4 and Lemma 5.8.5, factors through the isomorphism φ_J in (i). We will write the factorization as

$$S[1/f_J] \xrightarrow{\varphi_J} S_J \xrightarrow{\mu_J} \Gamma(X_J^{\mathrm{an}}, \mathcal{O}^{\mathrm{an}}) .$$

Remark 5.8.7. The set $X_J^{\mathrm{an}} = \mathbb{C}^n_{f_J}$ is the set of all $(z_1, z_2, \cdots, z_n) \in \mathbb{C}^n$ on which $f_J = \prod_{j \in J} x_j$ does not vanish. Rewriting this,

$$X_J^{\mathrm{an}} \quad = \quad \{(z_1, z_2, \cdots, z_n) \in \mathbb{C}^n \mid z_i \neq 0 \text{ if } i \in J\} .$$

Another way to write it is

$$X_J^{\mathrm{an}} = \prod_{i=1}^n U_i , \qquad \text{where} \quad U_i = \begin{cases} \mathbb{C} & \text{if } i \notin J, \\ \mathbb{C}^* = \mathbb{C} - \{0\} & \text{if } i \in J. \end{cases}$$

Lemma 5.8.5 establishes that the ring homomorphism $\mu_J : S_J \longrightarrow \Gamma(X_J^{\mathrm{an}}, \mathcal{O}^{\mathrm{an}})$ has a dense image. Next let us recall that we know the ring S_J; we studied it beginning with Exercise 3.3.5 and Remark 3.3.6, and then, more extensively, in Section 3.5. We remind the reader:

Reminder 5.8.8. Let J be the set $J = \{j_1, j_2, \ldots, j_\ell\} \subset \{1, 2, \ldots, n\}$. The ring

$$S_J \quad = \quad \mathbb{C}[x_1, x_2, \ldots, x_n, x_{j_1}^{-1}, x_{j_2}^{-1}, \ldots, x_{j_\ell}^{-1}]$$

has a basis, as a vector space over $\mathbb{C}$, consisting of the monomials

$$x_1^{a_1} x_2^{a_2} \cdots x_n^{a_n}$$

where $a_i \geq 0$ if $i \notin J$, while $a_i \in \mathbb{Z}$ is arbitrary if $i \in J$. This means, of course, that the elements of S_J can be expanded, uniquely, as finite sums

$$\sum \lambda_{(a_1, a_2, \ldots, a_n)} x_1^{a_1} x_2^{a_2} \cdots x_n^{a_n} .$$

We need a notation, to be able to cope with such expressions. We therefore introduce:

Notation 5.8.9. We will let boldface letters stand for vectors with n entries. More precisely:

(i) Letters late in the alphabet, such as $\mathbf{x}$, $\mathbf{y}$ and $\mathbf{z}$, will stand for vectors in $\mathbb{C}^n$; thus $\mathbf{x} = (x_1, x_2, \ldots, x_n)$ and $\mathbf{y} = (y_1, y_2, \ldots, y_n)$ will be points in $\mathbb{C}^n$.

(ii) Letters early in the alphabet, like $\mathbf{a}$ and $\mathbf{b}$, will stand for vectors of integers. Thus $\mathbf{a} = (a_1, a_2, \ldots, a_n)$ will have $a_i \in \mathbb{Z}$.

(iii) We will also reserve the notation $\mathbf{1} = (1, 1, \ldots, 1)$ for the vector in $\mathbb{Z}^n$, all of whose coordinates are 1.

(iv) For a vector $\mathbf{a} = (a_1, a_2, \ldots, a_n)$ in $\mathbb{Z}^n$, we will denote by $|\mathbf{a}|$ the integer

$$|\mathbf{a}| \quad = \quad |a_1| + |a_2| + \cdots + |a_n| \ .$$

(v) When we write $\mathbf{x}^{\mathbf{a}}$ we will mean the monomial

$$x_1^{a_1} x_2^{a_2} \cdots x_n^{a_n} \ .$$

Small variants should be self-explanatory; for example, $\mathbf{x}^{\mathbf{a}+\mathbf{1}}$ will stand for

$$x_1^{a_1+1} x_2^{a_2+1} \cdots x_n^{a_n+1} \ .$$

With this notation, the sum in Reminder 5.8.8 becomes more manageable to write; in the form $\sum \lambda_{\mathbf{a}} \mathbf{x}^{\mathbf{a}}$ it is far more compact than the right hand side in

$$\sum \lambda_{\mathbf{a}} \mathbf{x}^{\mathbf{a}} \quad = \quad \sum \lambda_{(a_1, a_2, \ldots, a_n)} x_1^{a_1} x_2^{a_2} \cdots x_n^{a_n} \ .$$

For this sum to lie in S_J it must be finite, and we must have $\lambda_{\mathbf{a}} = 0$ unless $\mathbf{x}^{\mathbf{a}} \in S_J$. The monomial $\mathbf{x}^{\mathbf{a}}$ lies in S_J if, in the vector $\mathbf{a} = (a_1, a_2, \ldots, a_n)$, we have $a_i \geq 0$ whenever $i \notin J$.

Remark 5.8.10. By Lemma 5.8.5 we know that the map $\mu_J : S_J \longrightarrow \Gamma(X_J^{\mathrm{an}}, \mathcal{O}^{\mathrm{an}})$ has a dense image. In other words, the ring $\Gamma(X_J^{\mathrm{an}}, \mathcal{O}^{\mathrm{an}})$ can be viewed as the completion of $[S_J, X_J^{\mathrm{an}}]$. The elements of S_J are the finite linear combinations $\sum_{\mathbf{x}^{\mathbf{a}} \in S_J} \lambda_{\mathbf{a}} \mathbf{x}^{\mathbf{a}}$. It turns out that the elements in the completion $\Gamma(X_J^{\mathrm{an}}, \mathcal{O}^{\mathrm{an}})$ are certain infinite linear combinations

$$\sum_{\mathbf{x}^{\mathbf{a}} \in S_J} \lambda_{\mathbf{a}} \mathbf{x}^{\mathbf{a}} \ ;$$

the requirement is that the series converge. In the remainder of this section we propose to go through this very explicitly, to remind the reader of the convergence properties of such series. As a byproduct, we will learn about the topology of the ring $S_J = [S_J, X_J^{\mathrm{an}}]$.

Remark 5.8.11. Remember that J can be any subset $J \subset \{1, 2, \ldots, n\}$. If $J \subset J' \subset \{1, 2, \ldots, n\}$, then there is an inclusion map $S_J \longrightarrow S_{J'}$. The minimal S_J occurs when $J = \emptyset$; in that case $S_J = S$. The maximal S_J is when $J = \{1, 2, \ldots, n\}$.

An inclusion of subsets $J \subset J' \subset \{1, 2, \ldots, n\}$ gives a reverse inclusion of open sets $X_{J'}^{\mathrm{an}} \subset X_J^{\mathrm{an}}$. Hence the holomorphic functions on X_J^{an} restrict to the holomorphic functions on $X_{J'}^{\mathrm{an}}$. It seems sensible to start by studying the holomorphic functions on the minimal X_J^{an}, and then analyze which ones are restrictions of holomorphic functions on larger X_J^{an}. The minimal X_J^{an}, which occurs when $J = \{1, 2, \ldots, n\}$, is the open set

$$X_{\{1,2,\ldots,n\}}^{\text{an}} = (\mathbb{C}^*)^n \subset \mathbb{C}^n ;$$

see Remark 5.8.7.

As suggested in Remark 5.8.11, we will remind the reader of the theory of holomorphic functions on $(\mathbb{C}^*)^n$. We begin with the case $n = 1$.

Reminder 5.8.12. Let f be a holomorphic function on $\mathbb{C}^* = \mathbb{C} - \{0\}$. Then f can be written as a convergent Laurent series

$$f(z) = \sum_{j=-\infty}^{\infty} a_i z^i .$$

Let us briefly recall the proof. The Cauchy integral formula says that

$$f(\zeta) = \frac{1}{2\pi i} \int_\gamma \frac{f(z)dz}{z - \zeta} ,$$

where γ can be taken as $\gamma = \gamma' \cup \gamma''$, with γ' and γ'' the circles below

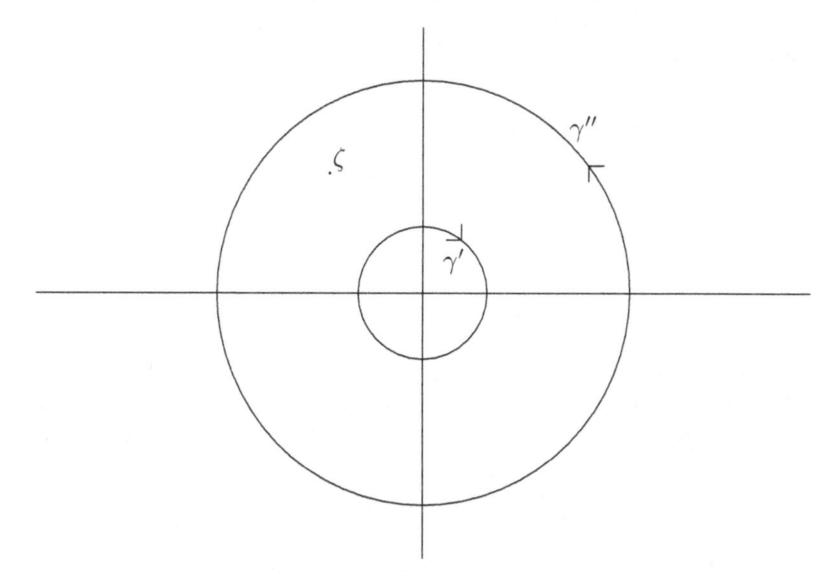

That is, ζ lies between the two circles, and we trace the outer circle γ'' counterclockwise, the inner circle γ' clockwise. If z lies on the circle γ' we have that $|z/\zeta| < 1$, while for z along γ'' the inequality to use is

$|\zeta/z| < 1$. The integral becomes

$$\begin{aligned} f(\zeta) &= \frac{1}{2\pi i}\int_{\gamma''}\frac{f(z)dz}{z-\zeta} + \frac{1}{2\pi i}\int_{\gamma'}\frac{f(z)dz}{z-\zeta} \\ &= \frac{1}{2\pi i}\int_{\gamma''}\frac{f(z)dz}{z(1-\zeta/z)} - \frac{1}{2\pi i}\int_{\gamma'}\frac{f(z)dz}{\zeta(1-z/\zeta)} \ . \end{aligned}$$

Now on γ'' we can expand

$$\frac{1}{z(1-\zeta/z)} = \frac{1}{z} + \frac{\zeta}{z^2} + \frac{\zeta^2}{z^3} + \frac{\zeta^3}{z^4} + \cdots$$

while on γ' we have

$$\frac{1}{\zeta(1-z/\zeta)} = \frac{1}{\zeta} + \frac{z}{\zeta^2} + \frac{z^2}{\zeta^3} + \frac{z^3}{\zeta^4} + \cdots$$

Expanding under the integral sign, and interchanging the order of summation and integration, we have

$$f(\zeta) = \frac{1}{2\pi i}\sum_{i=0}^{\infty}\zeta^i\int_{\gamma''}\frac{f(z)dz}{z^{i+1}} - \frac{1}{2\pi i}\sum_{i=-\infty}^{-1}\zeta^i\int_{\gamma'}\frac{f(z)dz}{z^{i+1}} \ .$$

This produces for us a Laurent series, converging to $f(\zeta)$, for ζ in the annulus between the circles γ' and γ''. For ζ in this annulus, we have produced an identity

$$f(\zeta) = \sum_{i=-\infty}^{\infty} a_i\zeta^i \ ,$$

where a_i is given by the formula

$$a_i = \begin{cases} \frac{1}{2\pi i}\int_{\gamma''}\frac{f(z)dz}{z^{i+1}} & \text{if } i \geq 0, \\ -\frac{1}{2\pi i}\int_{\gamma'}\frac{f(z)dz}{z^{i+1}} & \text{if } i < 0. \end{cases}$$

Now the curves γ' and γ'' have opposite orientations, and the function $f(z)/z^{i+1}$ has no singularity in the region between γ' and γ''. Taking the contour integral of $f(z)/z^{i+1}$ around this annulus, we obtain an equality

$$\int_{\gamma'}\frac{f(z)dz}{z^{i+1}} + \int_{\gamma''}\frac{f(z)dz}{z^{i+1}} = 0 \ .$$

The formula for the coefficients a_i can therefore be rewritten, more compactly, as

$$a_i = \int_{\gamma''}\frac{f(z)dz}{z^{i+1}} \ .$$

This argument also shows that the a_i are independent of the circles γ' and γ'' we happened to choose. Given two circles with the same orientation γ'' and $\widetilde{\gamma}''$, we have an identity

$$\int_{\gamma''} \frac{f(z)dz}{z^{i+1}} = \int_{\widetilde{\gamma}''} \frac{f(z)dz}{z^{i+1}} .$$

So far we have shown that there is a Laurent series, converging to $f(\zeta)$, for ζ in the annulus between the circles γ' and γ''. We have also shown that the coefficients a_i in this Laurent series are independent of γ' and γ''; by varying γ' and γ'', the reader can show that the series converges to f on all of $\mathbb{C}^*$.

This reminded us of the one-dimensional case. A very similar result holds in higher dimensions.

Exercise 5.8.13. We use the conventions of Notation 5.8.9. Let f be a holomorphic function on $(\mathbb{C}^*)^n$. The reader should show that f can be written as a convergent Laurent series

$$f(\mathbf{x}) = \sum_{\mathbf{a}\in\mathbb{Z}^n} \lambda_{\mathbf{a}} \mathbf{x}^{\mathbf{a}} ,$$

where

$$\lambda_{\mathbf{a}} = \frac{1}{(2\pi i)^n} \int_{\gamma_1''} \int_{\gamma_2''} \cdots \int_{\gamma_n''} \frac{f(\mathbf{z})dz_1 dz_2 \cdots dz_n}{\mathbf{z}^{\mathbf{a}+1}} ,$$

and where the γ_i'' are circles $|z_i| = r_i$. The hint is that, applying the ordinary, 1–dimensional Cauchy integral formula to each of the n variables, we obtain an identity

$$f(\mathbf{x}) = \frac{1}{(2\pi i)^n} \int_{\gamma_1} \int_{\gamma_2} \cdots \int_{\gamma_n} \frac{f(\mathbf{z})dz_1 dz_2 \cdots dz_n}{(z_1 - x_1)(z_2 - x_2) \cdots (z_n - x_n)} ,$$

where each γ_i is the union of two circles $\gamma_i = \gamma_i' \cup \gamma_i''$, just as in Reminder 5.8.12. Further, also as in Reminder 5.8.12, along the curves γ_i' (respectively γ_i'') we expand the expression $dz_i/(z_i - x_i)$ as a power series in z_i/x_i (respectively x_i/z_i).

We now know that, if f is holomorphic on $(\mathbb{C}^*)^n$, then f has a convergent Laurent series expansion. The fact that the series converges places restrictions on the coefficients $\lambda_{\mathbf{a}}$; they must decrease rapidly enough. Next we remind the reader how this goes. The way the estimates work is that, if we know that a Laurent series converges on a certain compact set, then we can bound the coefficients in terms of the size of the compact set. To make this precise we introduce the compact sets we will care about. Thinking ahead, we introduce all the compact sets we will consider in this section.

Notation 5.8.14. Let $B \geq 1$ be a real number. For each i in the range $1 \leq i \leq n$, consider the sets

(i) The disc $\Delta_i(B)$, of radius B, will be

$$\Delta_i(B) \quad = \quad \{x_i \in \mathbb{C} \mid |x_i| \leq B\} \ .$$

(ii) The annulus $A_i(B)$, of radius B, will be

$$A_i(B) \quad = \quad \{x_i \in \mathbb{C} \mid B^{-1} \leq |x_i| \leq B\} \ .$$

With this notation we define, for any $J \subset \{1, 2, \ldots, n\}$ and any $B \geq 1$, a compact region $K_J(B) \subset \mathbb{C}^n$ by the formula

$$K_J(B) = \prod_{i=1}^{n} K_i(B) \qquad \text{where} \qquad K_i(B) = \begin{cases} \Delta_i(B) & \text{if } i \notin J, \\ A_i(B) & \text{if } i \in J. \end{cases}$$

Remark 5.8.15. The increasing unions below clearly satisfy

$$\bigcup_{B \geq 1} \Delta_i(B) = \mathbb{C}, \qquad \bigcup_{B \geq 1} A_i(B) = \mathbb{C}^*.$$

Taking the product of these identities over i, more precisely taking the first identity if $i \notin J$ and the second if $i \in J$, we deduce

$$X_J^{\text{an}} \quad = \quad \bigcup_{B \geq 1} K_J(B) \ .$$

Here, X_J^{an} is as in Notation 5.8.6(iii) and Remark 5.8.7; it is a product of some copies of $\mathbb{C}$ with some copies of $\mathbb{C}^*$, indexed by J.

Note also that, if $J \subset J' \subset \{1, 2, \ldots, n\}$, then $K_{J'}(B) \subset K_J(B)$. The minimal $K_J(B)$ occurs when $J = \{1, 2, \ldots, n\}$; we will denote it $K^{\min}(B)$. We have that $K^{\min}(B)$ is the product of the annuli $A_i(B)$; it is the set of all

$$\mathbf{x} = (x_1, x_2, \cdots, x_n) \in \mathbb{C}^n \quad \text{satisfying} \quad B^{-1} \leq |x_i| \leq B \ .$$

Now we are ready to state the estimate, on the decay of the coefficients of a convergent Laurent series.

Lemma 5.8.16. *Suppose $B \geq 1$ is a real number, and suppose that we have a Laurent series*

$$\sum_{\mathbf{a} \in \mathbb{Z}^n} \lambda_{\mathbf{a}} \mathbf{x}^{\mathbf{a}} \ ,$$

which converges on the region $K^{\min}(B)$. Then there exists a real number $M = M(B)$, for which the following inequality holds:

$$|\lambda_{\mathbf{a}}| \le M/B^{|\mathbf{a}|} \ .$$

Reminder 5.8.17. We remind the reader of Notation 5.8.9(iv); for $\mathbf{a} \in \mathbb{Z}^n$, we declared that $|\mathbf{a}|$ is our shorthand for $|a_1| + |a_2| + \cdots + |a_n|$. Written less compactly, the inequality of Lemma 5.8.16 becomes

$$|\lambda_{\mathbf{a}}| \quad \le \quad \frac{M}{B^{|a_1|+|a_2|+\cdots+|a_n|}} \quad .$$

Proof. The Laurent series converges on all of $K^{\min}(B)$, and hence it must converge at the 2^n points, where some of the coordinates are B^{-1} while the others are B. That is, for any subset $J \subset \{0, 1, \ldots, n\}$, we can consider the point $P^J = (P_1^J, P_2^J, \ldots, P_n^J) \in \mathbb{C}^n$, with

$$P_i^J \quad = \quad \begin{cases} B & \text{if } i \notin J, \\ B^{-1} & \text{if } i \in J. \end{cases}$$

The points P^J all lie in $K^{\min}(B)$, and hence the Laurent series converges at P^J. This means that the individual terms in the series, the expressions

$$\lambda_{\mathbf{a}} \cdot \left(P_1^J\right)^{a_1} \cdot \left(P_2^J\right)^{a_2} \cdots \left(P_n^J\right)^{a_n} \ ,$$

must tend to zero. They must be bounded. For each J we have a bound, but there are only 2^n possible J's, so there has to be a bound that works for all J. Hence there must exist a number M with

$$|\lambda_{\mathbf{a}}| \left(P_1^J\right)^{a_1} \left(P_2^J\right)^{a_2} \cdots \left(P_n^J\right)^{a_n} \quad \le \quad M \ .$$

Putting

$$P_i^J \quad = \quad \begin{cases} B & \text{if } a_i \ge 0, \\ B^{-1} & \text{if } a_i < 0, \end{cases}$$

the inequality becomes $|\lambda_{\mathbf{a}}| \le M/B^{|\mathbf{a}|}$. □

Lemma 5.8.18. *Suppose f is a function on $(\mathbb{C}^*)^n$, having a convergent Laurent series expansion*

$$f(\mathbf{x}) \quad = \quad \sum_{\mathbf{a} \in \mathbb{Z}^n} \lambda_{\mathbf{a}} \mathbf{x}^{\mathbf{a}} \ .$$

Then this expansion is unique; in particular, it agrees with the expansion produced in Exercise 5.8.13.

Proof. For the uniqueness we do not even have to know that the series converges on all of $(\mathbb{C}^*)^n$. Given any real number $B > 1$, there is a formula for the coefficients $\lambda_{\mathbf{a}}$, which only uses the fact that the series converges on $K^{\min}(B)$.

We proceed as follows. Choose any circles $\gamma_1'', \gamma_2'', \ldots, \gamma_n''$, whose radii lie strictly between B^{-1} and B. Consider the integral

$$\frac{1}{(2\pi i)^n}\int_{\gamma_1''}\int_{\gamma_2''}\cdots\int_{\gamma_n''}\frac{f(\mathbf{z})dz_1dz_2\cdots dz_n}{\mathbf{z}^{\mathbf{a}+1}} .$$

Substituting the convergent Laurent series expansion for f, the integral becomes

$$\frac{1}{(2\pi i)^n}\int_{\gamma_1''}\int_{\gamma_2''}\cdots\int_{\gamma_n''}\left(\sum_{\mathbf{b}\in\mathbb{Z}}\frac{\lambda_{\mathbf{b}}\mathbf{z}^{\mathbf{b}}dz_1dz_2\cdots dz_n}{\mathbf{z}^{\mathbf{a}+1}}\right) .$$

The estimates of Lemma 5.8.16 allow us to interchange the order of summation and integration; the reader should check this. The integral is therefore equal to the sum

$$\frac{1}{(2\pi i)^n}\sum_{\mathbf{b}\in\mathbb{Z}^n}\int_{\gamma_1''}\int_{\gamma_2''}\cdots\int_{\gamma_n''}\frac{\lambda_{\mathbf{b}}\mathbf{z}^{\mathbf{b}}dz_1dz_2\cdots dz_n}{\mathbf{z}^{\mathbf{a}+1}} ,$$

and in this sum only one term is non-zero, namely

$$\lambda_{\mathbf{a}} = \frac{1}{(2\pi i)^n}\int_{\gamma_1''}\int_{\gamma_2''}\cdots\int_{\gamma_n''}\frac{\lambda_{\mathbf{a}}dz_1dz_2\cdots dz_n}{z_1z_2\cdots z_n} .$$

□

Remark 5.8.19. The uniqueness, proved in Lemma 5.8.18, allows us to speak of "the" Laurent series expansion of f. What we have learned, so far, is that there is a 1–to–1 correspondence between holomorphic functions on $(\mathbb{C}^*)^n$ and convergent Laurent series. Exercise 5.8.13 told us that any holomorphic function f has a Laurent series expansion, while Lemma 5.8.18 gives us the uniqueness of the expansion.

In Remark 5.8.19 we pointed out that holomorphic functions on $(\mathbb{C}^*)^n$ correspond, bijectively, to certain Laurent series. The next question we ask is: how do you recognize the Laurent series of a function on $(\mathbb{C}^*)^n$ which has an extension to X_J^{an}? We prove

Lemma 5.8.20. *Let f be a holomorphic function on X_J^{an}. Then in the Laurent series of f, that is in the series*

$$f(\mathbf{x}) = \sum_{\mathbf{a}\in\mathbb{Z}^n}\lambda_{\mathbf{a}}\mathbf{x}^{\mathbf{a}} ,$$

we have $\lambda_{\mathbf{a}} = 0$ unless $\mathbf{x}^{\mathbf{a}} \in S_J$.

Proof. Suppose $\mathbf{x}^{\mathbf{a}} \notin S_J$; we need to prove the vanishing of $\lambda_{\mathbf{a}}$. The fact that $\mathbf{x}^{\mathbf{a}} \notin S_J$ means that, in the vector $\mathbf{a} = (a_1, a_2, \cdots, a_n) \in \mathbb{Z}^n$, we must have $a_i < 0$ for some $i \notin J$; see Notation 5.8.9. This means that the function

$$\frac{f(z_1, z_2, \cdots, z_n)}{z_i^{a_i+1}} \quad ,$$

which we view as a function of a single variable z_i by holding constant all the variables $\{z_j, j \neq i\}$, is holomorphic on the entire complex plane. After all $f(z_1, z_2, \cdots, z_n)$ is an entire function of z_i, and $1/z_i^{a_i+1}$ is a non-negative power of the holomorphic function z_i. The integral

$$\int_{\gamma_i''} \frac{f(z_1, z_2, \cdots, z_n) dz_i}{z_i^{a_i+1}}$$

therefore vanishes, and hence so does the integral

$$\lambda_{\mathbf{a}} \quad = \quad \frac{1}{(2\pi i)^n} \int_{\gamma_1''} \int_{\gamma_2''} \cdots \int_{\gamma_n''} \frac{f(\mathbf{z}) dz_1 dz_2 \cdots dz_n}{z_1^{a_1+1} z_2^{a_2+1} \cdots z_n^{a_n+1}} \quad .$$

□

Remark 5.8.21. Let us take stock of where we are so far. To every holomorphic function f on X_J^{an}, Exercise 5.8.13 assigned a Laurent series, which converges to f on $(\mathbb{C}^*)^n \subset X_J^{\mathrm{an}}$. We know further that, in this series $f(\mathbf{x}) = \sum \lambda_{\mathbf{a}} \mathbf{x}^{\mathbf{a}}$, the coefficients $\lambda_{\mathbf{a}}$ must satisfy

(i) $\lambda_{\mathbf{a}} = 0$ unless $\mathbf{x}^{\mathbf{a}} \in S_J$; see Lemma 5.8.20. We will adopt the notation that such series will be written

$$\sum_{\mathbf{x}^{\mathbf{a}} \in S_j} \lambda_{\mathbf{a}} \mathbf{x}^{\mathbf{a}} \ .$$

(ii) For every $B \geq 1$ there exists a real number $M = M(B)$ so that, for all $\mathbf{a} \in \mathbb{Z}^n$, we have $|\lambda_{\mathbf{a}}| < M/B^{|\mathbf{a}|}$; see Lemma 5.8.16.

This gives us a map

$$\left\{ \begin{array}{c} \text{holomorphic functions} \\ \text{on the open set} \\ X_J^{\mathrm{an}} \subset \mathbb{C}^n \end{array} \right\} \xrightarrow{\quad \Phi \quad} \left\{ \begin{array}{c} \text{Laurent series, with} \\ \text{coefficients satisfying} \\ \text{(i) and (ii)} \end{array} \right\} ,$$

It is trivial that Φ is injective. For suppose $\Phi(f) = \Phi(f')$, that is f and f' are holomorphic on X_J^{an} and have the same Laurent series expansion. This common Laurent series converges to f and to f' on the open set $(\mathbb{C}^*)^n \subset X_J^{\mathrm{an}}$, hence f and f' must agree on $(\mathbb{C}^*)^n$. But $(\mathbb{C}^*)^n$ is a dense subset of X_J^{an}, and so we must have $f = f'$.

What we want to show next is that Φ is a bijection. This reduces to proving that any Laurent series, satisfying conditions (i) and (ii), converges to a holomorphic function f on X_J^{an}. The reduction is by the uniqueness of Lemma 5.8.18; if we define the function f as the sum of the convergent Laurent series

$$f(\mathbf{x}) \quad = \quad \sum_{\mathbf{a}\in\mathbb{Z}^n} \lambda_{\mathbf{a}}\mathbf{x}^{\mathbf{a}} \,,$$

then Lemma 5.8.18 establishes $\Phi(f)$ agrees with the series $\sum_{\mathbf{a}\in\mathbb{Z}^n} \lambda_{\mathbf{a}}\mathbf{x}^{\mathbf{a}}$.

It remains to prove a convergence statement. I leave this as an exercise to the reader:

Exercise 5.8.22. Suppose we are given a set of complex numbers $\{\lambda_{\mathbf{a}} \mid \mathbf{a} \in \mathbb{Z}^n\}$. Assume also that we are given two real numbers B, ε, with $B > 1$ and $\varepsilon > 0$. Suppose further that

(i) $\lambda_{\mathbf{a}} = 0$ unless $\mathbf{x}^{\mathbf{a}} \in S_J$.
(ii) There exists a real number $M = M(B+2\varepsilon)$ so that, for all $\mathbf{a} \in \mathbb{Z}^n$, we have $|\lambda_{\mathbf{a}}| \leq M/(B+2\varepsilon)^{|\mathbf{a}|}$.

The reader should check that the Laurent series $\sum_{\mathbf{a}\in\mathbb{Z}^n} \lambda_{\mathbf{a}}\mathbf{x}^{\mathbf{a}}$ converges, absolutely and uniformly, on the region $K_J(B+\varepsilon)$ of Notation 5.8.14. The limit has to be a holomorphic function, at least on $K_J(B)$.

In Remark 5.8.21(ii) we assumed that a constant $M = M(B+2\epsilon)$ exists for every $B > 1$ and $\epsilon > 0$. This means that the series converges, to a holomorphic function on $K_J(B)$, for every $B > 1$. But the union of the $K_J(B)$ is all of X_J^{an}; see Remark 5.8.15. Thus the series converges to a holomorphic function on X_J^{an}, completing the proof of all the assertions in Remark 5.8.21.

Construction 5.8.23. We define now a Fréchet space F_J as follows:

(i) The elements of the set F_J are the Laurent series

$$\sum_{\mathbf{x}^{\mathbf{a}}\in S_J} \lambda_{\mathbf{a}}\mathbf{x}^{\mathbf{a}} \,,$$

where $\lambda_{\mathbf{a}}$ are rapidly decreasing, in the sense of Remark 5.8.21(ii).
(ii) Addition and scalar multiplication are term-by-term; this makes F_J a vector space over $\mathbb{C}$.
(iii) The topology is given by a sequence of seminorms q_ℓ. The formula is

$$q_\ell\left(\sum_{\mathbf{x}^{\mathbf{a}}\in S_J} \lambda_{\mathbf{a}}\mathbf{x}^{\mathbf{a}}\right) \quad = \quad \sup_{\mathbf{a}\in\mathbb{Z}^n}\left(|\lambda_{\mathbf{a}}|\cdot \ell^{|\mathbf{a}|}\right) \,.$$

Note that the bound of Remark 5.8.21(ii), with $B = \ell$, guarantees the finiteness of the right hand side.

Remark 5.8.24. Observe that the seminorms q_ℓ, defined on F_J in Construction 5.8.23(iii), are independent of J. For any two subsets $J \subset J' \subset \{1, 2, \ldots, n\}$ there is an inclusion $F_J \subset F_{J'}$, and this inclusion is an isometry. The spaces F_J all embed, isometrically, into the largest one, which is $F_{\{1,2,\ldots,n\}}$. The space $F_{\{1,2,\ldots,n\}}$ consists of all Laurent series with rapidly decreasing coefficients, and the subspaces $F_J \subset F_{\{1,2,\ldots,n\}}$ contain the Laurent series where some coefficients are constrained to vanish.

Exercise 5.8.25. We ask the reader to check that F_J is a Fréchet space; in other words, that it is complete with respect to the collection of seminorms. We remind the reader of Exercise 5.4.7; what needs to be checked is the following. Given a sequence of elements of F_J, which is Cauchy with respect to each seminorm q_ℓ, we are asserting that it converges to an element of F_J, where the convergence is in every seminorm. The hint is that the following works:

$$\lim_{i\to\infty}\left(\sum_{\mathbf{x}^{\mathbf{a}}\in S_J}\lambda(i)_{\mathbf{a}}\mathbf{x}^{\mathbf{a}}\right) \quad = \quad \sum_{\mathbf{x}^{\mathbf{a}}\in S_J}\left(\lim_{i\to\infty}\lambda(i)_{\mathbf{a}}\right)\mathbf{x}^{\mathbf{a}} .$$

In Remark 5.8.21 we produced a map Φ, taking holomorphic functions on X_J^{an} to Laurent series. In our present notation, it can be written as

$$\Phi : \Gamma(X_J^{\mathrm{an}}, \mathcal{O}^{\mathrm{an}}) \longrightarrow F_J .$$

In Remark 5.8.21 we also observed that the map is bijective. It is a bijective map between Fréchet spaces. We assert:

Lemma 5.8.26. *The linear map*

$$\Phi : \Gamma(X_J^{\mathrm{an}}, \mathcal{O}^{\mathrm{an}}) \longrightarrow F_J$$

of Remark 5.8.21 is a homeomorphism.

Proof. We know that Φ is a bijective, linear map between Fréchet spaces; it is enough, by Corollary 5.4.12, to show that Φ is continuous. To do this we must briefly recall the definition of the Fréchet topology on $\Gamma(X_J^{\mathrm{an}}, \mathcal{O}^{\mathrm{an}})$.

In Example 5.4.8 we described this Fréchet topology. Given any open set $U \subset \mathbb{C}^n$, Example 5.4.8 constructed a sequence of compact sets $U_\ell \subset U$. The reader should check that, in the case where $U = X_J^{\mathrm{an}}$, the sets U_ℓ are precisely the $K_J(\ell)$ of Notation 5.8.14. Next, the recipe of

Example 5.4.8 provides us with a sequence of seminorms p_ℓ, given by the formula

$$p_\ell(f) = \sup_{z \in K_J(\ell)} |f(\mathbf{z})| \ .$$

The Fréchet topology on $\Gamma(X_J^{\mathrm{an}}, \mathcal{O}^{\mathrm{an}})$ is the topology given by these seminorms. To prove the continuity of Φ, it therefore suffices to establish

5.8.26.1. *Let f be an element of $\Gamma(X_J^{\mathrm{an}}, \mathcal{O}^{\mathrm{an}})$. Then*

$$q_\ell\big(\Phi(f)\big) \ \le \ p_\ell(f) \ .$$

That is, the q_ℓ seminorm, of the element $\Phi(f) \in F_J$, is bounded above by the p_ℓ seminorm, of $f \in \Gamma(X_J^{\mathrm{an}}, \mathcal{O}^{\mathrm{an}})$.

This is what we will now prove. We need to show that, if $p_\ell(f) \le M$, then $q_\ell\big(\Phi(f)\big) \le M$. Assume therefore that $p_\ell(f) \le M$. This means, concretely, that $|f(\mathbf{z})| \le M$ for all $z \in K_J(\ell)$. We need to estimate $q_\ell\big(\Phi(f)\big)$, that is, we need to find a bound on the growth of the coefficients in the Laurent series expansion of f.

As in Exercise 5.8.13, consider the Laurent series expansion of f

$$f(\mathbf{x}) \quad = \quad \sum_{\mathbf{a} \in \mathbb{Z}^n} \lambda_{\mathbf{a}} \mathbf{x}^{\mathbf{a}} \quad .$$

In Exercise 5.8.13 we learned a formula for $\lambda_{\mathbf{a}}$; we have

$$\lambda_{\mathbf{a}} \quad = \quad \frac{1}{(2\pi i)^n} \int_{\gamma_1''} \int_{\gamma_2''} \cdots \int_{\gamma_n''} \frac{f(\mathbf{z}) dz_1 dz_2 \cdots dz_n}{\mathbf{z}^{\mathbf{a}+1}} \qquad ,$$

where the γ_i'' are circles $|z_i| = r_i$. The radii r_i are arbitrary; let us choose them later. The only decision we make now is to have $\ell^{-1} \le r_i \le \ell$. This means that $\prod_{i=1}^n \gamma_i'' \subset K_J(\ell)$, and therefore $|f(\mathbf{z})| \le M$ for all $\mathbf{z} \in \prod_{i=1}^n \gamma_i''$.

Now we estimate

$$\begin{aligned}
|\lambda_{\mathbf{a}}| &\leq \left|\frac{1}{(2\pi i)^n}\right| \int_{\gamma_1''}\int_{\gamma_2''}\cdots\int_{\gamma_n''} \frac{|f(\mathbf{z})|\cdot|dz_1|\cdot|dz_2|\cdots|dz_n|}{|z_1^{a_1+1}|\cdot|z_2^{a_2+1}|\cdots|z_n^{a_n+1}|} \\
&\leq \frac{1}{(2\pi)^n}\int_{\gamma_1''}\int_{\gamma_2''}\cdots\int_{\gamma_n''} \frac{M\cdot|dz_1|\cdot|dz_2|\cdots|dz_n|}{r_1^{a_1+1}r_2^{a_2+1}\cdots r_n^{a_n+1}} \\
&= \frac{M}{(2\pi)^n r_1^{a_1+1}r_2^{a_2+1}\cdots r_n^{a_n+1}}\int_{\gamma_1''}|dz_1|\int_{\gamma_2''}|dz_2|\cdots\int_{\gamma_n''}|dz_n| \\
&= \frac{M}{(2\pi)^n r_1^{a_1+1}r_2^{a_2+1}\cdots r_n^{a_n+1}}(2\pi r_1)\cdot(2\pi r_2)\cdots(2\pi r_n) \\
&= \frac{M}{r_1^{a_1}r_2^{a_2}\cdots r_n^{a_n}}\ .
\end{aligned}$$

At this point we choose the r_i; we set

$$r_i \quad = \quad \begin{cases} \ell & \text{if } a_i \geq 0, \\ \ell^{-1} & \text{if } a_i < 0. \end{cases}$$

This immediately gives

$$|\lambda_{\mathbf{a}}|\cdot\ell^{|a_1|+|a_2|+\cdots+|a_n|} \quad\leq\quad M\quad ;$$

that is, $M \geq \sup\left(|\lambda_{\mathbf{a}}|\cdot\ell^{|\mathbf{a}|}\right)$. This is precisely the inequality

$$q_\ell(\Phi(f)) \quad\leq\quad M\ ;$$

see Construction 5.8.23(iii) for the definition of q_ℓ. □

Remark 5.8.27. The fact that Φ is a homeomorphism means that there exist bounds in the other direction; for each integer $\ell \in \mathbb{N}$, it is possible to find $M > 0$ and $\ell' \in \mathbb{N}$, so that

$$p_\ell(f) \quad\leq\quad M\cdot q_{\ell'}(\Phi(f))\ .$$

It is not difficult to obtain such estimates explicitly; the reader might wish to amuse herself by doing this.

Remark 5.8.28. The map Φ is a homeomorphism of $\Gamma(X_J^{\mathrm{an}}, \mathcal{O}^{\mathrm{an}})$ with F_J. We know that $\Gamma(X_J^{\mathrm{an}}, \mathcal{O}^{\mathrm{an}})$ is the completion of the image of $\mu_J : S_J \longrightarrow \Gamma(X_J^{\mathrm{an}}, \mathcal{O}^{\mathrm{an}})$, and the topology induced on S_J is $[S_J, X_J^{\mathrm{an}}]$; see Lemma 5.8.5 and Notation 5.8.6. We can compute this topology via the map to $F_J \cong \Gamma(X_J^{\mathrm{an}}, \mathcal{O}^{\mathrm{an}})$.

The composite

$$S_J \xrightarrow{\ \mu_J\ } \Gamma(X_J^{\mathrm{an}}, \mathcal{O}^{\mathrm{an}}) \xrightarrow{\ \Phi\ } F_J$$

is very straightforward, it is simply the map taking a Laurent polynomial to itself, viewed as a finite Laurent series. The reader can check directly that the Laurent polynomials are indeed dense in F_J; every Laurent series is the limit, with respect to the seminorms q_ℓ on F_J, of its finite truncations.

Since the composite $\Phi\mu_J : S_J \longrightarrow F_J$ is just the inclusion of the finite Laurent series into all Laurent series, we understand completely the restriction, to $S_J \subset F_J$, of the seminorms q_ℓ of Construction 5.8.23(iii). The formula simplifies slightly to

$$q_\ell\left(\sum_{\mathbf{x^a}\in S_J} \lambda_\mathbf{a}\mathbf{x^a}\right) \quad = \quad \max_{\mathbf{a}\in\mathbb{Z}^n}\left(|\lambda_\mathbf{a}|\cdot\ell^{|\mathbf{a}|}\right) \ ;$$

the supremum of finitely many terms is just the maximum. The completion of S_J, with respect to the system of seminorms q_ℓ, is isomorphic to $\Gamma(X_J^{\mathrm{an}}, \mathcal{O}^{\mathrm{an}})$.

Remark 5.8.29. If $J \subset J' \subset \{1, 2, \ldots, n\}$ are two subsets, we have an inclusion of rings $S_J \longrightarrow S_{J'}$. As in Remark 5.8.24 this map is an isometry. The subspace topology of $S_J \subset \Gamma(X_{J'}^{\mathrm{an}}, \mathcal{O}^{\mathrm{an}})$ is independent of $J' \supset J$.

It follows that the completion is also independent of J'; it is always $\Gamma(X_J^{\mathrm{an}}, \mathcal{O}^{\mathrm{an}})$. Hence S_J will not be dense in $\Gamma(X_{J'}^{\mathrm{an}}, \mathcal{O}^{\mathrm{an}})$ if the inclusion $J \subset J'$ is proper. We promised the reader such an example in Caution 5.7.7; one consequence is that $X_{J'}^{\mathrm{an}}$ cannot possibly be a polydisc in $X_J^{\mathrm{an}} = \left\{\operatorname{Spec}(S_J)\right\}^{\mathrm{an}}$.

Remark 5.8.30. What we will need most is the following observation. Suppose we partition the monomials in S_J into finitely many subsets. There is an induced direct sum decomposition of S_J, which we will write

$$S_J \quad = \quad V_1 \oplus V_2 \oplus \cdots \oplus V_t \ .$$

The formula for the seminorm q_ℓ, expressing it as the supremum of some function of the coefficients of monomials, tells us that, for any vector $v_1 + v_2 + \cdots + v_t$ in this direct sum, we have

$$q_\ell(v_1 + v_2 + \cdots + v_t) \quad = \quad \max\left(q_\ell(v_i)\right) \ .$$

From this two things follow:

(i) A sequence in S_J is Cauchy if and only if each of its t direct summands is Cauchy.

(ii) A sequence in S_J converges to zero if and only if each of its t direct summands does.

We can restate this, slightly differently, as follows. There is a natural map of completions

$$\widehat{V}_1 \oplus \widehat{V}_2 \oplus \cdots \oplus \widehat{V}_t \longrightarrow \widehat{S}_J = F_J .$$

The assertion (i) above shows that this map is surjective, while (ii) implies that it is injective. The map is obviously continuous; it is therefore a homeomorphism.

Remark 5.8.31. We know that $\Gamma(X_J^{\text{an}}, \mathcal{O}^{\text{an}})$ is a Fréchet ring, and F_J is a Fréchet space homeomorphic to it. It follows that F_J has a ring structure, where the multiplication is continuous. The simplest way to work out this ring structure, explicitly, is to observe that the inclusion $S_J \longrightarrow \Gamma(X_J^{\text{an}}, \mathcal{O}^{\text{an}})$ is a ring homomorphism, and hence the ring structure on F_J must make the inclusion $S_J \longrightarrow F_J$ a ring homomorphism. But S_J is dense in F_J; to multiply two Laurent series, just express each of them as the limit of its finite truncations, multiply the finite truncations, and take the limit. In other words, the rule for multiplication is to use the distributive law and multiply the series term by term.

From this description, it is not immediately clear why the product of two elements of F_J should lie in F_J. It is not clear why the infinite sums, which occur in the formula for the product, need to be summable. Accepting for an instant that it is possible to sum up the terms to obtain a product Laurent series, it is not obvious why the coefficients of this product series should decay rapidly enough for it to qualify as an element of F_J. And finally it is not immediate that the multiplication map $F_J \times F_J \longrightarrow F_J$, assuming it is well defined, need be continuous.

Because of the identification with $\Gamma(X_J^{\text{an}}, \mathcal{O}^{\text{an}})$, we know that the product is well defined and continuous. It is possible to prove this directly, estimating the various coefficients; we leave this to the interested reader.

5.9 Gluing it all

We now understand the affine case; the abstract theory was presented in Section 5.7, and in Section 5.8 we worked out, in some detail, a few simple examples. Now it is time to pass to the global case, where $(X, \mathcal{O})$ might not be affine. We have to understand how to glue the local data.

We begin with formal generalities. Let X be any topological space. We observe:

Lemma 5.9.1. *Suppose $X = \cup_{i \in I} U_i$ is an open cover of the topological space X. Suppose that on each U_i we have a sheaf $\mathscr{S}_i$, suppose we have* *isomorphisms* $\varphi_{ij} : \mathscr{S}_j|_{U_i \cap U_j} \longrightarrow \mathscr{S}_i|_{U_i \cap U_j}$, *and suppose the maps satisfy the following identities*

$$\varphi_{ij}\varphi_{ji} = 1\,, \qquad \varphi_{ij}\varphi_{jk}\varphi_{ki} = 1\,.$$

Then there exists a sheaf $\mathscr{S}$ *on* X *and isomorphisms*

$$\varphi_i : \mathscr{S}|_{U_i} \longrightarrow \mathscr{S}_i$$

so that $\varphi_{ij} = \varphi_i \varphi_j^{-1}$.

Proof. Let us begin by defining a basis of open subsets of X

$$B \quad = \quad \{V \subset X \mid V \text{ is open, and } V \subset U_i \text{ for some } i\}\,.$$

We should justify that B really is a basis. Every open set $W \subset X$ can be written as $W = \cup_i \{W \cap U_i\}$, and all $W \cap U_i \subset U_i$ are in B. It remains to be shown that the intersection of two elements of B lies in B. Let V, V' be open subsets of X belonging to B. Then $V \subset U_i$ for some i, and hence $V \cap V' \subset V \subset U_i$ must also lie in B.

For each V in the basis B choose and fix an $i \in I$ so that $V \subset U_i$. Call our chosen function $f : B \longrightarrow I$. Define

$$\Gamma(V, \mathscr{S}) \quad = \quad \Gamma\Big(V, \mathscr{S}_{f(V)}\Big)\,.$$

For $V \subset V'$ put $f(V) = i$, $f(V') = j$ and let the restriction map $\mathrm{res}_V^{V'} : \Gamma(V', \mathscr{S}) \longrightarrow \Gamma(V, \mathscr{S})$ be defined as the composite

$$\Gamma(V', \mathscr{S}) = \Gamma(V', \mathscr{S}_j) \xrightarrow{\mathrm{res}_V^{V'}} \Gamma(V, \mathscr{S}_j) \xrightarrow{\varphi_{ij}} \Gamma(V, \mathscr{S}_i) = \Gamma(V, \mathscr{S})\,.$$

It is easy to check that, for open sets $V \subset V' \subset V''$ in B, we have $\mathrm{res}_V^{V'}\, \mathrm{res}_{V'}^{V''} = \mathrm{res}_V^{V''}$. Lemma 3.4.10 tells us that to verify that $\mathscr{S}$ extends to a sheaf it is enough to show that, for every open cover of $V \in B$ by open subsets $V_k \in B$, the sequence

$$0 \longrightarrow \Gamma(V, \mathscr{S}) \xrightarrow{\ \alpha\ } \prod_k \Gamma(V_k, \mathscr{S}) \xrightarrow{\ \beta\ } \prod_{k,k' \in I} \Gamma(V_k \cap V_{k'}, \mathscr{S})$$

is exact. But V is in the basis, which means that $V \subset U_i$ for some $i \in I$. On U_i there is an obvious isomorphism of presheaves $\varphi_i : \mathscr{S}|_{U_i} \longrightarrow \mathscr{S}_i$, so the exactness of the sequence above comes from the fact that $\mathscr{S}_i$ is a sheaf on U_i. □

And we should also say something about gluing maps of sheaves.

Lemma 5.9.2. *Let* $f : X \longrightarrow Y$ *be a continuous map of topological spaces, let* $\mathcal{O}_X$ *be a sheaf of rings on* X *and* $\mathcal{O}_Y$ *a sheaf of rings on* Y. *Suppose we have a basis* B *of the open sets in* X *and a basis* B' *of the* *open sets in Y. Suppose further that, for any pair $U \in B$ and $V \in B'$ with $f(U) \subset V$, we are given a ring homomorphism*

$$\varphi_{U,V} : \Gamma(V, \mathcal{O}_Y) \longrightarrow \Gamma(U, \mathcal{O}_X) \ ,$$

and that the ring homomorphisms $\varphi_{U,V}$ are compatible with restrictions. The compatibility means that, for any $U \subset U'$ and $f(U') \subset V \subset V'$, the composite

$$\Gamma(V', \mathcal{O}_Y) \xrightarrow{\mathrm{res}_V^{V'}} \Gamma(V, \mathcal{O}_Y) \xrightarrow{\varphi_{U',V}} \Gamma(U', \mathcal{O}_X) \xrightarrow{\mathrm{res}_U^{U'}} \Gamma(U, \mathcal{O}_X)$$

must agree with $\varphi_{U,V'}$.

Then there is a unique way to extend the $\varphi_{U,V}$ to a map of ringed spaces.

Proof. Let V be any open set in the basis B'. We need to define a map

$$\Phi_U^* : \Gamma(V, \mathcal{O}_Y) \longrightarrow \Gamma(f^{-1}V, \mathcal{O}_X).$$

But the exact sequence

$$0 \longrightarrow \Gamma(f^{-1}V, \mathcal{O}_X) \xrightarrow{\alpha} \prod_{\substack{U \in B \\ f(U) \subset V}} \Gamma(U, \mathcal{O}_X) \xrightarrow{\beta} \prod_{\substack{U,U' \in B \\ f(U), f(U') \subset V}} \Gamma(U \cap U', \mathcal{O}_X)$$

says that it is enough to define the map from $\Gamma(V, \mathcal{O}_Y)$ to the kernel of β. And this map is defined to be the product of all the $\varphi_{U,V}$. □

We assert that all the global statements we made in Section 5.1, about the existence of an analytic space $(X^{\mathrm{an}}, \mathcal{O}^{\mathrm{an}})$ and of the various maps of ringed spaces, now follow from the local results by applying Lemmas 5.9.1 and 5.9.2.

6

The high road to analytification

Starting with a scheme $(X, \mathcal{O})$ locally of finite type over $\mathbb{C}$ we learned, in Chapters 4 and 5, how to construct an analytic space $(X^{\mathrm{an}}, \mathcal{O}^{\mathrm{an}})$. The constructions we gave were local, and as always with local constructions one needs to check that the local data glue well.

There is a high road, which mentions local information as little as possible. In this chapter I sketch this for the interested reader. None of this chapter is essential to what follows. It is most sensible to begin with the high road description of polydiscs.

6.1 A coordinate-free approach to polydiscs

If S is a finitely generated $\mathbb{C}$–algebra, and if $\{a_1, a_2, \ldots, a_n\} \subset S$ is a set of generators, we can embed $\left\{ \operatorname{Spec}(S) \right\}^{\mathrm{an}}$ in $\mathbb{C}^n$. The embedding allows us to form the open subsets $V = \Delta(g; w; r) \cap \left\{ \operatorname{Spec}(S) \right\}^{\mathrm{an}}$ of $\left\{ \operatorname{Spec}(S) \right\}^{\mathrm{an}}$, which are the intersections of $\left\{ \operatorname{Spec}(S) \right\}^{\mathrm{an}}$ with polydiscs $\Delta(g; w; r) \subset \mathbb{C}^n$. Proposition 5.6.4(i) told us that the subsets of $\left\{ \operatorname{Spec}(S) \right\}^{\mathrm{an}}$ obtained this way are independent of the choice of generators. But it would still be nice to have a definition, of these open sets in $\left\{ \operatorname{Spec}(S) \right\}^{\mathrm{an}}$, which does not mention generators anywhere. In this section we give such a definition. Remember that a polydisc $\Delta(g; w; r) \subset \mathbb{C}^n$ was defined as the set of points

$$\begin{aligned} \Delta(g; w; r) &= \Delta(g_1, g_2, \ldots, g_\ell; w_1, w_2, \ldots, w_\ell; r_1, r_2, \ldots, r_\ell) \\ &= \left\{ x \in \mathbb{C}^n \;\middle|\; |g_i(x) - w_i| < r_i \quad \forall 1 \le i \le \ell \right\}. \end{aligned}$$

We want to describe the open sets $\Delta(g; w; r) \cap \left\{ \operatorname{Spec}(S) \right\}^{\mathrm{an}}$ in a way that is clearly independent of the embedding $\left\{ \operatorname{Spec}(S) \right\}^{\mathrm{an}} \subset \mathbb{C}^n$. To do this we must start by explaining how to form the complex numbers $g_i(x)$ for $x \in \left\{ \operatorname{Spec}(S) \right\}^{\mathrm{an}}$.

Construction 6.1.1. Let $(X, \mathcal{O})$ be any scheme locally of finite type over $\mathbb{C}$. Let $U \subset X$ be any open subset, and let g be any element of $\Gamma(U, \mathcal{O})$. Let x be a closed point in U, which is closed in X by Corollary 4.2.9. We want to define a complex number $g(x)$.

The recipe is the following: Lemma 4.2.6 tells us that the closed point $x \in X$ corresponds to a unique map of ringed spaces over $\mathbb{C}$

$$(\Phi, \Phi^*) : \left(\operatorname{Spec}(\mathbb{C}), \widetilde{\mathbb{C}}\right) \longrightarrow (X, \mathcal{O}) ,$$

where $\Phi : \operatorname{Spec}(\mathbb{C}) \longrightarrow X$ takes the only point of $\operatorname{Spec}(\mathbb{C})$ to $x \in X$. The open set U contains the point x, and hence the map of ringed spaces gives a ring homomorphism

$$\Phi_U^* : \Gamma(U, \mathcal{O}) \longrightarrow \Gamma\left(\operatorname{Spec}(\mathbb{C}), \widetilde{\mathbb{C}}\right) \quad = \quad \mathbb{C}.$$

Definition 6.1.2. *Let the notation be as in Construction 6.1.1. The complex number $g(x) \in \mathbb{C}$ is defined to be $\Phi_U^*(g)$.*

Example 6.1.3. Consider the following special case:

(i) The ring $R = \mathbb{C}[x_1, x_2, \ldots, x_n]$ is the polynomial ring.
(ii) The scheme locally of finite type over $\mathbb{C}$ is $\left(\operatorname{Spec}(R), \widetilde{R}\right)$.
(iii) The open set $U \subset \operatorname{Spec}(R)$ is all of $\operatorname{Spec}(R)$.
(iv) The element $g \in \Gamma\left(\operatorname{Spec}(R), \widetilde{R}\right) = R = \mathbb{C}[x_1, x_2, \ldots, x_n]$ is any polynomial.
(v) The space $\left\{\operatorname{Spec}(R)\right\}^{\text{an}}$ identifies with $\mathbb{C}^n$ as in Lemma 4.4.1.

Then, given a closed point $a \in \operatorname{Spec}(R)$, Definition 6.1.2 produced a complex number $g(a)$ by the formula $g(a) = \Phi_X^*(g)$. We assert that this agrees with the number $g(a)$ obtained by evaluating the polynomial g at $a \in \mathbb{C}^n \cong \left\{\operatorname{Spec}(R)\right\}^{\text{an}}$.

The proof is an exercise in untangling the definitions. The closed point $a \in \operatorname{Spec}(R)$ corresponds, by Proposition 4.2.4, to a unique $\mathbb{C}$–algebra homomorphism $\varphi : R \longrightarrow \mathbb{C}$. The correspondence is such that the map

$$\operatorname{Spec}(\varphi) : \operatorname{Spec}(\mathbb{C}) \longrightarrow \operatorname{Spec}(R)$$

takes the unique point of $\operatorname{Spec}(\mathbb{C})$ to the point $a \in \operatorname{Spec}(R)$. The morphism

$$(\operatorname{Spec}(\varphi), \widetilde{\varphi}) : \left(\operatorname{Spec}(\mathbb{C}), \widetilde{\mathbb{C}}\right) \longrightarrow \left(\operatorname{Spec}(R), \widetilde{R}\right)$$

therefore satisfies the property which uniquely determines the morphism (Φ, Φ^*) of Lemma 4.2.6. We deduce that the ring homomorphism

$$\Phi_{\operatorname{Spec}(R)}^* : \Gamma\left(\operatorname{Spec}(R), \widetilde{R}\right) \longrightarrow \Gamma\left(\operatorname{Spec}(\mathbb{C}), \widetilde{\mathbb{C}}\right) \quad = \quad \mathbb{C}$$

is just the map $\widetilde{\varphi}_{\text{Spec}(R)} = \varphi$. This means that the complex number $\Phi^*_{\text{Spec}(R)}(g)$ is simply $\varphi(g)$.

But the point $a \in \mathbb{C}^n$ corresponds, under the bijection of Lemma 4.4.1, to the homomorphism $\varphi_a : R \longrightarrow \mathbb{C}$ which takes a polynomial g and evaluates it at a. Hence the value of $\Phi^*_{\text{Spec}(R)}(g) = \varphi_a(g)$ agrees with $g(a)$.

Discussion 6.1.4. Next we show that Definition 6.1.2 is compatible with restriction. Let us explain the relevance. In Example 6.1.3 we were given a polynomial $g \in R = \mathbb{C}[x_1, x_2, \ldots, x_n]$. We know that we can, using Definition 6.1.2, evaluate it at a closed point $a \in \left\{ \text{Spec}(R) \right\}^{\text{an}} \cong \mathbb{C}^n$, and that the somewhat abstruse way, in which we defined the evaluation in Definition 6.1.2, agrees with the common-sense way of just plugging a into the polynomial g.

So for $\mathbb{C}^n = \left\{ \text{Spec}(R) \right\}^{\text{an}}$ we are happy. But we really want to consider arbitrary finitely generated $\mathbb{C}$–algebras S. Now, as in Notation 5.3.2, we choose generators $b_1, b_2, \ldots, b_n$ for the ring S, and let $\theta : R \longrightarrow S$ be the surjection with $\theta(x_i) = b_i$. Given an element $g \in R$ we know how to evaluate it on $\mathbb{C}^n = \left\{ \text{Spec}(R) \right\}^{\text{an}}$. There was the abstruse formula of Definition 6.1.2, but it agrees with the concrete. The abstruse formula also makes sense if we want to evaluate an element of the ring

$$S \quad = \quad \Gamma\left(\text{Spec}(S), \widetilde{S}\right)$$

at a point $a \in \left\{ \text{Spec}(S) \right\}^{\text{an}}$. What we will show is that the evaluation map is compatible with the embedding. That is, given $g \in R$ we know (from Definition 6.1.2 or concretely) how to evaluate it at $a \in \mathbb{C}^n = \left\{ \text{Spec}(R) \right\}^{\text{an}}$. But Definition 6.1.2 also tells us how to evaluate $\theta(g) \in S$ at a point in $\left\{ \text{Spec}(S) \right\}^{\text{an}} \subset \mathbb{C}^n$. What we will see is that, if $a \in \mathbb{C}^n$ happens to lie in the subset $\left\{ \text{Spec}(S) \right\}^{\text{an}} \subset \mathbb{C}^n$, then the evaluation of $\theta(g)$ at a is just $g(a)$. In the special case, where S is also a polynomial ring, we have already seen such a result in Lemma 4.4.7; now we are after a very general version. Let us formulate our next results in glorious generality, and then return to the more concrete examples we have in mind.

Construction 6.1.5. Let $(\Psi, \Psi^*) : (X, \mathcal{O}_X) \longrightarrow (Y, \mathcal{O}_Y)$ be a morphism of schemes of finite type over $\mathbb{C}$. Let $a \in X$ be a closed point, and let $U \subset Y$ be an open set containing $\Psi(a)$. Let g be a section $g \in \Gamma(U, \mathcal{O}_Y)$.

Definition 6.1.2 allows us to construct two complex numbers from this data:

(i) We could evaluate $g \in \Gamma(U, \mathcal{O}_Y)$ at $\Psi(a) \in U$.

(ii) The ring homomorphism

$$\Psi_U^* : \Gamma(U, \mathcal{O}_Y) \longrightarrow \Gamma(\Psi^{-1}U, \mathcal{O}_X)$$

takes $g \in \Gamma(U, \mathcal{O}_Y)$ to $\Psi_U^* g \in \Gamma(\Psi^{-1}U, \mathcal{O}_X)$. We could evaluate $\Psi_U^* g$ at $a \in \Psi^{-1}U$.

Lemma 6.1.6. *Let the notation be as in Construction 6.1.5. Then*

$$g\big(\Psi(a)\big) = \{\Psi_U^* g\}(a).$$

Proof. Let a be a closed point in X. By Lemma 4.2.6 there is a morphism of ringed spaces over $\mathbb{C}$

$$(\Phi, \Phi^*) : \left(\operatorname{Spec}(\mathbb{C}), \widetilde{\mathbb{C}}\right) \longrightarrow (X, \mathcal{O}_X)$$

so that $\Phi\{\operatorname{Spec}(\mathbb{C})\} = a$. But then the composite morphism

$$\left(\operatorname{Spec}(\mathbb{C}), \widetilde{\mathbb{C}}\right) \xrightarrow{(\Phi,\Phi^*)} (X, \mathcal{O}_X) \xrightarrow{(\Psi,\Psi^*)} (Y, \mathcal{O}_Y)$$

satisfies $\Psi\Phi\{\operatorname{Spec}(\mathbb{C})\} = \Psi(a)$, that is the composite is the morphism corresponding under Lemma 4.2.6 with the closed point $\Psi(a) \in Y$.

Now let g be a section $g \in \Gamma(U, \mathcal{O}_Y)$ and assume $a \in \Psi^{-1}U \cap X^{\mathrm{an}}$. We need to compare $g\big(\Psi(a)\big)$ with $\{\Psi_U^* g\}(a)$. In terms of the maps of ringed spaces above, $g\big(\Psi(a)\big)$ is obtained by applying $\{\Psi\Phi\}_U^* : \Gamma(U, \mathcal{O}_Y) \longrightarrow \mathbb{C}$ to $g \in \Gamma(U, \mathcal{O}_Y)$. The definition of $\{\Psi_U^* g\}(a)$ is to apply the ring homomorphism $\Phi_{\Psi^{-1}U}^* : \Gamma(\Psi^{-1}U, \mathcal{O}_X) \longrightarrow \mathbb{C}$ to the element $\Psi_U^* g \in \Gamma(\Psi^{-1}U, \mathcal{O}_X)$. The equality is because $\{\Psi\Phi\}_U^* = \Phi_{\Psi^{-1}U}^* \Psi_U^*$. □

Remark 6.1.7. In the special case of Discussion 6.1.4 we considered the polynomial ring $R = \mathbb{C}[x_1, x_2, \ldots, x_n]$, and a surjective homomorphism $\theta : R \longrightarrow S$ of $\mathbb{C}$–algebras. There is an induced map of ringed spaces

$$\left(\operatorname{Spec}(\theta), \widetilde{\theta}\right) : \left(\operatorname{Spec}(S), \widetilde{S}\right) \longrightarrow \left(\operatorname{Spec}(R), \widetilde{R}\right) .$$

Lemma 6.1.6 tells us the following. Choose a closed point $a \in \operatorname{Spec}(S)$, and a section g in $R = \Gamma\left(\operatorname{Spec}(R), \widetilde{R}\right)$. Then the evaluation of g at $\{\operatorname{Spec}(\theta)\}(a) \in \big\{\operatorname{Spec}(R)\big\}^{\mathrm{an}}$ agrees with the evaluation of $\theta(g)$ at a. In the concrete case of the surjective map $\theta : R \longrightarrow S$, we have that $\theta(g)$ is simply the equivalence class of the polynomial g modulo the ideal $I = \ker(\theta)$. And the map

$$\big\{\operatorname{Spec}(\theta)\big\}^{\mathrm{an}} : \big\{\operatorname{Spec}(S)\big\}^{\mathrm{an}} \longrightarrow \big\{\operatorname{Spec}(R)\big\}^{\mathrm{an}} \quad \cong \quad \mathbb{C}^n$$

is, by Lemma 5.3.3, the embedding of the closed set of all solutions to the equation

$$f_1(x) = f_2(x) = \cdots = f_m(x) = 0\,.$$

If a is a solution to the equations, then $g(a)$ remains unchanged when g is changed to $g + \sum_{i=1}^m \lambda_i f_i$, that is $g(a)$ is a well-defined complex number when $g \in S$ and $a \in \big\{\operatorname{Spec}(S)\big\}^{\mathrm{an}} \subset \mathbb{C}^n$. And Lemma 6.1.6 tells us that this complex number agrees with the abstruse definition of $g(a)$.

Definition 6.1.8. *Let $(X, \mathcal{O})$ be a scheme locally of finite type over $\mathbb{C}$. An* enriched polydisc *$(U; g; w; r)$ is the following set of data:*

(i) *An open set $U \subset X$.*
(ii) *Sections $g_1, g_2, \ldots, g_\ell \in \Gamma(U, \mathcal{O})$.*
(iii) *Complex numbers $w_1, w_2, \ldots, w_\ell \in \mathbb{C}$ (the generalized center).*
(iv) *Positive real numbers $r_1, r_2, \ldots, r_\ell \in \mathbb{R}$ (the generalized polyradius).*

The data $(U; g; w; r)$ determines an open set $\Delta(U; g; w; r) \subset U^{\mathrm{an}} \subset X^{\mathrm{an}}$ as follows:

$$\Delta(U; g; w; r) \quad = \quad \Big\{ x \in U^{\mathrm{an}} \;\Big|\; |g_i(x) - w_i| < r_i \quad \forall 1 \le i \le \ell \Big\}\,.$$

Remark 6.1.9. Note that, unlike the traditional case of $\mathbb{C}^n$, we wish to remember not only the open set $\Delta(U; g; w; r) \subset U^{\mathrm{an}}$ but also the data $(U; g; w; r)$ that produced it. In the special case where $(U, \mathcal{O}_U) \cong \Big(\operatorname{Spec}(S), \widetilde{S}\Big)$ the sets $\Delta(U; g; w; r)$ are exactly the generalized polydiscs $\Delta(g; w; r) \cap \big\{\operatorname{Spec}(S)\big\}^{\mathrm{an}}$ of the embedding into $\mathbb{C}^n$. What is clear is that the sets $\Delta(U; g; w; r)$ are independent of the choice of the embedding into $\mathbb{C}^n$; that is, independent of the choice of generators for the algebra S.

6.2 The high road to the complex topology

Let $(X, \mathcal{O})$ be a scheme locally of finite type over $\mathbb{C}$. The set of closed points in X, denoted $\operatorname{Max}(X)$, was defined and studied in Section 4.2. To make it into a topological space we could declare that all the polydiscs $\Delta(U; g; w; r)$ are open. They do not form a basis for a topology, since finite intersections of polydiscs need not be polydiscs. But we can form a topology in which the finite intersections of polydiscs are a basis of open sets. This is the complex topology of X^{an}, which as a set is just $\operatorname{Max}(X)$. One could develop the subject starting with this definition; I leave it to the reader to see how the various facts can be proved.

6.3 The high road to the sheaf of analytic functions

Let $(X, \mathcal{O})$ be a scheme locally of finite type over $\mathbb{C}$. For a general polydisc $(U; g; w; r)$ the set $U \subset X$ need not be contained in any open affine subset of X. But if we are lucky then there will be such an inclusion. There will be an open immersion

$$(\Phi, \Phi^*) : \left(\operatorname{Spec}(R), \widetilde{R}\right) \longrightarrow (X, \mathcal{O})$$

with U contained in the image of Φ. When this happens then the open set $\Delta(U; g; w; r) \subset U$ is contained in $\left\{\operatorname{Spec}(R)\right\}^{\text{an}}$. There is a sheaf $\widetilde{R}^{\text{an}}$ of analytic functions on $\left\{\operatorname{Spec}(R)\right\}^{\text{an}}$, and, by Lemma 5.7.8, there is a map of ringed spaces from $\left(\left\{\operatorname{Spec}(R)\right\}^{\text{an}}, \widetilde{R}^{\text{an}}\right)$ to the ringed space $\left(\operatorname{Spec}(R), \widetilde{R}\right)$. In particular there will be a ring homomorphism

$$\Gamma(U, \mathcal{O}) \longrightarrow \Gamma(U^{\text{an}}, \mathcal{O}^{\text{an}}) \xrightarrow{\operatorname{res}^{U^{\text{an}}}_{\Delta(U;g;w;r)}} \Gamma(\Delta(U; g; w; r), \mathcal{O}^{\text{an}}) \,.$$

The space on the right is a Fréchet ring, and the homomorphism gives a pre-Fréchet topology on $\Gamma(U, \mathcal{O})$. The key to the high road approach to constructing the sheaf of analytic functions is

Lemma 6.3.1. *The pre-Fréchet topology on $\Gamma(U, \mathcal{O})$ is independent of the choice of open affine containing U. Furthermore, it does not depend on the open affine whether $\Gamma(U, \mathcal{O})$ is dense in $\Gamma(\Delta(U; g; w; r), \mathcal{O}^{\text{an}})$.*

Proof. Suppose we have two open affines $Y = \operatorname{Spec}(R)$ and $Z = \operatorname{Spec}(S)$ containing U. We must prove that the pre-Fréchet topologies, which they determine on $\Gamma(U, \mathcal{O})$, agree with each other and that, if $\Gamma(U, \mathcal{O})$ is dense in $\Gamma(\Delta(U; g; w; r), \mathcal{O}_Y^{\text{an}})$, then it is also dense $\Gamma(\Delta(U; g; w; r), \mathcal{O}_Z^{\text{an}})$.

By Proposition 3.10.9 we know that $U \subset Y \cap Z$ has an open cover $U = \cup W_i$, with each $W_i \cong Y_{f_i} \cong Z_{g_i}$ a basic open set both in $Y = \operatorname{Spec}(R)$ and in $Z = \operatorname{Spec}(S)$. By Corollary 5.2.4 we may assume the open cover finite.

Now in either Y or in Z we have a commutative diagram with exact

columns

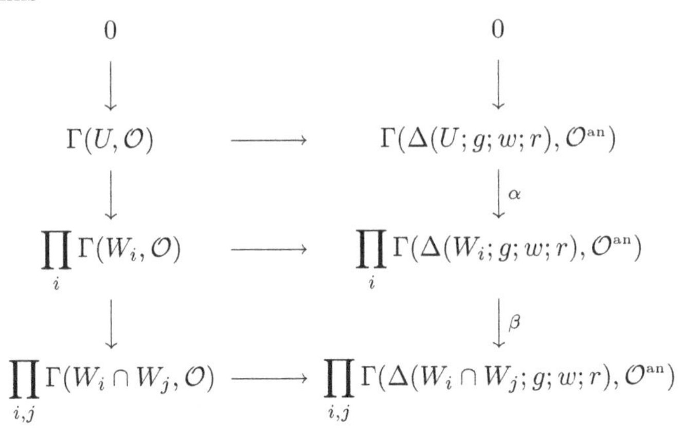

$$\begin{array}{ccc}
0 & & 0 \\
\downarrow & & \downarrow \\
\Gamma(U,\mathcal{O}) & \longrightarrow & \Gamma(\Delta(U;g;w;r),\mathcal{O}^{\mathrm{an}}) \\
\downarrow & & \downarrow{\scriptstyle\alpha} \\
\prod_i \Gamma(W_i,\mathcal{O}) & \longrightarrow & \prod_i \Gamma(\Delta(W_i;g;w;r),\mathcal{O}^{\mathrm{an}}) \\
\downarrow & & \downarrow{\scriptstyle\beta} \\
\prod_{i,j} \Gamma(W_i\cap W_j,\mathcal{O}) & \longrightarrow & \prod_{i,j} \Gamma(\Delta(W_i\cap W_j;g;w;r),\mathcal{O}^{\mathrm{an}})
\end{array} .$$

Note that the right column is deliberately vague; the sheaf $\mathcal{O}^{\mathrm{an}}$ could be either $\mathcal{O}_Y^{\mathrm{an}}$ or $\mathcal{O}_Z^{\mathrm{an}}$. There are two diagrams as above, one with Y and one with Z.

But the point is that the Fréchet rings $\Gamma(\Delta(W_i;g;w;r),\mathcal{O}^{\mathrm{an}})$ and $\Gamma(\Delta(W_i\cap W_j;g;w;r),\mathcal{O}^{\mathrm{an}})$ are independent of Y and Z. We learned in Lemma 5.7.4 that the ring homomorphisms $\alpha_f : R \longrightarrow R[1/f]$ and $\alpha_g : S \longrightarrow S[1/g]$ induce open immersions, so for any W which is simultaneously Y_f and Z_g the Fréchet rings

$$\Gamma(\Delta(W;g;w;r),\mathcal{O}_Y^{\mathrm{an}}) \quad\cong\quad \Gamma(\Delta(W;g;w;r),\mathcal{O}_Z^{\mathrm{an}})$$

can be identified as the completion of the ring $\Gamma(W,\mathcal{O}) \cong R[1/f] \cong S[1/g]$ with respect to the $\Delta(g;w;r)$ pre-Fréchet topology. The right column in the commutative diagram above identifies

$$\Gamma(\Delta(U;g;w;r),\mathcal{O}_Y^{\mathrm{an}}) \quad\text{and}\quad \Gamma(\Delta(U;g;w;r),\mathcal{O}_Z^{\mathrm{an}})$$

as the kernels of the homomorphism β, which is independent of Y and Z. So certainly they agree as rings.

Next we want to see that the Fréchet topologies also agree. We observe that the space $\prod_i \Gamma(\Delta(W_i;g;w;r),\mathcal{O}^{\mathrm{an}})$ is a finite product of Fréchet spaces, hence a Fréchet space (a countable product will also do, but not an arbitrary product). The kernel of β is a closed subspace of a Fréchet space, hence Fréchet. The map α is a bijective, continuous map into the kernel of β, and by Corollary 5.4.12 α must be a homeomorphism of $\Gamma(\Delta(U;g;w;r),\mathcal{O}^{\mathrm{an}})$ with the kernel of β. This is true whether by $\mathcal{O}^{\mathrm{an}}$ we understand $\mathcal{O}_Y^{\mathrm{an}}$ or $\mathcal{O}_Z^{\mathrm{an}}$. We conclude that these Fréchet spaces are homeomorphic. Therefore, independent of whether we use Y or Z, the topology on $\Gamma(U,\mathcal{O})$ will be the topology it inherits from its map into $\prod_i \Gamma(\Delta(W_i;g;w;r),\mathcal{O}^{\mathrm{an}})$, and whether it is dense in $\Gamma(\Delta(U;g;w;r),\mathcal{O}^{\mathrm{an}})$ will only depend on whether it is dense in the kernel of β. □

Definition 6.3.2. *An enriched polydisc $(U;g;w;r)$ will be called a* preferred polydisc *if it satisfies the following two hypotheses:*

(i) *The open set $U \subset X$ is contained in some affine open subset of X.*
(ii) *The ring $\Gamma(U,\mathcal{O})$ is dense in $\Gamma(\Delta(U;g;w;r),\mathcal{O}^{\mathrm{an}})$.*

Lemma 6.3.1 tells us that condition (ii) is independent of the particular open affine in which we embed U.

Remark 6.3.3. If $(U;g;w;r)$ is a preferred polydisc then the ring $\Gamma(U,\mathcal{O})$ is naturally a pre-Fréchet ring. The high road to constructing the ringed space $(X,\mathcal{O}^{\mathrm{an}})$ is to define the sheaf $\mathcal{O}^{\mathrm{an}}$ first on preferred polydiscs, as the completion of $\Gamma(U,\mathcal{O})$ with respect to its pre-Fréchet topology. To show that this works one needs to study the way the completion of $\Gamma(U,\mathcal{O})$ changes when we vary the preferred polydisc $(U;g;w;r)$. We leave this to the reader.

The high road approach is clearly intrinsic. There are several differences from the way we proceeded in Chapter 5. One is that the sets $\Delta(U;g;w;r)$ do not form a basis for the open sets. They do form a subbasis: every open set in X^{an} can be covered by preferred polydiscs. But finite intersections of preferred polydiscs need not be polydiscs.

7

Coherent sheaves

We have spent many chapters now worrying about ringed spaces. In Chapter 2 we used the example of $\mathcal{C}^k$–manifolds to motivate the definition and convince the reader that ringed spaces, that is spaces with sheaves of rings on them, are natural objects that are worth studying. In Chapter 3 we constructed the ringed spaces that form the object of study of this book, namely the schemes (locally) of finite type over $\mathbb{C}$. Chapters 4, 5 and 6 told us that, if $(X, \mathcal{O})$ is a scheme locally of finite type over $\mathbb{C}$, then there is a natural way to attach to it another ringed space $(X^{\text{an}}, \mathcal{O}^{\text{an}})$. Intuitively we think of $\mathcal{O}$ as the sheaf of polynomial functions on X, and $\mathcal{O}^{\text{an}}$ is the sheaf of holomorphic functions. For every open set $U \subset X$ one can speak of the polynomial functions on it, that is the ring $\Gamma(U, \mathcal{O})$. For any open subset $V \subset X^{\text{an}}$ (there are many more such open sets) one can talk of the holomorphic functions on V, that is the elements of $\Gamma(V, \mathcal{O}^{\text{an}})$.

If the reader glances back to the introduction she will discover that the results we want to prove are not only about spaces and functions on them, but also about vector bundles. It is only natural that we should again carefully define what we mean by an algebraic vector bundle, and what we mean by an analytic vector bundle. That is, for any scheme $(X, \mathcal{O})$, locally of finite type over $\mathbb{C}$, we need to disclose what are the algebraic vector bundles on it, and we also should tell the reader what are the holomorphic vector bundles on the analytification $(X^{\text{an}}, \mathcal{O}^{\text{an}})$.

It turns out that vector bundles are not general enough for us. The reason is the following. On a manifold M it is easy to define vector bundles, and to define maps of vector bundles $\mathscr{V} \longrightarrow \mathscr{V}'$. The problem is that the kernel, image and cokernel of maps of vector bundles are not usually vector bundles. On $\mathcal{C}^k$–manifolds we have no idea how to rectify this problem. The wonderful feature of both the complex analytic and algebraic framework is that this problem is easily solved. There is a very natural extension of the vector bundles, called the coherent sheaves. Coherent sheaves can be thought of as being locally quotients of maps of vector bundles. They behave well, and in particular the kernel, image and cokernel of a map of coherent sheaves are coherent sheaves. It is possible to speak about exact sequences of coherent sheaves.

In this chapter we explain the concept of coherent algebraic sheaf and, very briefly, the parallel notion of coherent analytic sheaf. We also explain the construction that takes a coherent algebraic sheaf $\mathscr{S}$, on a scheme $(X, \mathcal{O})$ locally of finite type over $\mathbb{C}$, and turns it into a coherent analytic sheaf $\mathscr{S}^{\mathrm{an}}$ on the analytic space $(X^{\mathrm{an}}, \mathcal{O}^{\mathrm{an}})$. Once all the vocabulary is in place we can state GAGA almost in complete generality (see Theorem 7.9.1) and say a little about the proof. The reader is referred particularly to Remark 7.9.9; the proof crucially depends on manipulating exact sequences of coherent sheaves. Even if one only cares about vector bundles, the proof unavoidably drags one into the world of coherent sheaves.

As we said in the introduction we will not prove GAGA for all the schemes $(X, \mathcal{O})$ for which it is known; see Remark 7.9.3. The general case follows from what we prove by relatively standard reductions, but the limited scope of the book does not allow us to develop the machinery necessary to perform these standard reductions.

7.1 Sheaves of modules on a ringed space

We learned, a long time ago, that a good way to study the properties of a ring R is to look at its modules. Not surprisingly there is an analogous notion for ringed spaces. If $(X, \mathcal{O})$ is a ringed space we want to define a reasonable notion of modules for the sheaf of rings $\mathcal{O}$.

Definition 7.1.1. *Let $(X, \mathcal{O})$ be a ringed space. A* sheaf of $\mathcal{O}$–modules *on X is a sheaf of abelian groups $\mathscr{S}$ so that, for every open set $U \subset X$, the abelian group $\Gamma(U, \mathscr{S})$ is a module over the ring $\Gamma(U, \mathcal{O})$, and so that the restriction maps*

$$\mathrm{res}_U^V : \Gamma(V, \mathscr{S}) \longrightarrow \Gamma(U, \mathscr{S})$$

respect the module structure.

Discussion 7.1.2. Perhaps some elaboration might help. For every open set $U \subset X$ we have an abelian group $\Gamma(U, \mathscr{S})$ and a ring $\Gamma(U, \mathcal{O})$. We want the ring to act on the group; that is, we want a multiplication map

$$m_U : \Gamma(U, \mathcal{O}) \times \Gamma(U, \mathscr{S}) \longrightarrow \Gamma(U, \mathscr{S}) ,$$

and this multiplication should satisfy the properties that make $\Gamma(U,\mathscr{S})$ a module over the ring $\Gamma(U,\mathcal{O})$. So far life is straightforward.

But both $\mathcal{O}$ and $\mathscr{S}$ are sheaves. Given open sets $U \subset V \subset X$ we can restrict sections, of either the sheaf $\mathcal{O}$ or the sheaf $\mathscr{S}$, from the larger open set V to the smaller open set U. What we ask is that this restriction should be compatible with multiplication. More concretely: suppose that on the open set V we are given a section of the sheaf $\mathcal{O}$ and a section of the sheaf $\mathscr{S}$. In symbols, we are given $\lambda \in \Gamma(V,\mathcal{O})$ and $s \in \Gamma(V,\mathscr{S})$. We could multiply these to get a section of the sheaf $\mathscr{S}$ on the open set V. That is we obtain $\lambda s \in \Gamma(V,\mathscr{S})$. Now we can restrict. We have $\mathrm{res}_U^V(\lambda)$, which is a section of the sheaf $\mathcal{O}$ on the open set U, and $\mathrm{res}_U^V(s), \mathrm{res}_U^V(\lambda s)$, both of which are sections of the sheaf $\mathscr{S}$ on U. To say that the multiplication is compatible with restriction is to insist on the identity

$$\mathrm{res}_U^V(\lambda s) = \mathrm{res}_U^V(\lambda)\cdot \mathrm{res}_U^V(s)\,.$$

One can also express this in terms of a diagram. To say that the multiplication is compatible with restriction is to say that the following square commutes:

$$\begin{array}{ccc}
\Gamma(V,\mathcal{O})\times\Gamma(V,\mathscr{S}) & \xrightarrow{\;m_V\;} & \Gamma(V,\mathscr{S}) \\
{\scriptstyle \mathrm{res}_U^V\times\mathrm{res}_U^V}\big\downarrow & & \big\downarrow{\scriptstyle \mathrm{res}_U^V} \\
\Gamma(U,\mathcal{O})\times\Gamma(U,\mathscr{S}) & \xrightarrow[\;m_U\;]{} & \Gamma(U,\mathscr{S})
\end{array} \quad .$$

Example 7.1.3. Let $(X,\mathcal{O})$ and $(X,\mathcal{O}')$ be two ringed spaces. That is we take the same topological space X but consider two sheaves of rings on it. Suppose

$$(1,\Phi^*):(X,\mathcal{O}') \longrightarrow (X,\mathcal{O})$$

is a morphism of ringed spaces. That is the underlying map of topological spaces is the identity, but the sheaves of rings map to each other. We know many examples. If X is a C^ℓ–manifold we can take $\mathcal{O}$ to be the sheaf of ℓ–times continuously differentiable functions on X. For any $k<\ell$ we can let $\mathcal{O}'$ be the sheaf of k–times continuously differentiable functions. For every open set $U\subset X$ there is a ring homomorphism

$$\Phi_U^*:\Gamma(U,\mathcal{O}) \longrightarrow \Gamma(U,\mathcal{O}')\ ,$$

including the ℓ–times continuously differentiable functions into the k–times continuously differentiable ones. This map is clearly a ring homomorphism and is compatible with restrictions. It gives a morphism of ringed spaces.

In the example above the maps Φ_U^* were all injective. But this is not necessarily the case, and for a general map of ringed spaces

$$(1, \Phi^*) : (X, \mathcal{O}') \longrightarrow (X, \mathcal{O})$$

we can define $\Gamma(U, \mathscr{S})$ to be the kernel of the ring homomorphism $\Phi_U^* : \Gamma(U, \mathcal{O}) \longrightarrow \Gamma(U, \mathcal{O}')$. This makes $\Gamma(U, \mathscr{S})$ an ideal in the ring $\Gamma(U, \mathcal{O})$, and in particular it is a module over $\Gamma(U, \mathcal{O})$. The fact that we have a map of ringed spaces means that the square below must commute

$$\begin{array}{ccc} \Gamma(V, \mathcal{O}) & \xrightarrow{\Phi_V^*} & \Gamma(V, \mathcal{O}') \\ {\scriptstyle \mathrm{res}_U^V}\downarrow & & \downarrow{\scriptstyle \mathrm{res}_U^V} \\ \Gamma(U, \mathcal{O}) & \xrightarrow[\Phi_U^*]{} & \Gamma(U, \mathcal{O}') \end{array} ,$$

and hence the restriction

$$\mathrm{res}_U^V : \Gamma(V, \mathcal{O}) \longrightarrow \Gamma(U, \mathcal{O})$$

takes the kernel of Φ_V^* to the kernel of Φ_U^*. We have a natural restriction map $\mathrm{res}_U^V : \Gamma(V, \mathscr{S}) \longrightarrow \Gamma(U, \mathscr{S})$ rendering commutative the following square

$$\begin{array}{ccc} \Gamma(V, \mathscr{S}) & \xrightarrow{i_V^*} & \Gamma(V, \mathcal{O}) \\ {\scriptstyle \mathrm{res}_U^V}\downarrow & & \downarrow{\scriptstyle \mathrm{res}_U^V} \\ \Gamma(U, \mathscr{S}) & \xrightarrow[i_U^*]{} & \Gamma(U, \mathcal{O}) \end{array} ,$$

where i^* is the inclusion of the ideal. Since $\mathcal{O}$ is a sheaf of rings the restriction map $\mathrm{res}_U^V : \Gamma(V, \mathcal{O}) \longrightarrow \Gamma(U, \mathcal{O})$ respects multiplication, and hence multiplying a section of $\mathcal{O}$ by a section of $\mathscr{S}$ will be compatible with restriction. We have definitely defined a presheaf of $\mathcal{O}$–modules on X. Our next observation is that the sheaf property comes for free. Since we want to prove this in slightly greater generality we first make a couple of definitions.

Definition 7.1.4. *Let X be a topological space, and let $\mathscr{S}$ and $\mathscr{T}$ be sheaves of abelian groups on X. A* morphism of sheaves $f : \mathscr{S} \longrightarrow \mathscr{T}$ *is, for every open set $U \subset X$, a homomorphism of abelian groups*

$$\Gamma(U, f) : \Gamma(U, \mathscr{S}) \longrightarrow \Gamma(U, \mathscr{T}) .$$

Whenever we have an inclusion of open sets $U \subset V \subset X$ the following *square must commute:*

$$\begin{array}{ccc} \Gamma(V,\mathscr{S}) & \xrightarrow{\Gamma(V,f)} & \Gamma(V,\mathscr{T}) \\ {\scriptstyle \mathrm{res}_U^V}\downarrow & & \downarrow{\scriptstyle \mathrm{res}_U^V} \\ \Gamma(U,\mathscr{S}) & \xrightarrow[\Gamma(U,f)]{} & \Gamma(U,\mathscr{T}) \end{array} .$$

Example 7.1.5. Given a morphism of ringed spaces $(1, \Phi^*) : (X, \mathcal{O}') \longrightarrow (X, \mathcal{O})$, as in Example 7.1.3, the maps

$$\Phi_U^* : \Gamma(U, \mathcal{O}) \longrightarrow \Gamma(U, \mathcal{O}')$$

give a morphism of sheaves of abelian groups on X.

Lemma 7.1.6. *Let X be a topological space, and let $f : \mathscr{S} \longrightarrow \mathscr{T}$ be a morphism of sheaves of abelian groups on X. The presheaf* $\ker(f)$, *which to every open set $U \subset X$ assigns the abelian group*

$$\Gamma\big(U, \ker(f)\big) \quad = \quad \ker\big\{\Gamma(U,f) : \Gamma(U,\mathscr{S}) \longrightarrow \Gamma(U,\mathscr{T})\big\} \quad ,$$

is a sheaf.

Proof. Let U be an open subset of X, and let $U = \cup_i U_i$ be an open cover. Consider the following commutative diagram

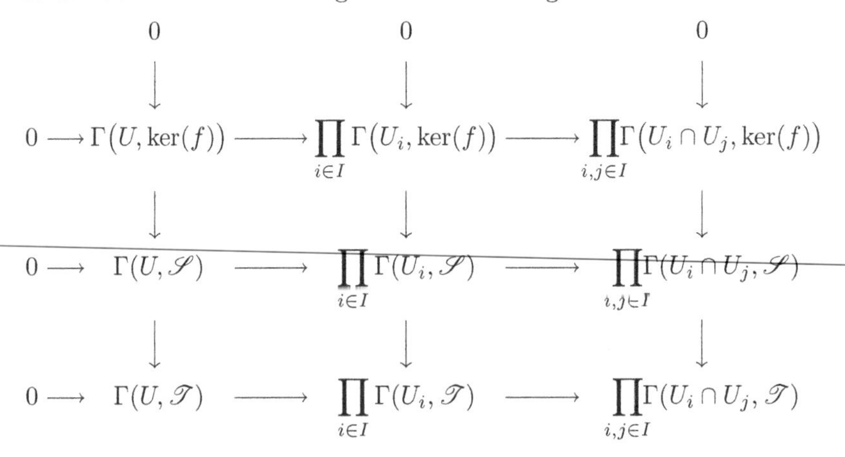

$$\begin{array}{ccccccc} & & 0 & & 0 & & 0 \\ & & \downarrow & & \downarrow & & \downarrow \\ 0 & \longrightarrow & \Gamma\big(U,\ker(f)\big) & \longrightarrow & \prod_{i\in I}\Gamma\big(U_i,\ker(f)\big) & \longrightarrow & \prod_{i,j\in I}\Gamma\big(U_i\cap U_j,\ker(f)\big) \\ & & \downarrow & & \downarrow & & \downarrow \\ 0 & \longrightarrow & \Gamma(U,\mathscr{S}) & \longrightarrow & \prod_{i\in I}\Gamma(U_i,\mathscr{S}) & \longrightarrow & \prod_{i,j\in I}\Gamma(U_i\cap U_j,\mathscr{S}) \\ & & \downarrow & & \downarrow & & \downarrow \\ 0 & \longrightarrow & \Gamma(U,\mathscr{T}) & \longrightarrow & \prod_{i\in I}\Gamma(U_i,\mathscr{T}) & \longrightarrow & \prod_{i,j\in I}\Gamma(U_i\cap U_j,\mathscr{T}) \end{array}$$

The columns are exact by the definition of the presheaf $\ker(f)$. The second and third rows are exact because $\mathscr{S}$ and $\mathscr{T}$ are sheaves; see Lemma 2.2.10. It follows by a diagram chase that the first row is exact and, using Lemma 2.2.10 again, we conclude that $\ker(f)$ is a sheaf. □

Remark 7.1.7. Next we want to discuss the obvious extensions of Definition 7.1.4 and of Lemma 7.1.6 to sheaves of modules on a ringed space. Suppose that $(X, \mathcal{O})$ is a ringed space and suppose $\mathscr{S}$ and $\mathscr{T}$ are sheaves of $\mathcal{O}$–modules, as in Definition 7.1.2. Then a morphism of sheaves of $\mathcal{O}$–modules $f : \mathscr{S} \longrightarrow \mathscr{T}$ is, first of all, a morphism of sheaves of abelian groups as in Definition 7.1.4, but we insist further that, for every open set $U \subset X$, the homomorphism of abelian groups

$$\Gamma(U, f) : \Gamma(U, \mathscr{S}) \longrightarrow \Gamma(U, \mathscr{T})$$

be a homomorphism of modules over the ring $\Gamma(U, \mathcal{O})$. The kernel sheaf, defined in Lemma 7.1.6, is already known to be a sheaf. Clearly every $\Gamma\big(U, \ker(f)\big)$ is a $\Gamma(U, \mathcal{O})$–submodule of $\Gamma(U, \mathscr{S})$. This gives it the structure of a $\Gamma(U, \mathcal{O})$–module. And the fact that the square

$$\begin{array}{ccc} \Gamma(V, \mathcal{O}) \times \Gamma(V, \mathscr{S}) & \xrightarrow{m_V} & \Gamma(V, \mathscr{S}) \\ {\scriptstyle \mathrm{res}^V_U \times \mathrm{res}^V_U}\downarrow & & \downarrow{\scriptstyle \mathrm{res}^V_U} \\ \Gamma(U, \mathcal{O}) \times \Gamma(U, \mathscr{S}) & \xrightarrow[m_U]{} & \Gamma(U, \mathscr{S}) \end{array}$$

commutes easily allows one to conclude that so does the square

$$\begin{array}{ccc} \Gamma(V, \mathcal{O}) \times \Gamma\big(V, \ker f\big) & \xrightarrow{m_V} & \Gamma\big(V, \ker f\big) \\ {\scriptstyle \mathrm{res}^V_U \times \mathrm{res}^V_U}\downarrow & & \downarrow{\scriptstyle \mathrm{res}^V_U} \\ \Gamma(U, \mathcal{O}) \times \Gamma\big(U, \ker f\big) & \xrightarrow[m_U]{} & \Gamma\big(U, \ker f\big) \end{array} .$$

Thus $\ker(f)$ is a sheaf of $\mathcal{O}$–modules on X, and the natural inclusion $\ker(f) \longrightarrow \mathscr{S}$ is a morphism of sheaves of $\mathcal{O}$–modules.

Discussion 7.1.8. We have learned what are sheaves of modules over a ringed space. We have learned what are $\mathcal{O}$–module homomorphisms $f : \mathscr{S} \longrightarrow \mathscr{T}$ of sheaves of $\mathcal{O}$–modules. And we have learned that the kernel $\ker(f)$ is naturally a sheaf of $\mathcal{O}$–modules. There is another very general process we will use, which manufactures new sheaves of $\mathcal{O}$–modules out of old ones. This is the Hom–sheaf construction. Given two sheaves of $\mathcal{O}$–modules $\mathscr{S}$ and $\mathscr{T}$ there is a natural way to construct another sheaf of $\mathcal{O}$–modules, denoted $\mathscr{H}om(\mathscr{S}, \mathscr{T})$. We will explain the construction.

Before we launch into the detail let us observe what *does not* work. We might be tempted to define

$$\Gamma\big(U, \mathscr{H}om(\mathscr{S}, \mathscr{T})\big) \quad = \quad \mathrm{Hom}\big[\Gamma(U, \mathscr{S})\ ,\ \Gamma(U, \mathscr{T})\big] .$$

This might look reasonable, but there is no obvious restriction map. That is given open sets $U \subset V \subset X$, we do not see any way of starting with a homomorphism

$$\Phi : \Gamma(V, \mathscr{S}) \longrightarrow \Gamma(V, \mathscr{T})$$

and, in some reasonably natural fashion, forming the restriction

$$\Phi|_U^V : \Gamma(U, \mathscr{S}) \longrightarrow \Gamma(U, \mathscr{T}) .$$

Instead we make the definition which does work.

Definition 7.1.9. *Let $\mathscr{S}$ and $\mathscr{T}$ be sheaves of $\mathcal{O}$–modules on the ringed space $(X, \mathcal{O})$. The presheaf $\mathscr{H}om(\mathscr{S}, \mathscr{T})$ is defined by*

$$\Gamma\big(U, \mathscr{H}om(\mathscr{S}, \mathscr{T})\big) = \mathrm{Hom}(\mathscr{S}|_U, \mathscr{T}|_U) .$$

Remark 7.1.10. In other words $\Gamma\big(V, \mathscr{H}om(\mathscr{S}, \mathscr{T})\big)$ is defined to be the set of all homomorphisms of sheaves of $\mathcal{O}|_V$–modules $f : \mathscr{S}|_V \longrightarrow \mathscr{T}|_V$. Now the restriction map is obvious: given a morphism of sheaves of $\mathcal{O}$–modules on V we have, for every open subset $U' \subset V$, a $\Gamma(U', \mathcal{O})$–module homomorphism

$$\Gamma(U', f) : \Gamma(U', \mathscr{S}) \longrightarrow \Gamma(U', \mathscr{T})$$

satisfying certain properties. If $U \subset V$ is an open subset then we certainly have all these $\Gamma(U', \mathcal{O})$–module homomorphisms for every $U' \subset U$; we just throw out much of the data that came with $f : \mathscr{S}|_V \longrightarrow \mathscr{T}|_V$ to restrict it to the morphism $f|_U : \mathscr{S}|_U \longrightarrow \mathscr{T}|_U$. Because all our rings are commutative, given an element $\lambda \in \Gamma(V, \mathcal{O})$, and a morphism of sheaves of $\mathcal{O}$–modules $f : \mathscr{S}|_V \longrightarrow \mathscr{T}|_V$, we can form λf by the formula

$$\Gamma(U', \lambda f) = \lambda \cdot \Gamma(U', f) .$$

In other words it is very straightforward to check that we have a presheaf of $\mathcal{O}$–modules. We need to verify:

Lemma 7.1.11. *The presheaf $\mathscr{H}om(\mathscr{S}, \mathscr{T})$ of Definition 7.1.9 is a sheaf.*

Proof. We have seen this fact many times before, under slightly different guises. Let us explain.

Suppose we are given an open set $U \subset X$ and an open cover $U = \cup U_i$. We want to show that a morphism of sheaves $f : \mathscr{S}|_U \longrightarrow \mathscr{T}|_U$ can be uniquely reassembled from its restrictions to the open subsets $U_i \subset U$. I assert that this can be reduced to the fact, which we have often seen before, that sheaves and morphisms of sheaves are uniquely specified by their restrictions to a basis of the open sets.

Let us therefore define a basis of open sets on U

$$B = \{V \subset X \mid V \text{ is open, and } V \subset U_i \text{ for some } i\} .$$

In the proof of Lemma 5.9.1 we saw that this is a basis. Let $U_i \subset U$ be open. Note that giving a morphism of sheaves $f_i : \mathscr{S}|_{U_i} \longrightarrow \mathscr{T}|_{U_i}$ is precisely the same as giving, for every open subset $U' \subset U_i$, a homomorphism

$$\Gamma(U', f_i) : \Gamma(U', \mathscr{S}) \longrightarrow \Gamma(U', \mathscr{T})$$

compatible with restrictions. Giving maps $f_i : \mathscr{S}|_{U_i} \longrightarrow \mathscr{T}|_{U_i}$ for every i is the same as giving, for every U' in our basis B and for every U_i containing it, a homomorphism

$$\Gamma(U', f_i) : \Gamma(U', \mathscr{S}) \longrightarrow \Gamma(U', \mathscr{T}) .$$

Saying that $\mathrm{res}^{U_i}_{U_i \cap U_j}(f_i) = \mathrm{res}^{U_j}_{U_i \cap U_j}(f_j)$ amounts to asking that there be exactly one well defined map

$$\Gamma(U', f) : \Gamma(U', \mathscr{S}) \longrightarrow \Gamma(U', \mathscr{T}) ,$$

which depends only on the open set U' in the basis B, not on the U_i in which we happen to choose to include it. But we know that a collection of maps $\Gamma(U', \mathscr{S}) \longrightarrow \Gamma(U', \mathscr{T})$, defined on a basis of the open sets and compatible with restriction, extends uniquely to a morphism of sheaves.

□

Remark 7.1.12. We have now defined $\mathscr{H}om(\mathscr{S}, \mathscr{T})$ and shown that it is a sheaf of $\mathcal{O}$–modules on X. If $f : \mathscr{S} \longrightarrow \mathscr{S}'$ and $g : \mathscr{T} \longrightarrow \mathscr{T}'$ are morphisms of sheaves of $\mathcal{O}$–modules, then there is an induced map

$$\mathscr{H}om(f, g) : \mathscr{H}om(\mathscr{S}', \mathscr{T}) \longrightarrow \mathscr{H}om(\mathscr{S}, \mathscr{T}') ;$$

on the open set $U \subset X$ the map

$$\Gamma\big(U, \mathscr{H}om(f, g)\big) : \Gamma\big(U, \mathscr{H}om(\mathscr{S}', \mathscr{T})\big) \longrightarrow \Gamma\big(U, \mathscr{H}om(\mathscr{S}, \mathscr{T}')\big)$$

$$\begin{array}{lcl} \text{takes} & \{\rho : \mathscr{S}'|_U \longrightarrow \mathscr{T}|_U\} & \in \Gamma\big(U, \mathscr{H}om(\mathscr{S}', \mathscr{T}\big) \\ \text{to} & \big\{(g|_U) \circ \rho \circ (f|_U) : \mathscr{S}|_U \longrightarrow \mathscr{T}'|_U\big\} & \in \Gamma\big(U, \mathscr{H}om(\mathscr{S}, \mathscr{T}'\big) . \end{array}$$

And now for the final general construction of new sheaves out of old.

Definition 7.1.13. *Let $f : X \longrightarrow Y$ be a continuous map of topological spaces, and let $\mathscr{S}$ be a sheaf of abelian groups on X. The sheaf $f_*\mathscr{S}$ is defined by the following:*

(i) *If $U \subset Y$ is an open set, then*

$$\Gamma(U, f_*\mathscr{S}) = \Gamma(f^{-1}U, \mathscr{S}) .$$

(ii) *If $U \subset U' \subset Y$ are open sets, then* $\mathrm{res}_U^{U'} : \Gamma(U', f_*\mathscr{S}) \longrightarrow \Gamma(U, f_*\mathscr{S})$ *is the map*

$$\mathrm{res}_{f^{-1}U}^{f^{-1}U'} : \Gamma(f^{-1}U', \mathscr{S}) \longrightarrow \Gamma(f^{-1}U, \mathscr{S}) .$$

Lemma 7.1.14. *Definition 7.1.13 does give a sheaf.*

Proof. It clearly gives a presheaf. Let V be an open subset of Y and let $V = \cup_{i\in I} U_i$ be an open cover. We need to prove the exactness of the sequence

$$0 \longrightarrow \Gamma(V, f_*\mathscr{S}) \xrightarrow{\ \alpha\ } \prod_i \Gamma(U_i, f_*\mathscr{S}) \xrightarrow{\ \beta\ } \prod_{i,j\in I} \Gamma(U_i \cap U_j, f_*\mathscr{S}) .$$

But this follows from the fact that $f^{-1}V = \cup_{i\in I} f^{-1}U_i$ is an open cover for $f^{-1}V$, and on X the sheaf $\mathscr{S}$ delivers the exact sequence

$$0 \longrightarrow \Gamma(f^{-1}V, \mathscr{S}) \xrightarrow{\alpha} \prod_i \Gamma(f^{-1}U_i, \mathscr{S}) \xrightarrow{\beta} \prod_{i,j\in I} \Gamma(f^{-1}U_i \cap f^{-1}U_j, \mathscr{S}) .$$

□

Example 7.1.15. If $(f, f^*) : (X, \mathcal{O}_X) \longrightarrow (Y, \mathcal{O}_Y)$ is a morphism of ringed spaces, then for every open set $U \subset Y$ the morphism gives a ring homomorphism

$$f_U^* : \Gamma(U, \mathcal{O}_Y) \longrightarrow \Gamma(f^{-1}U, \mathcal{O}_X) ,$$

compatible with restrictions. Another way to phrase this is that the data of a morphism of ringed spaces is a continuous map $f : X \longrightarrow Y$, together with a homomorphism of sheaves of rings on Y

$$f^* : \mathcal{O}_Y \longrightarrow f_*\mathcal{O}_X .$$

Exercise 7.1.16. It might help the reader to verify the following facts, which can be thought of as generalizations of [1, page 18, fifth paragraph]:

(i) If $(X, \mathcal{O})$ is a ringed space then $\mathcal{O}$ can be viewed as a sheaf of $\mathcal{O}$–modules on X.

(ii) Let $\mathscr{S}$ be any sheaf of $\mathcal{O}$–modules on X, and let $\mathcal{O}$ be viewed as a sheaf of modules over itself, as in (i). Then there is a natural bijection

$$\mathrm{Hom}(\mathcal{O}, \mathscr{S}) \quad = \quad \Gamma(X, \mathscr{S}) .$$

[Hint: any map of sheaves $f : \mathcal{O} \longrightarrow \mathscr{S}$ induces a homomorphism of $\Gamma(X, \mathcal{O})$–modules

$$\Gamma(X, f) \ : \ \Gamma(X, \mathcal{O}) \longrightarrow \Gamma(X, \mathscr{S}) .$$

This homomorphism takes $1 \in \Gamma(X, \mathcal{O})$ to an element $s \in \Gamma(X, \mathscr{S})$. Show that the map taking $f : \mathcal{O} \longrightarrow \mathscr{S}$ to $s \in \Gamma(X, \mathscr{S})$ is a bijection.]

(iii) Let $\mathcal{O}$ be viewed as a sheaf of $\mathcal{O}$–modules, as in (i). Suppose $f : \mathcal{O} \longrightarrow \mathcal{O}$ is a morphism of sheaves of $\mathcal{O}$–modules on X. Show that there exists a unique element $a \in \Gamma(X, \mathcal{O})$ so that, for any open set $U \subset X$ and any $s \in \Gamma(U, \mathcal{O})$, the map

$$\Gamma(U, f) \ : \ \Gamma(U, \mathcal{O}) \longrightarrow \Gamma(U, \mathcal{O})$$

takes s to $as \in \Gamma(U, \mathcal{O})$. [Hint: Work out explicitly what the correspondence of (ii) means when $\mathscr{S} = \mathcal{O}$.]

(iv) Show that, if $f : \mathcal{O} \longrightarrow \mathcal{O}$ is an isomorphism, then the element $a \in \Gamma(X, \mathcal{O})$ produced in (iii) is invertible. It is a unit in the ring $\Gamma(X, \mathcal{O})$.

(v) Show that the sheaf $\mathscr{H}om(\mathcal{O}, \mathscr{S})$ is canonically isomorphic to $\mathscr{S}$. [Hint: use (ii) to identify $\mathrm{Hom}(\mathcal{O}|_U, \mathscr{S}|_U)$ with $\Gamma(U, \mathscr{S})$.]

7.2 The sheaves $\widetilde{M}$

In this book we care about schemes. Schemes are built out of the building blocks $\left(\mathrm{Spec}(R), \widetilde{R}\right)$, and it seems natural to begin by studying sheaves of $\widetilde{R}$–modules on $\mathrm{Spec}(R)$. The ringed space $\left(\mathrm{Spec}(R), \widetilde{R}\right)$ might be an elaborate construction, but at the core it contains no information beyond the ring R from which it was built. A cruder way of saying this is that we obtained $\left(\mathrm{Spec}(R), \widetilde{R}\right)$ by starting from a ring R and torturing it. It seems reasonable to expect that, if we start with an R–module and torture it somewhat, then we should expect to end up with a sheaf of $\widetilde{R}$–modules over $\mathrm{Spec}(R)$. In this section we will explain how this is done. We will use a sleight-of-hand, called Nagata's trick of idealization.

Construction 7.2.1. Let R be a ring and let M be an R–module. We produce a ring $R \oplus M$ as follows. The addition is just the addition in the module $R \oplus M$. The multiplication is given by the rule

$$[r \oplus m] \cdot [r' \oplus m'] = [rr' \oplus (r'm + rm')] .$$

Remark 7.2.2. The ring $R \oplus M$ contains M as a square-zero ideal. That is, for any pair of elements m, m' in M the multiplication rule makes $mm' = 0$.

Definition 7.2.3. *For any R–module M we define ring homomorphisms*

$$R \xrightarrow{\psi_M} R \oplus M \xrightarrow{\varphi_M} R$$

by the rules

$$\psi_M(r) = r \oplus 0\,, \qquad\qquad \varphi_M(r \oplus m) = r\,.$$

Remark 7.2.4. The composite $\varphi_M \psi_M : R \longrightarrow R$ is clearly the identity. Proposition 3.6.5 tells us that the ring homomorphisms φ_M and ψ_M induce maps of ringed spaces

$$\Big(\mathrm{Spec}(R), \widetilde{R}\Big) \xrightarrow{\left(\mathrm{Spec}(\varphi_M), \widetilde{\varphi}_M\right)} \Big(\mathrm{Spec}(R \oplus M), \widetilde{R \oplus M}\Big) \xrightarrow{\left(\mathrm{Spec}(\psi_M), \widetilde{\psi}_M\right)} \Big(\mathrm{Spec}(R), \widetilde{R}\Big)$$

and Proposition 3.7.1 establishes that the composite is the identity. We now proceed to study these maps.

Lemma 7.2.5. *The continuous maps of topological spaces*

$$\mathrm{Spec}(R) \xrightarrow{\mathrm{Spec}(\varphi_M)} \mathrm{Spec}(R \oplus M) \xrightarrow{\mathrm{Spec}(\psi_M)} \mathrm{Spec}(R)$$

are both homeomorphisms.

Proof. Every prime ideal contains all the nilpotent elements of the ring, and every element of $M \subset R \oplus M$ is nilpotent. Hence the prime ideals in $R \oplus M$ all contain M, that is all prime ideals of $R \oplus M$ can be expressed as $\varphi_M^{-1}\mathfrak{p}$ with $\mathfrak{p}$ a prime ideal in R. This proves that the map $\mathrm{Spec}(\varphi_M) : \mathrm{Spec}(R) \longrightarrow \mathrm{Spec}(R \oplus M)$ is surjective. Since the composite $\mathrm{Spec}(\psi_M)\,\mathrm{Spec}(\varphi_M)$ is the identity, the map $\mathrm{Spec}(\varphi_M)$ must also be injective. It is bijective and $\mathrm{Spec}(\psi_M)$ must be its two-sided inverse.

But we know that both maps are continuous, and are inverse to each other by the above. It follows they must both be homeomorphisms. □

What this says is that we have only one topological space, namely $X = \mathrm{Spec}(R)$, and maps of ringed spaces

$$(X, \widetilde{R}) \xrightarrow{\left(1, \widetilde{\varphi}_M\right)} (X, \widetilde{R \oplus M}) \xrightarrow{\left(1, \widetilde{\psi}_M\right)} (X, \widetilde{R})\,.$$

To simplify the notation: on the space X we have morphisms of sheaves of rings

$$\widetilde{R} \xrightarrow{\widetilde{\psi}_M} \widetilde{R \oplus M} \xrightarrow{\widetilde{\varphi}_M} \widetilde{R}\,.$$

This is precisely the situation of Example 7.1.3. On every open set we have two rings, and we also have ring homomorphisms

$$\Gamma(U, \widetilde{R}) \xrightarrow{\{\widetilde{\psi}_M\}_U} \Gamma\big(U, \widetilde{R \oplus M}\big) \xrightarrow{\{\widetilde{\varphi}_M\}_U} \Gamma(U, \widetilde{R}) .$$

Definition 7.2.6. *We define*

$$\Gamma(U, \widetilde{M}) \quad = \quad \ker\big(\{\widetilde{\varphi}_M\}_U\big) .$$

The discussion in Example 7.1.3 assures us that this is a presheaf of $\widetilde{R \oplus M}$–modules, but then the collection of ring homomorphisms $\{\widetilde{\psi}_M\}_U : \Gamma(U, \widetilde{R}) \longrightarrow \Gamma\big(U, \widetilde{R \oplus M}\big)$ makes every presheaf of $\widetilde{R \oplus M}$–modules into a presheaf of $\widetilde{R}$–modules. And Lemma 7.1.6 guarantees that $\widetilde{M}$ is a sheaf.

For every R–module M we have constructed on $\mathrm{Spec}(R)$ a sheaf $\widetilde{M}$ of $\widetilde{R}$–modules.

7.3 Localization for modules

In Section 7.2 we learned how, for every R–module M, one can construct on $\mathrm{Spec}(R)$ a sheaf $\widetilde{M}$ of $\widetilde{R}$–modules. The construction was based on Nagata's trick; we looked at the ring $R \oplus M$, at the two ring homomorphisms between R and $R \oplus M$, and then we studied the induced maps of spectra. Now recall that, for any ring S and any basic open set Y_f in $Y = \mathrm{Spec}(S)$, the ring $\Gamma(Y_f, \widetilde{S})$ is $S[1/f]$. Since we care about the case where $S = R \oplus M$ it might help to remind ourselves briefly about the localization of the module M at an element $f \in R$.

Reminder 7.3.1. Let M be an R–module, and let f be an element of R. The module $M[1/f]$ is the module of all elements m/f^n with $m \in M$ and $n \geq 0$. There is a homomorphism $\alpha_f : M \longrightarrow M[1/f]$, sending $m \in M$ to $m/1 \in M[1/f]$. This homomorphism satisfies the universal property that any map $\rho : M \longrightarrow N$, where f acts invertibly on N, factors uniquely through

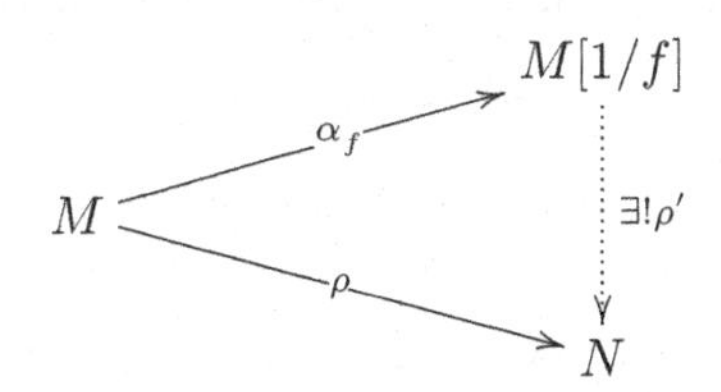

Let $\eta : M \longrightarrow N$ be a homomorphism of R–modules. There is an

induced map $\eta[1/f] : M[1/f] \longrightarrow N[1/f]$; it is the map given by the formula

$$\eta[1/f]\left(\frac{m}{f^n}\right) = \frac{\eta(m)}{f^n} .$$

We know further that the following square of R–module homomorphisms commutes:

$$\begin{array}{ccc} M & \xrightarrow{\eta} & N \\ {\scriptstyle \alpha_f}\downarrow & & \downarrow{\scriptstyle \alpha_f} \\ M[1/f] & \xrightarrow[\eta[1/f]]{} & N[1/f] \end{array} .$$

By the universal property of $\alpha_f : M \longrightarrow M[1/f]$, the commutative square uniquely defines $\eta[1/f]$. Furthermore, given any exact sequence of R–modules

$$0 \longrightarrow M' \longrightarrow M \longrightarrow M'' \longrightarrow 0$$

the localized sequence

$$0 \longrightarrow M'[1/f] \longrightarrow M[1/f] \longrightarrow M''[1/f] \longrightarrow 0$$

is also exact. For the proof see [1, Proposition 3.3].

Observation 7.3.2. The above gives us a map of R–modules

$$\beta_f : \mathrm{Hom}(M,N) \longrightarrow \mathrm{Hom}(M[1/f], N[1/f]) .$$

It is the map sending $\eta : M \longrightarrow N$ to the homomorphism $\beta_f(\eta) = \eta[1/f] : M[1/f] \longrightarrow N[1/f]$. The element $f \in R$ acts invertibly on the R–module $\mathrm{Hom}(M[1/f], N[1/f])$, and by the universal property of Reminder 7.3.1 we have a unique factorization of β_f as

$$\mathrm{Hom}(M,N) \xrightarrow{\alpha_f} \mathrm{Hom}(M,N)[1/f] \xrightarrow{\theta} \mathrm{Hom}(M[1/f], N[1/f]) .$$

In general the map θ is neither injective nor surjective. Nevertheless in the cases we care about it is an isomorphism. In preparation for the next lemma we should remind the reader that an R–module M is called *finitely presented* if it is the quotient of one finitely generated free module by the image of another. That is, if there is an exact sequence

$$R^m \longrightarrow R^n \longrightarrow M \longrightarrow 0 .$$

Lemma 7.3.3. *Suppose M is a finitely presented R–module and suppose that N is any R–module. Then the natural homomorphism*

$$\theta : \mathrm{Hom}(M,N)[1/f] \longrightarrow \mathrm{Hom}(M[1/f], N[1/f])$$

of Observation 7.3.2 is an isomorphism.

Proof. M will be finitely presented if and only if there exists an exact sequence

$$R^m \longrightarrow R^n \longrightarrow M \longrightarrow 0\,.$$

This will give us a commutative diagram with exact columns

$$\begin{array}{ccc} 0 & & 0 \\ \downarrow & & \downarrow \\ \mathrm{Hom}(M,N)[1/f] & \xrightarrow{\alpha} & \mathrm{Hom}\big(M[1/f], N[1/f]\big) \\ \downarrow & & \downarrow \\ \mathrm{Hom}(R^n,N)[1/f] & \xrightarrow{\beta} & \mathrm{Hom}\big(R^n[1/f], N[1/f]\big) \\ \downarrow & & \downarrow \\ \mathrm{Hom}(R^m,N)[1/f] & \xrightarrow{\gamma} & \mathrm{Hom}\big(R^m[1/f], N[1/f]\big) \end{array} .$$

The exactness of the columns is by [1, Proposition 2.9(i)]. The maps β and γ are isomorphisms. To see this note that, by [1, Page 18, fifth paragraph], we have

$$\mathrm{Hom}(R,N) = N\,, \qquad \mathrm{Hom}\big(R[1/f], N[1/f]\big) = N[1/f]\,.$$

Hence the natural map

$$\mathrm{Hom}(R,N)[1/f] \xrightarrow{\delta} \mathrm{Hom}\big(R[1/f], N[1/f]\big)$$

is the identity map $1 : N[1/f] \longrightarrow N[1/f]$. The maps and β and γ are just direct sums of the map δ with itself. An easy diagram chase allows us to conclude that α is an isomorphism. □

Reminder 7.3.4. It is important to remember that, if R is a finitely generated algebra over a field, then every finitely generated R–module is finitely presented. For the proof see [1, Proposition 6.5 and Corollary 7.7]. In this book we care almost only about rings finitely generated over $\mathbb{C}$. Their finitely generated modules, or equivalently finitely presented modules, will simply be called *finite modules.*

7.4 The sheaf of modules more explicitly

Now the time has come to work out concretely what the sheaves $\widetilde{M}$ actually are.

Lemma 7.4.1. *Let R be a ring, let M be an R–module and let $\widetilde{M}$ be the* *sheaf of $\widetilde{R}$–modules of Definition 7.2.6. Put $X = \mathrm{Spec}(R)$. If $X_f \subset X$ is a basic open set then*

$$\Gamma(X_f, \widetilde{M}) = M[1/f]\,,$$

and if $X_g \subset X_f$ is an inclusion of basic open sets then f acts invertibly on $M[1/g]$, hence the map $\alpha_g : M \longrightarrow M[1/g]$ factors uniquely through $\alpha_g^f : M[1/f] \longrightarrow M[1/g]$, and the restriction map

$$\mathrm{res}^{X_f}_{X_g} : \Gamma(X_f, \widetilde{M}) \longrightarrow \Gamma(X_g, \widetilde{M})$$

identifies with $\alpha_g^f : M[1/f] \longrightarrow M[1/g]$.

Remark 7.4.2. In particular the special case $f = 1$ tells us that $\Gamma(X, \widetilde{M}) = M$.

Proof. The maps of ringed spaces

$$(X, \widetilde{R}) \xrightarrow{\left(1, \widetilde{\varphi}_M\right)} \left(X, \widetilde{R \oplus M}\right) \xrightarrow{\left(1, \widetilde{\psi}_M\right)} (X, \widetilde{R})$$

are quite explicit; in the proof of Proposition 3.6.5 we computed them. In particular

$$\Gamma(X_f, \widetilde{R}) \xrightarrow{\{\widetilde{\psi}_M\}_{X_f}} \Gamma\big(X_f, \widetilde{R \oplus M}\big) \xrightarrow{\{\widetilde{\varphi}_M\}_{X_f}} \Gamma(X_f, \widetilde{R})$$

identify as

$$R[1/f] \xrightarrow{\psi[1/f]} R[1/f] \oplus M[1/f] \xrightarrow{\varphi[1/f]} R[1/f]\,.$$

This means that $\Gamma(X_f, \widetilde{M}) = \ker(\varphi[1/f])$ must be $M[1/f]$. Also the commutative diagram

$$\begin{array}{ccccc}
\Gamma(X_f, \widetilde{R}) & \xrightarrow{\{\widetilde{\psi}_M\}_{X_f}} & \Gamma\big(X_f, \widetilde{R \oplus M}\big) & \xrightarrow{\{\widetilde{\varphi}_M\}_{X_f}} & \Gamma(X_f, \widetilde{R}) \\
{\scriptstyle \mathrm{res}^{X_f}_{X_g}}\downarrow & & {\scriptstyle \mathrm{res}^{X_f}_{X_g}}\downarrow & & \downarrow{\scriptstyle \mathrm{res}^{X_f}_{X_g}} \\
\Gamma(X_g, \widetilde{R}) & \xrightarrow{\{\widetilde{\psi}_M\}_{X_g}} & \Gamma\big(X_g, \widetilde{R \oplus M}\big) & \xrightarrow{\{\widetilde{\varphi}_M\}_{X_g}} & \Gamma(X_g, \widetilde{R})
\end{array}$$

is completely explicit and identifies as

$$\begin{array}{ccccc}
R[1/f] & \xrightarrow{\psi[1/f]} & R[1/f] \oplus M[1/f] & \xrightarrow{\varphi[1/f]} & R[1/f] \\
{\scriptstyle \alpha_g^f}\downarrow & & {\scriptstyle \alpha_g^f}\downarrow & & \downarrow{\scriptstyle \alpha_g^f} \\
R[1/g] & \xrightarrow{\psi[1/g]} & R[1/g] \oplus M[1/g] & \xrightarrow{\varphi[1/g]} & R[1/g]
\end{array}\,,$$

and this means the map $\mathrm{res}^{X_f}_{X_g} : \Gamma(X_f, \widetilde{M}) \longrightarrow \Gamma(X_g, \widetilde{M})$ must agree with α_g^f. □

Lemma 7.4.3. *Let R be a ring, let M be an R–module and let $\widetilde{M}$ be the sheaf of $\widetilde{R}$–modules of Definition 7.2.6. Put $X = \mathrm{Spec}(R)$. If $X_f \subset X$ is a basic open set then the restriction of the sheaf $\widetilde{M}$ to X_f is the sheaf $\widetilde{M[1/f]}$ on $X_f = \mathrm{Spec}(R[1/f])$.*

Proof. On the space X we have morphisms of sheaves of rings

$$\widetilde{R} \xrightarrow{\ \widetilde{\psi}_M\ } \widetilde{R \oplus M} \xrightarrow{\ \widetilde{\varphi}_M\ } \widetilde{R}\,.$$

Now Proposition 3.7.3 tells us that, for any ring R and any $f \in R$, the map

$$\left(\mathrm{Spec}(\alpha_f), \widetilde{\alpha_f}\right) : \left(\mathrm{Spec}(R[1/f]), \widetilde{R[1/f]}\right) \longrightarrow \left(\mathrm{Spec}(R), \widetilde{R}\right)$$

in an open immersion, in other words on the space X_f the natural map of sheaves of rings is an isomorphism

$$\widetilde{R}|_{X_f} \quad = \quad \widetilde{R[1/f]}\,.$$

Applying Proposition 3.7.3 to $f \oplus 0 \in R \oplus M$ we deduce that

$$\widetilde{R \oplus M}|_{X_f} \quad = \quad \widetilde{\{R \oplus M\}[1/f]}\,.$$

This means that the restriction to the open set $X_f \subset X$, of the morphisms of sheaves of rings

$$\widetilde{R} \xrightarrow{\ \widetilde{\psi}_M\ } \widetilde{R \oplus M} \xrightarrow{\ \widetilde{\varphi}_M\ } \widetilde{R}\,,$$

identifies as

$$\widetilde{R[1/f]} \xrightarrow{\ \widetilde{\psi[1/f]_M}\ } \widetilde{\{R \oplus M\}[1/f]} \xrightarrow{\ \widetilde{\varphi[1/f]_M}\ } \widetilde{R[1/f]}\,.$$

Hence the sheaf kernel of $\widetilde{\varphi[1/f]_M}$ agrees with the sheaf kernel of $\widetilde{\varphi}_M$, restricted to $X_f \subset X$. □

7.5 Morphisms of sheaves

So far we have explained how, given an R–module M, we can construct a sheaf $\widetilde{M}$ of $\widetilde{R}$–modules on $\mathrm{Spec}(R)$. We all know about homomorphisms of R–modules, and it seems natural to generalize to homomorphisms of sheaves of $\widetilde{R}$–modules on $\mathrm{Spec}(R)$.

Proposition 7.5.1. *Let R be a ring, let M be an R–module, and let $\mathscr{T}$ be a sheaf of $\widetilde{R}$–modules on $X = \mathrm{Spec}(R)$. For any R–module homomorphism $\eta : M \longrightarrow \Gamma(X, \mathscr{T})$ there is a unique map of sheaves of $\widetilde{R}$–modules on X*

$$\widetilde{\eta} : \widetilde{M} \longrightarrow \mathscr{T}$$

so that, on global sections, the induced map

$$\Gamma(X,\widetilde{\eta}) : \Gamma(X,\widetilde{M}) \longrightarrow \Gamma(X,\mathscr{T})$$

agrees with $\eta : M \longrightarrow \Gamma(X,\mathscr{T})$.

Proof. Note that Remark 7.4.2 tells us $M = \Gamma(X,\widetilde{M})$, so the statement of the proposition is possible. We are given an R–module homomorphism $\eta : M \longrightarrow \Gamma(X,\mathscr{T})$. If $X_f \subset X$ is a basic open subset we form the composite

$$M \xrightarrow{\ \eta\ } \Gamma(X,\mathscr{T}) \xrightarrow{\ \mathrm{res}^X_{X_f}\ } \Gamma(X_f,\mathscr{T}) .$$

This composite is a homomorphism of R–modules, and the module $\Gamma(X_f,\mathscr{T})$ is a module over $R[1/f] = \Gamma(X_f,\widetilde{R})$. In other words, f must act invertibly on $\Gamma(X_f,\mathscr{T})$. By Reminder 7.3.1 there is a unique map $\eta[1/f] : M[1/f] \longrightarrow \Gamma(X_f,\mathscr{T})$ rendering commutative the square

$$\begin{array}{ccc}
M & \xrightarrow{\ \eta\ } & \Gamma(X,\mathscr{T}) \\
{\scriptstyle \alpha_f}\downarrow & & \downarrow{\scriptstyle \mathrm{res}^X_{X_f}} \\
M[1/f] & \xrightarrow[\eta[1/f]]{} & \Gamma(X_f,\mathscr{T}) .
\end{array}$$

This already proves the uniqueness of $\widetilde{\eta}$; if we assume that a morphism $\widetilde{\eta} : \widetilde{M} \longrightarrow \mathscr{T}$ exists, as in the statement of the proposition, then the commutative diagram

$$\begin{array}{ccccc}
M & = & \Gamma(X,\widetilde{M}) & \xrightarrow{\ \Gamma(X,\widetilde{\eta})\ } & \Gamma(X,\mathscr{T}) \\
{\scriptstyle \alpha_f}\downarrow & & {\scriptstyle \mathrm{res}^X_{X_f}}\downarrow & & \downarrow{\scriptstyle \mathrm{res}^X_{X_f}} \\
M[1/f] & = & \Gamma(X_f,\widetilde{M}) & \xrightarrow[\Gamma(X_f,\widetilde{\eta})]{} & \Gamma(X_f,\mathscr{T}) \quad .
\end{array}$$

forces $\Gamma(X_f,\eta)$ to agree with $\eta[1/f] : M[1/f] \longrightarrow \Gamma(X_f,\mathscr{T})$. Next we show that this works. Therefore define, for each basic open set $X_f \subset X$, the map $\Gamma(X_f,\widetilde{\eta})$ to be $\eta[1/f]$.

Let $X_g \subset X_f \subset X$ be open subsets. We consider the larger diagram

$$\begin{array}{ccccc}
M & = & \Gamma(X,\widetilde{M}) & \xrightarrow{\ \Gamma(X,\widetilde{\eta})\ } & \Gamma(X,\mathscr{T}) \\
{\scriptstyle \alpha_f}\downarrow & & {\scriptstyle \mathrm{res}^X_{X_f}}\downarrow & & \downarrow{\scriptstyle \mathrm{res}^X_{X_f}} \\
M[1/f] & = & \Gamma(X_f,\widetilde{M}) & \xrightarrow[\Gamma(X_f,\widetilde{\eta})]{} & \Gamma(X_f,\mathscr{T}) \\
{\scriptstyle \alpha^f_g}\downarrow & & {\scriptstyle \mathrm{res}^{X_f}_{X_g}}\downarrow & & \downarrow{\scriptstyle \mathrm{res}^{X_f}_{X_g}} \\
M[1/g] & = & \Gamma(X_g,\widetilde{M}) & \xrightarrow[\Gamma(X_g,\widetilde{\eta})]{} & \Gamma(X_g,\mathscr{T}) \quad .
\end{array}$$

What we know about the diagram, from the definition of $\Gamma(X_f, \widetilde{\eta})$ as $\eta[1/f]$, is that two of the squares commute. It follows that the three composites from top left to bottom right agree. Forgetting the irrelevant part of the diagram we have

$$\begin{array}{ccccc} M & & & & \\ \downarrow{\scriptstyle \alpha_f} & & & & \\ M[1/f] & = & \Gamma(X_f, \widetilde{M}) & \xrightarrow{\Gamma(X_f,\widetilde{\eta})} & \Gamma(X_f, \mathscr{T}) \\ \downarrow{\scriptstyle \alpha_g^f} & & \downarrow{\scriptstyle \mathrm{res}^{X_f}_{X_g}} & & \downarrow{\scriptstyle \mathrm{res}^{X_f}_{X_g}} \\ M[1/g] & = & \Gamma(X_g, \widetilde{M}) & \xrightarrow[\Gamma(X_g,\widetilde{\eta})]{} & \Gamma(X_g, \mathscr{T}) \end{array}$$

and the two composites from top left to bottom right agree. This common map has two factorizations through $\alpha_f : M \longrightarrow M[1/f]$, but, by the universal property of Reminder 7.3.1, these two factorizations must agree. Hence the square

$$\begin{array}{ccc} \Gamma(X_f, \widetilde{M}) & \xrightarrow{\Gamma(X_f,\widetilde{\eta})} & \Gamma(X_f, \mathscr{T}) \\ \downarrow{\scriptstyle \mathrm{res}^{X_f}_{X_g}} & & \downarrow{\scriptstyle \mathrm{res}^{X_f}_{X_g}} \\ \Gamma(X_g, \widetilde{M}) & \xrightarrow[\Gamma(X_g,\widetilde{\eta})]{} & \Gamma(X_g, \mathscr{T}) \end{array}$$

commutes. We have defined the morphism of sheaves on a basis for the open sets, compatibly with the restriction homomorphisms. The map extends to all open sets. □

Discussion 7.5.2. Consider the special case where $\mathscr{T} = \widetilde{N}$. In this case $\Gamma(X, \widetilde{N}) = N$, and an R–module homomorphism $\eta : M \longrightarrow \Gamma(X, \widetilde{N}) = N$ is simple enough. We learn from Proposition 7.5.1 that all morphisms of sheaves of $\widetilde{R}$–modules $\widetilde{M} \longrightarrow \widetilde{N}$ are of the form $\widetilde{\eta}$ for a unique $\eta : M \longrightarrow N$. The proposition delivers an explicit construction of $\widetilde{\eta}$, given $\eta : M \longrightarrow N$.

Of course we have another way we might wish to construct $\widetilde{\eta}$. Recall that the sheaf $\widetilde{M}$ was constructed from M by applying Nagata's trick; we considered the ring homomorphisms

$$R \xrightarrow{\psi_M} R \oplus M \xrightarrow{\varphi_M} R\ ,$$

and studied the corresponding maps of ringed spaces. Now that we have a map of R–modules $\eta : M \longrightarrow N$ we can apply Nagata's trick to it. We can look at the longer sequence of ring homomorphisms

$$R \xrightarrow{\psi_M} R \oplus M \xrightarrow{1 \oplus \eta} R \oplus N \xrightarrow{\varphi_N} R\ .$$

On the space $X = \mathrm{Spec}(R)$ there is an induced sequence of maps of sheaves of rings

$$\widetilde{R} \xrightarrow{\widetilde{\psi}_M} \widetilde{R \oplus M} \xrightarrow{\widetilde{1 \oplus \eta}} \widetilde{R \oplus N} \xrightarrow{\widetilde{\varphi}_N} \widetilde{R}\ .$$

The sheaf $\widetilde{M}$, which is the kernel of $\widetilde{\varphi}_M = \widetilde{\varphi}_N \circ \{\widetilde{1 \oplus \eta}\}$, maps to the sheaf $\widetilde{N}$, which is the kernel of $\widetilde{\varphi}_N$. This produces another map of sheaves of $\widetilde{R}$–modules $E : \widetilde{M} \longrightarrow \widetilde{N}$. Not surprisingly the two agree. We prove

Corollary 7.5.3. *With the notation as in Discussion 7.5.2, the two maps $E, \widetilde{\eta} : \widetilde{M} \longrightarrow \widetilde{N}$ agree.*

Proof. By Proposition 7.5.1 we need only check that the maps agree on global sections. The map E is induced by the morphisms of sheaves of rings

$$\widetilde{R} \xrightarrow{\widetilde{\psi}_M} \widetilde{R \oplus M} \xrightarrow{\widetilde{1 \oplus \eta}} \widetilde{R \oplus N} \xrightarrow{\widetilde{\varphi}_N} \widetilde{R}$$

and, taking global sections, this yields

$$R \xrightarrow{\psi_M} R \oplus M \xrightarrow{1 \oplus \eta} R \oplus N \xrightarrow{\varphi_N} R\ .$$

The map from the kernel M of $\varphi_M = \varphi_N \circ \{1 \oplus \eta\}$ to the kernel N of φ_N is precisely $\eta : M \longrightarrow N$. □

Corollary 7.5.4. *Let $\varepsilon : L \longrightarrow M$ and $\eta : M \longrightarrow N$ be homomorphisms of R–modules. Then they induce maps $\widetilde{\varepsilon} : \widetilde{L} \longrightarrow \widetilde{M}$, $\widetilde{\eta} : \widetilde{M} \longrightarrow \widetilde{N}$ and $\widetilde{\eta\varepsilon} : \widetilde{L} \longrightarrow \widetilde{N}$. We assert*

$$\widetilde{\eta\varepsilon} = \widetilde{\eta} \circ \widetilde{\varepsilon}.$$

Proof. Proposition 7.5.1 tells us that a morphism $\widetilde{L} \longrightarrow \widetilde{N}$ is uniquely determined by the map it induces on global sections. It therefore suffices to show that on global sections the maps agree, and this is obvious. □

Remark 7.5.5. By Proposition 7.5.1 and Discussion 7.5.2 we know that any map $\Phi : \widetilde{M} \longrightarrow \widetilde{N}$ is of the form $\widetilde{\eta}$, where $\eta : M \longrightarrow N$ is the homomorphism

$$M \quad = \quad \Gamma(X, \widetilde{M}) \xrightarrow{\Gamma(X,\Phi)} \Gamma(X, \widetilde{N}) \quad = \quad N\ .$$

In particular Φ will be an isomorphism if and only if $\Gamma(X, \Phi)$ is. Isomorphisms $\widetilde{M} \longrightarrow \widetilde{N}$ are in bijection with isomorphisms $M \longrightarrow N$. If a sheaf $\mathscr{S}$ is isomorphic to $\widetilde{M}$ then the module $M \cong \Gamma(X, \mathscr{S})$ and the isomorphism are unique up to canonical isomorphism.

Discussion 7.5.6. In Definition 7.1.9 we introduced the Hom–sheaves. Given two sheaves of $\mathcal{O}$–modules $\mathscr{S}$ and $\mathscr{T}$, Definition 7.1.9 tells us what we mean by the sheaf $\mathscr{H}om(\mathscr{S}, \mathscr{T})$. We end this section by explicitly working this out in the case where $\mathscr{S} = \widetilde{M}$ and $\mathscr{T} = \widetilde{N}$. The general definition of Hom–sheaves gives the recipe

$$\Gamma\big(U, \mathscr{H}om(\mathscr{S}, \mathscr{T})\big) \quad = \quad \mathrm{Hom}(\mathscr{S}|_U, \mathscr{T}|_U) \ .$$

Now let R be a ring, let $X = \mathrm{Spec}(R)$ and let M and N be R–modules. Suppose $U = X_f$ is a basic open set. Lemma 7.4.3 tells us that the sheaf $\widetilde{M}|_{X_f}$ is $\widetilde{M[1/f]}$, and similarly $\widetilde{N}|_{X_f}$ is $\widetilde{N[1/f]}$. Thus

$$\Gamma\big(X_f, \mathscr{H}om(\widetilde{M}, \widetilde{N})\big) \quad = \quad \mathrm{Hom}\Big(\widetilde{M[1/f]}, \widetilde{N[1/f]}\Big) \ .$$

But Proposition 7.5.1 now guarantees that every morphism of sheaves $\eta : \widetilde{M[1/f]} \longrightarrow \widetilde{N[1/f]}$ is uniquely determined by the induced map on global sections; that is

$$\mathrm{Hom}\Big(\widetilde{M[1/f]}, \widetilde{N[1/f]}\Big) \quad = \quad \mathrm{Hom}\big(M[1/f], N[1/f]\big) \ .$$

If $f = 1$ this tells us that $\Gamma\big(X, \mathscr{H}om(\widetilde{M}, \widetilde{N})\big) = \mathrm{Hom}(M, N)$. The identity map, from the R–module $\mathrm{Hom}(M, N)$ to the global sections of the sheaf $\mathscr{H}om(\widetilde{M}, \widetilde{N})$, induces by Proposition 7.5.1 a morphism of sheaves of $\widetilde{R}$–modules

$$\sigma : \widetilde{\mathrm{Hom}(M, N)} \longrightarrow \mathscr{H}om(\widetilde{M}, \widetilde{N}) \ .$$

We prove:

Lemma 7.5.7. *Let R be a ring, M a finitely presented R–module and N an arbitrary R–module. Then the map σ constructed above is an isomorphism.*

Proof. It suffices to prove that σ induces an isomorphism on every basic open set. Let X_f be a basic open set. The commutative square

$$\begin{array}{ccc}
\Gamma\Big(X, \widetilde{\mathrm{Hom}(M, N)}\Big) & \xrightarrow{\Gamma(X,\sigma)} & \Gamma\Big(X, \mathscr{H}om(\widetilde{M}, \widetilde{N})\Big) \\
{\scriptstyle \mathrm{res}^X_{X_f}}\Big\downarrow & & \Big\downarrow{\scriptstyle \mathrm{res}^X_{X_f}} \\
\Gamma\Big(X_f, \widetilde{\mathrm{Hom}(M, N)}\Big) & \xrightarrow[\Gamma(X_f,\sigma)]{} & \Gamma\Big(X_f, \mathscr{H}om(\widetilde{M}, \widetilde{N})\Big)
\end{array}$$

comes down to a commutative square

$$\begin{array}{ccc} \mathrm{Hom}(M,N) & \xrightarrow{\ 1\ } & \mathrm{Hom}(M,N) \\ {\scriptstyle \alpha_f}\downarrow & & \downarrow{\scriptstyle \beta_f} \\ \mathrm{Hom}(M,N)[1/f] & \xrightarrow[\Gamma(X_f,\sigma)]{} & \mathrm{Hom}\big(M[1/f],N[1/f]\big) \end{array} .$$

The map $\Gamma(X_f,\sigma)$ is therefore the unique map factoring β_f through α_f; it is the map which in Observation 7.3.2 we called θ. The fact that, under the hypotheses above, the map θ is an isomorphism may be found in Lemma 7.3.3. □

7.6 Coherent algebraic sheaves

Now it is time to globalize. In this entire section $(X, \mathcal{O})$ will be a fixed scheme locally of finite type over $\mathbb{C}$. We define

Definition 7.6.1. *A sheaf $\mathscr{S}$ of $\mathcal{O}$–modules over X is called* coherent *if it is locally isomorphic to $\widetilde{M}$, with M a finite R–module.*

Discussion 7.6.2. To say that $\mathscr{S}$ is coherent is to require, first of all, that there should be a cover of X by open immersions

$$(\Phi_i, \Phi_i^*) : \left(\mathrm{Spec}(R_i), \widetilde{R_i}\right) \longrightarrow (X, \mathcal{O}) ,$$

with R_i finitely generated $\mathbb{C}$–algebras. Further, for each i there should be a finite R_i–module M_i and an isomorphism

$$\widetilde{M_i} \quad \cong \quad \mathscr{S}|_{\mathrm{Spec}(R_i)} .$$

Recall that a finite R_i–module means a module with finitely many generators, which, for our rings R_i, is the same as a finitely presented module.

When we speak about coherent sheaves the term only makes sense on a ringed space, and the sheaves must always be sheaves of $\mathcal{O}$–modules. For this reason we will abbreviate "coherent sheaves of $\mathcal{O}$–modules over the ringed space $(X, \mathcal{O})$" to either "coherent sheaves over X" or "coherent sheaves over $(X, \mathcal{O})$".

Remark 7.6.3. By Remark 7.5.5 we note that, if $\mathscr{S}|_U \cong \widetilde{M}$, then the module M and the isomorphism are canonically unique.

Remark 7.6.4. It turns out to be true that, if $\mathscr{S}$ is coherent and $U \subset X$ is affine, then $\mathscr{S}|_U \cong \widetilde{M}$ for a finite module M. It also turns out that we will need a more general version of this fact, about coherent sheaves with a G–action for a suitable group G. Since I do not want to repeat the same proof twice we leave the proof for now. We prove the more general fact in Section 9.8, and deduce the claim we just made in Corollary 9.9.1.

Because we are postponing the proof of this claim, we have to spend a little time worrying about whether $\mathscr{S}|_U$ is isomorphic to $\widetilde{M}$ on any given open affine subset $U \subset X$. The next lemma deals with the difficulty in a way which suffices for now.

Lemma 7.6.5. *Given any finite number of coherent sheaves on $(X, \mathcal{O})$ there is an affine open cover $X = \cup U_i$ of X so that, simultaneously, on each of the open sets U_i all the sheaves are isomorphic to $\widetilde{M_j}$, where the M_j are all finite modules.*

Proof. We will prove this by induction on the number of coherent sheaves. If there is only one coherent sheaf there is nothing to do; the existence of the cover is guaranteed by the definition. Assume the statement true for all sets of $< n$ coherent sheaves. Let $\{\mathscr{S}_j, 1 \le j \le n\}$ be a finite set of coherent sheaves. By induction we know that every point $p \in X$ has an affine neighborhood $(U, \mathcal{O}|_U) \cong \left(\text{Spec}(R), \widetilde{R}\right)$ so that, for all $1 \le j \le n-1$, there is an isomorphism $\mathscr{S}_j|_U \cong \widetilde{M_j}$ with M_j a finite R–module. Since $\mathscr{S}_n$ is coherent there is also a neighborhood $(V, \mathcal{O}|_V) \cong \left(\text{Spec}(S), \widetilde{S}\right)$ of the point $p \in X$, and a finite S–module M_n with $\widetilde{M_n} \cong \mathscr{S}_n|_V$.

But Proposition 3.10.9 guarantees that $U \cap V = \cup_k W_k$, with W_k basic open affines in both $U \cong \text{Spec}(R)$ and $V \cong \text{Spec}(S)$. Choose a $W = W_k$ containing the point $p \in U \cap V$. There exists an $f \in R$ and a $g \in S$ with

$$\left(\text{Spec}(R[1/f]), \widetilde{R[1/f]}\right) \cong (W, \mathcal{O}|_W) \cong \left(\text{Spec}(S[1/g]), \widetilde{S[1/g]}\right) .$$

Lemma 7.4.3 tells us that $\mathscr{S}_j|_W \cong \widetilde{M_j[1/f]}$ if $1 \le j < n$, while $\mathscr{S}_n|_W \cong \widetilde{M_n[1/g]}$. □

Lemma 7.6.6. *Let $\mathscr{S}$ and $\mathscr{T}$ be coherent sheaves on $(X, \mathcal{O})$. Then the sheaf $\mathscr{H}om(\mathscr{S}, \mathscr{T})$ is coherent.*

Proof. The question is local. Choose a point $p \in X$; by Lemma 7.6.5 the point p has an open affine neighborhood $U \subset X$, so that

(i) $(U, \mathcal{O}|_U) \cong \left(\text{Spec}(R), \widetilde{R}\right)$, with R a finitely generated $\mathbb{C}$–algebra.
(ii) There are finite R–modules M and N and isomorphisms of sheaves of $\widetilde{R}$–modules over $\text{Spec}(R)$

$$\widetilde{M} \cong \mathscr{S}|_U , \qquad \widetilde{N} \cong \mathscr{T}|_U ,$$

and Lemma 7.5.7 tells us that $\mathscr{H}om(\widetilde{M},\widetilde{N})$ is isomorphic to the sheaf $\widetilde{\mathrm{Hom}(M,N)}$. □

When dealing with homomorphisms $\eta : M \longrightarrow N$ of modules over a ring, one of the most important facts is that we can associate to η its kernel, image and cokernel. This allows us to speak of exact sequences; the whole subject of homological algebra begins there. In the world of sheaves this is a little complicated, but for coherent sheaves the complications disappear.

Construction 7.6.7. Let $\eta : \mathscr{S} \longrightarrow \mathscr{T}$ be a morphism of coherent sheaves of $(X,\mathcal{O})$–modules. Then the kernel, image and cokernel of η are most naturally defined on a subbasis for the open sets. First we define the subbasis: B will be given by

$$B \quad = \quad \left\{ U \subset X \;\middle|\; \begin{array}{c} U \text{ is open in } X,\;\; (U,\mathcal{O}|_U) \cong \left(\mathrm{Spec}(R),\widetilde{R}\right), \\ R \text{ is a finitely generated algebra over } \mathbb{C}, \\ \text{and there exist finite } R\text{–modules } M \text{ and } N \\ \text{with } \widetilde{M} \cong \mathscr{S}|_U \text{ and } \widetilde{N} \cong \mathscr{T}|_U \end{array} \right\}.$$

First we observe that B is a subbasis. We need to show that every open set $V \subset X$ can be covered by open sets in B. That is, given any point $p \in V$, we need to find a $U \in B$ so that $p \in U \subset V$.

By Lemma 7.6.5 the point $p \in X$ is contained in some open set $U \in B$. The set $U \cap V$ is an open subset of $U \cong \mathrm{Spec}(R)$, and therefore contains a basic open subset U_f containing p. Lemma 7.4.3 tells us that, if U lies in B, then every basic open subset $U_f \subset U$ also lies in B. We have found a $U_f \in B$ with $p \in U_f \subset V$.

Definition 7.6.8. *On $U \in B$ we define*

$$\begin{array}{rcr} \Gamma\big(U,\ker(\eta)\big) & = & \ker\big\{\Gamma(U,\eta) : \Gamma(U,\mathscr{S}) \longrightarrow \Gamma(U,\mathscr{T})\big\} \\ \Gamma\big(U,\mathrm{im}(\eta)\big) & = & \mathrm{im}\big\{\Gamma(U,\eta) : \Gamma(U,\mathscr{S}) \longrightarrow \Gamma(U,\mathscr{T})\big\} \\ \Gamma\big(U,\mathrm{coker}(\eta)\big) & = & \mathrm{coker}\big\{\Gamma(U,\eta) : \Gamma(U,\mathscr{S}) \longrightarrow \Gamma(U,\mathscr{T})\big\} \end{array}$$

and the restriction maps are obvious.

Proposition 7.6.9. *The three formulas of Definition 7.6.8, giving* $\ker(\eta)$, $\mathrm{im}(\eta)$ *and* $\mathrm{coker}(\eta)$ *on open sets in B, extend to three coherent sheaves on $(X,\mathcal{O})$. More precisely: let U be an open set in B. By virtue of U belonging to B there must be an isomorphism $(U,\mathcal{O}|_U) \cong \left(\mathrm{Spec}(R),\widetilde{R}\right)$ for a finitely generated $\mathbb{C}$–algebra R. Further, $\mathscr{S}|_U \cong \widetilde{M}$, $\mathscr{T}|_U \cong \widetilde{N}$, and $\eta|_U : \mathscr{S}|_U \longrightarrow \mathscr{T}|_U$ is the map $\widetilde{\theta}$ for some $\theta : M \longrightarrow N$. If $K = \ker(\theta)$,*

$I = \mathrm{im}(\theta)$ and $Q = \mathrm{coker}(\theta)$ then, on the open set $U \subset X$, the three sheaves, whose existence we asserted above, are precisely $\widetilde{K}$, $\widetilde{I}$ and $\widetilde{Q}$.

Remark 7.6.10. In the case of the sheaf $\ker(\eta)$, Lemma 7.1.6 tells us there is an extension to a sheaf; only the coherence needs proof. For the other two even the existence of a sheaf is not clear.

Before launching into the proof of Proposition 7.6.9 let us discuss the implications, for exact sequences of coherent sheaves. First we define them.

Definition 7.6.11. *Let $(X, \mathcal{O})$ be a scheme, locally of finite type over $\mathbb{C}$. Suppose $\mathscr{R} \xrightarrow{\eta} \mathscr{S} \xrightarrow{\varepsilon} \mathscr{T}$ are morphisms of coherent sheaves of $(X, \mathcal{O})$–modules with $\varepsilon\eta = 0$. The sequence is* exact at $\mathscr{S}$ *if and only if* $\mathrm{im}(\eta) = \ker(\varepsilon)$, *with the subsheaves* $\mathrm{im}(\eta) \subset \mathscr{S}$ *and* $\ker(\varepsilon) \subset \mathscr{S}$ *as in Proposition 7.6.9.*

Remark 7.6.12. Exactness is local. This means the following. Suppose we have a sequence $\mathscr{R} \xrightarrow{\eta} \mathscr{S} \xrightarrow{\varepsilon} \mathscr{T}$ of morphisms of coherent sheaves of $(X, \mathcal{O})$–modules, and suppose the composite $\varepsilon\eta : \mathscr{R} \longrightarrow \mathscr{T}$ vanishes. Let $X = \cup_{i \in I} U_i$ be an open cover of X. The natural map of sheaves

$$\mathrm{im}(\eta) \longrightarrow \ker(\varepsilon)$$

is an isomorphism if and only if its restrictions, to every U_i in the cover, are all isomorphisms. This means that $\mathscr{R} \xrightarrow{\eta} \mathscr{S} \xrightarrow{\varepsilon} \mathscr{T}$ is exact if and only if all its restrictions

$$\mathscr{R}|_{U_i} \xrightarrow{\eta|_{U_i}} \mathscr{S}|_{U_i} \xrightarrow{\varepsilon|_{U_i}} \mathscr{T}|_{U_i}$$

are exact.

Remark 7.6.13. Let us figure out, concretely, what exactness is about. Let $\mathscr{R} \xrightarrow{\eta} \mathscr{S} \xrightarrow{\varepsilon} \mathscr{T}$ be morphisms of coherent sheaves of $(X, \mathcal{O})$–modules. For each of the two morphisms η, ε, Construction 7.6.7 produced a subbasis of open sets; we have two subbases, $B(\eta)$ and $B(\varepsilon)$. Lemma 7.6.5 guarantees that every point $p \in X$ is contained in some open set belonging to the intersection $B = B(\eta) \cap B(\varepsilon)$; the open sets in B cover X. Let $B' \subset B$ be any subcover. We assert that the sequence is exact at $\mathscr{S}$ precisely if, for every $U \in B'$, the sequence of vector spaces

$$\Gamma(U, \mathscr{R}) \xrightarrow{\Gamma(U,\eta)} \Gamma(U, \mathscr{S}) \xrightarrow{\Gamma(U,\varepsilon)} \Gamma(U, \mathscr{T})$$

is exact.

Let us prove this assertion. Choose any $U \in B'$. Let $I_U = \mathrm{im}\big[\Gamma(U, \eta)\big]$

and $K_U = \ker\left[\Gamma(U,\varepsilon)\right]$. The sequence

$$\Gamma(U,\mathscr{R}) \xrightarrow{\Gamma(U,\eta)} \Gamma(U,\mathscr{S}) \xrightarrow{\Gamma(U,\varepsilon)} \Gamma(U,\mathscr{T})$$

is exact if and only if the natural map $I_U \longrightarrow K_U$ is an isomorphism. The natural map $I_U \longrightarrow K_U$ is an isomorphism if and only if the map of sheaves $\widetilde{I}_U \longrightarrow \widetilde{K}_U$ is an isomorphism. Proposition 7.6.9 tells us that, on the open set $U \in B(\eta) \cap B(\varepsilon)$, we have

$$\mathrm{im}(\eta)|_U = \widetilde{I}_U\ , \qquad \ker(\varepsilon)|_U = \widetilde{K}_U\ .$$

The map of sheaves $\widetilde{I}_U \longrightarrow \widetilde{K}_U$ identifies with the restriction to U of the natural map $\mathrm{im}(\eta) \longrightarrow \ker(\varepsilon)$. It is an isomorphism if and only if the sequence

$$\mathscr{R}|_U \xrightarrow{\eta|_U} \mathscr{S}|_U \xrightarrow{\varepsilon|_U} \mathscr{T}|_U$$

is exact. Summarizing: we have proved that, for $U \in B'$, the sequence

$$(*) \qquad \Gamma(U,\mathscr{R}) \xrightarrow{\Gamma(U,\eta)} \Gamma(U,\mathscr{S}) \xrightarrow{\Gamma(U,\varepsilon)} \Gamma(U,\mathscr{T})$$

is exact if and only if the sequence

$$(**) \qquad \mathscr{R}|_U \xrightarrow{\eta|_U} \mathscr{S}|_U \xrightarrow{\varepsilon|_U} \mathscr{T}|_U$$

is.

In Remark 7.6.12 we showed that $\mathscr{R} \xrightarrow{\eta} \mathscr{S} \xrightarrow{\varepsilon} \mathscr{T}$ is exact if and only if, for all $U \in B'$, the sequence (**) is exact. Because the exactness of (**) is equivalent to the exactness of (*), our assertion follows.

In the special case, where $X = \mathrm{Spec}(R)$ is affine, this simplifies further. We have

Lemma 7.6.14. *Let R be a finitely generated $\mathbb{C}$–algebra. Let $M' \xrightarrow{\eta} M \xrightarrow{\varepsilon} M''$ be maps of finite R–modules, composing to zero. The sequence $\widetilde{M'} \xrightarrow{\widetilde{\eta}} \widetilde{M} \xrightarrow{\widetilde{\varepsilon}} \widetilde{M''}$, of coherent sheaves on $X = \mathrm{Spec}(R)$, is exact if and only if the sequence $M' \xrightarrow{\eta} M \xrightarrow{\varepsilon} M''$ is.*

Proof. The point here is that, under the hypotheses, X belongs to the set $B = B(\widetilde{\eta}) \cap B(\widetilde{\varepsilon})$ of Remark 7.6.13. Let $B' \subset B$ consist of the singleton $\{X\}$; clearly B' is a subcover, and the lemma follows from Remark 7.6.13. □

Caution 7.6.15. The reader is warned; even when the sequence $\mathscr{R} \longrightarrow \mathscr{S} \longrightarrow \mathscr{T}$ is exact, there will usually exist open subsets $U \subset X$ for which the sequence

$$\Gamma(U,\mathscr{R}) \xrightarrow{\Gamma(U,\eta)} \Gamma(U,\mathscr{S}) \xrightarrow{\Gamma(U,\varepsilon)} \Gamma(U,\mathscr{T})$$

is *not* exact. The open sets $U \in B$ are quite special.

It is time to begin the proof of Proposition 7.6.9.

Proof. We assert first that it suffices to prove Proposition 7.6.9 in the special case where X belongs to the subbasis B of open sets, defined in Construction 7.6.7. For now let us assume the special case, and show how to deduce the general statement.

Let X be general. If U is an open set belonging to B, then we can apply the special case of Proposition 7.6.9, to the map $\eta|_U : \mathscr{S}|_U \longrightarrow \mathscr{T}|_U$, to obtain, on U, sheaves $\ker(\eta|_U)$, $\mathrm{im}(\eta|_U)$ and $\mathrm{coker}(\eta|_U)$. If U and U' are in B, then the sheaves constructed from U and from U' visibly agree on a subbasis of $U \cap U'$, namely on the open sets belonging to B; therefore the restrictions must agree on all open sets in $U \cap U'$. Now Lemma 5.9.1 establishes that they must extend uniquely to sheaves on all of X. The coherence is local, and on open sets in B the sheaves are of the form $\widetilde{K}$, $\widetilde{I}$ and $\widetilde{Q}$. □

Discussion 7.6.16. It therefore remains to prove the special case. Assume $X \in B$; we have on X sheaves $\widetilde{K}$, $\widetilde{I}$ and $\widetilde{Q}$, given to us in the statement of Proposition 7.6.9. On the open sets $V \subset X, V \in B$ we need to produce isomorphisms

$$\begin{array}{ccc} \Gamma(V,\widetilde{K}) & \longrightarrow & \Gamma\big(V,\ker(\eta)\big) \\ \Gamma(V,\widetilde{I}) & \longrightarrow & \Gamma\big(V,\mathrm{im}(\eta)\big) \\ \Gamma(V,\widetilde{Q}) & \longrightarrow & \Gamma\big(V,\mathrm{coker}(\eta)\big) \end{array}$$

compatible with the restriction maps. This would establish that $\widetilde{K}$, $\widetilde{I}$ and $\widetilde{Q}$ are sheaves on X extending, to all open sets $V \subset X$, the formulas given in Definition 7.6.8 for open sets $V \subset X, V \in B$.

Since $X \in B$, we know that X is affine; that is $(X, \mathcal{O}) = \Big(\mathrm{Spec}(R), \widetilde{R}\Big)$. Furthermore, $\mathscr{S} \cong \widetilde{M}$ and $\mathscr{T} \cong \widetilde{N}$. The map $\eta : \mathscr{S} \longrightarrow \mathscr{T}$ is a morphism $\widetilde{M} \longrightarrow \widetilde{N}$. By Proposition 7.5.1 η must be of the form $\widetilde{\theta}$ for some $\theta : M \longrightarrow N$. As in the statement of Proposition 7.6.9 we let $K = \ker(\theta)$, $I = \mathrm{im}(\theta)$ and $Q = \mathrm{coker}(\theta)$. They are finite R–modules by [1, Proposition 6.3(i), Proposition 6.5 and Corollary 7.7]. There is a sequence of maps of finite R–modules

$$K \longrightarrow M \longrightarrow I \longrightarrow N \longrightarrow Q$$

and hence a sequence of morphisms of sheaves on $(X, \mathcal{O}) = \Big(\mathrm{Spec}(R), \widetilde{R}\Big)$

$$\widetilde{K} \longrightarrow \widetilde{M} \longrightarrow \widetilde{I} \longrightarrow \widetilde{N} \longrightarrow \widetilde{Q}\,.$$

For every open set $V \subset X$ there are homomorphisms

$$\Gamma(V,\widetilde{K}) \longrightarrow \Gamma(V,\widetilde{M}) \longrightarrow \Gamma(V,\widetilde{I}) \longrightarrow \Gamma(V,\widetilde{N}) \longrightarrow \Gamma(V,\widetilde{Q})\ .$$

To show that $\widetilde{K}$, $\widetilde{I}$ and $\widetilde{Q}$ agree, on open sets $V \in B$, with the formulas of Definition 7.6.8 amounts to showing that, for any open subset $V \subset X, V \in B$, the modules $\Gamma(V,\widetilde{K})$, $\Gamma(V,\widetilde{I})$ and $\Gamma(V,\widetilde{Q})$ are, respectively, the kernel, image and cokernel of the map $\Gamma(V,\widetilde{\theta}) : \Gamma(V,\widetilde{M}) \longrightarrow \Gamma(V,\widetilde{N})$. The next lemma does this.

Lemma 7.6.17. *Let $(X,\mathcal{O}) = \left(\mathrm{Spec}(R),\widetilde{R}\right)$ be an affine scheme of finite type over $\mathbb{C}$. Let $\theta : M \longrightarrow N$ be a homomorphism of R–modules, and let $K = \ker(\theta)$, $I = \mathrm{im}(\theta)$ and $Q = \mathrm{coker}(\theta)$. For every open subset $V \subset X$ we have a sequence*

$$\Gamma(V,\widetilde{K}) \longrightarrow \Gamma(V,\widetilde{M}) \longrightarrow \Gamma(V,\widetilde{I}) \longrightarrow \Gamma(V,\widetilde{N}) \longrightarrow \Gamma(V,\widetilde{Q})\ .$$

If the open set V belongs to the subbasis B of Construction 7.6.7 then the modules $\Gamma(V,\widetilde{K})$, $\Gamma(V,\widetilde{I})$ and $\Gamma(V,\widetilde{Q})$ are, respectively, the kernel, image and cokernel of the map $\Gamma(V,\widetilde{\theta}) : \Gamma(V,\widetilde{M}) \longrightarrow \Gamma(V,\widetilde{N})$.

Proof. Let us start with the easiest case, namely $\widetilde{K}$. The maps of modules $K \longrightarrow M \longrightarrow N$ compose to zero, and hence the induced maps of sheaves $\widetilde{K} \longrightarrow \widetilde{M} \longrightarrow \widetilde{N}$ must also compose to zero. We deduce that, for every open set $U \subset X$, the composite

$$\Gamma(U,\widetilde{K}) \longrightarrow \Gamma(U,\widetilde{M}) \xrightarrow{\Gamma(U,\widetilde{\theta})} \Gamma(U,\widetilde{N})$$

vanishes, and the homomorphism $\Gamma(U,\widetilde{K}) \longrightarrow \Gamma(U,\widetilde{M})$ must therefore factor through the kernel of $\Gamma(U,\widetilde{\theta})$. There is, for every U, a map

$$\Gamma(U,\widetilde{K}) \longrightarrow \Gamma(U,\ker(\theta))\ ,$$

where $\ker(\theta)$ is the sheaf of Lemma 7.1.6. This is clearly compatible with restrictions; we have produced a morphism of sheaves $\widetilde{K} \longrightarrow \ker(\theta)$. We want to show it to be an isomorphism; it is enough to check this on a basis for X.

But on the sets $X_f \subset X$ this is easy; Lemma 7.4.1 informs us that the sequence

$$0 \longrightarrow \Gamma(X_f,\widetilde{K}) \longrightarrow \Gamma(X_f,\widetilde{M}) \xrightarrow{\Gamma(X_f,\widetilde{\theta})} \Gamma(X_f,\widetilde{N})$$

is simply

$$0 \longrightarrow K[1/f] \longrightarrow M[1/f] \xrightarrow{\theta[1/f]} N[1/f]\ ,$$

which is exact by Reminder 7.3.1.

The two remaining cases are $\widetilde{I}$ and $\widetilde{Q}$. We can view them as the same by regarding I as the cokernel of the map $K \longrightarrow M$; in the remainder of the proof we therefore focus on proving the assertion for $\widetilde{Q}$. We have an exact sequence of R–modules

$$M \xrightarrow{\theta} N \xrightarrow{\psi} Q \longrightarrow 0\,,$$

which induces maps $\widetilde{M} \longrightarrow \widetilde{N} \longrightarrow \widetilde{Q}$ of sheaves on X. Restricting to the open set $V \subset X$, $V \in B$ we have a sequence

$$\Gamma(V,\widetilde{M}) \xrightarrow{\Gamma(V,\widetilde{\theta})} \Gamma(V,\widetilde{N}) \xrightarrow{\Gamma(V,\widetilde{\psi})} \Gamma(V,\widetilde{Q}) \longrightarrow 0$$

which we need to prove exact. Define Q' to be the cokernel of $\Gamma(V,\widetilde{\theta})$; that is we have an exact sequence

$$\Gamma(V,\widetilde{M}) \xrightarrow{\Gamma(V,\widetilde{\theta})} \Gamma(V,\widetilde{N}) \longrightarrow Q' \longrightarrow 0$$

and a homomorphism $Q' \longrightarrow \Gamma(V,\widetilde{Q})$, which we need to prove an isomorphism.

But now recall that $V \in B$. This means two things:

(i) $(V, \mathcal{O}|_V) = \left(\mathrm{Spec}(S), \widetilde{S}\right)$ is an affine scheme of finite type over $\mathbb{C}$.

(ii) The sheaves $\widetilde{M}|_V$, $\widetilde{N}|_V$ are isomorphic to $\widetilde{M'}$, $\widetilde{N'}$ for some finite S–modules M' and N'.

The isomorphisms $\widetilde{M'} \cong \widetilde{M}|_V$ and $\widetilde{N'} \cong \widetilde{N}|_V$ give, on the open set V, isomorphisms

$$M' = \Gamma(V,\widetilde{M'}) \cong \Gamma(V,\widetilde{M})\,, \qquad N' = \Gamma(V,\widetilde{N'}) \cong \Gamma(V,\widetilde{N})\,.$$

Put more simply we have a commutative diagram

$$\begin{array}{ccccccc}
M' & \xrightarrow{\theta'} & N' & \xrightarrow{\psi'} & Q' & \longrightarrow & 0 \\
{\scriptstyle a}\downarrow & & {\scriptstyle b}\downarrow & & {\scriptstyle c}\downarrow & & \\
\Gamma(V,\widetilde{M}) & \xrightarrow[\Gamma(V,\widetilde{\theta})]{} & \Gamma(V,\widetilde{N}) & \xrightarrow[\Gamma(V,\widetilde{\psi})]{} & \Gamma(V,\widetilde{Q}) & &
\end{array}\;.$$

We know that the top row is exact and that a and b are isomorphisms, and want to prove that so is c. Proposition 7.5.1, applied to $\left(\mathrm{Spec}(S), \widetilde{S}\right)$, tells us that there is a bijective correspondence between maps of S–modules $Q' \longrightarrow \Gamma(V,\widetilde{Q})$ and morphisms of sheaves $\widetilde{Q'} \longrightarrow \widetilde{Q}|_V$. The commutative diagram above therefore gives a commutative diagram of sheaves of modules on V

$$\begin{array}{ccccc} \widetilde{M'} & \xrightarrow{\widetilde{\theta}'} & \widetilde{N'} & \xrightarrow{\widetilde{\psi}'} & \widetilde{Q'} \\ \alpha\downarrow & & \beta\downarrow & & \downarrow\gamma \\ \widetilde{M}|_V & \xrightarrow[\widetilde{\theta}|_V]{} & \widetilde{N}|_V & \xrightarrow[\widetilde{\psi}|_V]{} & \widetilde{Q}|_V \end{array},$$

where α and β are isomorphisms by hypothesis. We want to prove that the map $c : Q' \longrightarrow \Gamma(V, \widetilde{Q})$, which is also the map

$$\Gamma(V, \gamma) \; : \; \Gamma(V, \widetilde{Q'}) \longrightarrow \Gamma(V, \widetilde{Q})$$

ia an isomorphism. It certainly suffices to show that the map of sheaves $\gamma : \widetilde{Q'} \longrightarrow \widetilde{Q}|_V$ is an isomorphism.

We have two open affine subsets of X, namely $V \subset X$ and $X \subset X$. As it happens one of them is all of X, but Proposition 3.10.9 still applies. Since $(X, \mathcal{O}|_X) \cong \left(\text{Spec}(R), \widetilde{R}\right)$ and $(V, \mathcal{O}|_V) \cong \left(\text{Spec}(S), \widetilde{S}\right)$ Proposition 3.10.9 tells us that $V = V \cap X$ can be covered by open sets W_i which are basic open affines in both $X = \text{Spec}(R)$ and $V = \text{Spec}(S)$. It suffices to prove that the map of sheaves $\gamma : \widetilde{Q'} \longrightarrow \widetilde{Q}|_V$ is an isomorphism when restricted to each W_i. Fix i, and put $W_i = W = X_f = V_g$ for some $f \in R$ and $g \in S$. We restrict the morphisms of sheaves above to W, obtaining a commutative diagram

$$\begin{array}{ccccc} \widetilde{M'}|_W & \xrightarrow{\widetilde{\theta}'|_W} & \widetilde{N'}|_W & \xrightarrow{\widetilde{\psi}'|_W} & \widetilde{Q'}|_W \\ \alpha|_W\downarrow & & \beta|_W\downarrow & & \downarrow\gamma|_W \\ \widetilde{M}|_W & \xrightarrow[\widetilde{\theta}|_W]{} & \widetilde{N}|_W & \xrightarrow[\widetilde{\psi}|_W]{} & \widetilde{Q}|_W \end{array}.$$

Because $W = X_f$ Lemma 7.4.3 tells us that $\widetilde{Q}|_W \cong \widetilde{Q[1/f]}$. Because $W = V_g$ Lemma 7.4.3 tells us that $\widetilde{Q'}|_W = \widetilde{Q'[1/g]}$. Without being so explicit about the precise modules involved, on the affine open set W the sheaves $\widetilde{Q'}|_W$ and $\widetilde{Q}|_W$ are both of the form $\widetilde{(-)}$ for appropriate modules $(-)$. Remark 7.5.5 tells us that to show the map $\gamma|_W$ an isomorphism it suffices to establish that

$$\Gamma(W, \gamma) \; : \; \Gamma(W, \widetilde{Q'}|_W) \longrightarrow \Gamma(W, \widetilde{Q}|_W)$$

is an isomorphism. To do so we consider the commutative diagram

$$\begin{array}{ccccccc}
\Gamma(W,\widetilde{M'}) & \xrightarrow{\Gamma(W,\widetilde{\theta'})} & \Gamma(W,\widetilde{N'}) & \xrightarrow{\Gamma(W,\widetilde{\psi'})} & \Gamma(W,\widetilde{Q'}) & \longrightarrow & 0 \\
\downarrow{\scriptstyle \Gamma(W,\alpha)} & & \downarrow{\scriptstyle \Gamma(W,\beta)} & & \downarrow{\scriptstyle \Gamma(W,\gamma)} & & \\
\Gamma(W,\widetilde{M}) & \xrightarrow[\Gamma(W,\widetilde{\theta})]{} & \Gamma(W,\widetilde{N}) & \xrightarrow[\Gamma(W,\widetilde{\psi})]{} & \widetilde{\Gamma}(W,Q) & \longrightarrow & 0 \, .
\end{array}$$

We know that $\Gamma(W,\alpha)$ and $\Gamma(W,\beta)$ are isomorphisms. To prove that so is $\Gamma(W,\gamma)$ it suffices to prove that both rows are exact. We now show this.

For the bottom row, observe that we started with the exact sequence of R–modules $M \longrightarrow N \longrightarrow Q \longrightarrow 0$. On $X = \mathrm{Spec}(R)$ we deduced maps of sheaves $\widetilde{M} \longrightarrow \widetilde{N} \longrightarrow \widetilde{Q}$ which, by Lemma 7.4.1, evaluate on the open set $W = X_f$ to the sequence

$$M[1/f] \xrightarrow{\theta[1/f]} N[1/f] \longrightarrow Q[1/f] \longrightarrow 0 \, .$$

This sequence is exact by Reminder 7.3.1. Similarly for the top row: the exact sequence of S–modules $M' \longrightarrow N' \longrightarrow Q' \longrightarrow 0$ gives, on $V = \mathrm{Spec}(S)$, a sequence of sheaves $\widetilde{M'} \longrightarrow \widetilde{N'} \longrightarrow \widetilde{Q'}$, and, taking sections on the open set $W = V_g$, we find an exact sequence

$$M'[1/g] \xrightarrow{\theta'[1/g]} N'[1/g] \longrightarrow Q'[1/g] \longrightarrow 0 \, .$$

□

This finishes the results we will need in the remainder of this chapter. What follows is a technical lemma which will be useful in Section 9.8. We include it here because it is a very easy consequence of what we have just proved.

Lemma 7.6.18. *Let R be a finitely generated $\mathbb{C}$–algebra and let Q be a finite R–module. Put $(X,\mathcal{O}) = \left(\mathrm{Spec}(R),\widetilde{R}\right)$, and let $V \subset X$ be an open affine subset; that is, $(V,\mathcal{O}|_V) \cong \left(\mathrm{Spec}(S),\widetilde{S}\right)$ for a finitely generated algebra S. Then two things hold:*

(i) *The sheaf $\widetilde{Q}|_V$ is isomorphic to $\widetilde{Q'}$, for some finite module Q' over the ring S.*

(ii) *The canonical map*

$$S \otimes_R \Gamma(X,\widetilde{Q}) \longrightarrow \Gamma(V,\widetilde{Q})$$

is an isomorphism.

Proof. Since Q is finite it has a finite presentation. That is, there is an exact sequence

$$R^m \xrightarrow{\ \theta\ } R^n \longrightarrow Q \longrightarrow 0\,.$$

Put $M = R^m$, $N = R^n$ and $\theta : M \longrightarrow N$ as above. Then $\widetilde{M} = \mathcal{O}^m$ and $\widetilde{N} = \mathcal{O}^n$, and the restrictions to the open set $V \subset X$ are $\widetilde{S^m}$ and $\widetilde{S^n}$. In other words V belongs to the subbasis B of Construction 7.6.7. To prove (i) note that Proposition 7.6.9 tells us that $\widetilde{Q}$ is the cokernel sheaf, and that its restriction to V is a $\widetilde{Q'}$. To prove (ii) observe that, in the commutative diagram

$$\begin{array}{ccccccc}
\Gamma(X,\widetilde{M}) & \longrightarrow & \Gamma(X,\widetilde{N}) & \longrightarrow & \Gamma(X,\widetilde{Q}) & \longrightarrow & 0 \\
\Big\downarrow{\scriptstyle \mathrm{res}^X_V} & & \Big\downarrow{\scriptstyle \mathrm{res}^X_V} & & \Big\downarrow{\scriptstyle \mathrm{res}^X_V} & & \\
\Gamma(V,\widetilde{M}) & \longrightarrow & \Gamma(V,\widetilde{N}) & \longrightarrow & \Gamma(V,\widetilde{Q}) & \longrightarrow & 0
\end{array}\ ,$$

both rows are exact by Lemma 7.6.17; both the open set X and the open set V belong to B. But now the entire diagram is a commutative diagram of $R = \Gamma(X, \mathcal{O})$–modules, where the R–algebra $S = \Gamma(V, \mathcal{O})$ acts on the bottom row. There is an induced commutative diagram

$$\begin{array}{ccccccc}
S \otimes_R \Gamma(X,\widetilde{M}) & \longrightarrow & S \otimes_R \Gamma(X,\widetilde{N}) & \longrightarrow & S \otimes_R \Gamma(X,\widetilde{Q}) & \longrightarrow & 0 \\
\Big\downarrow{\scriptstyle \alpha} & & \Big\downarrow{\scriptstyle \beta} & & \Big\downarrow{\scriptstyle \gamma} & & \\
\Gamma(V,\widetilde{M}) & \longrightarrow & \Gamma(V,\widetilde{N}) & \longrightarrow & \Gamma(V,\widetilde{Q}) & \longrightarrow & 0
\end{array}\ .$$

The top row is still exact since the tensor product is right exact. To prove that γ is an isomorphism it therefore suffices to prove that α and β are. But α and β are, respectively, the natural maps $\alpha : S \otimes_R R^m \longrightarrow S^m$ and $\beta : S \otimes_R R^n \longrightarrow S^n$. □

7.7 Coherent analytic sheaves

In the previous sections we developed some of the theory of coherent algebraic sheaves on a scheme $(X, \mathcal{O})$ locally of finite type over $\mathbb{C}$. There is a parallel analytic theory. It is a little difficult to give an honest treatment of this theory here, since we did not ever define what an analytic space is. We produced examples: given a scheme $(X, \mathcal{O})$, locally of finite type over $\mathbb{C}$, then the analytification $(X^{\mathrm{an}}, \mathcal{O}^{\mathrm{an}})$ is an analytic space. As it happens these are the only analytic spaces we care about in this book. For further reading we have already referred the reader to the many fine treatments of the theory to be found elsewhere.

On any analytic space (whatever this is) there is a notion of coherent analytic sheaves. If the ringed space $(Y, \mathcal{O})$ is an analytic space then a coherent analytic sheaf on it is, first of all, a sheaf of $\mathcal{O}$–modules. But there is also the hypothesis that, on all sufficiently small polydiscs, it is finitely presented. This means the following: for $\mathscr{S}$ to be a coherent analytic sheaf every point p must have a neighborhood U, as well as a sequence of maps of sheaves of $\mathcal{O}$–modules

$$\mathcal{O}^m|_U \longrightarrow \mathcal{O}^n|_U \longrightarrow \mathscr{S}|_U ,$$

which, for any polydisc $V \subset U$, gives an exact sequence

$$\Gamma\big(V, \mathcal{O}^m|_U\big) \longrightarrow \Gamma\big(V, \mathcal{O}^n|_U\big) \longrightarrow \Gamma\big(V, \mathscr{S}|_U\big) \longrightarrow 0 .$$

Coherent analytic sheaves on analytic spaces have many of the same formal properties as coherent algebraic sheaves on schemes locally of finite type over $\mathbb{C}$. The two facts we want are

Lemma 7.7.1. *If $\mathscr{S}$ and $\mathscr{T}$ are coherent analytic sheaves, on an analytic space $(Y, \mathcal{O})$, then the sheaf $\mathscr{H}om(\mathscr{S}, \mathscr{T})$ is also a coherent analytic sheaf.*

Proof. The argument of Lemma 7.3.3 works in the analytic category, virtually unchanged. □

Lemma 7.7.2. *If $\eta : \mathscr{S} \longrightarrow \mathscr{T}$ is a homomorphism of coherent analytic sheaves, on an analytic space $(Y, \mathcal{O})$, then the kernel, image and cokernel of η are coherent analytic sheaves, and on polydiscs $U \subset Y$ they are obtained by the simple-minded formulas*

$$\begin{array}{rcr}
\Gamma\big(U, \ker(\eta)\big) & = & \ker\big\{\Gamma(U,\eta) : \Gamma(U,\mathscr{S}) \longrightarrow \Gamma(U,\mathscr{T})\big\} \\
\Gamma\big(U, \mathrm{im}(\eta)\big) & = & \mathrm{im}\big\{\Gamma(U,\eta) : \Gamma(U,\mathscr{S}) \longrightarrow \Gamma(U,\mathscr{T})\big\} \\
\Gamma\big(U, \mathrm{coker}(\eta)\big) & = & \mathrm{coker}\big\{\Gamma(U,\eta) : \Gamma(U,\mathscr{S}) \longrightarrow \Gamma(U,\mathscr{T})\big\}
\end{array}$$

Proof. See [3, Proposition 13 on page 132, Theorem 1 on page 134 and Theorem 14 on page 243]. □

7.8 The analytification of coherent algebraic sheaves

Let $(X, \mathcal{O})$ be a scheme locally of finite type over $\mathbb{C}$. We understand what is meant by coherent algebraic sheaves on $(X, \mathcal{O})$. We know, from the previous chapters, how to attach to $(X, \mathcal{O})$ an analytic space $(X^{\mathrm{an}}, \mathcal{O}^{\mathrm{an}})$. In Section 7.7 we briefly outlined the notion of coherent analytic sheaves on $(X^{\mathrm{an}}, \mathcal{O}^{\mathrm{an}})$. We now want to explain the analytification functor, which takes coherent algebraic sheaves to coherent analytic sheaves.

Construction 7.8.1. Let $(X, \mathcal{O})$ be a scheme locally of finite type over $\mathbb{C}$. Let $\mathscr{S}$ be a coherent algebraic sheaf on $(X, \mathcal{O})$. We do Nagata's trick globally. Define a sheaf of rings $\mathcal{O} \oplus \mathscr{S}$ on X; the rule is that

$$\Gamma(U, \mathcal{O} \oplus \mathscr{S}) \quad = \quad \Gamma(U, \mathcal{O}) \oplus \Gamma(U, \mathscr{S})$$

and the ring structure is still $(r \oplus s) \cdot (r' \oplus s') = \big(rr' \oplus (rs' + r's)\big)$. We have global maps of sheaves of rings on X

$$\mathcal{O} \xrightarrow{\quad \psi \quad} \mathcal{O} \oplus \mathscr{S} \xrightarrow{\quad \varphi \quad} \mathcal{O}$$

which, as in Definition 7.2.3, are given by the formulas $\psi(r) = r \oplus 0$ and $\varphi(r \oplus s) = r$.

Lemma 7.8.2. *The ringed space $(X, \mathcal{O} \oplus \mathscr{S})$ is a scheme locally of finite type over $\mathbb{C}$.*

Proof. Because the sheaf $\mathscr{S}$ is coherent X has an open cover $X = \cup U_i$, with $(U_i, \mathcal{O}|_{U_i}) \cong \Big(\mathrm{Spec}(R_i), \widetilde{R_i}\Big)$, so that

(i) Each R_i is finitely generated over $\mathbb{C}$.
(ii) The restriction of $\mathscr{S}$ to $(U_i, \mathcal{O}|_{U_i}) \cong \Big(\mathrm{Spec}(R_i), \widetilde{R_i}\Big)$ is $\widetilde{M_i}$, for some finite R_i–module M_i.

But then the ringed space $\big(U_i, \{\mathcal{O} \oplus \mathscr{S}\}|_{U_i}\big)$ is isomorphic, by the definition of $\widetilde{M_i}$, to $\Big(\mathrm{Spec}(R_i \oplus M_i), \widetilde{R_i \oplus M_i}\Big)$. We know that R_i is finitely generated as an algebra over $\mathbb{C}$, and M_i is finitely generated as a module over R_i. The union of a set of algebra generators for R_i and module generators for M_i generates the ring $R_i \oplus M_i$. Hence it is finitely generated over $\mathbb{C}$. □

Remark 7.8.3. We have maps of schemes locally of finite type over $\mathbb{C}$

$$(X, \mathcal{O}) \xrightarrow{(1,\varphi)} (X, \mathcal{O} \oplus \mathscr{S}) \xrightarrow{(1,\psi)} (X, \mathcal{O}) \ ,$$

and the results of Section 5.1 permit us to analytify. We deduce a commutative diagram of maps of ringed spaces

$$\begin{array}{ccccc}
(X^{\mathrm{an}}, \mathcal{O}^{\mathrm{an}}) & \xrightarrow{(\Phi,\Phi^*)} & (Y^{\mathrm{an}}, \{\mathcal{O} \oplus \mathscr{S}\}^{\mathrm{an}}) & \xrightarrow{(\Psi,\Psi^*)} & (X^{\mathrm{an}}, \mathcal{O}^{\mathrm{an}}) \\
{\scriptstyle (\lambda_X,\lambda_X^*)}\downarrow & & {\scriptstyle (\lambda_Y,\lambda_Y^*)}\downarrow & & {\scriptstyle (\lambda_X,\lambda_X^*)}\downarrow \\
(X, \mathcal{O}) & \xrightarrow[(1,\varphi)]{} & (X, \mathcal{O} \oplus \mathscr{S}) & \xrightarrow[(1,\psi)]{} & (X, \mathcal{O})
\end{array} .$$

Note that we are still being very careful with the notation. We know that X^{an} and Y^{an} have the same set of points, namely the closed points of X. But we are not yet sure of the complex topology of Y^{an}, which is determined by complex embeddings using the rings of functions.

But the maps

$$X^{\mathrm{an}} \xrightarrow{\ \Phi\ } Y^{\mathrm{an}} \xrightarrow{\ \Psi\ } X^{\mathrm{an}}$$

are continuous in the complex topology, and as maps of sets they are just the identity on the set $\mathrm{Max}(X)$ of closed points in X. It follows that they induce inverse homeomorphisms in the complex topology. There is only one space $X^{\mathrm{an}} = Y^{\mathrm{an}}$, and two sheaves of rings on it. The diagram of maps of ringed spaces simplifies to

$$\begin{array}{ccccc}
(X^{\mathrm{an}}, \mathcal{O}^{\mathrm{an}}) & \xrightarrow{(1,\varphi^{\mathrm{an}})} & (X^{\mathrm{an}}, \{\mathcal{O} \oplus \mathscr{S}\}^{\mathrm{an}}) & \xrightarrow{(1,\psi^{\mathrm{an}})} & (X^{\mathrm{an}}, \mathcal{O}^{\mathrm{an}}) \\
{\scriptstyle (\lambda_X,\lambda_X^*)}\downarrow & & {\scriptstyle (\lambda_X,\lambda_X^*)}\downarrow & & {\scriptstyle (\lambda_X,\lambda_X^*)}\downarrow \\
(X, \mathcal{O}) & \xrightarrow[(1,\varphi)]{} & (X, \mathcal{O} \oplus \mathscr{S}) & \xrightarrow[(1,\psi)]{} & (X, \mathcal{O})
\end{array} ,$$

and on the space X^{an} we have maps of sheaves of rings

$$\mathcal{O}^{\mathrm{an}} \xrightarrow{\ \psi^{\mathrm{an}}\ } \{\mathcal{O} \oplus \mathscr{S}\}^{\mathrm{an}} \xrightarrow{\ \varphi^{\mathrm{an}}\ } \mathcal{O}^{\mathrm{an}}$$

composing to the identity. The kernel of φ^{an} is a sheaf of ideals for the sheaf of rings $\{\mathcal{O} \oplus \mathscr{S}\}^{\mathrm{an}}$, but the map ψ^{an} makes it a sheaf of modules for $\mathcal{O}^{\mathrm{an}}$. We define

$$\mathscr{S}^{\mathrm{an}} \quad = \quad \ker(\varphi^{\mathrm{an}}) .$$

Remark 7.8.4. The diagram above, of maps of ringed spaces, gives on X a commutative diagram of homomorphisms of sheaves of rings

$$\begin{array}{ccccc}
\mathcal{O} & \xrightarrow{\ \psi\ } & \mathcal{O} \oplus \mathscr{S} & \xrightarrow{\ \varphi\ } & \mathcal{O} \\
{\scriptstyle \lambda_X^*}\downarrow & & {\scriptstyle \lambda_X^*}\downarrow & & {\scriptstyle \lambda_X^*}\downarrow \\
\{\lambda_X\}_*\mathcal{O}^{\mathrm{an}} & \xrightarrow[\{\lambda_X\}_*\psi]{} & \{\lambda_X\}_*\{\mathcal{O} \oplus \mathscr{S}\}^{\mathrm{an}} & \xrightarrow[\{\lambda_X\}_*\varphi]{} & \{\lambda_X\}_*\mathcal{O}^{\mathrm{an}}
\end{array} ;$$

see Example 7.1.15. We conclude that the kernel of φ maps to the kernel of φ^{an}; we have a map of sheaves of $(X, \mathcal{O})$–modules $\lambda_X^* : \mathscr{S} \longrightarrow \{\lambda_X\}_*\mathscr{S}^{\mathrm{an}}$.

Notation 7.8.5. Until now we have been denoting the map of ringed spaces, from the analytification $(X^{\mathrm{an}}, \mathcal{O}^{\mathrm{an}})$ to $(X, \mathcal{O})$, by the symbol

$$(\lambda_X, \lambda_X^*) : (X^{\mathrm{an}}, \mathcal{O}^{\mathrm{an}}) \longrightarrow (X, \mathcal{O}) .$$

In this section and the next the notation is a little silly; the space X

is fixed, and all that changes is the sheaf of rings $\mathcal{O}$. We will write the map as

$$(\lambda, \lambda^*) : (X^{\mathrm{an}}, \mathcal{O}^{\mathrm{an}}) \longrightarrow (X, \mathcal{O}) \; .$$

In our compactified notation Example 7.1.15 tells us that this map of ringed spaces consists of the following data

(i) A continuous map $\lambda : X^{\mathrm{an}} \longrightarrow X$, which is fixed since X is fixed.
(ii) A homomorphism $\lambda^* : \mathcal{O} \longrightarrow \lambda_* \mathcal{O}^{\mathrm{an}}$ of sheaves of rings over the topological space X.

What we have learned in Remark 7.8.4 is that maps λ^*, which were originally defined for sheaves of rings on X, can be extended to any coherent sheaf of $\mathcal{O}$–modules $\mathscr{S}$; there is a map which we now denote $\lambda^* : \mathscr{S} \longrightarrow \lambda_* \mathscr{S}^{\mathrm{an}}$.

Remark 7.8.6. Furthermore, given a map of coherent algebraic sheaves $\eta : \mathscr{S} \longrightarrow \mathscr{T}$, the commutative diagram

$$\begin{array}{ccccc}
(X, \mathcal{O}) & \xrightarrow{(1,\varphi_{\mathscr{T}})} & (X, \mathcal{O} \oplus \mathscr{T}) & \xrightarrow{(1,\psi_{\mathscr{T}})} & (X, \mathcal{O}) \\
{\scriptstyle (1,1)}\big\downarrow & & {\scriptstyle (1,1\oplus\eta)}\big\downarrow & & {\scriptstyle (1,1)}\big\downarrow \\
(X, \mathcal{O}) & \xrightarrow[(1,\varphi_{\mathscr{S}})]{} & (X, \mathcal{O} \oplus \mathscr{S}) & \xrightarrow[(1,\psi_{\mathscr{S}})]{} & (X, \mathcal{O})
\end{array} \quad ,$$

analytifies to give on X^{an} a commutative diagram of homomorphisms of sheaves of rings

$$\begin{array}{ccccc}
\mathcal{O}^{\mathrm{an}} & \xrightarrow{\psi^{\mathrm{an}}_{\mathscr{S}}} & \{\mathcal{O} \oplus \mathscr{S}\}^{\mathrm{an}} & \xrightarrow{\varphi^{\mathrm{an}}_{\mathscr{S}}} & \mathcal{O}^{\mathrm{an}} \\
{\scriptstyle 1}\big\downarrow & & {\scriptstyle \{1\oplus\eta\}^{\mathrm{an}}}\big\downarrow & & \big\downarrow{\scriptstyle 1} \\
\mathcal{O}^{\mathrm{an}} & \xrightarrow{\psi^{\mathrm{an}}_{\mathscr{T}}} & \{\mathcal{O} \oplus \mathscr{T}\}^{\mathrm{an}} & \xrightarrow{\varphi^{\mathrm{an}}_{\mathscr{T}}} & \mathcal{O}^{\mathrm{an}}
\end{array}$$

It follows that the kernel $\mathscr{S}^{\mathrm{an}}$ of the map $\varphi^{\mathrm{an}}_{\mathscr{S}}$ maps to the kernel $\mathscr{T}^{\mathrm{an}}$ of the map $\varphi^{\mathrm{an}}_{\mathscr{T}}$, and that the homomorphism is a homomorphism of sheaves of $\mathcal{O}^{\mathrm{an}}$–modules. We denote this map $\eta^{\mathrm{an}} : \mathscr{S}^{\mathrm{an}} \longrightarrow \mathscr{T}^{\mathrm{an}}$.

Summary 7.8.7. Summarizing, we have learnt

(i) Every coherent algebraic sheaf $\mathscr{S}$ produces a sheaf $\mathscr{S}^{\mathrm{an}}$ of $\mathcal{O}^{\mathrm{an}}$–modules on X^{an}. See Remark 7.8.3 for the construction.
(ii) For every coherent algebraic sheaf $\mathscr{S}$ there is a map of sheaves of $\mathcal{O}$–modules $\lambda^* : \mathscr{S} \longrightarrow \lambda_* \mathscr{S}^{\mathrm{an}}$. See Remark 7.8.4 for the construction.

(iii) Any map of coherent algebraic sheaves $\eta : \mathscr{S} \longrightarrow \mathscr{T}$ induces a map $\eta^{\mathrm{an}} : \mathscr{S}^{\mathrm{an}} \longrightarrow \mathscr{T}^{\mathrm{an}}$ of sheaves of $\mathcal{O}^{\mathrm{an}}$–modules on X^{an}. See Remark 7.8.6 for the construction.

Exercise 7.8.8. We leave to the reader to check the following easy statements.

(i) The analytification commutes with restriction. This means the following. Suppose $\mathscr{S}$ is a coherent algebraic sheaf and $U \subset X$ is an open subset. Then there is a canonical identification $\{\mathscr{S}|_U\}^{\mathrm{an}} = \mathscr{S}^{\mathrm{an}}|_{U^{\mathrm{an}}}$. Suppose $\eta : \mathscr{S} \longrightarrow \mathscr{T}$ is a morphism of coherent algebraic sheaves on $(X, \mathcal{O})$. Then $\eta^{\mathrm{an}}|_{U^{\mathrm{an}}} = \{\eta|_U\}^{\mathrm{an}}$.

(ii) For every morphism of coherent algebraic sheaves $\eta : \mathscr{S} \longrightarrow \mathscr{T}$ the following square commutes

$$\begin{array}{ccc} \mathscr{S} & \xrightarrow{\quad\eta\quad} & \mathscr{T} \\ {\scriptstyle\lambda^*}\big\downarrow & & \big\downarrow{\scriptstyle\lambda^*} \\ \lambda_*\mathscr{S}^{\mathrm{an}} & \xrightarrow[\lambda_*\eta^{\mathrm{an}}]{} & \lambda_*\mathscr{T}^{\mathrm{an}} \end{array} .$$

(iii) Given composable morphisms of coherent algebraic sheaves

$$\mathscr{R} \xrightarrow{\quad\eta\quad} \mathscr{S} \xrightarrow{\quad\varepsilon\quad} \mathscr{T} ,$$

we have $\{\varepsilon\eta\}^{\mathrm{an}} = \varepsilon^{\mathrm{an}}\eta^{\mathrm{an}}$.

All the facts we have explained so far are immediate consequences of the construction, and chasing some commutative diagrams of maps of ringed spaces. Not quite so trivial are the following.

Facts 7.8.9. The following are true:

(i) If $\mathscr{S}$ is a coherent algebraic sheaf then $\mathscr{S}^{\mathrm{an}}$ is a coherent analytic sheaf.

(ii) For any coherent algebraic sheaf $\mathscr{S}$ on X, and for any open subset $U \subset X$, the map

$$\lambda_U^* : \Gamma(U, \mathscr{S}) \longrightarrow \Gamma(U^{\mathrm{an}}, \mathscr{S}^{\mathrm{an}})$$

is injective.

(iii) The map λ_U^* is not usually surjective, but something weaker is true. Suppose $U \subset X$ is affine, and suppose $V \subset U^{\mathrm{an}}$ is a polydisc. Let $\mathscr{S}$ be a coherent algebraic sheaf over $(X, \mathcal{O})$, and suppose that on the affine open set $U \subset X$ there is an isomorphism $\mathscr{S}|_U \cong \widetilde{M}$ (in Corollary 9.9.1 we will learn that this is true for every affine

$U \subset X$). The map $\lambda^* : \mathscr{S} \longrightarrow \lambda_* \mathscr{S}^{\mathrm{an}}$ induces a homomorphism of $\Gamma(U, \mathcal{O})$–modules

$$\Gamma(U, \mathscr{S}) \xrightarrow{\lambda_U^*} \Gamma(U^{\mathrm{an}}, \mathscr{S}^{\mathrm{an}}) \xrightarrow{\mathrm{res}_V^{U^{\mathrm{an}}}} \Gamma(V, \mathscr{S}^{\mathrm{an}}) .$$

We assert two things about the composite map $\Gamma(U, \mathscr{S}) \longrightarrow \Gamma(V, \mathscr{S}^{\mathrm{an}})$:

(a) The image of $\Gamma(U, \mathscr{S})$ in $\Gamma(V, \mathscr{S}^{\mathrm{an}})$ is dense in the Fréchet topology.

(b) The image generates $\Gamma(V, \mathscr{S}^{\mathrm{an}})$ as a module over $\Gamma(V, \mathcal{O}^{\mathrm{an}})$.

(iv) If $\mathscr{R} \longrightarrow \mathscr{S} \longrightarrow \mathscr{T}$ is an exact sequence of coherent algebraic sheaves, then $\mathscr{R}^{\mathrm{an}} \longrightarrow \mathscr{S}^{\mathrm{an}} \longrightarrow \mathscr{T}^{\mathrm{an}}$ is an exact sequence of coherent analytic sheaves.

(v) Let $\mathscr{T}$ be a sheaf of $\mathcal{O}^{\mathrm{an}}$–modules on X^{an}. Then $\lambda_* \mathscr{T}$ is a sheaf of $\mathcal{O}$–modules on X. If $\mathscr{S}$ is a coherent algebraic sheaf, and if $\eta : \mathscr{S} \longrightarrow \lambda_* \mathscr{T}$ is any morphism of sheaves of $\mathcal{O}$–modules on X, then there is a unique map of $\mathcal{O}^{\mathrm{an}}$–modules $E : \mathscr{S}^{\mathrm{an}} \longrightarrow \mathscr{T}$ so that η is the composite

$$\mathscr{S} \xrightarrow{\lambda^*} \lambda_* \mathscr{S}^{\mathrm{an}} \xrightarrow{\lambda_* E} \lambda_* \mathscr{T} .$$

(vi) Let $\mathscr{S}$ and $\mathscr{T}$ be coherent algebraic sheaves. There is a natural map

$$\theta : \mathscr{H}om(\mathscr{S}, \mathscr{T}) \longrightarrow \lambda_* \mathscr{H}om(\mathscr{S}^{\mathrm{an}}, \mathscr{T}^{\mathrm{an}}) .$$

This map sends $\eta \in \Gamma\big(U, \mathscr{H}om(\mathscr{S}, \mathscr{T})\big)$, that is a morphism $\eta : \mathscr{S}|_U \longrightarrow \mathscr{T}|_U$, to $\eta^{\mathrm{an}} : \mathscr{S}^{\mathrm{an}}|_U \longrightarrow \mathscr{T}^{\mathrm{an}}|_U$. By (v) the map θ factorizes as

$$\mathscr{H}om(\mathscr{S}, \mathscr{T}) \xrightarrow{\lambda^*} \lambda_* \big\{\mathscr{H}om(\mathscr{S}, \mathscr{T})\big\}^{\mathrm{an}} \xrightarrow{\lambda_* \Theta} \lambda_* \mathscr{H}om(\mathscr{S}^{\mathrm{an}}, \mathscr{T}^{\mathrm{an}})$$

for a unique

$$\Theta : \big\{\mathscr{H}om(\mathscr{S}, \mathscr{T})\big\}^{\mathrm{an}} \longrightarrow \mathscr{H}om(\mathscr{S}^{\mathrm{an}}, \mathscr{T}^{\mathrm{an}}) .$$

We assert that Θ is an isomorphism.

Remark 7.8.10. The six statements in Facts 7.8.9 are not especially difficult, they simply are not totally trivial consequences of the construction. The commutative algebra used in the proofs is ever so slightly above the level of the commutative algebra we have used so far. We will include the proofs in Appendix 1.

7.9 The statement of GAGA

We are now ready to state GAGA in almost the generality in which we will prove it.

Theorem 7.9.1. *Let $(X, \mathcal{O})$ be a scheme of finite type over $\mathbb{C}$, and assume the complex space X^{an} is a compact, Hausdorff topological space. Then*

(i) *If $\mathscr{S}$ and $\mathscr{T}$ are two coherent algebraic sheaves on $(X, \mathcal{O})$, and if $f : \mathscr{S}^{\mathrm{an}} \longrightarrow \mathscr{T}^{\mathrm{an}}$ is a map of sheaves of $\mathcal{O}^{\mathrm{an}}$–modules between them, then there exists a unique map of sheaves of $\mathcal{O}$–modules $\varphi : \mathscr{S} \longrightarrow \mathscr{T}$ with $f = \varphi^{\mathrm{an}}$.*

(ii) *If $\mathscr{R}$ is a coherent analytic sheaf of $\mathcal{O}^{\mathrm{an}}$–modules over X^{an}, then there exists a coherent algebraic sheaf $\mathscr{S}$ of $\mathcal{O}$–modules and an isomorphism $\mathscr{R} \cong \mathscr{S}^{\mathrm{an}}$.*

Remark 7.9.2. Until now we have been working in the generality of schemes locally of finite type over $\mathbb{C}$, but the theorem assumes that X^{an} is compact. The compactness assumption forces any open cover of X^{an} to have a finite subcover, and one easily deduces that any open cover of X has a finite subcover.

Remark 7.9.3. We will not prove the theorem in the generality stated above, but only for projective space. As explained in the introduction the reduction of the general case to the case of projective space is standard, although beyond this book.

Next we should explain to the reader why the theorems of the introduction are consequences of Theorem 7.9.1. We begin with the easiest. We show

Lemma 7.9.4. *Theorem 1.1.5 follows from Theorem 7.9.1.*

Proof. Let $\mathscr{V}$ and $\mathscr{V}'$ be algebraic vector bundles on the scheme $(X, \mathcal{O})$. We need to prove that any analytic map $f : \mathscr{V}^{\mathrm{an}} \longrightarrow \{\mathscr{V}'\}^{\mathrm{an}}$ is equal to φ^{an} for a unique algebraic map $\varphi : \mathscr{V} \longrightarrow \mathscr{V}'$. But vector bundles are particularly simple special cases of coherent sheaves, and hence the assertion is immediate from Theorem 7.9.1(i). □

Lemma 7.9.5. *Theorem 1.1.3 follows from Theorem 7.9.1.*

Proof. We need to show that, if $(X, \mathcal{O})$ is a scheme of finite type over $\mathbb{C}$, X^{an} is compact and Hausdorff and Y is a closed analytic subspace (which is a closed subspace locally given by the vanishing of a finite number of holomorphic functions), then Y is algebraic.

Consider the ideal $\mathscr{I} \subset \mathcal{O}^{\mathrm{an}}$ of all holomorphic functions vanishing on Y. By assumption this ideal is locally finitely generated. It is locally the image of a map $\{\mathcal{O}^{\mathrm{an}}\}^n|_V \longrightarrow \mathcal{O}^{\mathrm{an}}|_V$. By Lemma 7.7.2 the image of a map of coherent analytic sheaves is coherent, and hence the ideal $\mathscr{I}|_V$ is a coherent analytic sheaf on V. But coherence is local, hence $\mathscr{I}$ is a coherent analytic sheaf on all of X^{an}. It follows from Theorem 7.9.1(ii) that there is a coherent algebraic sheaf $\mathscr{S}$ on $(X, \mathcal{O})$ and an isomorphism $\mathscr{I} \cong \mathscr{S}^{\mathrm{an}}$. The inclusion $\mathscr{I} \longrightarrow \mathcal{O}^{\mathrm{an}}$, of the ideal $\mathscr{I}$ into the analytic functions, is a map of $\mathcal{O}^{\mathrm{an}}$–modules $f : \mathscr{S}^{\mathrm{an}} \longrightarrow \mathcal{O}^{\mathrm{an}}$, and Theorem 7.9.1(i) tells us that $f = \varphi^{\mathrm{an}}$ for some map of $\mathcal{O}$–modules $\varphi : \mathscr{S} \longrightarrow \mathcal{O}$. Now consider the diagram of sheaves on X

$$\begin{array}{ccccc} & & \mathscr{S} & \xrightarrow{\varphi} & \mathcal{O} \\ & & \lambda^*\downarrow & & \downarrow\lambda^* \\ \lambda_*\mathscr{I} & = & \lambda_*\mathscr{S}^{\mathrm{an}} & \xrightarrow[\lambda_*\varphi^{\mathrm{an}}=\lambda_* f]{} & \lambda_*\mathcal{O}^{\mathrm{an}} \end{array} .$$

Maybe it is more illuminating to say this in terms of modules over open sets: for every open affine $U \subset X$ and for every polydisc $V \subset U^{\mathrm{an}}$ we have a commutative diagram

$$\begin{array}{ccccc} & & \Gamma(U,\mathscr{S}) & \xrightarrow{\Gamma(U,\varphi)} & \Gamma(U,\mathcal{O}) \\ & & \downarrow & & \downarrow \\ \Gamma(V,\mathscr{I}) & = & \Gamma(V,\mathscr{S}^{\mathrm{an}}) & \xrightarrow[\Gamma(V,\varphi^{\mathrm{an}})]{} & \Gamma(V,\mathcal{O}^{\mathrm{an}}) \end{array} .$$

Fact 7.8.9(iii)(b) tells us that the image of $\Gamma(U,\mathscr{S}) \longrightarrow \Gamma(V,\mathscr{S}^{\mathrm{an}})$ generates $\Gamma(V,\mathscr{S}^{\mathrm{an}})$ as a module over the ring $\Gamma(V,\mathcal{O}^{\mathrm{an}})$. The image of the composite

$$\begin{array}{ccccc} & & \Gamma(U,\mathscr{S}) & & \\ & & \downarrow & & \\ \Gamma(V,\mathscr{I}) & = & \Gamma(V,\mathscr{S}^{\mathrm{an}}) & \xrightarrow[\Gamma(V,\varphi^{\mathrm{an}})]{} & \Gamma(V,\mathcal{O}^{\mathrm{an}}) \end{array}$$

must therefore generate the ideal $\Gamma(V,\mathscr{I}) \subset \Gamma(V,\mathcal{O}^{\mathrm{an}})$. But the commutativity of the square says that this is also the image of

$$\begin{array}{ccc} \Gamma(U,\mathscr{S}) & \xrightarrow{\Gamma(U,\varphi)} & \Gamma(U,\mathcal{O}) \\ & & \downarrow \\ & & \Gamma(V,\mathcal{O}^{\mathrm{an}}) \end{array} ,$$

and the image of $\Gamma(U,\varphi)$ is an ideal of $\Gamma(U,\mathcal{O})$ which generates $\mathcal{I}\subset\mathcal{O}^{\mathrm{an}}$ in the ring $\Gamma(V,\mathcal{O}^{\mathrm{an}})$. □

It remains to prove

Lemma 7.9.6. *Theorem 1.1.4 follows from Theorem 7.9.1.*

Proof. Strangely enough this is slightly harder than the previous statements. Given an analytic vector bundle $\mathscr{V}'$ on $(X^{\mathrm{an}},\mathcal{O}^{\mathrm{an}})$ we need to find an algebraic vector bundle $\mathscr{V}$ on $(X,\mathcal{O})$ with $\mathscr{V}^{\mathrm{an}}=\mathscr{V}'$. We know that $\mathscr{V}'$ is a coherent analytic sheaf, and Theorem 7.9.1(ii) tells us that there exists a coherent algebraic sheaf $\mathscr{V}$ with $\mathscr{V}^{\mathrm{an}}\cong\mathscr{V}'$. The problem is to show that the coherent algebraic sheaf $\mathscr{V}$ is a vector bundle. Although not difficult, this is another of the statements that we leave to Appendix 1. It again involves commutative algebra slightly above the level we have seen so far. □

Discussion 7.9.7. We end the chapter with just a little hint about the proof of Theorem 7.9.1(i). We first remind the reader of Fact 7.8.9(vi). We constructed a morphism of sheaves

$$\theta:\mathscr{H}om(\mathscr{S},\mathscr{T})\longrightarrow\lambda_*\mathscr{H}om(\mathscr{S}^{\mathrm{an}},\mathscr{T}^{\mathrm{an}})\,.$$

If we only look at what the map does to global sections, that is if we consider

$$\Gamma\big(X,\mathscr{H}om(\mathscr{S},\mathscr{T})\big)\xrightarrow{\Gamma(X,\theta)}\Gamma\big(X,\lambda_*\mathscr{H}om(\mathscr{S}^{\mathrm{an}},\mathscr{T}^{\mathrm{an}})\big)\,,$$

then, by construction, the map takes $\eta:\mathscr{S}\longrightarrow\mathscr{T}$, which is an element of $\Gamma\big(X,\mathscr{H}om(\mathscr{S},\mathscr{T})\big)$, to its analytification $\eta^{\mathrm{an}}:\mathscr{S}^{\mathrm{an}}\longrightarrow\mathscr{T}^{\mathrm{an}}$. Theorem 7.9.1(i) is the assertion that $\Gamma(X,\theta)$ is an isomorphism, provided X^{an} is compact and Hausdorff. So far we have merely reformulated Theorem 7.9.1(i).

In Fact 7.8.9(v) we saw that the morphism of sheaves θ factorizes as

$$\mathscr{H}om(\mathscr{S},\mathscr{T})\xrightarrow{\lambda^*}\lambda_*\big\{\mathscr{H}om(\mathscr{S},\mathscr{T})\big\}^{\mathrm{an}}\xrightarrow{\lambda_*\Theta}\lambda_*\mathscr{H}om(\mathscr{S}^{\mathrm{an}},\mathscr{T}^{\mathrm{an}})\,.$$

Fact 7.8.9(vi) tells us that, in this composition, the map Θ is an isomorphism; it induces an isomorphism on every open set in X^{an}, and hence $\lambda_*\Theta$ induces an isomorphism on every open set in X (of which there are fewer). In particular $\Gamma(X,\lambda_*\Theta)$ is an isomorphism. It suffices therefore to prove that, in the factorisation above, the map

$$\Gamma\big(X,\mathscr{H}om(\mathscr{S},\mathscr{T})\big)\xrightarrow{\Gamma(X,\lambda^*)}\Gamma\Big(X,\lambda_*\big\{\mathscr{H}om(\mathscr{S},\mathscr{T})\big\}^{\mathrm{an}}\Big)$$

is an isomorphism. Theorem 7.9.1(i) certainly follows from

Theorem 7.9.8. *Let $(X, \mathcal{O})$ be a scheme of finite type over $\mathbb{C}$, and assume the complex space X^{an} is a compact, Hausdorff topological space. Then, for any coherent algebraic sheaf $\mathscr{R}$, the natural map of global sections*

$$\Gamma(X, \lambda^*) \; : \; \Gamma(X, \mathscr{R}) \longrightarrow \Gamma(X, \lambda_* \mathscr{R}^{\mathrm{an}}) \quad = \quad \Gamma(X^{\mathrm{an}}, \mathscr{R}^{\mathrm{an}})$$

is an isomorphism.

Remark 7.9.9. In Discussion 7.9.7 we noted that Theorem 7.9.1(i) follows from Theorem 7.9.8 by putting $\mathscr{R} = \mathscr{H}om(\mathscr{S}, \mathscr{T})$. Since every coherent sheaf $\mathscr{R}$ can be expressed as $\mathscr{H}om(\mathcal{O}, \mathscr{R})$, we can deduce Theorem 7.9.8 from Theorem 7.9.1(i) by putting $\mathscr{S} = \mathcal{O}$, $\mathscr{T} = \mathscr{R}$. There is not a very deep difference between Theorem 7.9.1(i) and Theorem 7.9.8. But conceptually the statement of Theorem 7.9.8 is closer to the proof. The idea will be to prove the assertion for some sheaves $\mathscr{R}$, and then play around with exact sequences of sheaves to deduce it for more $\mathscr{R}$'s. A clear understanding of the way exact sequences of sheaves behave under analytification is therefore key. Note also that, even if the reader thinks coherent sheaves are for the birds and only vector bundles are natural objects worth studying, the proof forces one to consider coherent sheaves. The exact sequences we form in the proof inevitably will take honest, God-fearing vector bundles and make out of them Godless coherent sheaves.

8

Projective space: the statements

The version of GAGA we plan to prove is about projective space, and it is high time that we disclose to the reader what projective space is. The differential geometer would think of projective space $\mathbb{CP}^n$ as the quotient of $\mathbb{C}^{n+1} - \{0\}$ by the action of $\mathbb{C}^*$. Let us remind the reader.

The group $\mathbb{C}^*$ is the group of non-zero complex numbers under multiplication. The group $\mathbb{C}^*$ acts on $\mathbb{C}^{n+1}$ by the following rule: the action sends the pair (λ, x), where $\lambda \in \mathbb{C}^*$ and $x \in \mathbb{C}^{n+1}$, to $\lambda^{-1}x \in \mathbb{C}^{n+1}$. The point $0 \in \mathbb{C}^{n+1}$ is a fixed point for this action, and therefore $\mathbb{C}^*$ acts on the complement $\mathbb{C}^{n+1} - \{0\}$ of $0 \in \mathbb{C}^{n+1}$. The orbit space is the space of lines through the origin in $\mathbb{C}^{n+1}$, and this is what a differential geometer would understand by projective space $\mathbb{CP}^n$.

We have become more sophisticated by now. We would expect to define a ringed space $(\mathbb{P}^n, \mathcal{O})$, which if we are lucky should be a scheme of finite type over $\mathbb{C}$, and so that $\{\mathbb{P}\}^{\text{an}} = \mathbb{CP}^n$. The object of the next two chapters is to give the construction. The idea is simple enough. The topological space $\mathbb{C}^{n+1}$ can most definitely be viewed as X^{an} for a scheme $(X, \mathcal{O}_X)$ of finite type over $\mathbb{C}$; nothing could be easier. Just take $(X, \mathcal{O}_X) = \left(\operatorname{Spec}(S), \widetilde{S}\right)$, where $S = \mathbb{C}[x_0, x_1, \ldots, x_n]$ is the polynomial ring in $(n+1)$ variables. Lemma 4.4.1 tells us that X^{an} is naturally identified with $\mathbb{C}^{n+1}$. The point $0 \in \mathbb{C}^{n+1} = X^{\text{an}}$ is a closed point in X, hence its complement $U = X - \{0\}$ is a Zariski open subset of X. Corollary 5.2.4 says that $(U, \{\mathcal{O}_X\}|_U)$ is a scheme of finite type over $\mathbb{C}$. All would be well if we understood how to take quotients.

There is a branch of algebraic geometry that deals precisely with this. It concerns itself with taking quotients of schemes by the action of groups. This subject is called geometric invariant theory. We will introduce just enough geometric invariant theory to be able to construct projective space as a scheme of finite type over $\mathbb{C}$, more specifically as the quotient of U by the action of a suitable group scheme $(G, \mathcal{O}_G)$. The group scheme will have the property that $G^{\mathrm{an}} = \mathbb{C}^*$, and the action of G^{an} on U^{an} will be the usual action of $\mathbb{C}^*$ on $\mathbb{C}^{n+1} - \{0\}$.

Geometric invariant theory is a much deeper and more extensive subject than the small portion that will be presented here. As has often been the case we refer the reader to the literature for more about the field. In the case of geometric invariant theory there is one obvious choice of an excellent book to recommend. The reader might wish to look at [6].

In this chapter we set up the little corner of geometric invariant theory that we will use in the proof. First we define what it means for an affine scheme $(G, \mathcal{O}_G)$ over $\mathbb{C}$ to be a group scheme. Then we define what it means for it to act on the affine scheme $(X, \mathcal{O}_X)$. Following that, we proceed to define a ringed space $(X/G, \mathcal{O}_{X/G})$ and a morphism

$$(\pi, \pi^*) : (X, \mathcal{O}_X) \longrightarrow (X/G, \mathcal{O}_{X/G})$$

of ringed spaces over $\mathbb{C}$. We also explain how to take a coherent sheaf $\mathscr{S}$ on $(X, \mathcal{O}_X)$, which admits an action by the group $(G, \mathcal{O}_G)$, and form a sheaf $\{\pi_*\mathscr{S}\}^{G^{\mathrm{an}}}$ over $(X/G, \mathcal{O}_{X/G})$. In general the spaces $(X/G, \mathcal{O}_{X/G})$ are dreadful, as are the sheaves $\{\pi_*\mathscr{S}\}^{G^{\mathrm{an}}}$. But under some reasonable hypotheses there is an open subset $V \subset X/G$ which is a scheme of finite type over $\mathbb{C}$, and, over this open set V, the restrictions of the sheaves $\{\pi_*\mathscr{S}\}^{G^{\mathrm{an}}}$ are coherent sheaves. Furthermore all coherent sheaves on V can be obtained as $\{\pi_*\mathscr{S}\}^{G^{\mathrm{an}}}|_V$ for some coherent $\mathscr{S}$ on X.

We denote projective space, meaning the scheme version, by $(\mathbb{P}^n, \mathcal{O}_{\mathbb{P}^n})$. It can be obtained by the construction above, and so can the coherent sheaves on it. It is an important feature, of the construction of projective space $\mathbb{P}^n$ as a quotient, that we obtain a good understanding of the coherent algebraic sheaves. GAGA is a theorem about coherent sheaves, and the proof we will give relies heavily on the fact that we can express $(\mathbb{P}^n, \mathcal{O}_{\mathbb{P}^n})$ as an open set in $(X/G, \mathcal{O}_{X/G})$, and, more importantly, that the coherent sheaves on $(\mathbb{P}^n, \mathcal{O}_{\mathbb{P}^n})$ are all of the form $\{\pi_*\mathscr{S}\}^{G^{\mathrm{an}}}|_{\mathbb{P}^n}$ for $\mathscr{S}$ on X. In our case $(X, \mathcal{O}_X) = \left(\mathrm{Spec}(S), \widetilde{S}\right)$ is affine and S is particularly simple, being the polynomial ring in $(n+1)$ variables. We will show that one can take $\mathscr{S} = \widetilde{M}$ where M is a finite S–module with a G–action. It is easier to understand finite S–modules with a G–action than sheaves on X/G, or on the open subset $\mathbb{P}^n \subset X/G$.

This chapter builds up the notation for geometric invariant theory, and contains the statements of the major results we will prove. The majority of the proofs are in Chapter 9.

8.1 Products of affine schemes

Suppose we are given two affine schemes of finite type over $\mathbb{C}$. That is, we have ringed spaces $(X, \mathcal{O}_X) = \left(\mathrm{Spec}(R), \widetilde{R}\right)$ and $(Y, \mathcal{O}_Y) = \left(\mathrm{Spec}(S), \widetilde{S}\right)$ over $\mathbb{C}$. It is natural to wonder how to form the product. We want to form a ringed space $(X \times Y, \mathcal{O}_{X\times Y})$. What should it be?

From a category-theoretical perspective, the product should have the property that any pair of maps

$$(Z, \mathcal{O}_Z) \xrightarrow{\ f'\ } (X, \mathcal{O}_X), \qquad (Z, \mathcal{O}_Z) \xrightarrow{\ g'\ } (Y, \mathcal{O}_Y)$$

should factor uniquely through a map φ' as below:

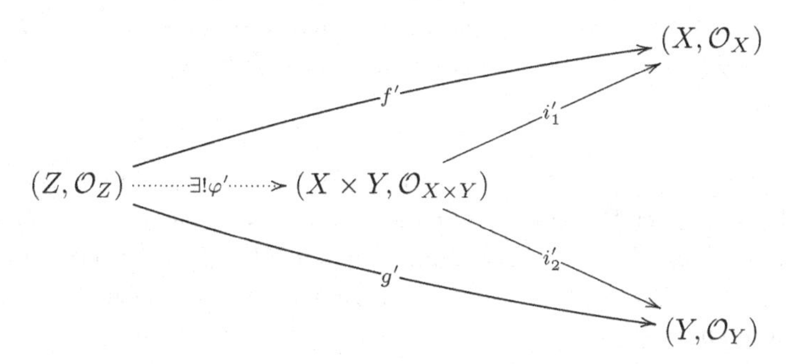

We can pursue this perspective. Assume that all the ringed spaces in the diagram are affine schemes of finite type over $\mathbb{C}$, that is each of them is of the form $\left(\mathrm{Spec}(A), \widetilde{A}\right)$ for some finitely generated $\mathbb{C}$–algebra A, and that all the maps are maps of ringed spaces over $\mathbb{C}$. Theorem 3.9.4 then tells us that all the maps are induced by $\mathbb{C}$–algebra homomorphisms. That is, the diagram of ringed spaces above must come from a diagram of $\mathbb{C}$–algebras

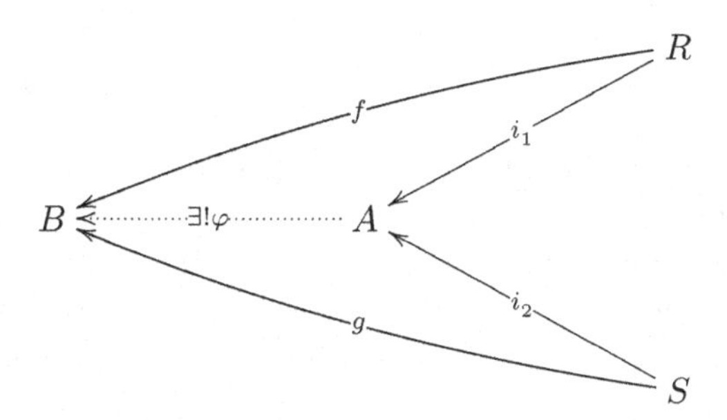

Put concretely this means the following. We are given two finitely generated $\mathbb{C}$–algebras R and S. We want to form the product. Category theory would lead us to look for a ring A, with two homomorphisms $i_1 : R \longrightarrow A$ and $i_2 : S \longrightarrow A$, so that any pair of homomorphisms $f : R \longrightarrow B$ and $g : S \longrightarrow B$ factors uniquely through a map $\varphi : A \longrightarrow B$ as above. Category theory turns out to give precisely the right intuition. There is such a ring A, and it is the correct definition of the product.

Definition 8.1.1. *Let R and S be finitely generated $\mathbb{C}$–algebras. The product of the two affine schemes $\left(\text{Spec}(R), \widetilde{R}\right)$ and $\left(\text{Spec}(S), \widetilde{S}\right)$ is defined to be $\left(\text{Spec}(R \otimes_{\mathbb{C}} S), \widetilde{R \otimes_{\mathbb{C}} S}\right)$. The two ring homomorphisms $i_1 : R \longrightarrow R \otimes_{\mathbb{C}} S$ and $i_2 : S \longrightarrow R \otimes_{\mathbb{C}} S$ are the natural inclusions:*

$$i_1(r) = r \otimes 1, \qquad i_2(s) = 1 \otimes s.$$

Lemma 8.1.2. *Any two $\mathbb{C}$–algebra homomorphisms $f : R \longrightarrow B$ and $g : S \longrightarrow B$ define a unique map $\varphi : R \otimes_{\mathbb{C}} S \longrightarrow B$ with*

$$f = \varphi i_1, \qquad g = \varphi i_2.$$

Proof. We have a map $R \times S \longrightarrow B$ taking $(r, s) \in R \times S$ to $f(r)g(s) \in B$. This map is $\mathbb{C}$ bilinear, and hence factors uniquely through the tensor product. This gives a unique homomorphism of $\mathbb{C}$–vector spaces $\varphi : R \otimes_{\mathbb{C}} S \longrightarrow B$, and the ring structure on $R \otimes_{\mathbb{C}} S$ is such that φ is a ring homomorphism. □

Remark 8.1.3. To see that the definition is half-way reasonable, look at the analytic spaces $\{\text{Spec}(R)\}^{\text{an}}$, $\{\text{Spec}(S)\}^{\text{an}}$ and $\{\text{Spec}(R \otimes_{\mathbb{C}} S)\}^{\text{an}}$. The two homomorphisms $i_1 : R \longrightarrow R \otimes_{\mathbb{C}} S$ and $i_2 : S \longrightarrow R \otimes_{\mathbb{C}} S$ induce continuous maps

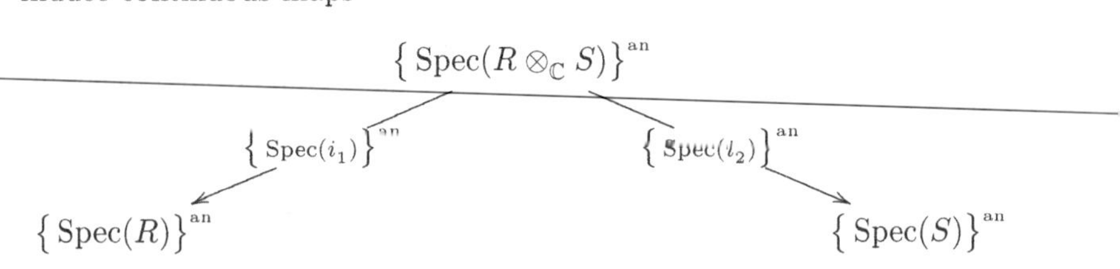

What are the maps, and what are the spaces?

To simplify the notation put $X = \text{Spec}(R)$, $Y = \text{Spec}(S)$ and $Z = \text{Spec}(R \otimes_{\mathbb{C}} S)$. Also, let us abbreviate $\pi_1 = \{\text{Spec}(i_1)\}^{\text{an}}$ and $\pi_2 = \{\text{Spec}(i_2)\}^{\text{an}}$. The points in $X^{\text{an}} = \{\text{Spec}(S)\}^{\text{an}}$ are the closed points in X, which by Proposition 4.2.4 are in bijective correspondence with $\mathbb{C}$–algebra homomorphisms $f : R \longrightarrow \mathbb{C}$. Similarly the points in Y^{an} are in bijective correspondence with $\mathbb{C}$–algebra homomorphisms $g : S \longrightarrow \mathbb{C}$, and the points of Z^{an} are in bijective correspondence with $\mathbb{C}$–algebra homomorphisms $\varphi : R \otimes_{\mathbb{C}} S \longrightarrow \mathbb{C}$. Lemma 4.3.2 tells us that the map $\left\{\operatorname{Spec}(i_1)\right\}^{\text{an}} : Z^{\text{an}} \longrightarrow X^{\text{an}}$, which we decided to shorten to $\pi_1 : Z^{\text{an}} \longrightarrow X^{\text{an}}$, takes the $\mathbb{C}$–algebra homomorphism $\varphi : R \otimes_{\mathbb{C}} S \longrightarrow \mathbb{C}$ to $\varphi i_1 : R \longrightarrow \mathbb{C}$. Similarly $\pi_2 : Z^{\text{an}} \longrightarrow Y^{\text{an}}$ takes $\varphi : R \otimes_{\mathbb{C}} S \longrightarrow \mathbb{C}$ to $\varphi i_2 : S \longrightarrow \mathbb{C}$. Lemma 8.1.2, in the special case where $B = \mathbb{C}$, establishes that any pair of homomorphisms (f, g) as above factor uniquely as

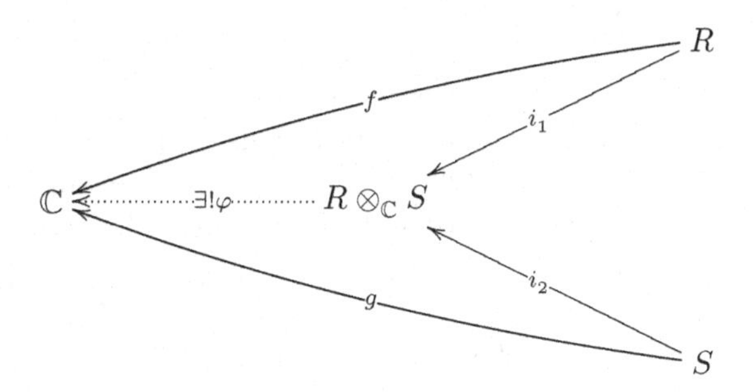

This means that any pair $(f, g) \in X^{\text{an}} \times Y^{\text{an}}$ can be written uniquely as $(f, g) = \big(\pi_1(\varphi), \pi_2(\varphi)\big)$ for $\varphi \in Z^{\text{an}}$. The map $(\pi_1, \pi_2) : Z^{\text{an}} \longrightarrow X^{\text{an}} \times Y^{\text{an}}$ is a bijection. What about the topology?

The complex topology can be defined by choosing generators. We know it to be independent of the choice, so let us make a convenient choice. Choose generators $\{r_1, r_2, \ldots, r_m\} \subset R$ for the $\mathbb{C}$–algebra R, and $\{s_1, s_2, \ldots, s_n\} \subset S$ for the $\mathbb{C}$–algebra S. Together the set of generators

$$r_1 \otimes 1, r_2 \otimes 1, \ldots, r_m \otimes 1, 1 \otimes s_1, 1 \otimes s_2, \ldots, 1 \otimes s_n$$

generate the $\mathbb{C}$–algebra $R \otimes_{\mathbb{C}} S$. If A and B are the polynomial rings

$$A = \mathbb{C}[x_1, x_2, \ldots, x_m], \qquad B = \mathbb{C}[y_1, y_2, \ldots, y_n],$$

then $A \otimes_{\mathbb{C}} B$ is also a polynomial ring

$$A \otimes_{\mathbb{C}} B \quad = \quad \mathbb{C}[x_1, x_2, \ldots, x_m, y_1, y_2, \ldots, y_n].$$

Let $\theta : A \longrightarrow R$, $\theta' : B \longrightarrow S$ be defined by $\theta(x_i) = r_i$ and $\theta'(y_j) = s_j$. Then we have a commutative diagram of ring homomorphisms

$$\begin{array}{ccccc} A & \longrightarrow & A \otimes_{\mathbb{C}} B & \longleftarrow & B \\ \big\downarrow{\scriptstyle \theta} & & \big\downarrow{\scriptstyle \theta \otimes \theta'} & & \big\downarrow{\scriptstyle \theta'} \\ R & \longrightarrow & R \otimes_{\mathbb{C}} S & \longleftarrow & S \end{array}$$

where all three vertical maps are surjections. Applying $\left\{\operatorname{Spec}(-)\right\}^{\text{an}}$ to this diagram we obtain the following commutative diagram of spaces

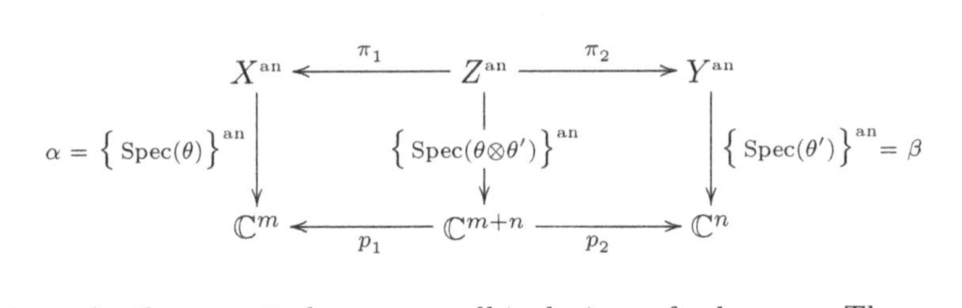

where the three vertical maps are all inclusions of subspaces. The maps $p_1 : \mathbb{C}^{m+n} \longrightarrow \mathbb{C}^m$ and $p_2 : \mathbb{C}^{m+n} \longrightarrow \mathbb{C}^n$ are easily computed, using Lemma 4.4.5, to be just the natural projections. The commutativity means that the embedding $\{\operatorname{Spec}(\theta \otimes \theta')\}^{\text{an}}$ takes the unique point $\varphi \in Z^{\text{an}}$ corresponding to $(f,g) \in X^{\text{an}} \times Y^{\text{an}}$ to the point $\big(\alpha(f), \beta(g)\big) \in \mathbb{C}^m \times \mathbb{C}^n$. That is, under the identification $Z^{\text{an}} \cong X^{\text{an}} \times Y^{\text{an}}$, the map $\{\operatorname{Spec}(\theta \otimes \theta')\}^{\text{an}}$ agrees with $\alpha \times \beta : X^{\text{an}} \times Y^{\text{an}} \longrightarrow \mathbb{C}^m \times \mathbb{C}^n$. The complex topology on $\mathbb{C}^{m+n} = \mathbb{C}^m \times \mathbb{C}^n$ is the product topology, and hence so is the topology on the subspace $X^{\text{an}} \times Y^{\text{an}}$.

Caution 8.1.4. Let $X = \operatorname{Spec}(R)$, $Y = \operatorname{Spec}(S)$ and $Z = \operatorname{Spec}(R \otimes_{\mathbb{C}} S)$ as above. There is always a map $Z \longrightarrow X \times Y$, even in the Zariski topology. The reader should be warned that this map is not an isomorphism. It is not even a bijection of sets, and the topologies are quite different. Only in the complex topology is it true that $Z^{\text{an}} = X^{\text{an}} \times Y^{\text{an}}$.

8.2 Affine group schemes

A group is a set G together with: (i) a multiplication map $\mu : G \times G \longrightarrow G$, (ii) an identity element $e \in G$, and (iii) a map $i : G \longrightarrow G$ taking $g \in G$ to $i(g) = g^{-1} \in G$. If we want to describe the group purely in terms of maps, we would say that there are three maps giving the group structure:

$$\mu : G \times G \longrightarrow G\,, \qquad e : \{*\} \longrightarrow G\,, \qquad i : G \longrightarrow G\,,$$

where $\{*\}$ is the 1-point set, and where $e : \{*\} \longrightarrow G$ takes the unique element $* \in \{*\}$ to the identity in G. The fact that G is a group means that these maps must satisfy certain compatibility conditions, which can be described by requiring that some diagrams commute. Maybe we should remind the reader. The conditions are

8.2.1. Associativity means that the following square commutes

$$\begin{array}{ccc} G \times G \times G & \xrightarrow{1\times\mu} & G \times G \\ {\scriptstyle \mu\times 1}\downarrow & & \downarrow{\scriptstyle \mu} \\ G \times G & \xrightarrow[\mu]{} & G \end{array} .$$

8.2.2. The fact that e is a two-sided identity can be phrased as saying that the two composites

$$\begin{array}{ccc} \{*\} \times G \times \{*\} & \xrightarrow{1\times 1\times e} & \{*\} \times G \times G \\ {\scriptstyle e\times 1\times 1}\downarrow & & \downarrow{\scriptstyle \mu} \\ G \times G \times \{*\} & \xrightarrow[\mu]{} & G \end{array}$$

are the identity on G, where of course we feel free to identify, whenever we can do so without risk of confusion, $\{*\} \times X = X \times \{*\} = X$.

8.2.3. The fact that $i : G \longrightarrow G$ takes $g \in G$ to its two-sided inverse can be rephrased, as saying that the following square commutes

$$\begin{array}{ccc} G & \xrightarrow{(1,i)} & G \times G \\ {\scriptstyle (i,1)}\downarrow & & \downarrow{\scriptstyle \mu} \\ G \times G & \xrightarrow[\mu]{} & G \end{array} ,$$

and that the equal composites agree with the composite

$$G \longrightarrow \{*\} \xrightarrow{e} G .$$

In Section 8.1 we defined the product of two affine schemes of finite type over $\mathbb{C}$. We can now use this to define what a group scheme should be.

Definition 8.2.4. *An affine scheme $(G, \mathcal{O})$, of finite type over $\mathbb{C}$, is called an* affine group scheme *if it has three maps*

$$\mu : (G, \mathcal{O}) \times (G, \mathcal{O}) \longrightarrow (G, \mathcal{O}) , \qquad e : \left(\operatorname{Spec}(\mathbb{C}), \widetilde{\mathbb{C}}\right) \longrightarrow (G, \mathcal{O}) ,$$
$$i : (G, \mathcal{O}) \longrightarrow (G, \mathcal{O})$$

satisfying the obvious analogues of 8.2.1, 8.2.2, and 8.2.3.

Remark 8.2.5. It is important to note that the topological space G^{an} is a topological group. As we saw in Remark 8.1.3 the product of affine schemes, confusing as it might be, induces the ordinary product on the analytic spaces. We therefore have continuous maps

$$\mu^{an} : G^{an} \times G^{an} \longrightarrow G^{an} , \qquad e^{an} : \{*\} \longrightarrow G^{an} , \qquad i^{an} : G^{an} \longrightarrow G^{an}$$

and the commutative diagrams ensure precisely that G^{an} satisfies the axioms of a topological group.

Remark 8.2.6. An affine scheme of finite type over $\mathbb{C}$ is little more than a badly tortured finitely generated algebra over $\mathbb{C}$. It should therefore be possible to rephrase Definition 8.2.4 in terms of algebras over $\mathbb{C}$. Suppose $(G, \mathcal{O}) = \left(\mathrm{Spec}(R), \widetilde{R}\right)$, for R a finitely generated $\mathbb{C}$–algebra. What does it mean for it to be an affine group scheme?

The three structure maps μ, e and i must correspond to $\mathbb{C}$–algebra homomorphisms

$$\mu' : R \longrightarrow R \otimes_{\mathbb{C}} R \qquad e' : R \longrightarrow \mathbb{C} \qquad i' : R \longrightarrow R$$

satisfying compatibilities. Associativity says that the square of $\mathbb{C}$–algebra homomorphisms

$$\begin{array}{ccc} R & \xrightarrow{\mu'} & R \otimes_{\mathbb{C}} R \\ {\scriptstyle \mu'}\downarrow & & \downarrow{\scriptstyle 1\otimes\mu'} \\ R \otimes_{\mathbb{C}} R & \xrightarrow[\mu'\otimes 1]{} & R \otimes_{\mathbb{C}} R \otimes_{\mathbb{C}} R \end{array}$$

should commute. The fact that e is a two-sided identity means that the two composites

$$\begin{array}{ccc} R & \xrightarrow{\mu'} & R \otimes_{\mathbb{C}} R \otimes_{\mathbb{C}} \mathbb{C} \\ {\scriptstyle \mu'}\downarrow & & \downarrow{\scriptstyle e'\otimes 1\otimes 1} \\ \mathbb{C} \otimes_{\mathbb{C}} R \otimes_{\mathbb{C}} R & \xrightarrow[1\otimes 1\otimes e']{} & \mathbb{C} \otimes_{\mathbb{C}} R \otimes_{\mathbb{C}} \mathbb{C} \end{array}$$

are both equal to the identity on R. And the fact that i gives a two-sided inverse means that the following square commutes

$$\begin{array}{ccc} R & \xrightarrow{\mu'} & R \otimes_{\mathbb{C}} R \\ {\scriptstyle \mu'}\downarrow & & \downarrow{\scriptstyle (i',1)} \\ R \otimes_{\mathbb{C}} R & \xrightarrow[(1,i')]{} & R \end{array} ,$$

and the two equal composites agree with the composite

$$R \xrightarrow{e'} \mathbb{C} \xrightarrow{\rho} R ,$$

where $\rho : \mathbb{C} \longrightarrow R$ is the map giving R its structure as a $\mathbb{C}$–algebra.

The ring theoretic data described above has been studied extensively in several other contexts. A ring with this structure is called a *Hopf algebra* over $\mathbb{C}$. The reader is encouraged to read further in the rich and extensive literature on Hopf algebras.

There are two examples the reader should keep in mind.

Example 8.2.7. The trivial group is obtained by letting $R = \mathbb{C}$. To make R into a Hopf algebra we need $\mathbb{C}$–algebra homomorphisms

$$\mu' : R \longrightarrow R \otimes_{\mathbb{C}} R \qquad e' : R \longrightarrow \mathbb{C} \qquad i' : R \longrightarrow R$$

satifying the compatibilities of Remark 8.2.6. In the case where $R = \mathbb{C}$ is the trivial $\mathbb{C}$–algebra, the maps μ', e' and i' can only be the identity map $1 : \mathbb{C} \longrightarrow \mathbb{C}$, and the diagrams in Remark 8.2.6 all commute just for the simple reason that all composites are $\mathbb{C}$–algebra homomorphisms $\mathbb{C} \longrightarrow \mathbb{C}$, which means they must all be the identity. This gives us the trivial affine group scheme. Note that $G^{\mathrm{an}} = \{*\}$ is the 1–point space. The only element of the group G^{an} is the identity.

Example 8.2.8. The non-trivial example we care about is the group $\mathbb{C}^*$. Recall that, for the polynomial ring in one variable $R = \mathbb{C}[x]$, the space $X^{\mathrm{an}} = \{\operatorname{Spec}(R)\}^{\mathrm{an}}$ identifies naturally with $\mathbb{C}$. The open subset X_x^{an} identifies with the subset where $x \neq 0$, that is it identifies with $\mathbb{C}^*$. This is naturally a group. I would like to say that it is an affine group scheme as above. It is certainly affine; Proposition 3.7.3 asserts that $(X_f, \mathcal{O}|_{X_f})$ is always $\left(\operatorname{Spec}(R[1/f]), \widetilde{R[1/f]}\right)$. The special case, where $R = \mathbb{C}[x]$ and $f \in R$ is $x \in \mathbb{C}[x]$, was already considered in Example 3.3.3; in the example we showed that the ring $R[1/x] = \mathbb{C}[x][1/x]$ is isomorphic, via a map we called φ', with the Laurent polynomial ring $\mathbb{C}[x, x^{-1}]$. We have that, with the isomorphism induced by φ', $(X_x, \mathcal{O}|_{X_x}) = \left(\operatorname{Spec}(\mathbb{C}[x, x^{-1}]), \widetilde{\mathbb{C}[x, x^{-1}]}\right)$. We need to establish that the ring $\mathbb{C}[x, x^{-1}]$ is a Hopf algebra in such a way that the induced group structure, on $\{\operatorname{Spec}(\mathbb{C}[x, x^{-1}])\}^{\mathrm{an}}$, is precisely the usual multiplicative group structure on $\mathbb{C}^*$.

It is perhaps simpler to start with $\mathbb{C} = \{\operatorname{Spec}(R)\}^{\mathrm{an}}$, with R the polynomial ring $R = \mathbb{C}[x]$. The ring $R \otimes_{\mathbb{C}} R$ is isomorphic to the polynomial ring in two variables $\mathbb{C}[x, y]$. Consider the map $\nu' : R \longrightarrow R \otimes_{\mathbb{C}} R$ given by the formula

$$\nu'(x) = xy.$$

Lemma 4.4.5 allows us to compute that the induced map $\nu : \mathbb{C} \times \mathbb{C} \longrightarrow \mathbb{C}$

is given by the formula $\nu(a,b) = ab$. Now consider the commutative diagram of ring homomorphisms

$$\begin{array}{ccccc}
R & \xrightarrow{\nu'} & R \otimes_{\mathbb{C}} R & \xrightarrow{\alpha_x \otimes \alpha_x} & R[1/x] \otimes_{\mathbb{C}} R[1/x] \\
 & & \| \wr \downarrow & & \downarrow \varphi' \otimes \varphi' \\
 & & \mathbb{C}[x] \otimes_{\mathbb{C}} \mathbb{C}[y] & \xrightarrow{\varphi \otimes \varphi} & \mathbb{C}[x, x^{-1}] \otimes_{\mathbb{C}} \mathbb{C}[y, y^{-1}] \\
 & & \tau \downarrow & & \downarrow \tau \\
 & & \mathbb{C}[x, y] & \xrightarrow[\text{inclusion}]{} & \mathbb{C}[x, y, x^{-1}, y^{-1}]
\end{array} ,$$

where the top square comes from Example 3.3.3, while the bottom square is as in Remark 3.3.6; in particular, the vertical maps are all isomorphisms. The composite

$$\begin{array}{cccc}
R & \xrightarrow{\nu'} & R \otimes_{\mathbb{C}} R & \\
 & & \| \wr \downarrow & \\
 & & \mathbb{C}[x] \otimes_{\mathbb{C}} \mathbb{C}[y] & \\
 & & \tau \downarrow & \\
 & & \mathbb{C}[x, y] & \xrightarrow[\text{inclusion}]{} \mathbb{C}[x, y, x^{-1}, y^{-1}]
\end{array}$$

takes $x \in R = \mathbb{C}[x]$ to $xy \in \mathbb{C}[x, y, x^{-1}, y^{-1}]$, which is an invertible element. The fact that τ is an isomorphism means that the image of $x \in R$ becomes invertible already under the composite

$$\begin{array}{cccc}
R & \xrightarrow{\nu'} & R \otimes_{\mathbb{C}} R & \\
 & & \| \wr \downarrow & \\
 & & \mathbb{C}[x] \otimes_{\mathbb{C}} \mathbb{C}[y] & \xrightarrow{\varphi \otimes \varphi} \mathbb{C}[x, x^{-1}] \otimes_{\mathbb{C}} \mathbb{C}[y, y^{-1}]
\end{array} .$$

This composite must therefore factor through $R[1/x] \cong \mathbb{C}[x, x^{-1}]$. We conclude that there is a unique map $\mu' : \mathbb{C}[x, x^{-1}] \longrightarrow \mathbb{C}[x, x^{-1}] \otimes_{\mathbb{C}} \mathbb{C}[y, y^{-1}]$ rendering commutative the square

$$\begin{array}{ccc}
R & \xrightarrow{\nu'} & R \otimes_{\mathbb{C}} R \\
\varphi \downarrow & & \downarrow \varphi \otimes \varphi \\
\mathbb{C}[x, x^{-1}] & \xrightarrow[\mu']{} & \mathbb{C}[x, x^{-1}] \otimes_{\mathbb{C}} \mathbb{C}[y, y^{-1}]
\end{array} .$$

If we apply $\{\operatorname{Spec}(-)\}^{\mathrm{an}}$ to the diagram, we obtain a commutative square

$$\begin{array}{ccc} \mathbb{C}^* \times \mathbb{C}^* & \xrightarrow{\mu} & \mathbb{C}^* \\ {\scriptstyle \{\text{inc}\}\times\{\text{inc}\}}\big\downarrow & & \big\downarrow{\scriptstyle \text{inc}} \\ \mathbb{C} \times \mathbb{C} & \xrightarrow[\nu]{} & \mathbb{C} \end{array},$$

where the vertical maps are induced by the inclusion $\{\text{inc}\} : \mathbb{C}^* \longrightarrow \mathbb{C}$. Since we know that ν takes the pair $(a, b) \in \mathbb{C} \times \mathbb{C}$ to ab, so does μ, acting on the open set where $a \neq 0 \neq b$.

The map $e' : \mathbb{C}[x, x^{-1}] \longrightarrow \mathbb{C}$ takes a Laurent polynomial in x and evaluates it at $x = 1 \in \mathbb{C}$. The map $i' : \mathbb{C}[x, x^{-1}] \longrightarrow \mathbb{C}[x, x^{-1}]$ is the homomorphism with $i'(x) = x^{-1}$, $i'(x^{-1}) = x$. We leave it to the reader to check that the axioms of a Hopf algebra are satisfied. We do have an affine group scheme of finite type over $\mathbb{C}$.

8.3 Affine group schemes acting on affine schemes

Whenever we meet a group we expect it to act on something. What should it mean to say that an affine group scheme $(G, \mathcal{O}_G)$ acts on a scheme $(X, \mathcal{O}_X)$? Without getting too sophisticated, what should it mean in the special case where $(X, \mathcal{O}_X) = \left(\text{Spec}(S), \widetilde{S}\right)$ is affine? What we should expect is to have a map

$$a : (G, \mathcal{O}_G) \times (X, \mathcal{O}_X) \longrightarrow (X, \mathcal{O}_X)$$

giving the action, and some diagrams should commute. Of course any statement about maps of affine schemes of finite type over $\mathbb{C}$ translates, immediately, to a statement about $\mathbb{C}$–algebra homomorphisms. In the interest of keeping our commutative diagrams smaller we will sometimes formulate our conditions in terms of homomorphisms of $\mathbb{C}$–algebras; we ask the reader to make the translation to maps of schemes.

Definition 8.3.1. *Suppose*

$$(G, \mathcal{O}_G) = \left(\text{Spec}(R), \widetilde{R}\right), \qquad (X, \mathcal{O}_X) = \left(\text{Spec}(S), \widetilde{S}\right)$$

for some finitely generated $\mathbb{C}$–algebras R and S. Suppose further that $(G, \mathcal{O}_G)$ is an affine group scheme, as in Section 8.2. An action *of $(G, \mathcal{O}_G)$ on $(X, \mathcal{O}_X)$ is given by a homomorphism of $\mathbb{C}$–algebras*

$$a' : S \longrightarrow R \otimes_{\mathbb{C}} S\,.$$

This homomorphism must satisfy two conditions:

(i) *The following square of $\mathbb{C}$–algebra homomorphisms commutes*

$$\begin{array}{ccc} S & \xrightarrow{a'} & R\otimes_{\mathbb{C}} S \\ {\scriptstyle a'}\downarrow & & \downarrow{\scriptstyle 1\otimes a'} \\ R\otimes_{\mathbb{C}} S & \xrightarrow[\mu'\otimes 1]{} & R\otimes_{\mathbb{C}} R\otimes_{\mathbb{C}} S \end{array} .$$

(ii) *The following triangle commutes*

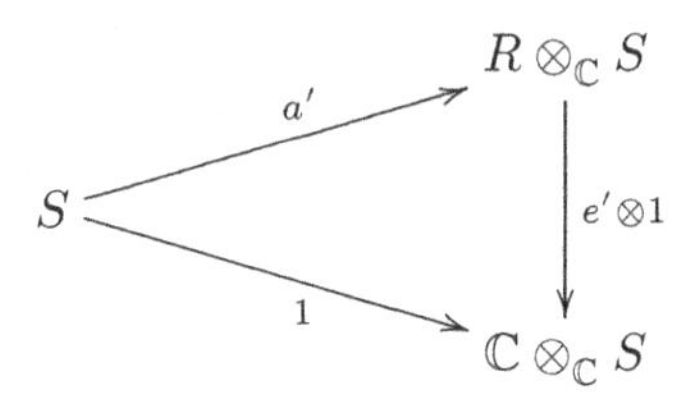

Remark 8.3.2. The fact that the square in Definition 8.3.1(i) commutes says that the action is associative; for the induced map $a : G^{\mathrm{an}} \times X^{\mathrm{an}} \longrightarrow X^{\mathrm{an}}$ it says $a\big(g, a(h,x)\big) = a\big(\mu(g,h), x\big)$, which one usually writes $g(hx) = (gh)x$. The commutativity of the triangle in Definition 8.3.1(ii) implies that, for $e \in G^{\mathrm{an}}$ the image of the map $\mathrm{Spec}(e') : \mathrm{Spec}(\mathbb{C}) \longrightarrow G^{\mathrm{an}}$, the identity $a(e,x) = x$ holds; this identity is normally written $ex = x$.

Notation 8.3.3. We will say that S is a G–ring if the affine group scheme $(G, \mathcal{O}_G)$ acts on the affine scheme $(X, \mathcal{O}_X) = \Big(\mathrm{Spec}(S), \widetilde{S}\Big)$ as in Definition 8.3.1. For us G–rings are always finitely generated algebras over $\mathbb{C}$, and the group scheme $(G, \mathcal{O}_G)$ is always an affine group scheme of finite type over $\mathbb{C}$.

Example 8.3.4. If $(G, \mathcal{O}_G)$ is the trivial group of Example 8.2.7, that is if R is the trivial Hopf algebra $R = \mathbb{C}$, then any finitely generated $\mathbb{C}$–algebra S can be given the structure of a G–ring. We let $a' : S \longrightarrow R\otimes_{\mathbb{C}} S$ be the identity map

$$1 \;:\; S \longrightarrow \mathbb{C}\otimes_{\mathbb{C}} S \quad = \quad S\;.$$

This satisfies the axioms for an action; the commutativity of the diagrams, in Definition 8.3.1(i) and (ii), comes down to the fact that all the maps in the diagrams are $1 : S \longrightarrow S$

Example 8.3.5. Once again, there is basically only one non-trivial example we care about in this book. Let $S = \mathbb{C}[x_0, x_1, \ldots, x_n]$ be the polynomial ring in $(n+1)$ variables, and let $R = \mathbb{C}[t, t^{-1}]$ be the Hopf algebra of Example 8.2.8. The example that interests us the most is the action $a' : S \longrightarrow R \otimes_{\mathbb{C}} S$ given by the formula

$$a'(x_i) = t^{-1} \otimes x_i \,.$$

We leave it to the reader to check that the conditions of Definition 8.3.1(i) and (ii) are satisfied. When we apply $\{\, \mathrm{Spec}(-)\}^{\mathrm{an}}$ to the $\mathbb{C}$–algebra homomorphism $a' : S \longrightarrow R \otimes_{\mathbb{C}} S$ we get a map $a : \mathbb{C}^* \times \mathbb{C}^{n+1} \longrightarrow \mathbb{C}^{n+1}$. Lemma 4.4.5 tells us how to compute this map. It is easy to check that the map a takes a pair (λ, b), with $\lambda \in \mathbb{C}^*$ and $b \in \mathbb{C}^{n+1}$, to $a(\lambda, b) = \lambda^{-1} b$.

Perhaps we should make more precise what we mean when we say that this is the only non-trivial example we care about. We will also look at the G–invariant open subsets of $\left(\mathrm{Spec}(S), \widetilde{S}\right)$, and we will permit ourselves to do Nagata's trick and add to S some square-zero ideal. But, up to these qualifications, the action of $\left(\mathrm{Spec}(R), \widetilde{R}\right)$ on $\left(\mathrm{Spec}(S), \widetilde{S}\right)$ above really is the only case that interests us here.

Definition 8.3.6. *Suppose $(G, \mathcal{O}_G)$, $(X, \mathcal{O}_X)$ and $(Y, \mathcal{O}_Y)$ are all affine schemes of finite type over $\mathbb{C}$. Suppose $(G, \mathcal{O}_G)$ is an affine group scheme, and suppose $(G, \mathcal{O}_G)$ acts on $(X, \mathcal{O}_X)$ and $(Y, \mathcal{O}_Y)$, as in Definition 8.3.1. A morphism $\beta : (X, \mathcal{O}_X) \longrightarrow (Y, \mathcal{O}_Y)$, of schemes of finite type over $\mathbb{C}$, is called a* G–morphism *if the following square commutes*

$$\begin{array}{ccc}
(G, \mathcal{O}_G) \times (X, \mathcal{O}_X) & \xrightarrow{a_X} & (X, \mathcal{O}_X) \\
{\scriptstyle 1\times\beta}\big\downarrow & & \big\downarrow{\scriptstyle \beta} \\
(G, \mathcal{O}_G) \times (Y, \mathcal{O}_Y) & \xrightarrow[a_Y]{} & (Y, \mathcal{O}_Y)
\end{array} \,.$$

Remark 8.3.7. Definition 8.3.6 can of course be rephrased in terms of rings. If R, S and S' are finitely generated $\mathbb{C}$–algebras with

$$(G, \mathcal{O}_G) = \left(\mathrm{Spec}(R), \widetilde{R}\right) , \qquad (X, \mathcal{O}_X) = \left(\mathrm{Spec}(S), \widetilde{S}\right) ,$$
$$(Y, \mathcal{O}_Y) = \left(\mathrm{Spec}(S'), \widetilde{S'}\right)$$

then the map $\beta : (X, \mathcal{O}_X) \longrightarrow (Y, \mathcal{O}_Y)$ is equal to $\left(\mathrm{Spec}(\beta'), \widetilde{\beta'}\right)$ for a unique $\mathbb{C}$–algebra homomorphism $\beta' : S' \longrightarrow S$, and for β to be a G–morphism means that the following square of ring homomorphisms must commute:

$$\begin{array}{ccc} S' & \xrightarrow{a'_{S'}} & R \otimes_{\mathbb{C}} S' \\ {\scriptstyle \beta'}\downarrow & & \downarrow{\scriptstyle 1\otimes\beta'} \\ S & \xrightarrow[a'_S]{} & R \otimes_{\mathbb{C}} S \end{array} .$$

Notation 8.3.8. A $\mathbb{C}$–algebra homomorphism $\beta' : S' \longrightarrow S$ as in Remark 8.3.7 will be called a G–homomorphism of G–rings.

Example 8.3.9. Let $(G, \mathcal{O}_G)$ be trivial, that is $R = \mathbb{C}$ as in Example 8.2.7. Every finitely generated $\mathbb{C}$–algebra S was made into a G–ring in Example 8.3.4, by declaring the action $a' : S \longrightarrow \mathbb{C} \otimes_{\mathbb{C}} S$ to be the identity map $1 : S \longrightarrow S$. Any $\mathbb{C}$–algebra homomorphism $\beta : S' \longrightarrow S$ commutes with this action. That is, all $\mathbb{C}$–algebra homomorphisms $\beta : S' \longrightarrow S$ are G–homomorphisms of G–rings for the trivial group G.

Before proceeding to construct non-trivial examples we should note the little fact.

Lemma 8.3.10. *Let S be a G–ring, with $(G, \mathcal{O}_G) = \big(\mathrm{Spec}(R), \widetilde{R}\big)$. Suppose $f \in S$ is a non-zero element such that $a'(f) = r \otimes f$. Then $r \in R$ is invertible.*

Proof. Consider the diagram

$$\begin{array}{ccccc} S & \xrightarrow{a'} & R \otimes_{\mathbb{C}} S & \xrightarrow{e'\otimes 1} & \mathbb{C} \otimes_{\mathbb{C}} S \\ {\scriptstyle a'}\downarrow & & \downarrow{\scriptstyle \mu'\otimes 1} & & \downarrow{\scriptstyle \rho\otimes 1} \\ R \otimes_{\mathbb{C}} S & \xrightarrow[1\otimes a']{} & R \otimes_{\mathbb{C}} R \otimes_{\mathbb{C}} S & \xrightarrow[(1,i')\otimes 1]{} & R \otimes_{\mathbb{C}} S \end{array} .$$

In this diagram the two small squares commute. We know that the left square commutes by Definition 8.3.1(i), and the right square comes from taking the third diagram of Remark 8.2.6 and tensoring with S. The two composites from top left to bottom right must agree, and the idea is to check what they do to the element $f \in S$.

Let us start with the composite

$$\begin{array}{ccccc} S & & & & \\ {\scriptstyle a'}\downarrow & & & & \\ R \otimes_{\mathbb{C}} S & \xrightarrow[1\otimes a']{} & R \otimes_{\mathbb{C}} R \otimes_{\mathbb{C}} S & \xrightarrow[(1,i')\otimes 1]{} & R \otimes_{\mathbb{C}} S \end{array} .$$

The fact that $a'(f) = r \otimes f$ means that $(1 \otimes a')a'(f) = r \otimes r \otimes f$, and applying $(1, i') \otimes 1$ to this we obtain $\big(r \cdot i'(r)\big) \otimes f$. To compute the other composite note first that

$$S \xrightarrow{\ a'\ } R \otimes_{\mathbb{C}} S \xrightarrow{\ e' \otimes 1\ } \mathbb{C} \otimes_{\mathbb{C}} S$$

composes to the identity by Definition 8.3.1(ii). This means that

$$\begin{array}{ccccc} S & \xrightarrow{\ a'\ } & R \otimes_{\mathbb{C}} S & \xrightarrow{\ e' \otimes 1\ } & \mathbb{C} \otimes_{\mathbb{C}} S \\ & & & & \big\downarrow {\scriptstyle \rho \otimes 1} \\ & & & & R \otimes_{\mathbb{C}} S \end{array}$$

takes f to $1 \otimes f$. Since $f \neq 0$ we conclude that $r \cdot i'(r) = 1$, that is r is invertible in R. □

Remark 8.3.11. One way to get examples of G–homomorphisms of G–rings is the following. Suppose S is a G–ring, and suppose $f \in S$ is an element such that $a'_S(f) = r \otimes f$. Lemma 8.3.10 tells us that $r \in R$ is invertible. Thus the composite

$$S \xrightarrow{\ a'_S\ } R \otimes_{\mathbb{C}} S \xrightarrow{\ 1 \otimes \alpha_f\ } R \otimes_{\mathbb{C}} S[1/f]$$

takes $f \in S$ to the invertible element $r \otimes f \in R \otimes_{\mathbb{C}} S[1/f]$. We conclude that the composite must factor uniquely through $\alpha_f : S \longrightarrow S[1/f]$; there exists a unique map $a'_{S[1/f]} : S[1/f] \longrightarrow R \otimes_{\mathbb{C}} S[1/f]$ rendering commutative the square

$$\begin{array}{ccc} S & \xrightarrow{\ a'_S\ } & R \otimes_{\mathbb{C}} S \\ {\scriptstyle \alpha_f}\big\downarrow & & \big\downarrow {\scriptstyle 1 \otimes \alpha_f} \\ S[1/f] & \xrightarrow[\ a'_{S[1/f]}\ ]{} & R \otimes_{\mathbb{C}} S[1/f] \end{array} .$$

Exercise 8.3.12. With the notation as in Remark 8.3.11, we leave it to the reader to verify the following three assertions:

(i) Let the notation be as above. The map $a'_{S[1/f]} : S[1/f] \longrightarrow R \otimes_{\mathbb{C}} S[1/f]$ satisfies conditions (i) and (ii) of Definition 8.3.1; that is $S[1/f]$ is a G–ring.

(ii) The map $\alpha_f : S \longrightarrow S[1/f]$ is a G–homomorphism.

(iii) If $f, f' \in S$ are two elements, with $a'_S(f) = r \otimes f$ and $a'_S(f') = r' \otimes f'$, and if in $X = \mathrm{Spec}(S)$ we have an inclusion $X_{f'} \subset X_f$ of basic open sets, then the natural map $\alpha^f_{f'} : S[1/f] \longrightarrow S[1/f']$, of Lemma 3.4.3, is a G–homomorphism.

Example 8.3.13. Back to our favorite Example 8.3.5. The Hopf algebra R is $\mathbb{C}[t, t^{-1}]$, the ring $S = \mathbb{C}[x_0, x_1, \ldots, x_n]$ is the polynomial ring in $(n+1)$ variables and the action $a'_S : S \longrightarrow R \otimes_{\mathbb{C}} S$ is given by $a'(x_i) = t^{-1} \otimes x_i$. There are many elements $f \in S$ for which $a'_S(f) = r \otimes f$. All the homogeneous polynomials are of this form. If $f \in S$ is a homogeneous polynomial, of degree N in the variables $x_0, x_1, \ldots, x_n$, then $a'_S(f) = t^{-N} \otimes f$. Thus we learn that all the rings $S[1/f]$, where f is any homogeneous polynomial, are naturally G–rings. Furthermore all the maps $\alpha_f : S \longrightarrow S[1/f]$ and $\alpha^f_{f'} : S[1/f] \longrightarrow S[1/f']$ are G–homomorphisms.

Remark 8.3.14. Suppose $(G, \mathcal{O}_G)$ is an affine group scheme of finite type over $\mathbb{C}$. Suppose $(X, \mathcal{O}_X)$ is an affine scheme of finite type over $\mathbb{C}$, and suppose that $(G, \mathcal{O}_G)$ acts on $(X, \mathcal{O}_X)$ as in Definition 8.3.1. What do we mean by an action of $(G, \mathcal{O}_G)$ on a coherent sheaf of modules $\widetilde{M}$ over the affine scheme $(X, \mathcal{O}_X)$?

Here life is simple: recall Nagata's trick. Suppose R and S are finitely generated $\mathbb{C}$–algebras with

$$(G, \mathcal{O}_G) = \left(\mathrm{Spec}(R), \widetilde{R}\right), \qquad (X, \mathcal{O}_X) = \left(\mathrm{Spec}(S), \widetilde{S}\right).$$

Let M be a finite S–module. Then $S \oplus M$ is a finitely generated $\mathbb{C}$–algebra, and we have ring homomorphisms

$$S \xrightarrow{\quad\psi\quad} S \oplus M \xrightarrow{\quad\varphi\quad} S .$$

The ring S is a G–ring, as in Notation 8.3.3. We will say that $(G, \mathcal{O}_G)$ acts on $\widetilde{M}$ if the ring $S \oplus M$ is also a G–ring, and if the ring homomorphisms ψ and φ are G–homomorphisms. In this case we will call M a G–module for the G–ring S. An S–module homomorphism $\eta : M \longrightarrow N$ is called a G–homomorphism if the induced ring homomorphism $1 \oplus \eta : S \oplus M \longrightarrow S \oplus N$ is a G–homomorphism of G–rings

Example 8.3.15. Let $(G, \mathcal{O}_G)$ be the trivial group scheme, that is suppose $R = \mathbb{C}$ as in Example 8.2.7. In Example 8.3.9 we learned that any homomorphism $\beta : S \longrightarrow S'$ of finitely generated $\mathbb{C}$–algebras is a G–homomorphism of G–rings. In particular it follows that, for any finite S–module M, the homomorphisms

$$S \xrightarrow{\quad\psi\quad} S \oplus M \xrightarrow{\quad\varphi\quad} S$$

are G–homomorphisms of G–rings. Thus all finite modules M are G–modules. The reader can easily check that any map $\eta : M \longrightarrow N$ of finite S–modules is a G–homomorphism.

Remark 8.3.16. Let us give a less compact way of rephrasing the definition of a G–module over the G–ring S; the terse version was given in Remark 8.3.14. Let us actually generalize a little, and consider S–modules that are not necessarily finite. We define a *large G–module* for the G–ring S to be an S–module M (not necessarily finitely generated) for which there is a map

$$a'_M : M \longrightarrow R \otimes_{\mathbb{C}} M$$

satisfying the following hypotheses:

(i) The following square commutes

$$\begin{array}{ccc} M & \xrightarrow{a'_M} & R \otimes_{\mathbb{C}} M \\ {\scriptstyle a'_M}\downarrow & & \downarrow{\scriptstyle 1\otimes a'_M} \\ R \otimes_{\mathbb{C}} M & \xrightarrow[\mu'\otimes 1]{} & R \otimes_{\mathbb{C}} R \otimes_{\mathbb{C}} M \end{array} .$$

(ii) The following triangle commutes

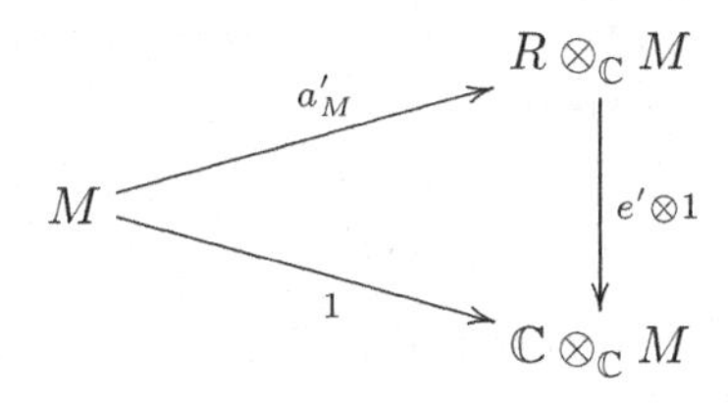

(iii) The maps a'_S and a'_M are compatible with the module structure of M as an S–module. This means the following. Let $A : S \otimes_{\mathbb{C}} M \longrightarrow M$ be the map giving M its structure as an S–module, and let $m : R \otimes_{\mathbb{C}} R \longrightarrow R$ be the multiplication. Then the following diagram commutes

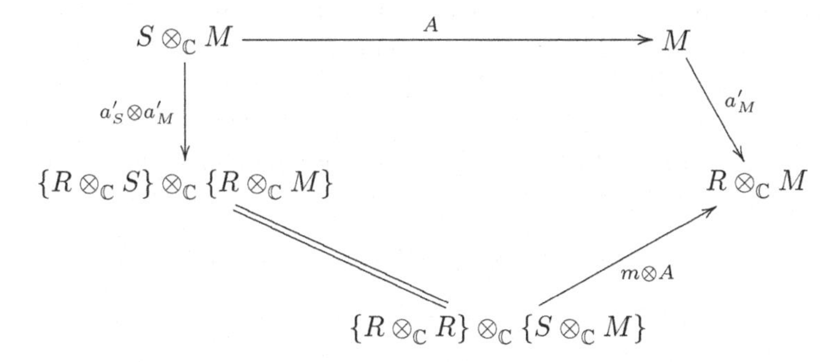

We are not insisting that M should be finite, which means that the ring

$S \oplus M$ need not be a finitely generated $\mathbb{C}$–algebra. Let us nevertheless see that with (i), (ii) and (iii) above the ring $S \oplus M$ admits an action

$$a'_{S\oplus M} : S \oplus M \longrightarrow R \otimes_{\mathbb{C}} \{S \oplus M\} ,$$

and that

$$S \xrightarrow{\psi} S \oplus M \xrightarrow{\varphi} S$$

are $\mathbb{C}$–algebra homomorphisms compatible with the $(G, \mathcal{O}_G)$–action. Conditions (i) and (ii) establish that the map

$$S \oplus M \xrightarrow{a'_S \oplus a'_M} \{R \otimes_{\mathbb{C}} S\} \oplus \{R \otimes_{\mathbb{C}} M\} \quad = \quad R \otimes_{\mathbb{C}} \{S \oplus M\}$$

satisfies the commutativity requirements of Definition 8.3.1(i) and (ii). The fact that $\psi : S \longrightarrow S \oplus M$ and $\varphi : S \oplus M \longrightarrow S$ are G–homomorphisms is guaranteed by the definition of $a'_{S\oplus M} = a'_S \oplus a'_M$ as a sum. The requirement (iii) above is what is needed to make sure that $a'_{S\oplus M} = a'_S \oplus a'_M$ is a ring homomorphism. More precisely, to make sure that $a'_{S\oplus M}$ respects the action of S on the ideal $M \subset S \oplus M$.

Remark 8.3.17. The last paragraph of Remark 8.3.16 says that, if M is a finite S–module, then the definition of a G–action given in Remark 8.3.16 agrees with that of Remark 8.3.14. The main difference is that Remark 8.3.16 is more detailed and less succinct. We will only very occasionally consider large G–modules.

Remark 8.3.18. The more pedestrian, less compact version of the definition makes it clear that, given a G–homomorphism $\eta : M \longrightarrow N$ of G–modules over the G–ring S, the kernel, image and cokernel of η are also G–modules.

8.4 The action of the group of closed points

We have noted that the action of the affine group scheme $(G, \mathcal{O}_G)$ on the affine scheme $(X, \mathcal{O}_X)$ gives a map

$$a : (G, \mathcal{O}_G) \times (X, \mathcal{O}_X) \longrightarrow (X, \mathcal{O}_X) .$$

By Proposition 4.2.4, or else by the fancier Lemma 4.2.6, we know that any closed point $g \in G^{\mathrm{an}}$ corresponds to a morphism of ringed spaces

$$\Psi_g : \left(\mathrm{Spec}(\mathbb{C}), \widetilde{\mathbb{C}}\right) \longrightarrow (G, \mathcal{O}_G) .$$

This gives a composite

$$\left(\mathrm{Spec}(\mathbb{C}), \widetilde{\mathbb{C}}\right) \times (X, \mathcal{O}_X) \xrightarrow{\Psi_g \times 1} (G, \mathcal{O}_G) \times (X, \mathcal{O}_X) \xrightarrow{a} (X, \mathcal{O}_X) ;$$

since $\left(\mathrm{Spec}(\mathbb{C}), \widetilde{\mathbb{C}}\right) \times (X, \mathcal{O}_X) = (X, \mathcal{O}_X)$ we deduce a morphism of ringed spaces

$$(T_g, T_g^*) : (X, \mathcal{O}_X) \longrightarrow (X, \mathcal{O}_X) .$$

We have a continuous map $T_g : X \longrightarrow X$. When restricted to $X^{\mathrm{an}} \subset X$ this map is just the translation by $g \in G^{\mathrm{an}}$; it is the map $T_g^{\mathrm{an}} : X^{\mathrm{an}} \longrightarrow X^{\mathrm{an}}$ which takes x to gx. But the above allows us to extend the action to all of X. We note this as a lemma.

Lemma 8.4.1. *The map sending $g \in G^{\mathrm{an}}$ to $T_g : X \longrightarrow X$ is a homomorphism from the group G^{an} to the group of homeomorphisms from X to itself.*

Proof. The axioms of the group action guarantee that $(T_g, T_g^*)(T_h, T_h^*) = (T_{gh}, T_{gh}^*)$; in particular $T_{gh} = T_g T_h$. □

Since (T_g, T_g^*) is a morphism of ringed spaces it induces maps on rings. Given any open set $U \subset X$ there is a map

$$\{T_g\}_U^* \ : \Gamma(U, \mathcal{O}_X) \longrightarrow \Gamma(T_g^{-1}U, \mathcal{O}_X) .$$

In the light of Lemma 8.4.1 it is probably simpler to denote $T_g^{-1}U$ by $g^{-1}U$. The most interesting case of this is when $g^{-1}U = U$. We make a definition:

Definition 8.4.2. *The open set $U \subset X$ will be called G–*invariant *if, for all $g \in G^{\mathrm{an}}$, we have $g^{-1}U = T_g^{-1}U = U$.*

Construction 8.4.3. If $U \subset X$ is a G–invariant open set then, to every $g \in G^{\mathrm{an}}$, we can associate a ring homomorphism $\{T_{g^{-1}}\}_U^* : \Gamma(U, \mathcal{O}) \longrightarrow \Gamma(U, \mathcal{O})$.

Lemma 8.4.4. *If U is a G–invariant open set in X then the map, taking $g \in G^{\mathrm{an}}$ to $\{T_{g^{-1}}\}_U^*$, is a homomorphism from the group G^{an} into the group of automorphisms of the ring $\Gamma(U, \mathcal{O})$. Further, this map $\Phi : G^{\mathrm{an}} \longrightarrow \mathrm{Aut}\big\{\Gamma(U, \mathcal{O})\big\}$ is compatible with restriction. If $U \subset V$ are G–invariant open subsets of X then the restriction homomorphism*

$$\mathrm{res}_U^V : \Gamma(V, \mathcal{O}) \longrightarrow \Gamma(U, \mathcal{O})$$

respects the action of G^{an}.

Proof. Let g, h be two elements of the group G^{an}. The axioms of a group action guarantee that the composite of the two maps

$$(X, \mathcal{O}_X) \xrightarrow{(T_{g^{-1}}, T_{g^{-1}}^*)} (X, \mathcal{O}_X) \xrightarrow{(T_{h^{-1}}, T_{h^{-1}}^*)} (X, \mathcal{O}_X)$$

agrees with

$$(T_{h^{-1}g^{-1}}, T^*_{h^{-1}g^{-1}}) : (X, \mathcal{O}_X) \longrightarrow (X, \mathcal{O}_X) .$$

For U any G–invariant open set this means that the ring homomorphisms

$$\Gamma(U, \mathcal{O}_X) \xrightarrow{\{T_{h^{-1}}\}^*_U} \Gamma(U, \mathcal{O}_X) \xrightarrow{\{T_{g^{-1}}\}^*_U} \Gamma(U, \mathcal{O}_X)$$

composes to

$$\{T_{\{gh\}^{-1}}\}^*_U : \Gamma(U, \mathcal{O}_X) \longrightarrow \Gamma(U, \mathcal{O}_X) .$$

In other words, we obtain the identity

$$\{T_{\{gh\}^{-1}}\}^*_U \quad = \quad \{T_{g^{-1}}\}^*_U \{T_{h^{-1}}\}^*_U ;$$

the map sending $g \in G^{\mathrm{an}}$ to $\{T_{g^{-1}}\}^*_U$ is a group homomorphism. The fact that it commutes with restriction is obvious, just because $(T_{g^{-1}}, T^*_{g^{-1}})$ is a morphism of ringed spaces. □

Notation 8.4.5. In the light of Lemma 8.4.4 we now know that, under the hypotheses of the Lemma, the group G^{an} acts on the ring $\Gamma(U, \mathcal{O})$ by automorphisms. We will feel free to denote by $g(f)$, or even by gf, the image of $f \in \Gamma(U, \mathcal{O})$ under the action of $g \in G^{\mathrm{an}}$. The notation $\{T_{g^{-1}}\}^*_U f$ might be closer to the definition of this action, but we will expect the reader to forgive us for using the less cumbersome notation. This will be especially true in later sections, when it is not so relevant to remember the definition of the action.

Remark 8.4.6. As in Lemma 8.4.4, let $(G, \mathcal{O}_G) = \left(\mathrm{Spec}(R), \widetilde{R}\right)$ be an affine group scheme of finite type over $\mathbb{C}$, and suppose it acts on an affine scheme $(X, \mathcal{O}_X) = \left(\mathrm{Spec}(S), \widetilde{S}\right)$, of finite type over $\mathbb{C}$. From Lemma 8.4.4 we know that the group G^{an} acts on $\Gamma(U, \mathcal{O}_X)$, for every G–invariant open subset $U \subset X$. What about the special case $U = X$? Each element $g \in G^{\mathrm{an}}$ gives a ring automorphism $\{T_{g^{-1}}\}^*_X$ of the ring $\Gamma(X, \mathcal{O}_X) = S$. What is this map $\{T_{g^{-1}}\}^*_X : S \longrightarrow S$?

Let us trace our way through the definitions. The element $g^{-1} \in G^{\mathrm{an}}$ gives a map of affine schemes

$$(T_{g^{-1}}, T^*_{g^{-1}}) : (X, \mathcal{O}_X) \longrightarrow (X, \mathcal{O}_X) .$$

This map was defined to be the composite

$$\left(\mathrm{Spec}(\mathbb{C}), \widetilde{\mathbb{C}}\right) \times (X, \mathcal{O}_X) \xrightarrow{\Psi_{g^{-1}} \times 1} (G, \mathcal{O}_G) \times (X, \mathcal{O}_X) \xrightarrow{a} (X, \mathcal{O}_X) .$$

Of course, it can also be described as

$$\left(\mathrm{Spec}(\theta), \widetilde{\theta}\right) : \left(\mathrm{Spec}(S), \widetilde{S}\right) \longrightarrow \left(\mathrm{Spec}(S), \widetilde{S}\right)$$

for some ring homomorphism $\theta : S \longrightarrow S$. It is not difficult to work out what θ is. The map

$$(G, \mathcal{O}_G) \times (X, \mathcal{O}_X) \xrightarrow{\ a\ } (X, \mathcal{O}_X)$$

is $\left(\mathrm{Spec}(a'), \widetilde{a'}\right)$, where $a' : S \longrightarrow R \otimes_{\mathbb{C}} S$ is the action. The map $\Psi_{g^{-1}} : \left(\mathrm{Spec}(\mathbb{C}), \widetilde{\mathbb{C}}\right) \longrightarrow (G, \mathcal{O}_G)$ is $\left(\mathrm{Spec}(\varphi_{g^{-1}}), \widetilde{\varphi_{g^{-1}}}\right)$, where $\varphi_{g^{-1}}$ is the $\mathbb{C}$–algebra homomorphism $\varphi_{g^{-1}} : R \longrightarrow \mathbb{C}$ corresponding, under the bijection of Proposition 4.2.4, with the closed point $g^{-1} \in G = \mathrm{Spec}(R)$. The composite

$$\left(\mathrm{Spec}(\mathbb{C}), \widetilde{\mathbb{C}}\right) \times (X, \mathcal{O}_X) \xrightarrow{\ \Psi_{g^{-1}} \times 1\ } (G, \mathcal{O}_G) \times (X, \mathcal{O}_X) \xrightarrow{\ a\ } (X, \mathcal{O}_X)$$

is therefore obtained by taking the composite ring homomorphism

$$S \xrightarrow{\ a'\ } R \otimes_{\mathbb{C}} S \xrightarrow{\ \varphi_{g^{-1}} \otimes 1\ } \mathbb{C} \otimes_{\mathbb{C}} S \quad = \quad S$$

and forming the map of schemes it induces. We have found the map $\theta : S \longrightarrow S$, for which

$$(T_{g^{-1}}, T^*_{g^{-1}}) : (X, \mathcal{O}_X) \longrightarrow (X, \mathcal{O}_X)$$

is just $\left(\mathrm{Spec}(\theta), \widetilde{\theta}\right) : \left(\mathrm{Spec}(S), \widetilde{S}\right) \longrightarrow \left(\mathrm{Spec}(S), \widetilde{S}\right)$.

The action of $g \in G^{\mathrm{an}}$ on $\Gamma(X, \mathcal{O}_X) = \Gamma\left(\mathrm{Spec}(S), \widetilde{S}\right) = S$ is, by definition, whatever the morphism $(T_{g^{-1}}, T^*_{g^{-1}}) = \left(\mathrm{Spec}(\theta), \widetilde{\theta}\right)$ does to global sections. That is, it is the map

$$\widetilde{\theta}_{\mathrm{Spec}(S)} : \Gamma\left(\mathrm{Spec}(S), \widetilde{S}\right) \longrightarrow \Gamma\left(\mathrm{Spec}(S), \widetilde{S}\right) .$$

Proposition 3.6.5 tells us that this is simply the ring homomorphism $\theta : S \longrightarrow S$.

Example 8.4.7. Let $f \in S$ be an element with $a'(f) = r \otimes f$. In Lemma 8.3.10 we proved that r is invertible in R. It follows that, for all ring homomorphisms $\varphi : R \longrightarrow \mathbb{C}$, the complex number $\varphi(r)$ will be non-zero. If we consider the map $\theta : S \longrightarrow S$, given in Remark 8.4.6 as the composite

$$S \xrightarrow{\ a'\ } R \otimes_{\mathbb{C}} S \xrightarrow{\ \varphi_{g^{-1}} \otimes 1\ } \mathbb{C} \otimes_{\mathbb{C}} S \quad = \quad S ,$$

then $\theta(f) = \varphi_{g^{-1}}(r) \cdot f$ is a unit times f. But we know that the map

$g^{-1} : X \longrightarrow X$ is just $\text{Spec}(\theta) : X \longrightarrow X$. The inverse image of X_f under $\text{Spec}(\theta) : X \longrightarrow X$ is, by Remark 3.6.4, precisely $X_{\theta(f)} = X_f$. That is, $gX_f = X_f$. This being true for every $g \in G^{\text{an}}$, we deduce that X_f is a G–invariant open subset of X.

Lemma 8.4.4 therefore tells us that the group G^{an} acts on $\Gamma(X_f, \mathcal{O}) = S[1/f]$. In this case, the action is easy to compute. We know that the action is by ring homomorphisms, and we also know that the restriction map

$$S \quad = \quad \Gamma(X, \mathcal{O}) \longrightarrow \Gamma(X_f, \mathcal{O}) \quad = \quad S[1/f]$$

intertwines the action of G^{an}. Since every element of $S[1/f]$ can be written as x/f^n, with $x \in S$, we conclude that $g(x/f^n) = g(x)/g(f)^n$. And the action of g on S is by $\theta : S \longrightarrow S$; see Remark 8.4.6. Hence

$$g(x/f^n) \quad = \quad \theta(x)/\theta(f)^n \ .$$

Lemma 8.4.8. *Suppose $(G, \mathcal{O}_G)$ is an affine group scheme, of finite type over $\mathbb{C}$, acting on the affine schemes $(X, \mathcal{O}_X)$ and $(Y, \mathcal{O}_Y)$, also both of finite type over $\mathbb{C}$. Let $(\beta, \beta^*) : (X, \mathcal{O}_X) \longrightarrow (Y, \mathcal{O}_Y)$ be a G–morphism, as in Definition 8.3.6. Then we know the following:*

(i) *The continuous map $\beta : X \longrightarrow Y$ intertwines the action of G^{an}.*
(ii) *If $U \subset Y$ is a G–invariant open set then the set $\beta^{-1}U \subset X$ is G–invariant, and the ring homomorphism*

$$\beta_U^* : \Gamma(U, \mathcal{O}_Y) \longrightarrow \Gamma(\beta^{-1}U, \mathcal{O}_X)$$

respects the action of G^{an}.

Proof. Let $g \in G^{\text{an}}$ be any closed point. By either Proposition 4.2.1 or Lemma 4.2.6 we know that there is a corresponding morphism $(\Psi_g, \Psi_g^*) : \left(\text{Spec}(\mathbb{C}), \widetilde{\mathbb{C}}\right) \longrightarrow (G, \mathcal{O}_G)$ of ringed spaces over $\mathbb{C}$. In the diagram below

$$\begin{array}{ccccc}
\left(\text{Spec}(\mathbb{C}), \widetilde{\mathbb{C}}\right) \times (X, \mathcal{O}_X) & \xrightarrow{(\Psi_g, \Psi_g^*) \times 1} & (G, \mathcal{O}_G) \times (X, \mathcal{O}_X) & \xrightarrow{a_X} & (X, \mathcal{O}_X) \\
{\scriptstyle 1 \times (\beta, \beta^*)} \downarrow & & {\scriptstyle 1 \times (\beta, \beta^*)} \downarrow & & {\scriptstyle (\beta, \beta^*)} \downarrow \\
\left(\text{Spec}(\mathbb{C}), \widetilde{\mathbb{C}}\right) \times (Y, \mathcal{O}_Y) & \xrightarrow[(\Psi_g, \Psi_g^*) \times 1]{} & (G, \mathcal{O}_G) \times (Y, \mathcal{O}_Y) & \xrightarrow[a_Y]{} & (Y, \mathcal{O}_Y)
\end{array}$$

the square on the left commutes trivially, while the square on the right commutes because $(\beta, \beta^*) : (X, \mathcal{O}_X) \longrightarrow (Y, \mathcal{O}_Y)$ is a G–morphism. By the definition of (T_g, T_g^*) we have a commutative square

$$\begin{array}{ccc} (X, \mathcal{O}_X) & \xrightarrow{(T_g, T_g^*)} & (X, \mathcal{O}_X) \\ {\scriptstyle (\beta,\beta^*)}\downarrow & & \downarrow{\scriptstyle (\beta,\beta^*)} \\ (Y, \mathcal{O}_Y) & \xrightarrow[(T_g, T_g^*)]{} & (Y, \mathcal{O}_Y) \end{array} .$$

If we take a point $x \in X$ and look at its image under the two composites we learn that $\beta(gx) = g\beta(x)$, that is $\beta : X \longrightarrow Y$ intertwines the action of G^{an}. Hence, for any open set U, we have $\beta^{-1}g^{-1}U = g^{-1}\beta^{-1}U$; if U is G–invariant then so is $\beta^{-1}U$. On the level of rings we learn that, for any G–invariant U, the following square of ring homomorphisms commutes

$$\begin{array}{ccc} \Gamma(U, \mathcal{O}_Y) & \xrightarrow{\{T_g\}_U^*} & \Gamma(U, \mathcal{O}_Y) \\ {\scriptstyle \beta_U^*}\downarrow & & \downarrow{\scriptstyle \beta_U^*} \\ \Gamma(\beta^{-1}U, \mathcal{O}_X) & \xrightarrow[\{T_g\}_{\beta^{-1}U}^*]{} & \Gamma(\beta^{-1}U, \mathcal{O}_X) \end{array} .$$

This says that the ring homomorphism $\beta_U^* : \Gamma(U, \mathcal{O}_Y) \longrightarrow \Gamma(\beta^{-1}U, \mathcal{O}_X)$ intertwines the action of $g^{-1} \in G^{\text{an}}$. □

Remark 8.4.9. There are two cases of the above that interest us particularly. The first is where the topological spaces X and Y agree; that is, only the sheaf of rings changes. In particular if S is a G–ring, and the S–module M is a G–module, then we have homomorphisms of G–rings

$$S \xrightarrow{\psi} S \oplus M \xrightarrow{\varphi} S .$$

These correspond to G–morphisms of affine schemes

$$(X, \widetilde{S}) \xrightarrow{(\text{Spec}(\varphi), \widetilde{\varphi})} (Y, \widetilde{S \oplus M}) \xrightarrow{(\text{Spec}(\psi), \widetilde{\psi})} (X, \widetilde{S}) .$$

We know that $\text{Spec}(\varphi) : X \longrightarrow Y$ and $\text{Spec}(\psi) : Y \longrightarrow X$ are inverse homeomorphisms, and Lemma 8.4.8(i) tells us that they also respect the action of G^{an}. Most important for us is that the G–invariant open sets are the same, and that ring homomorphisms induced by the maps of the ringed spaces respect the G^{an}–action. On the space $X = Y$ we have a sequence of sheaves of rings

$$\widetilde{S} \xrightarrow{\widetilde{\psi}} \widetilde{S \oplus M} \xrightarrow{\widetilde{\varphi}} \widetilde{S} .$$

If U is any open set we deduce a sequence of ring homomorphisms composing to the identity

$$\Gamma(U, \widetilde{S}) \xrightarrow{\Gamma(U, \widetilde{\psi})} \Gamma(U, \widetilde{S \oplus M}) \xrightarrow{\Gamma(U, \widetilde{\varphi})} \Gamma(U, \widetilde{S}) .$$

What Lemma 8.4.8(ii) buys us is that, if $U \subset X$ happens to be G–invariant, then these ring homomorphisms intertwine the action of G^{an}. It means that $\Gamma(U,\widetilde{M})$, which is defined to be the kernel of $\Gamma(U,\widetilde{\varphi})$, is naturally acted on by the group G^{an}. Summarizing, we have:

Corollary 8.4.10. *Let $(G,\mathcal{O}_G)$ be an affine group scheme of finite type over $\mathbb{C}$. Let S be a G–ring, let M be a G–module, and let $U \subset \mathrm{Spec}(S)$ be a G–invariant open set. Then the group G^{an} acts on $\Gamma(U,\widetilde{M})$, the multiplication map*

$$\Gamma(U,\widetilde{S}) \otimes \Gamma(U,\widetilde{M}) \longrightarrow \Gamma(U,\widetilde{M})$$

intertwines the G^{an}–action, and, if $U \subset V$ are both G–invariant open sets, then the restriction map

$$\mathrm{res}^V_U : \Gamma(V,\widetilde{M}) \longrightarrow \Gamma(U,\widetilde{M})$$

also intertwines the G^{an}–action.

Remark 8.4.11. In the special case, where the G–invariant open set U is all of X, we get an action of G^{an} on $M = \Gamma(X,\widetilde{M})$. As in Remark 8.4.6, the action of $g \in G^{\mathrm{an}}$ on the module M is by the composite

$$M \xrightarrow{\;a'\;} R \otimes_{\mathbb{C}} M \xrightarrow{\;\varphi_{g^{-1}}\otimes 1\;} \mathbb{C} \otimes_{\mathbb{C}} M \quad = \quad M\,.$$

The formula above works for rings, by Remark 8.4.6. To show it for modules, we only have to recall that the action on $M = \Gamma(X,\widetilde{M})$ is the action on the ideal M, where M is viewed as the kernel of the ring homomorphism $\varphi : S \oplus M \longrightarrow S$. The induced maps, on the kernels of the vertical maps in the commutative diagram of ring homomorphisms

$$\begin{array}{ccccccc}
S \oplus M & \xrightarrow{\;a'\;} & R \otimes_{\mathbb{C}} \{S \oplus M\} & \xrightarrow{\;\varphi_{g^{-1}}\otimes 1\;} & \mathbb{C} \otimes_{\mathbb{C}} \{S \oplus M\} & = & S \oplus M \\
\Big\downarrow{\scriptstyle \varphi} & & \Big\downarrow{\scriptstyle 1\otimes\varphi} & & \Big\downarrow{\scriptstyle 1\otimes\varphi} & & \Big\downarrow{\scriptstyle \varphi} \\
S & \xrightarrow{\;a'\;} & R \otimes_{\mathbb{C}} S & \xrightarrow{\;\varphi_{g^{-1}}\otimes 1\;} & \mathbb{C} \otimes_{\mathbb{C}} S & = & S
\end{array},$$

therefore compute the action.

Remark 8.4.12. Let $f \in R$ be an element satisfying $a'(f) = r \otimes f$. In Example 8.4.7 we saw that the open set $X_f \subset X$ is G–invariant, and Corollary 8.4.10 says that the group G^{an} acts on $\Gamma(X_f,\widetilde{M}) = M[1/f]$. What is this action?

As in Example 8.4.7, there is nothing mysterious about this action. We know that the multiplication map

$$S[1/f] \otimes M[1/f] = \Gamma(X_f,\widetilde{S}) \otimes \Gamma(X_f,\widetilde{M}) \xrightarrow{\;A\;} \Gamma(X_f,\widetilde{M}) = M[1/f]$$

intertwines the action of G^{an}, as does the restriction homomorphism

$$M \quad = \quad \Gamma(X, \widetilde{M}) \xrightarrow{\mathrm{res}^X_{X_f}} \Gamma(X_f, \widetilde{M}) \quad = \quad M[1/f] \, .$$

Hence so does the composite

$$S[1/f] \otimes M \xrightarrow{1 \otimes \mathrm{res}^X_{X_f}} S[1/f] \otimes M[1/f] \xrightarrow{A} M[1/f] \, .$$

But any element of $M[1/f]$ lies in the image of this composite; we can write m/f^n as the image of $\{1/f^n\} \otimes m$. Therefore

$$g(m/f^n) \quad = \quad g(1/f^n) \cdot g(m) \, ,$$

where $g(1/f^n)$ is computed using the action of g on $S[1/f]$, while $g(m)$ comes from the action of g on M.

We will leave to the reader to check:

Exercise 8.4.13. Let $(G, \mathcal{O}_G)$ be an affine group scheme of finite type over $\mathbb{C}$. Let S be a G–ring, let $U \subset \mathrm{Spec}(S)$ be a G–invariant open set, and let $\eta : M \longrightarrow N$ be a G–homomorphism of G–modules. Then the induced map

$$\Gamma(U, \widetilde{\eta}) : \Gamma(U, \widetilde{M}) \longrightarrow \Gamma(U, \widetilde{N})$$

intertwines the G^{an}–action.

Remark 8.4.14. At the beginning of Remark 8.4.9 we said there were two cases of Lemma 8.4.8 that we will find particularly important. So far we have only discussed one, namely the case where the underlying space X does not change, only the sheaf of rings varies. What is the mysterious second case I have in mind?

Suppose we are given a G–ring S, and an element $f \in S$ with $a'(f) = r \otimes f$. In Exercise 8.3.12 we saw that $S[1/f]$ is a G–ring, and that $\alpha_f : S \longrightarrow S[1/f]$ is a G–homomorphism of G–rings. If we let $(X, \mathcal{O}_X) = \left(\mathrm{Spec}(S), \widetilde{S}\right)$ and $(X_f, \mathcal{O}_{X_f}) = \left(\mathrm{Spec}(S[1/f]), \widetilde{S[1/f]}\right)$ then we have a G–morphism

$$\left(\mathrm{Spec}(\alpha_f), \widetilde{\alpha_f}\right) : (X_f, \mathcal{O}_{X_f}) \longrightarrow (X, \mathcal{O}_X) \, .$$

Proposition 3.7.3 tells us that this map is an open immersion, identifying the image with the basic open set X_f. What we learn now is that the map also respects the G^{an}–action. The G–invariant open sets in X_f are precisely the G–invariant open subsets of X which happen to be contained in X_f, and, for a G–invariant open subset $U \subset X_f$, the isomorphism

$$\Gamma(U, \mathcal{O}_X) \longrightarrow \Gamma(U, \mathcal{O}_{X_f})$$

respects the G^{an}–action. That is, the isomorphism of sheaves of rings $\mathcal{O}_{X_f} \cong \{\mathcal{O}_X\}|_{X_f}$ respects the G^{an}–action.

Example 8.4.15. Let us consider the special case where $(G, \mathcal{O}_G)$ is the trivial group, that is where $R = \mathbb{C}$ as in Example 8.2.7. Every finitely generated algebra S was made into a G–ring in Example 8.3.4. Since the group $G^{\mathrm{an}} = \{*\}$ is trivial, all actions of G^{an} must be trivial. For every open set $U \subset X = \mathrm{Spec}(S)$ and all $g \in G^{\mathrm{an}}$ (of which there is only the identity) we have $gU = U$; that is every open set is G–invariant. The group G^{an} acts on $\Gamma(U, \mathcal{O}_X)$, and the only possible action is trivial.

8.5 Back to the world of the concrete

In all the abstraction the reader should not lose sight of the fact that the non-trivial $(G, \mathcal{O}_G)$ we care about corresponds to the Hopf algebra $\mathbb{C}[t, t^{-1}]$, and the G–ring we want to remember is $S = \mathbb{C}[x_0, x_1, \ldots, x_n]$. Then $(G, \mathcal{O}_G)$ acts on $(X, \mathcal{O}_X) = \left(\mathrm{Spec}(S), \widetilde{S}\right)$, and the theory applies. For example, the open set $U = X$ is clearly G–invariant, and Remark 8.4.6 provides a recipe for computing the action of $G^{\mathrm{an}} = \mathbb{C}^*$ on the ring $\Gamma(X, \mathcal{O}_X) = S$. What is this action, expressed concretely? Let us work this out in detail.

Example 8.5.1. The general rule is the following. Given $g \in G^{\mathrm{an}} = \mathbb{C}^*$, the action of g on S is via the homomorphism

$$S \xrightarrow{\ a'\ } R \otimes_{\mathbb{C}} S \xrightarrow{\ \varphi_{g^{-1}} \otimes 1\ } \mathbb{C} \otimes_{\mathbb{C}} S \quad = \quad S\,.$$

In our particular case, the action $a' : S \longrightarrow R \otimes_{\mathbb{C}} S$ is given by the formula $a'(x_i) = t^{-1} \otimes x_i$. The map $g : S \longrightarrow S$ will take $x_i \in S = \mathbb{C}[x_0, x_1, \ldots, x_n]$ to

$$\begin{aligned} \{\varphi_{g^{-1}} \otimes 1\} a'(x_i) &= \{\varphi_{g^{-1}} \otimes 1\}(t^{-1} \otimes x_i) \\ &= \varphi_{g^{-1}}(t^{-1}) \cdot x_i\,. \end{aligned}$$

But $\varphi_{g^{-1}}(t^{-1})$ is computed by evaluating the Laurent polynomial $t^{-1} \in \mathbb{C}[t, t^{-1}]$ at the non-zero complex number $t = g^{-1}$. This gives us the formula

$$g(x_i) = g x_i\,,$$

where on the left we mean the group element g acting on $x_i \in S$, and on the right we mean the product of the complex number g with x_i.

Now the map $g : S \longrightarrow S$ is a ring homomorphism. Hence

$$\begin{aligned} g\left(\sum \lambda_{\mathbf{a}} x_0^{a_0} x_1^{a_1} \cdots x_n^{a_n}\right) &= \sum \lambda_{\mathbf{a}} g(x_0)^{a_0} g(x_1)^{a_1} \cdots g(x_n)^{a_n} \\ &= \sum \lambda_{\mathbf{a}} g^{a_0+a_1+\cdots+a_n} x_0^{a_0} x_1^{a_1} \cdots x_n^{a_n} \,. \end{aligned}$$

This completely describes the action, on any polynomial.

Notation 8.5.2. We want to return to the conventions of Notation 5.8.9, ever so slightly modified; we will now be in dimension $(n+1)$ instead of n. We remind the reader: vectors are denoted by boldface letters, with late letters in the alphabet, like $\mathbf{x}$, standing for vectors $\mathbf{x} = (x_0, x_1, \ldots, x_n)$ in $\mathbb{C}^{n+1}$, while early letters, like $\mathbf{a}$, will stand for vectors $\mathbf{a} = (a_0, a_1, \ldots, a_n)$ in $\mathbb{Z}^{n+1}$. The symbol $\mathbf{x}^{\mathbf{a}}$ is our shorthand for the monomial

$$\mathbf{x}^{\mathbf{a}} = x_0^{a_0} x_1^{a_1} \cdots x_n^{a_n} \,.$$

In Notation 5.8.9 we declared that $|\mathbf{a}|$ was to stand for

$$|\mathbf{a}| = |a_0| + |a_1| + \cdots + |a_n| \,.$$

We now also need a symbol for $a_0 + a_1 + \cdots + a_n$; our abbreviation will be

$$\langle \mathbf{a} \rangle = a_0 + a_1 + \cdots + a_n \,.$$

In this notation, the formula of Example 8.5.1 becomes

$$g\left(\sum_{\mathbf{x}^{\mathbf{a}} \in S} \lambda_{\mathbf{a}} \mathbf{x}^{\mathbf{a}}\right) = \sum_{\mathbf{x}^{\mathbf{a}} \in S} \lambda_{\mathbf{a}} g^{\langle \mathbf{a} \rangle} \mathbf{x}^{\mathbf{a}} \,.$$

Problem 8.5.3. Recall that for any subset $J \subset \{0, 1, \ldots, n\}$ we defined, in Notation 3.5.3, an element $f_J \in S$ by the formula

$$f_J = \prod_{i \in J} x_i \,.$$

Then f_J is a homogeneous polynomial of degree $|J|$, and Example 8.3.13 established that $a'(f_J) = t^{-|J|} \otimes f_J$. Example 8.4.7 tells us that the open set $X_J = X_{f_J}$ is G–invariant, and Lemma 8.4.4 asserts that the ring $\Gamma(X_J, \mathcal{O}) = S[1/f_J]$ admits an action of G^{an}. What is this action?

Reminder 8.5.4. Before we answer Problem 8.5.3, let us remind ourselves of some results from Section 3.5. In Exercise 3.5.4(i) and in Remark 3.5.6 we constructed an isomorphism $\varphi_J^{-1} : S_J \longrightarrow \Gamma(X_J, \mathcal{O})$, where S_J is the ring of Laurent polynomials

$$S_J = \mathbb{C}[x_0, x_1, \ldots, x_n, x_{j_1}^{-1}, x_{j_2}^{-1}, \ldots, x_{j_\ell}^{-1}] \,.$$

Recall that a basis for the ring S_J consists of the Laurent monomials

$$\mathbf{x}^{\mathbf{a}} \quad = \quad x_0^{a_0} x_1^{a_1} \cdots x_n^{a_n} ,$$

where $a_i \geq 0$ if $i \notin J$, while $a_i \in \mathbb{Z}$ is unrestricted if $i \in J$. Finally in Remark 3.5.6 we noticed that, whenever we have an inclusion of subsets $J \subset J' \subset \{0, 1, \ldots, n\}$, the following square commutes

$$\begin{array}{ccc} S_J & \xrightarrow{\text{inclusion}} & S_{J'} \\ \varphi_J^{-1} \big\downarrow & & \big\downarrow \varphi_{J'}^{-1} \\ \Gamma(X_J, \mathcal{O}) & \xrightarrow[\operatorname{res}^{X_J}_{X_{J'}}]{} & \Gamma(X_{J'}, \mathcal{O}) \end{array} .$$

In the special case where $J = \emptyset$ we have $f_J = 1$, and hence $X_J = X_{f_J} = X$. The commutative square above becomes

$$\begin{array}{ccc} S & \xrightarrow{\text{inclusion}} & S_{J'} \\ \varphi_\emptyset^{-1} \big\downarrow & & \big\downarrow \varphi_{J'}^{-1} \\ \Gamma(X, \mathcal{O}) & \xrightarrow[\operatorname{res}^{X}_{X_{J'}}]{} & \Gamma(X_{J'}, \mathcal{O}) \end{array} .$$

The reader can verify that the map $\varphi_\emptyset^{-1} : S \longrightarrow \Gamma(X, \mathcal{O})$, defined in Exercise 3.5.4(i), is the natural identification.

Remark 8.5.5. It is now time to answer Problem 8.5.3. From Example 8.5.1 we know the action of $G^{\mathrm{an}} = \mathbb{C}^*$ on the ring $S = \Gamma(X, \mathcal{O})$. By Lemma 8.4.4 we know that the restriction map

$$\operatorname{res}^X_{X_{J'}} : \Gamma(X, \mathcal{O}) \longrightarrow \Gamma(X_{J'}, \mathcal{O})$$

is a ring homomorphism respecting the G^{an}–action. It follows that, if we give the ring $S_{J'}$ an action of G^{an} by way of the isomorphism $\varphi_{J'}^{-1} : S_{J'} \longrightarrow \Gamma(X_{J'}, \mathcal{O})$, then the inclusion $S \longrightarrow S_{J'}$ must also be a homomorphism of rings, intertwining the action of G^{an}. We know the action on S, so we should be able to figure out the action on $S_{J'}$.

Now a basis for S consists of monomials $\mathbf{x}^{\mathbf{a}}$, with $a_i \geq 0$ for all $0 \leq i \leq n$. The action of $g \in G^{\mathrm{an}}$ was given, in Example 8.5.1, by the formula

$$g(\mathbf{x}^{\mathbf{a}}) \quad = \quad g^{\langle \mathbf{a} \rangle} \mathbf{x}^{\mathbf{a}} .$$

Our basis for $S_{J'}$ consists of monomials $\mathbf{x}^{\mathbf{c}}$, where some of the c_i may be negative. We may certainly write the vector $\mathbf{c} \in \mathbb{Z}^{n+1}$ as $\mathbf{c} = \mathbf{a} - \mathbf{b}$, where $\mathbf{a}$ and $\mathbf{b}$ are integer vectors with non-negative entries. This expresses $\mathbf{x}^{\mathbf{c}}$

as

$$\mathbf{x}^{\mathbf{c}} \quad = \quad \mathbf{x}^{\mathbf{a}-\mathbf{b}} \quad = \quad \mathbf{x}^{\mathbf{a}}/\mathbf{x}^{\mathbf{b}} \ ,$$

with $\mathbf{x}^{\mathbf{a}}$ and $\mathbf{x}^{\mathbf{b}}$ both in $S \subset S_{J'}$. It follows that

$$g(\mathbf{x}^{\mathbf{a}-\mathbf{b}}) \quad = \quad g\left(\frac{\mathbf{x}^{\mathbf{a}}}{\mathbf{x}^{\mathbf{b}}}\right) \quad = \quad \frac{g^{\langle \mathbf{a} \rangle}\mathbf{x}^{\mathbf{a}}}{g^{\langle \mathbf{b} \rangle}\mathbf{x}^{\mathbf{b}}} \quad = \quad g^{\langle \mathbf{a}-\mathbf{b} \rangle}\mathbf{x}^{\mathbf{a}-\mathbf{b}} \ .$$

Put more succinctly, the formula $g(\mathbf{x}^{\mathbf{c}}) = g^{\langle \mathbf{c} \rangle}\mathbf{x}^{\mathbf{c}}$ extends to all monomials in $S_{J'}$, even ones where some of the c_i are negative.

8.6 Quotients of affine schemes

Let $(G, \mathcal{O}_G)$ and $(X, \mathcal{O}_X)$ be affine schemes of finite type over $\mathbb{C}$. Suppose $(G, \mathcal{O}_G)$ is a group scheme, and suppose it acts on $(X, \mathcal{O}_X)$. Next we want to form the quotient $(X/G, \mathcal{O}_{X/G})$. In this section we define the topological space X/G, and in Section 8.7 we turn it into a ringed space by producing on it a sheaf of rings $\mathcal{O}_{X/G}$.

Remark 8.6.1. The construction we will give is not particularly sensible in general. It works well only if the group G is connected and reductive (whatever this means), and the space X is particularly nice. As we will see, the ringed space $(X/G, \mathcal{O}_{X/G})$ need not, in general, be a scheme; usually only an open subset $V \subset X/G$ is a scheme of finite type over $\mathbb{C}$. Even for the example we have in mind this is what happens; what will interest us is the open subset $V \subset X/G$ which happens to be a scheme.

Definition 8.6.2. *Let the notation be as above. We declare:*

(i) *The points of X/G are the fixed points in X under the action of the group G^{an}; see Lemma 8.4.1 for the fact that G^{an} acts on X.*
(ii) *The topology on $X/G \subset X$ is the subspace topology.*

Remark 8.6.3. Recall Example 8.3.5, where $(G, \mathcal{O}_G) = \left(\mathrm{Spec}(R), \widetilde{R}\right)$ with $R = \mathbb{C}[t, t^{-1}]$, and where $S = \mathbb{C}[x_0, x_1, \ldots, x_n]$ is the polynomial ring. In this case $G^{\mathrm{an}} = \mathbb{C}^*$, and it acts on $X^{\mathrm{an}} = \mathbb{C}^{n+1}$ by dilation. The only fixed point for the action of G^{an} on X^{an} is the point $0 \in \mathbb{C}^{n+1}$.

The reader should not be disheartened; there are plenty more fixed points, only none of them is closed. That is, most of the fixed points lie in $X - X^{\mathrm{an}}$. For the first time in the book we have here a construction which makes heavy use of the non-closed points in X.

The next step is to construct a map $\pi : X \longrightarrow X/G$. To define it we need a lemma.

Lemma 8.6.4. *Assume* $(G, \mathcal{O}_G) = \left(\operatorname{Spec}(R), \widetilde{R}\right)$ *is a connected affine group scheme of finite type over* $\mathbb{C}$. *Let* S *be a* G*–ring. If* $\mathfrak{p}$ *is any prime ideal in* S *then so is the ideal*

$$\pi(\mathfrak{p}) \quad = \quad \bigcap_{g \in G^{\mathrm{an}}} g\mathfrak{p} \; .$$

Remark 8.6.5. In the two special cases we care about, where the Hopf algebra R is either the trivial $\mathbb{C}$ or the non-trivial $\mathbb{C}[t, t^{-1}]$, the proof is very simple. The more general argument involves commutative algebra which is above the level we have had in the book. For this reason the proof will proceed as follows. The beginning of the proof is the same, whether one considers the general or the special G. When the argument begins to become difficult I will set out what is needed to complete the proof, and show that it is trivially satisfied for $R = \mathbb{C}$ or $\mathbb{C}[t, t^{-1}]$. Then, in an aside to the experts, we explain how to proceed in general.

Proof. The action of G is given by a ring homomorphism $a' : S \longrightarrow R \otimes_{\mathbb{C}} S$, and we are given a prime ideal $\mathfrak{p} \subset S$. We first need to understand what $g \in G^{\mathrm{an}}$ does to the prime ideal $\mathfrak{p}$. The element $g \in G^{\mathrm{an}}$ corresponds to a homomorphism $\varphi_g : R \longrightarrow \mathbb{C}$. In Remark 8.4.6 we observed that the action g on $X = \operatorname{Spec}(S)$ is given by a map $\operatorname{Spec}(\theta) : X \longrightarrow X$, where $\theta : S \longrightarrow S$ is the composite of the ring homomorphisms

$$S \xrightarrow{\quad a' \quad} R \otimes_{\mathbb{C}} S \xrightarrow{\varphi_g \otimes 1} \mathbb{C} \otimes_{\mathbb{C}} S \quad = \quad S \; .$$

The prime ideal $g\mathfrak{p}$ is the inverse image of $\mathfrak{p} \subset S$ by this composite. Of course we could view $\mathfrak{p}$ as the kernel of the map $\psi : S \longrightarrow S/\mathfrak{p} \longrightarrow k(\mathfrak{p})$, where $k(\mathfrak{p})$ is the quotient field of the integral domain $S/\mathfrak{p}$. This makes $g\mathfrak{p}$ the kernel of

$$S \xrightarrow{\quad a' \quad} R \otimes_{\mathbb{C}} S \xrightarrow{\varphi_g \otimes 1} \mathbb{C} \otimes_{\mathbb{C}} S \xrightarrow{1 \otimes \psi} \mathbb{C} \otimes_{\mathbb{C}} k(\mathfrak{p}) \; ,$$

and we could rewrite this composite as

$$S \xrightarrow{\quad a' \quad} R \otimes_{\mathbb{C}} S \xrightarrow{1 \otimes \psi} R \otimes_{\mathbb{C}} k(\mathfrak{p}) \xrightarrow{\varphi_g \otimes 1} \mathbb{C} \otimes_{\mathbb{C}} k(\mathfrak{p}) \; .$$

As we vary the element $g \in G^{\mathrm{an}}$ only the map $\varphi_g \otimes 1 : R \otimes_{\mathbb{C}} k(\mathfrak{p}) \longrightarrow \mathbb{C} \otimes_{\mathbb{C}} k(\mathfrak{p})$ changes. The kernel of this map is a prime ideal $\mathfrak{q}_g$ in $R \otimes_{\mathbb{C}} k(\mathfrak{p})$, the inverse image of $\mathfrak{q}_g$ by the composite

$$S \xrightarrow{\quad a' \quad} R \otimes_{\mathbb{C}} S \xrightarrow{1 \otimes \psi} R \otimes_{\mathbb{C}} k(\mathfrak{p})$$

is $g\mathfrak{p}$, and the composite $\{1 \otimes \psi\} a'$ is independent of $g \in G^{\mathrm{an}}$. It therefore suffices to prove that the intersection of the prime ideals $\mathfrak{q}_g \subset R \otimes_{\mathbb{C}} k(\mathfrak{p})$ is a prime ideal. Note that from this we already see that the result is largely independent of the ring S and of the action a'.

It suffices then to prove the following two assertions:

(i) For any field $k(\mathfrak{p})$ the ring $R \otimes_{\mathbb{C}} k(\mathfrak{p})$ is an integral domain.
(ii) The intersection, in the integral domain $R \otimes_{\mathbb{C}} k(\mathfrak{p})$, of the kernels of the maps $\varphi_g \otimes 1 : R \otimes_{\mathbb{C}} k(\mathfrak{p}) \longrightarrow \mathbb{C} \otimes_{\mathbb{C}} k(\mathfrak{p})$ is the ideal $\{0\} \subset R \otimes_{\mathbb{C}} k(\mathfrak{p})$, which is prime by (i).

In the case where $R = \mathbb{C}$ we have $R \otimes_{\mathbb{C}} k(\mathfrak{p}) = k(\mathfrak{p})$ is a field, and (i) and (ii) are trivial. If $R = \mathbb{C}[t, t^{-1}]$ both assertions are also quite easy. The ring $R \otimes_{\mathbb{C}} k(\mathfrak{p}) = k(\mathfrak{p})[t, t^{-1}]$ is clearly an integral domain. Further, if $f(t, t^{-1}) \in k(\mathfrak{p})[t, t^{-1}]$ is a polynomial, which is annihilated by all the maps $\varphi_g : R \longrightarrow \mathbb{C}$, then it vanishes at all the points $t = g$ with $g \in \mathbb{C}^* \subset k(\mathfrak{p})$. There are infinitely many such points, and a polynomial vanishing at infinitely many points is zero. □

Aside for the Experts 8.6.6. The general argument goes as follows. From [5, page 101, Theorem] we know that group schemes over fields of characteristic zero are smooth and, as we are assuming G connected, both R and $R \otimes_{\mathbb{C}} k(\mathfrak{p})$ must be integral domains. By the strong Nullstellensatz the prime ideal $\{0\} \subset R$ is the intersection of all the maximal ideals containing it; see Reminder 3.8.1(ii). The maximal ideals in R, that is the points in $G^{\mathrm{an}} = \{\mathrm{Spec}(R)\}^{\mathrm{an}}$, are precisely the kernels of the homomorphisms $\varphi_g : R \longrightarrow \mathbb{C}$. That is, the intersection of the kernels is $\{0\} \subset R$, and tensoring with $k(\mathfrak{p})$ we have that the intersection over all $g \in G^{\mathrm{an}}$ of the kernels of the homomorphisms

$$R \otimes_{\mathbb{C}} k(\mathfrak{p}) \xrightarrow{\varphi_g \otimes 1} \mathbb{C} \otimes_{\mathbb{C}} k(\mathfrak{p})$$

is the ideal $\{0\} \subset R \otimes_{\mathbb{C}} k(\mathfrak{p})$. This proves (i) and (ii) for a general connected G.

Definition 8.6.7. *The map $\pi : X \longrightarrow X/G$ is the map taking a prime ideal $\mathfrak{p} \in X = \mathrm{Spec}(S)$ to $\pi(\mathfrak{p})$, which is the intersection of all $g\mathfrak{p}$ over $g \in G^{\mathrm{an}}$.*

Note that Lemma 8.6.4 proved that the ideal $\pi(\mathfrak{p})$ is prime, and the fact that it is fixed by any $g \in G^{\mathrm{an}}$ is obvious. Hence it is a well-defined point of X/G.

Remark 8.6.8. A more topological description of the point $\pi(\mathfrak{p})$ runs as follows. Let $T(\mathfrak{p}) \subset \mathrm{Spec}(S)$ be the set defined by

$$T(\mathfrak{p}) \quad = \quad \left\{ \mathfrak{q} \;\middle|\; \begin{array}{c} \mathfrak{q} \text{ is a point in } X = \mathrm{Spec}(S) \\ \text{and } g\mathfrak{p} \in \overline{\mathfrak{q}} \text{ for all } g \in G^{\mathrm{an}} \end{array} \right\} .$$

Here $\overline{\mathfrak{q}}$ means the closure of the point $\mathfrak{q}$ in the topological space X.

Remark 3.1.10 tells us that $\mathfrak{p} \in \overline{\mathfrak{q}}$ if and only if $\mathfrak{q} \subset \mathfrak{p}$. Thus the set $T(\mathfrak{p})$ is simply the set of all prime ideals $\mathfrak{q}$ contained in every $g\mathfrak{p}$. Then $\pi(\mathfrak{p})$ is the maximal element of the set $T(\mathfrak{p})$, which one can say topologically by

$$\{\pi(\mathfrak{p}) = \mathfrak{q}\} \quad \Longleftrightarrow \quad \left\{ \begin{array}{c} \mathfrak{q} \in T(\mathfrak{p}), \text{ and} \\ \mathfrak{q} \in \overline{\mathfrak{q}'} \text{ for all } \mathfrak{q}' \in T(\mathfrak{p}) \end{array} \right\} .$$

If $U \subset X$ is any open set and $\mathfrak{p} \in U$, then the set $T(\mathfrak{p})$ is contained in U by Remark 3.1.10; the primes in $T(\mathfrak{p})$ are all smaller than $\mathfrak{p}$ and therefore lie in every open set containing $\mathfrak{p}$. Among other things this means that $\pi(\mathfrak{p}) \in U$; after all $\pi(\mathfrak{p})$ is one of the points of $T(\mathfrak{p})$. But if U happens to be not just any open subset but a G–invariant one, then we can say more. If $\mathfrak{p}$ is a point of U and U is G–invariant then U must contain all the points $g\mathfrak{p}$. The definition of the set $T(\mathfrak{p})$ can be formulated, as above, purely in terms of topological data about the the open set U. And so can the definition of $\pi(\mathfrak{p})$.

Lemma 8.6.9. *The map $\pi : X \longrightarrow X/G$ is continuous and surjective. For any $x \in X$ and any $g \in G^{\mathrm{an}}$ we have $\pi(x) = \pi(gx)$. Furthermore, the inverse images of the open sets in X/G are precisely all the G–invariant open subsets of X.*

Proof. The fact that $\pi(x) = \pi(gx)$ is obvious: both are defined to be the intersection over all $h \in G^{\mathrm{an}}$ of the ideals $hx \subset S$. The fact that the map is surjective is also obvious; every $x \in X/G$ is a fixed-point for the action of G^{an}, that is a prime ideal $x \subset S$ with $gx = x$ for all $g \in G^{\mathrm{an}}$. The formula for $\pi(x)$ immediately tells us that $\pi(x) = x$ for all $x \in X/G$.

Now for the continuity, or the stronger assertion that the inverse images of open sets in X/G are precisely all the G–invariant open subsets of X. Recall that the topology on X/G is the subspace topology via its inclusion in X. Any open subset in X/G is of the form $V \cap \{X/G\}$, where V is an open subset of X. Since every point of X/G is fixed by every $g \in G^{\mathrm{an}}$, we have $V \cap \{X/G\} = gV \cap \{X/G\}$ for all g. This makes

$$V \cap \{X/G\} \quad = \quad \bigcup_{g \in G^{\mathrm{an}}} [gV \cap \{X/G\}] \quad = \quad \{X/G\} \cap \bigcup_{g \in G^{\mathrm{an}}} gV \, .$$

Thus any open subset of X/G can be written as $U \cap \{X/G\}$ with U a G–invariant open subset of X. Next we will prove that $\pi^{-1}[U \cap \{X/G\}] = U$. This would show what we want: it would establish that $\pi^{-1}[U \cap \{X/G\}]$ is open, and exhibit every G–invariant U as an inverse image $\pi^{-1}V$ with V open in X/G.

To show $\pi^{-1}[U \cap \{X/G\}] = U$ we need to show that the map π takes $U \subset X$ to $U \cap \{X/G\}$, and takes the complement $X - U$ to $X - U$. We already know that πU is contained in $U \cap \{X/G\}$. Clearly $\pi U \subset X/G$, and Remark 8.6.8 tells us that $\pi U \subset U$. It remains to show that $\pi(X - U) \subset X - U$.

But $X - U$ is Zariski closed, and is therefore $V(I)$ for some ideal $I \subset S$. We are given that $\mathfrak{p} \in V(I)$ and that $V(I)$ is G^{an}–invariant (because its complement U is invariant under the action of G^{an}). Hence $g\mathfrak{p}$ is in $V(I)$ for all $g \in G^{\mathrm{an}}$. This means that $I \subset g\mathfrak{p}$ for all $g \in G^{\mathrm{an}}$. Therefore I is contained in the intersection, that is $I \subset \pi(\mathfrak{p})$. □

Remark 8.6.10. Note that the construction is local. Suppose $f \in S$ is as in Remarks 8.3.11 and 8.4.14. That is, the map $a' : S \longrightarrow R \otimes_{\mathbb{C}} S$ takes $f \in S$ to $a'(f) = r \otimes f$. The open set $X_f \subset X$ is G–invariant and affine. We could define X_f/G, and compare it to $X_f \cap \{X/G\}$. The assertion is that they are the same.

The set X_f/G is the set of fixed points for the action of G^{an} on $X_f \subset X$, and this is also the set of fixed points in X which happen to lie in the open subset X_f. That is, as subsets of X_f we have that

$$X_f/G \quad = \quad X_f \cap \{X/G\} \ .$$

Since the topology on $X_f = \mathrm{Spec}(S[1/f])$ is the subspace topology, the topologies also agree.

It remains to worry about the continuous map $\pi : X \longrightarrow X/G$, and how it compares to the continuous map $\pi_f : X_f \longrightarrow X_f/G$. By Remark 8.6.8, for any point $\mathfrak{p} \in X_f$ there is a topological description of the point $\pi(\mathfrak{p})$, a description which never mentions any of the rings involved, and is the same in the open set X_f as in X. Hence the continuous map $\pi_f : X_f \longrightarrow X_f/G$ is just the restriction to X_f of the continuous map $\pi : X \longrightarrow X/G$.

Example 8.6.11. Suppose $(G, \mathcal{O}_G)$ is trivial, that is $R = \mathbb{C}$ as in Example 8.2.7. Any finitely generated $\mathbb{C}$–algebra S was made into a G–ring in Example 8.3.4. The induced action of $G^{\mathrm{an}} = \{*\}$ on $X = \mathrm{Spec}(S)$ must be trivial; every point of X is fixed by the action of G. The subspace $X/G \subset X$ is therefore all of X. The map $\pi : X \longrightarrow X/G$ presents no problem; given any prime ideal $\mathfrak{p} \in \mathrm{Spec}(S) = X$ the group $G^{\mathrm{an}} = \{*\}$ fixes it, so the formula $\pi(\mathfrak{p}) = \cap_{g \in G^{\mathrm{an}}} g\mathfrak{p}$ gives the prime ideal $\mathfrak{p}$. The map $\pi : X \longrightarrow X/G$ is the identity.

Example 8.6.12. We should never lose sight of the important Example 8.3.5. The Hopf algebra is $R = \mathbb{C}[t, t^{-1}]$ and S is the polynomial ring $S = \mathbb{C}[x_0, x_1, \ldots, x_n]$. What are the fixed points of the action of $G^{\mathrm{an}} = \mathbb{C}^*$? They are the prime ideals $\mathfrak{p} \subset S$ which are invariant under the action of $\mathbb{C}^*$. Let us therefore work out what are the ideals invariant under the action of $\mathbb{C}^*$.

The action of $\mathbb{C}^*$ on the ring S was computed in Example 8.5.1. With the conventions as in Notation 8.5.2, the element $\lambda \in \mathbb{C}^*$ takes the mononomial $\mathbf{x}^{\mathbf{a}}$ to $\lambda^{\langle \mathbf{a} \rangle}\mathbf{x}^{\mathbf{a}}$. Let f be any element of $S = \mathbb{C}[x_0, x_1, \ldots, x_n]$. We can write f as a sum of its homogeneous components, that is

$$f \quad = \quad f_0 + f_1 + f_2 + \cdots + f_m$$

with $f_i(x_0, x_1, \ldots, x_n)$ a homogeneous polynomial of degree i. The formula becomes

$$\lambda(f) \quad = \quad f_0 + \lambda f_1 + \lambda^2 f_2 + \cdots + \lambda^m f_m \,.$$

Any G–invariant ideal containing f must contain $f_0 + \lambda f_1 + \lambda^2 f_2 + \cdots + \lambda^m f_m$ for every complex number λ, and hence will have to contain all the homogeneous components f_i. The G–invariant ideals are therefore precisely the homogeneous ideals.

Now the fixed points for the action of $G^{\mathrm{an}} = \mathbb{C}^*$ on X are the invariant ideals which happen to be prime; by the above these are precisely the homogeneous prime ideals in the ring $S = \mathbb{C}[x_0, x_1, \ldots, x_n]$. The reader familiar with the construction of projective space will note that these are the points of projective space, except that I have not yet explicitly removed the maximal ideal corresponding to the closed point $0 \in \mathbb{C}^{n+1}$. The topology is the subspace topology from the embedding in $\mathrm{Spec}(S)$. But we noticed, in the proof of Lemma 8.6.9, that every open subset in X/G is the intersection of $X/G \subset X$ with a G–invariant open subset of X. Equivalently every closed subset is the intersection of $X/G \subset X$ with a G–invariant closed subset.

Now every closed subset of X is $V(I)$ for some ideal I. If $V(I)$ is G–invariant then $V(I) = V(gI)$ for all $g \in G^{\mathrm{an}}$. But then by Lemma 3.1.9 we have an equality of sets

$$V(I) \quad = \quad \bigcap_{g \in G^{\mathrm{an}}} V(gI) \quad = \quad V\left(\sum_{g \in G^{\mathrm{an}}} gI \right) \,.$$

This exhibits the closed set $V(I)$ as $V(J)$ for an invariant ideal J, and invariant ideals are homogeneous by the above. In other words the closed sets in projective space are the sets of homogeneous prime ideals $\mathfrak{p}$ containing a given homogeneous ideal J.

Finally the map π takes a prime ideal in S to the largest homogeneous ideal it contains. This largest homogeneous ideal must be prime by Lemma 8.6.4.

8.7 Sheaves on the quotient

In this section let the notation be as in Section 8.6. We have a connected affine group scheme $(G, \mathcal{O}_G) = \left(\text{Spec}(R), \widetilde{R}\right)$ acting on an affine scheme $(X, \mathcal{O}_X) = \left(\text{Spec}(S), \widetilde{S}\right)$, and we have constructed a continuous map $\pi : X \longrightarrow X/G$. Now we want to construct on X/G some sheaves; we want to turn it into a ringed space over $\mathbb{C}$, and have some sheaves of modules over it. Of course some sheaves come for free. These are the pushforwards of sheaves on X. We remind the reader.

Reminder 8.7.1. We have a continuous map $\pi : X \longrightarrow X/G$. Given any sheaf $\mathscr{S}$ on X, Definition 7.1.13 permits us to define a sheaf $\pi_*\mathscr{S}$ on X/G. We remind the reader of the recipe:

$$\Gamma(V, \pi_*\mathscr{S}) \quad = \quad \Gamma(\pi^{-1}V, \mathscr{S}) \ .$$

On X we know many sheaves. The ringed space $(X, \mathcal{O}_X) = \left(\text{Spec}(S), \widetilde{S}\right)$ came equipped with the sheaf $\mathcal{O}_X = \widetilde{S}$. Also, given any S–module M there is the sheaf $\widetilde{M}$ of $\mathcal{O}_X$–modules on X. And we have recently spent a great deal of time studying the special case where M is not merely an S–module, but is a G–module as well.

In the interesting case, where M is a G–module for the G–ring S, Corollary 8.4.10 tells us that the group G^{an} acts on $\Gamma(U, \widetilde{M})$, whenever $U \subset X$ is a G–invariant open set. But the open sets $\pi^{-1}V$ are all G–invariant; see Lemma 8.6.9. For $\mathscr{S} = \widetilde{M}$, where M is a G–module, we have that the sheaf $\pi_*\mathscr{S} = \pi_*\widetilde{M}$ is a sheaf with a G^{an}–action. Every $\Gamma(V, \pi_*\mathscr{S})$ is acted on by G^{an}, and restriction maps respect the action. The following construction, and the couple of lemmas that succeed it, study arbitrary topological spaces Y and sheaves of abelian groups on Y with the action of a group H. Then we return to the situation where the space is $Y = X/G$, the group is $H = G^{\text{an}}$, and the sheaves with the G^{an}–action are the $\pi_*\widetilde{M}$.

Construction 8.7.2. Let Y be a topological space and let H be a group. Suppose $\mathscr{S}$ is a sheaf of abelian groups on Y, and suppose the group H acts on $\mathscr{S}$. By this we mean that, for each open set $V \subset Y$, the group H acts on $\Gamma(V, \mathscr{S})$, and furthermore the restriction maps

$$\text{res}^V_U : \Gamma(V, \mathscr{S}) \longrightarrow \Gamma(U, \mathscr{S})$$

all intertwine the H–action. We will call these H–sheaves on Y. For the H–sheaf $\mathscr{S}$ we define a presheaf $\mathscr{S}^H$ on Y by the formula

$$\Gamma(V, \mathscr{S}^H) \quad = \quad \Gamma(V, \mathscr{S})^H .$$

This means that the sections of the presheaf $\mathscr{S}^H$, on the open set $V \subset Y$, are precisely the H–invariant sections of $\mathscr{S}$, that is the fixed points for the action of H on $\Gamma(V, \mathscr{S})$.

Lemma 8.7.3. *Let the notation be as in Construction 8.7.2. Then the presheaf $\mathscr{S}^H$ is a sheaf.*

Proof. We begin with the observation that, for any open set $V \subset Y$, there is an exact sequence

$$0 \longrightarrow \Gamma(V, \mathscr{S})^H \stackrel{i}{\longrightarrow} \Gamma(V, \mathscr{S}) \stackrel{j}{\longrightarrow} \prod_{h \in H} \Gamma(V, \mathscr{S}) ,$$

where $i : \Gamma(V, \mathscr{S})^H \longrightarrow \Gamma(V, \mathscr{S})$ is the inclusion of the H–invariant sections of $\mathscr{S}$ into all the sections, and the map j takes a section $s \in \Gamma(V, \mathscr{S})$ to the product, over all $h \in H$, of the terms $\{hs - s\}$. Now let $V \subset Y$ be an open set, and let $V = \cup_i V_i$ be an open cover. Consider the following commutative diagram

$$\begin{array}{ccccccc}
 & & 0 & & 0 & & 0 \\
 & & \downarrow & & \downarrow & & \downarrow \\
0 & \longrightarrow & \Gamma(V, \mathscr{S})^H & \longrightarrow & \prod_{i \in I} \Gamma(V_i, \mathscr{S})^H & \longrightarrow & \prod_{i,j \in I} \Gamma(V_i \cap V_j, \mathscr{S})^H \\
 & & \downarrow & & \downarrow & & \downarrow \\
0 & \longrightarrow & \Gamma(V, \mathscr{S}) & \longrightarrow & \prod_{i \in I} \Gamma(V_i, \mathscr{S}) & \longrightarrow & \prod_{i,j \in I} \Gamma(V_i \cap V_j, \mathscr{S}) \\
 & & \downarrow & & \downarrow & & \downarrow \\
0 & \longrightarrow & \prod_{h \in H} \Gamma(V, \mathscr{S}) & \longrightarrow & \prod_{\substack{i \in I, \\ h \in H}} \Gamma(V_i, \mathscr{S}) & \longrightarrow & \prod_{\substack{i,j \in I, \\ h \in H}} \Gamma(V_i \cap V_j, \mathscr{S})
\end{array}$$

The columns are exact by the first sentence of the proof. Lemma 2.2.10, coupled with the fact that $\mathscr{S}$ is a sheaf, implies that the second and third rows are exact. An easy diagram chase shows that the first row is exact; hence, again by Lemma 2.2.10, we deduce that $\mathscr{S}^H$ is a sheaf. □

While we are at it, let us prove a similar result.

Lemma 8.7.4. *The functor taking $\mathscr{S}$ to $\mathscr{S}^H$ is left-exact: that is, given a morphism of H–sheaves $\mathscr{S} \longrightarrow \mathscr{T}$ on Y with kernel $\mathscr{R}$, which we denote by a short exact sequence*

$$0 \longrightarrow \mathscr{R} \longrightarrow \mathscr{S} \longrightarrow \mathscr{T} ,$$

then the following sequence is also exact

$$0 \longrightarrow \mathscr{R}^H \longrightarrow \mathscr{S}^H \longrightarrow \mathscr{T}^H ,$$

meaning $\mathscr{R}^H$ is the kernel of the map $\mathscr{S}^H \longrightarrow \mathscr{T}^H$.

Proof. Obvious: the H–invariant elements of the kernel of $\Gamma(U, \mathscr{S}) \longrightarrow \Gamma(U, \mathscr{T})$ are the H–invariant elements of $\Gamma(U, \mathscr{S})$ which happen to lie in the kernel. □

Remark 8.7.5. We return to our specific case of the map $\pi : X \longrightarrow X/G$. With $Y = X/G$ and $H = G^{\mathrm{an}}$, Reminder 8.7.1 showed us that we know many examples of G^{an}–sheaves on the topological space $Y = X/G$. Given any G–module M for the G–ring S, the sheaf $\pi_* \widetilde{M}$ is a G^{an}–sheaf on X/G. We define

Definition 8.7.6. *The ringed space $(X/G, \mathcal{O}_{X/G})$ is given as follows:*

(i) *The topological space X/G is as in Definition 8.6.2.*
(ii) *The sheaf of rings $\mathcal{O}_{X/G}$ is defined to be $\{\pi_* \mathcal{O}_X\}^{G^{\mathrm{an}}}$.*

Remark 8.7.7. The space $(X/G, \mathcal{O}_{X/G})$ is naturally a ringed space over $\mathbb{C}$. The ring

$$\Gamma(X/G, \mathcal{O}_{X/G}) \quad = \quad \Gamma\left(X/G, \{\pi_* \mathcal{O}_X\}^{G^{\mathrm{an}}}\right) \quad = \quad \Gamma(X, \mathcal{O}_X)^{G^{\mathrm{an}}}$$

clearly contains the image of the map $\rho : \mathbb{C} \longrightarrow \Gamma(X, \mathcal{O}_X)$. After all, the group G^{an} acts by $\mathbb{C}$–algebra homomorphisms.

Remark 8.7.8. The map $\pi : X \longrightarrow X/G$ now extends to a morphism of ringed spaces over $\mathbb{C}$. In Example 7.1.15 we saw that to give a map of ringed spaces $(\pi, \pi^*) : (X, \mathcal{O}_X) \longrightarrow (X/G, \mathcal{O}_{X/G})$ is the same as giving a continuous map $\pi : X \longrightarrow X/G$ (which we have), as well as a morphism $\pi^* : \mathcal{O}_{X/G} \longrightarrow \pi_* \mathcal{O}_X$ of sheaves of rings on X/G. In our case we define π^* to be the inclusion of the sheaves of rings on X/G

$$\pi^* : \{\pi_* \mathcal{O}_X\}^{G^{\mathrm{an}}} \longrightarrow \pi_* \mathcal{O}_X .$$

Remark 8.7.9. Not only do we have a ringed space $(X/G, \mathcal{O}_{X/G})$, we also know many sheaves of $\mathcal{O}_{X/G}$–modules on it, namely all the sheaves $\{\pi_*\widetilde{M}\}^{G^{\mathrm{an}}}$, where M is a G–module for the G–ring S. Given an exact sequence of G–modules

$$0 \longrightarrow M' \longrightarrow M \longrightarrow M'' \longrightarrow 0\,,$$

Lemma 7.6.14 tells us that the sequence of coherent sheaves on X

$$0 \longrightarrow \widetilde{M'} \longrightarrow \widetilde{M} \longrightarrow \widetilde{M''} \longrightarrow 0$$

is exact. In particular for any open set $U \subset X$ we have that $\Gamma(U, \widetilde{M'})$ is the kernel of the map $\Gamma(U, \widetilde{M}) \longrightarrow \Gamma(U, \widetilde{M''})$. Taking the special case where $U = \pi^{-1}V$, with V open in X/G, we learn that the sequence

$$0 \longrightarrow \Gamma(\pi^{-1}V, \widetilde{M'}) \longrightarrow \Gamma(\pi^{-1}V, \widetilde{M}) \longrightarrow \Gamma(\pi^{-1}V, \widetilde{M''})$$

is exact. This says that the sequence of sheaves

$$0 \longrightarrow \pi_*\widetilde{M'} \longrightarrow \pi_*\widetilde{M} \longrightarrow \pi_*\widetilde{M''}$$

is exact on X/G, and Lemma 8.7.4 now tells us that the sequence

$$0 \longrightarrow \{\pi_*\widetilde{M'}\}^{G^{\mathrm{an}}} \longrightarrow \{\pi_*\widetilde{M}\}^{G^{\mathrm{an}}} \longrightarrow \{\pi_*\widetilde{M''}\}^{G^{\mathrm{an}}}$$

is also exact.

Example 8.7.10. For the trivial group $(G, \mathcal{O}_G)$ all the above simplifies. If $R = \mathbb{C}$, as in Example 8.2.7, then every finitely generated $\mathbb{C}$–algebra S was made into a G–ring in Example 8.3.4. If M is a finite S–module Example 8.3.15 teaches us that M is a G–module. The map $\pi : X \longrightarrow X/G$ is the identity map $1 : X \longrightarrow X$ by Example 8.6.11. The sheaf $\pi_*\widetilde{M}$ simplifies to $\widetilde{M}$, and since the group $G^{\mathrm{an}} = \{*\}$ is trivial it acts trivially on $\Gamma(U, \pi_*\widetilde{M}) = \Gamma(U, \widetilde{M})$ for every open set $U \subset X$. Therefore $\{\pi_*\widetilde{M}\}^{G^{\mathrm{an}}} = \widetilde{M}$.

8.8 The main results

So far our statements were true with hardly any assumption on X and G. In coming sections we will prove more precise statements, but the cost will be that we have more restrictive hypotheses. The theorem we will prove will have the following hypotheses:

Hypothesis 8.8.1. *Let* $(G, \mathcal{O}_G) = \left(\mathrm{Spec}(R), \widetilde{R}\right)$ *be a connected, reductive group acting on an affine scheme* $(X, \mathcal{O}_X) = \left(\mathrm{Spec}(S), \widetilde{S}\right)$ *of finite* *type over $\mathbb{C}$. Suppose $U \subset X$ is a G–invariant open set satisfying the following:*

(i) *There is an open cover for U by G–invariant open affines X_f, with $f \in S$ as in Remarks 8.3.11 and 8.4.14. That is, the map $a' : S \longrightarrow R \otimes_{\mathbb{C}} S$ takes $f \in S$ to $a'(f) = r \otimes f$.*

(ii) *The closed points in U have closed orbits. Since I have only defined how the group G^{an} of closed points acts, I had better make this precise. Given a closed point $x \in U^{\mathrm{an}}$ we can look at the orbit $G^{\mathrm{an}}x = \{gx \mid g \in G^{\mathrm{an}}\}$. We require that there exist a Zariski closed set $Z \subset U$ with $Z \cap U^{\mathrm{an}} = G^{\mathrm{an}}x$. In other words we require that the orbit $G^{\mathrm{an}}x$ should be closed in the Zariski topology of U^{an}.*

Remark 8.8.2. The continuous map $U^{\mathrm{an}} \longrightarrow U$ tells us that any Zariski closed set $Z \subset U$ will intersect U^{an} in a set which is closed in the complex topology. In general many sets, which are closed in the complex topology on U^{an}, need not be closed in the Zariski topology. Nevertheless it turns out to be true that, if the orbit $G^{\mathrm{an}}x \subset U^{\mathrm{an}}$ of Hypothesis 8.8.1(ii) is closed in the complex topology, then it is also closed in the Zariski topology. This is true, but the commutative algebra required for the proof is beyond our book.

Remark 8.8.3. We have not defined, so far in this book, what it means for an affine group scheme $(G, \mathcal{O}_G)$ to be reductive and connected. We will give one of the equivalent characterizations of a reductive group in Definition 9.3.11; connectedness is too complicated for us to discuss. In any case, the reader can comfortably focus on the only two examples that interest us, the trivial group where $R = \mathbb{C}$ and the non-trivial group where $R = \mathbb{C}[t, t^{-1}]$. These two groups are reductive and connected, whatever the terms mean.

Example 8.8.4. Suppose $(G, \mathcal{O}_G)$ is the trivial group, that is $R = \mathbb{C}$ as in Example 8.2.7. The group $(G, \mathcal{O}_G)$ is reductive and connected by Remark 8.8.3. Any finite $\mathbb{C}$–algebra S is a G–ring by Example 8.3.4. This means that $(G, \mathcal{O}_G)$ acts on $(X, \mathcal{O}_X) = \left(\mathrm{Spec}(S), \widetilde{S}\right)$. In Example 8.4.15 we learned that every open subset $U \subset X$ is G–invariant. We would like to assert that the conditions of Hypothesis 8.8.1 hold for every $U \subset X$.

First we prove Hypothesis 8.8.1(i). Proposition 3.2.4 tells us that the basic open sets X_f form a basis for the topology of X, and hence U may be covered by sets X_f with $f \in S$. Our action $a' : S \longrightarrow \mathbb{C} \otimes_{\mathbb{C}} S = S$ sends $f \in S$ to $a'(f) = 1 \otimes f \in \mathbb{C} \otimes_{\mathbb{C}} S$. This proves Hypothesis 8.8.1(i).

Now for Hypothesis 8.8.1(ii). Let $x \in U^{\mathrm{an}}$ be a closed point. The group $G^{\mathrm{an}} = \{*\}$ is trivial, hence it fixes x. The G^{an}–orbit of $x \in U^{\mathrm{an}}$ is x, and by virtue of belonging to U^{an} it is a Zariski closed subset of $U \subset X$.

Example 8.8.5. We return to the key Example 8.3.5, with $R = \mathbb{C}[t, t^{-1}]$ acting on $S = \mathbb{C}[x_0, x_1, \ldots, x_n]$. The group $(G, \mathcal{O}_G)$ is reductive and connected (whatever this means). In fact some arguments simplify for our particular group G, and when this is the case we will present the simplified argument first and then, in an aside to the expert, discuss the more general case.

The open set $U \subset X$, to which we wish to apply our theorems, will be the set $X - \{0\}$. Recall that $X^{\mathrm{an}} = \mathbb{C}^{n+1}$, and $0 \in \mathbb{C}^{n+1}$ is most definitely a closed point, and is also fixed by the action of G^{an}. Hence the set $X - \{0\}$, being the complement of a closed point, is open in X and is invariant under the action of G^{an}; it is what we have been calling a G–invariant open set. It is the example to keep in mind, and now we check that Hypothesis 8.8.1 is satisfied in this case.

First of all we produce an open cover of U by open sets X_f, as in Hypothesis 8.8.1(i). The closed point $0 \in \mathbb{C}^{n+1}$ corresponds to a maximal ideal $\mathfrak{m} \subset S$. The correspondence concretely takes the closed point 0 to the kernel of the map $\varphi_0 : S \longrightarrow \mathbb{C}$ which evaluates a polynomial $f \in S = \mathbb{C}[x_0, x_1, \ldots, x_n]$ at the point $0 \in \mathbb{C}^{n+1}$. The ideal $\mathfrak{m} = \ker(\varphi_0)$ is the ideal of all polynomials with a zero constant term; that is polynomials $f = f_1 + f_2 + \cdots + f_m$, with each f_i homogeneous of degree $i > 0$. This means that $\mathfrak{m}$ is generated by all the homogeneous polynomials of degree > 0. Another way of saying this is

$$\{\mathfrak{m}\} \quad = \bigcap_{\substack{f_i \text{ homogeneous} \\ \text{of degree } i > 0}} V(Sf_i)\ ;$$

the above formula simply says that only the prime ideal $\mathfrak{m}$ contains all the principal ideals Sf_i, with f_i homogeneous of degree $i > 0$. Taking complements we have

$$X - \{0\} \quad = \bigcup_{\substack{f_i \text{ homogeneous} \\ \text{of degree } i > 0}} X_{f_i}\ .$$

And now Example 8.3.13 tells us that the homogeneous $f_i \in S$ satisfy $a'(f_i) = t^{-i} \otimes f_i$, as in Remark 8.3.11.

It remains to establish Hypothesis 8.8.1(ii), that is that the orbits of G^{an} in U^{an} are Zariski closed. The group is $G^{\text{an}} = \mathbb{C}^*$, the space is $X^{\text{an}} = \mathbb{C}^{n+1}$, and the open set is $U^{\text{an}} = \mathbb{C}^{n+1} - \{0\}$. In Example 8.3.5 we saw that the action of $\mathbb{C}^*$ on $\mathbb{C}^{n+1}$ was by dilation; $\lambda \in \mathbb{C}^*$ takes the point $b \in \mathbb{C}^{n+1}$ to $\lambda^{-1}b$. If we take $b \in U^{\text{an}} = \mathbb{C}^{n+1} - \{0\}$ then its orbit is the straight line of all non-zero multiples of b. We want to show this to be a Zariski closed set in U^{an}. Let I be the ideal generated by all homogeneous polynomials vanishing at b. Of course if f is homogeneous of degree i and vanishes at b then $f(\lambda^{-1}b) = \lambda^{-i}f(b) = 0$, and hence f vanishes on the entire G^{an}–orbit of b. This means that $V(I)$ is a Zariski closed set containing the G^{an}–orbit of b. We need to show that $V(I) \cap U^{\text{an}}$ contains no other point.

The argument is very simple; given a point $b' \in U^{\text{an}} - G^{\text{an}}b$, that is a point b' not in the orbit of b, we want to produce an element of the ideal I which does not vanish at b'. The fact that $b' \neq 0$ is not a multiple of b means that b and b' are linearly independent vectors in $\mathbb{C}^{n+1}$. Linear algebra tells us that there is a linear function $f : \mathbb{C}^{n+1} \longrightarrow \mathbb{C}$ which vanishes at b but not at b'. Thus we have produced a homogeneous polynomial f, of degree 1, vanishing at b but not b'.

What we will prove is

Theorem 8.8.6. *As in the rest of this section let $(G, \mathcal{O}_G)$ be a connected, reductive affine group scheme, of finite type over $\mathbb{C}$, acting on an affine scheme $(X, \mathcal{O}_X) = \left(\text{Spec}(S), \widetilde{S}\right)$, also of finite type over $\mathbb{C}$. Let $U \subset X$ be a G–invariant open subset, which by Lemma 8.6.9 we know to be $\pi^{-1}V$ for a unique open set $V \subset X/G$. Let us assume Hypothesis 8.8.1 for $U \subset X$. Then the following assertions are true.*

(i) *The ringed space $(V, \mathcal{O}_V)$ is a scheme of finite type over $\mathbb{C}$, where the sheaf of rings $\mathcal{O}_V$ is the restriction to V of the sheaf $\mathcal{O}_{X/G}$ of Definition 8.7.6. In symbols*

$$\mathcal{O}_V \quad = \quad \{\mathcal{O}_{X/G}\}|_V \ .$$

(ii) *If M is a G–module over the G–ring S then the sheaf $\{\pi_*\widetilde{M}\}^{G^{\text{an}}}$, constructed in Remark 8.7.9, restricts on V to a coherent sheaf of $\mathcal{O}_V$–modules.*

(iii) *Any exact sequence of G–modules for the G–ring S*

$$0 \longrightarrow M' \longrightarrow M \longrightarrow M'' \longrightarrow 0$$

gives, on $V \subset X/G$, an exact sequence of coherent sheaves

$$0 \longrightarrow \{\pi_*\widetilde{M'}\}^{G^{\text{an}}} \longrightarrow \{\pi_*\widetilde{M}\}^{G^{\text{an}}} \longrightarrow \{\pi_*\widetilde{M''}\}^{G^{\text{an}}} \longrightarrow 0 \ .$$

(iv) *Every coherent sheaf on V is isomorphic to $\{\pi_*\widetilde{M}\}^{G^{\mathrm{an}}}$, for some G–module M over the G–ring S.*

Remark 8.8.7. Most of Chapter 9 will be devoted to the proof of Theorem 8.8.6. Note that three of the assertions, namely Theorem 8.8.6(i), (ii) and (iii), are local in V. Hypothesis 8.8.1 tells us that the open set $U = \pi^{-1}V$ has an open cover by G–invariant open sets X_f which are themselves affine schemes with a G–action. Corollary 5.2.3 asserts that there is a finite subcover. This means that X/G has a cover by finitely many open sets $X_f \cap X/G$, and it is enough to prove the assertions of Theorem 8.8.6(i), (ii) and (iii) for the open sets $X_f \cap X/G$. Remark 8.6.10 tells us that the continuous map $\pi_f : X_f \longrightarrow X_f/G$ agrees with the restriction to X_f of the continuous map $\pi : X \longrightarrow X/G$. The sheaves $\pi_*\widetilde{M}$ (the sheaf $\pi_*\mathcal{O}_X = \pi_*\widetilde{S}$ is a special case, where $M = S$) are defined by the formula

$$\Gamma(W, \pi_*\widetilde{M}) \quad = \quad \Gamma(\pi^{-1}W, \widetilde{M}) \ .$$

Since the map π_f is the restriction to X_f of π, on open sets $W \subset X_f/G$ we have

$$\Gamma(\pi^{-1}W, \widetilde{M}) \quad = \quad \Gamma(\pi_f^{-1}W, \widetilde{M}) \quad \cong \quad \Gamma\big(\pi_f^{-1}W, \widetilde{M[1/f]}\big) \ ,$$

where the last isomorphism is because $\widetilde{M}|_{X_f} = \widetilde{M[1/f]}$; see Lemma 7.4.3. This rewrites as

$$\Gamma(W, \pi_*\widetilde{M}) \quad \cong \quad \Gamma\big(W, \{\pi_f\}_*\widetilde{M[1/f]}\big) \ .$$

Remark 8.4.14 tells us that action of the group G^{an} on $\Gamma(W, \pi_*\widetilde{M})$ agrees with its action on the isomorphic $\Gamma\big(W, \{\pi_f\}_*\widetilde{M[1/f]}\big)$. This means that the submodules of G^{an}–invariant elements agree, that is

$$\Gamma(W, \pi_*\widetilde{M})^{G^{\mathrm{an}}} \quad \cong \quad \Gamma\big(W, \{\pi_f\}_*\widetilde{M[1/f]}\big)^{G^{\mathrm{an}}} \ .$$

But, by Construction 8.7.2, this says

$$\Gamma\left(W, \{\pi_*\widetilde{M}\}^{G^{\mathrm{an}}}\right) \quad \cong \quad \Gamma\left(W, \left\{\{\pi_f\}_*\widetilde{M[1/f]}\right\}^{G^{\mathrm{an}}}\right) \ .$$

Therefore the restriction to $X_f/G \subset X/G$ of the invariant subsheaf $\{\pi_*\widetilde{M}\}^{G^{\mathrm{an}}}$ agrees with $\left\{\{\pi_f\}_*\widetilde{M[1/f]}\right\}^{G^{\mathrm{an}}}$. It suffices to prove parts (i), (ii) and (iii) of Theorem 8.8.6 for the sheaves $\left\{\{\pi_f\}_*\widetilde{M[1/f]}\right\}^{G^{\mathrm{an}}}$ on X_f/G. In other words we may replace X by X_f and $U \subset X$ by $U \cap X_f = X_f$. This buys us that, in the proof of the first three parts of

Theorem 8.8.6, we may assume $U = X$, in which case Hypothesis 8.8.1 simplifies. Part (i) becomes redundant; we can always cover $U = X$ by the open affine $X = X_1$, and $1 \in S$ certainly satisfies $a'(1) = 1 \otimes 1$. The only non-trivial part left, to the conditions imposed in Theorem 8.8.6, is Hypothesis 8.8.1(ii), the statement about orbits of closed points being Zariski closed.

8.9 What it all means, in a concrete example

We have just stated the main theorem, and in Chapter 9 we will be launching into the proof. Before we plunge in, it might help to look at our favorite example and see, in down-to-earth terms, what the theorem means for the example. In order to do that, we need a little technical refinement to the statement of Theorem 8.8.6(i). The statement, as it appeared above, asserted that $(V, \mathcal{O}_V)$ is a scheme of finite type over $\mathbb{C}$. This means it can be covered by (finitely many) affine schemes of finite type over $\mathbb{C}$. The refinement we want is

Refinement 8.9.1. *Let $f \in S$ be any element satisfying the conditions in Hypothesis 8.8.1(i); that is, X_f is contained in $U \subset X$, and $a'(f) = r \otimes f$. Then the open set $X_f/G \subset V$ is an* affine *scheme of finite type over $\mathbb{C}$.*

We have already seen, in Remark 8.8.7, that Theorem 8.8.6(i) can be deduced from the local case. To prove that V is a scheme of finite type over $\mathbb{C}$, it suffices to show that all the X_f/G are. The refinement is genuinely finer; it asserts, moreover, that each X_f/G is affine. It turns out that the refinement comes for free, in the course of the proof.

Now we return to our favorite Example 8.8.5; the Hopf algebra is $R = \mathbb{C}[t, t^{-1}]$, it acts on the ring $S = \mathbb{C}[x_0, x_1, \ldots, x_n]$ by the formula $a'(x_i) = t^{-1} \otimes x_i$, and the open set $U \subset X = \operatorname{Spec}(S)$ was chosen to be $U = X - \{0\}$. Let $\pi : X \longrightarrow X/G$ be the projection. The open set $V \subset X/G$ is, as always, the unique open set for which $\pi^{-1}V = U$. Since this will be our key example, we fix some notation for it.

Definition 8.9.2. *The* projective space $(\mathbb{P}^n, \mathcal{O}_{\mathbb{P}^n})$ *is defined to be the ringed space $\big(V, \mathcal{O}_V\big)$, with $V \subset X/G$ as above (which is the same as Example 8.8.5).*

Remark 8.9.3. For any subset $J \subset \{0, 1, \ldots, n\}$ we have, on occasion, considered the monomials

$$f_J \quad = \quad \prod_{i \in J} x_i \; .$$

These monomials are always homogeneous polynomials; the action of the group G is by $a'(f_J) = t^{-|J|} \otimes f_J$. As long as the set J is non-empty, we have that f_J vanishes at $0 \in \mathbb{C}^{n+1}$. For this section we will only consider non-empty $J \subset \{0, 1, \ldots, n\}$. For such J we know that $X_J = X_{f_J} \subset U$, and hence Refinement 8.9.1 applies. It follows that the open subset $X_J/G \subset \mathbb{P}^n$ is affine. In this section we propose to discuss these open sets, and make everything as explicit as possible. To facilitate the discussion we introduce some notation.

Notation 8.9.4. Let X, G, $U \subset X$ and $V = \mathbb{P}^n$ be as above.

(i) Let $f \in S$ be an arbitrary homogeneous polynomial of degree > 0. Then $\mathbb{P}^n_f$ will be our shorthand for the open set $X_f/G \subset \mathbb{P}^n$.

(ii) In the special case, where $f = f_J = \prod_{i \in J} x_i$ for some non-empty $J \subset \{0, 1, \ldots, n\}$, we will write $\mathbb{P}^n_J$ for the open set

$$\mathbb{P}^n_J \quad = \quad \mathbb{P}^n_{f_J} \quad = \quad X_J/G \quad = \quad X_{f_J}/G\,.$$

In Remark 8.9.3 we noted that the open sets $\mathbb{P}^n_J \subset \mathbb{P}^n$ are all affine.

(iii) In the case where $J = \{i\}$ contains exactly one element, we will feel free to write $\mathbb{P}^n_i$ for $\mathbb{P}^n_{\{i\}} = \mathbb{P}^n_{x_i}$.

Remark 8.9.5. If $f, g \in S$ are homogeneous polynomials of positive degree, then we have

$$X_f = \pi^{-1}\mathbb{P}^n_f\,, \qquad X_g = \pi^{-1}\mathbb{P}^n_g\,, \qquad X_{fg} = \pi^{-1}\mathbb{P}^n_{fg}\,.$$

Lemma 3.2.5 says that $X_f \cap X_g = X_{fg}$, and hence

$$\pi^{-1}\mathbb{P}^n_{fg} \quad = \quad \pi^{-1}\mathbb{P}^n_f \cap \pi^{-1}\mathbb{P}^n_g \quad = \quad \pi^{-1}\{\mathbb{P}^n_f \cap \mathbb{P}^n_g\}\,.$$

Since the map π is surjective we conclude that $\mathbb{P}^n_{fg} = \mathbb{P}^n_f \cap \mathbb{P}^n_g$.

Lemma 8.9.6. *Let $J \subset \{0, 1, \ldots, n\}$ be a non-empty subset. Then*

$$\mathbb{P}^n_J \quad = \quad \bigcap_{i \in J} \mathbb{P}^n_i\,.$$

Proof. Apply Remark 8.9.5 to $f_J = \prod_{i \in J} x_i$. □

Remark 8.9.7. We know that the open sets $\mathbb{P}^n_i$ are all affine, as are their finite intersections $\mathbb{P}^n_J$. Next we observe:

Lemma 8.9.8. *The open sets $\mathbb{P}^n_i$, of Notation 8.9.4(iii), cover $\mathbb{P}^n$.*

Proof. We need to show that the open sets $\mathbb{P}^n_i = \mathbb{P}^n_{x_i}$ cover $\mathbb{P}^n$, or equivalently that their inverse images by $\pi : X \longrightarrow X/G$ cover $U = X - \{0\}$, which is the inverse image of $\mathbb{P}^n \subset X/G$. That is we need to show that the basic open sets X_{x_i} cover U. Taking complements, we must show that the closed sets $X - X_{x_i} = V(Sx_i)$ intersect in the closed point $0 \in \mathbb{C}^{n+1} = X^{\mathrm{an}}$. This means we must show that the only prime ideal $\mathfrak{p} \in \operatorname{Spec}(S) = X$, lying in every $V(Sx_i)$, is the prime ideal corresponding to $0 \in \mathbb{C}^{n+1}$. We should remind ourselves of the correspondence between maximal ideals $\mathfrak{m}_a \in \operatorname{Spec}(S)$ and points $a \in \mathbb{C}^{n+1}$.

The correspondence takes $a \in \mathbb{C}^{n+1}$ to the kernel of the homomorphism $\varphi_a : S \longrightarrow \mathbb{C}$, which evaluates $f \in S$ at the point a. Thus $0 \in \mathbb{C}^{n+1}$ corresponds to the kernel $\mathfrak{m}$ of $\varphi_0 : S \longrightarrow \mathbb{C}$. This maximal ideal $\mathfrak{m} = \ker(\varphi_0)$ is the ideal of all polynomials vanishing at 0, that is all polynomials with a zero constant term. The ideal is generated by $\{x_0, x_1, \ldots, x_n\} \subset S$.

Any prime ideal lying in every $V(Sx_i)$ must contain all the elements $x_i \in S$. It therefore contains the ideal $\mathfrak{m}$ which they generate. This ideal happens to be $\mathfrak{m} = \ker(\varphi_0)$, which is maximal. There is only one point $\mathfrak{m} = \ker(\varphi_0)$ in the intersection of the $V(Sx_i)$, and this point is $0 \in \mathbb{C}^{n+1} = X^{\mathrm{an}}$. □

Remark 8.9.9. Lemma 8.9.8 tells us that the affine open sets $\mathbb{P}^n_i$, with $0 \le i \le n$, cover projective space $\mathbb{P}^n$. In Remark 8.9.7 we saw that all their intersections $\mathbb{P}^n_J$ are also affine. Now it is time to figure out what these affines are.

Remark 8.9.10. We are assuming Theorem 8.8.6(i); in fact, we are assuming the sharper Refinement 8.9.1. This tells us that the scheme $(\mathbb{P}^n_J, \mathcal{O}_{\mathbb{P}^n_J})$ is affine, making the map $(\pi, \pi^*) : (X_J, \mathcal{O}_{X_J}) \longrightarrow (\mathbb{P}^n_J, \mathcal{O}_{\mathbb{P}^n_J})$ a morphism of affine schemes of finite type over $\mathbb{C}$. Theorem 3.9.4 establishes that (π, π^*) must be of the form $\left(\operatorname{Spec}(\theta), \widetilde{\theta}\right)$, for some ring homomorphism θ. Proposition 3.6.5 informs us how to compute this ring homomorphism θ; it is simply the map on global sections

$$\pi^*_{\mathbb{P}^n_J} : \Gamma\left(\mathbb{P}^n_J, \mathcal{O}_{\mathbb{P}^n_J}\right) \longrightarrow \Gamma\left(X_J, \mathcal{O}_{X_J}\right).$$

If there is anything we understand, it is this particular map. We know that $\mathbb{P}^n_J = X_J/G$. Definition 8.7.6(ii) tells us that the sheaf $\mathcal{O}_{\mathbb{P}^n_J}$ is simply the subsheaf of invariants $\{\pi_*\mathcal{O}_{X_J}\}^{G^{\mathrm{an}}} \subset \pi_*\mathcal{O}_{X_J}$. This means that, for any open subset $W \subset \mathbb{P}^n_J$, we defined

$$\Gamma(W, \mathcal{O}_{\mathbb{P}^n_J}) \quad = \quad \Gamma(\pi^{-1}W, \mathcal{O}_{X_J})^{G^{\mathrm{an}}};$$

in the special case where $W = \mathbb{P}^n_J$, we obtain

$$\Gamma\Big(\mathbb{P}^n_J\,,\,\mathcal{O}_{\mathbb{P}^n_J}\Big) \quad = \quad \Gamma\Big(\pi^{-1}\mathbb{P}^n_J\,,\,\mathcal{O}_{X_J}\Big)^{G^{\mathrm{an}}} \quad = \quad \Gamma\Big(X_J\,,\,\mathcal{O}_{X_J}\Big)^{G^{\mathrm{an}}} .$$

Remark 8.7.8 goes on to tell us that the map of sheaves $\pi^* : \mathcal{O}_{\mathbb{P}^n_J} \longrightarrow \pi_*\mathcal{O}_{X_J}$ is the inclusion of the subsheaf of invariants. This means that, for any open set $W \subset \mathbb{P}^n_J$, the ring homomorphism $\pi^*_W : \Gamma(W, \mathcal{O}_{\mathbb{P}^n_J}) \longrightarrow \Gamma(\pi^{-1}W\,,\,\mathcal{O}_{X_J})$ is simply the inclusion

$$\Gamma(\pi^{-1}W\,,\,\mathcal{O}_{X_J})^{G^{\mathrm{an}}} \longrightarrow \Gamma(\pi^{-1}W\,,\,\mathcal{O}_{X_J})\ ;$$

taking the special case of the open set $W = \mathbb{P}^n_J$, we conclude that

$$\theta = \pi^*_{\mathbb{P}^n_J} : \Gamma\Big(\mathbb{P}^n_J, \mathcal{O}_{\mathbb{P}^n_J}\Big) \longrightarrow \Gamma\Big(X_J, \mathcal{O}_{X_J}\Big)$$

identifies with the inclusion

$$\Gamma\Big(X_J, \mathcal{O}_{X_J}\Big)^{G^{\mathrm{an}}} \longrightarrow \Gamma\Big(X_J, \mathcal{O}_{X_J}\Big)\ .$$

Hence the map θ, which we are after, is just the inclusion of the subring of G^{an}–invariants in $\Gamma(X_J, \mathcal{O}_{X_J})$.

In Exercise 3.5.4(i) and in Remark 3.5.6 we produced an isomorphism $\varphi_J^{-1} : S_J \longrightarrow \Gamma(X_J, \mathcal{O}_{X_J})$, and in Remark 8.5.5 we worked out, explicitly, how the group $G^{\mathrm{an}} = \mathbb{C}^*$ acts on $S_J \cong \Gamma(X_J, \mathcal{O}_{X_J})$. Surely we should be in a position to put it all together. To understand the map $(\pi, \pi^*) : (U, \mathcal{O}_U) \longrightarrow (\mathbb{P}^n, \mathcal{O}_{\mathbb{P}^n})$, over the open affine $\mathbb{P}^n_J \subset \mathbb{P}^n$, we need to compute the inclusion $\theta : S_J^{G^{\mathrm{an}}} \longrightarrow S_J$, of the subring of G^{an}–invariant elements in the ring S_J.

Reminder 8.9.11. Let us briefly remind ourselves of the notation; the reader is referred to Sections 3.5, 5.8 and 8.5 for more detail. As above, let $J \subset \{0, 1, \ldots, n\}$ be a non-empty subset. The ring

$$S_J \quad = \quad \mathbb{C}[x_0, x_1, \ldots, x_n, x_{j_1}^{-1}, x_{j_2}^{-1}, \ldots, x_{j_\ell}^{-1}]$$

is a ring of Laurent polynomials; its elements are finite linear combinations of Laurent monomials

$$\mathbf{x}^{\mathbf{a}} \quad = \quad x_0^{a_0} x_1^{a_1} \cdots x_n^{a_n}\ ,$$

where $a_i \geq 0$ if $i \notin J$, while $a_i \in \mathbb{Z}$ is unrestricted if $i \in J$. In Remark 8.5.5 we learned the action of G^{an}; given an element $g \in G^{\mathrm{an}} = \mathbb{C}^*$, the formula was

$$g(\mathbf{x}^{\mathbf{a}}) \quad = \quad g^{\langle \mathbf{a} \rangle}\mathbf{x}^{\mathbf{a}}\ .$$

Lemma 8.9.12. *Let $J \subset \{0,1,\ldots,n\}$ be a non-empty subset. The subring $S_J^{G^{\mathrm{an}}}$, of G^{an}–invariant elements in S_J, has a basis consisting of all the monomials $\mathbf{x}^{\mathbf{a}}$ with $\langle \mathbf{a} \rangle = 0$. That is, its basis is all the monomials*

$$\mathbf{x}^{\mathbf{a}} \quad = \quad x_0^{a_0} x_1^{a_1} \cdots x_n^{a_n} \; ,$$

where

$$\langle \mathbf{a} \rangle \quad = \quad a_0 + a_1 + \cdots + a_n \quad = \quad 0 \; .$$

Proof. Any element of S_J can be written as a finite linear combination

$$\sum_{\mathbf{a}} \lambda_{\mathbf{a}} \mathbf{x}^{\mathbf{a}} \, .$$

Applying $g \in G^{\mathrm{an}}$ to this sum, we have

$$g\left(\sum_{\mathbf{a}} \lambda_{\mathbf{a}} \mathbf{x}^{\mathbf{a}}\right) \quad = \quad \sum_{\mathbf{a}} \lambda_{\mathbf{a}} g^{\langle \mathbf{a} \rangle} \mathbf{x}^{\mathbf{a}} \, .$$

If this is to be invariant we must have, for every $g \in G^{\mathrm{an}} = \mathbb{C}^*$,

$$\sum_{\mathbf{a}} \lambda_{\mathbf{a}} g^{\langle \mathbf{a} \rangle} \mathbf{x}^{\mathbf{a}} \quad = \quad \sum_{\mathbf{a}} \lambda_{\mathbf{a}} \mathbf{x}^{\mathbf{a}} \, .$$

This means $\left(g^{\langle \mathbf{a} \rangle} - 1\right) \cdot \lambda_{\mathbf{a}} = 0$ for all $\mathbf{a}$ and for all g. If we fix $\mathbf{a} \in \mathbb{Z}^{n+1}$, then either $\lambda_{\mathbf{a}} = 0$, or else $g^{\langle \mathbf{a} \rangle} = 1$ for every $g \in \mathbb{C}^*$. The second possibility can only happen if $\langle \mathbf{a} \rangle = 0$. □

Remark 8.9.13. It is common to refer to the integer $\mathbf{a} = a_0 + a_1 + \cdots + a_n$ as the *degree* of the Laurent monomial $\mathbf{x}^{\mathbf{a}}$. A Laurent polynomial is said to be *homogeneous of degree m* if it is a finite linear combination of monomials $\mathbf{x}^{\mathbf{a}}$, each of degree m. The elements of the subring $S_J^{G^{\mathrm{an}}} \subset S_J$ are precisely the homogeneous Laurent polynomials of degree 0.

Remark 8.9.14. Suppose we are given non-empty subsets $J \subset J' \subset \{0,1,\ldots,n\}$. Then the open immersion $X_{J'} \longrightarrow X_J$, together with the induced open immersion of the quotients $X_{J'}/G \longrightarrow X_J/G$, gives a commutative square of morphisms of affine schemes

$$\begin{array}{ccc} \left(X_{J'}, \mathcal{O}_{X_{J'}}\right) & \longrightarrow & \left(X_J, \mathcal{O}_{X_J}\right) \\ \downarrow & & \downarrow \\ \left(\mathbb{P}^N_{J'}, \mathcal{O}_{\mathbb{P}^N_{J'}}\right) & \longrightarrow & \left(\mathbb{P}^N_J, \mathcal{O}_{\mathbb{P}^N_J}\right) \end{array} \quad .$$

On global sections, this gives a commutative diagram of ring homomorphisms

$$\begin{array}{ccc}
\Gamma\Big(\mathbb{P}^N_J, \mathcal{O}_{\mathbb{P}^N_J}\Big) & \xrightarrow{\operatorname{res}^{\mathbb{P}^N_J}_{\mathbb{P}^N_{J'}}} & \Gamma\Big(\mathbb{P}^N_{J'}, \mathcal{O}_{\mathbb{P}^N_{J'}}\Big) \\
\downarrow & & \downarrow \\
\Gamma\Big(X_J, \mathcal{O}_{X_J}\Big) & \longrightarrow & \Gamma\Big(X_{J'}, \mathcal{O}_{X_{J'}}\Big)
\end{array} .$$

In this diagram, we understand the part

$$\begin{array}{ccc}
\Gamma\Big(\mathbb{P}^N_J, \mathcal{O}_{\mathbb{P}^N_J}\Big) & & \Gamma\Big(\mathbb{P}^N_{J'}, \mathcal{O}_{\mathbb{P}^N_{J'}}\Big) \\
\downarrow & & \downarrow \\
\Gamma\Big(X_J, \mathcal{O}_{X_J}\Big) & \longrightarrow & \Gamma\Big(X_{J'}, \mathcal{O}_{X_{J'}}\Big)
\end{array} .$$

The horizontal map $\operatorname{res}^{X_J}_{X_{J'}} : \Gamma(X_J, \mathcal{O}_{X_J}) \longrightarrow \Gamma(X_{J'}, \mathcal{O}_{X_{J'}})$ was identified, in Remark 3.5.6, with the inclusion $S_J \longrightarrow S_{J'}$. The vertical maps were described in Remark 8.9.10. With these identifications the diagram becomes

$$\begin{array}{ccc}
S_J^{G^{\text{an}}} & & S_{J'}^{G^{\text{an}}} \\
\downarrow & & \downarrow \\
S_J & \longrightarrow & S_{J'}
\end{array} ,$$

where all three maps are the natural inclusions. There is only one possible way to complete to a commutative square; the map

$$\operatorname{res}^{\mathbb{P}^N_J}_{\mathbb{P}^N_{J'}} : \Gamma\Big(\mathbb{P}^N_J, \mathcal{O}_{\mathbb{P}^N_J}\Big) \longrightarrow \Gamma\Big(\mathbb{P}^N_{J'}, \mathcal{O}_{\mathbb{P}^N_{J'}}\Big)$$

must identify with the natural inclusion $\psi^J_{J'} : S_J^{G^{\text{an}}} \longrightarrow S_{J'}^{G^{\text{an}}}$. We deduce:

Lemma 8.9.15. *As in Remark 8.9.14, suppose we are given non-empty subsets $J \subset J' \subset \{0, 1, \ldots, n\}$. Then the open immersion of schemes*

$$\Big(\mathbb{P}^N_{J'}, \mathcal{O}_{\mathbb{P}^N_{J'}}\Big) \longrightarrow \Big(\mathbb{P}^N_J, \mathcal{O}_{\mathbb{P}^N_J}\Big)$$

must identify with $\Big(\operatorname{Spec}(\psi^J_{J'}), \widetilde{\psi^J_{J'}}\Big)$, where $\psi^J_{J'} : S_J^{G^{\text{an}}} \longrightarrow S_{J'}^{G^{\text{an}}}$ is the inclusion.

Proof. In Remark 8.9.14 we computed the map induced by the open immersion $\mathbb{P}^N_{J'} \subset \mathbb{P}^N_J$, at the level of global sections. We showed that this map, which we would write in symbols as

$$\mathrm{res}^{\mathbb{P}^N_J}_{\mathbb{P}^N_{J'}} : \Gamma\left(\mathbb{P}^N_J, \mathcal{O}_{\mathbb{P}^N_J}\right) \longrightarrow \Gamma\left(\mathbb{P}^N_{J'}, \mathcal{O}_{\mathbb{P}^N_{J'}}\right),$$

must identify with the inclusion $\psi^J_{J'} : S^{G^{\mathrm{an}}}_J \longrightarrow S^{G^{\mathrm{an}}}_{J'}$. Now Theorem 3.9.4 says that the open immersion of affine schemes

$$\left(\mathbb{P}^N_{J'}, \mathcal{O}_{\mathbb{P}^N_{J'}}\right) \longrightarrow \left(\mathbb{P}^N_J, \mathcal{O}_{\mathbb{P}^N_J}\right)$$

must be of the form $(\mathrm{Spec}(\sigma), \widetilde{\sigma})$, for some ring homomorphism $\sigma : S^{G^{\mathrm{an}}}_J \longrightarrow S^{G^{\mathrm{an}}}_{J'}$. And Proposition 3.6.5 informs us that, to compute σ, all we have to do is find the map induced on global sections. □

Remark 8.9.16. Let us put together what we know so far. Lemma 8.9.8 told us that the open affines $\mathbb{P}^n_i = \mathbb{P}^n_{\{i\}}$ cover $\mathbb{P}^n$. Remark 8.9.10 identified the ringed space $(\mathbb{P}^n_i, \mathcal{O}_{\mathbb{P}^n_i})$ with $\left(\mathrm{Spec}(S^{G^{\mathrm{an}}}_{\{i\}}), \widetilde{S^{G^{\mathrm{an}}}_{\{i\}}}\right)$. In case this was not explicit enough, Lemma 8.9.12 lets us know that the ring $S^{G^{\mathrm{an}}}_{\{i\}}$ is the ring of homogeneous Laurent polynomials of degree 0, where only the variable x_i is allowed to have a negative exponent.

Now consider a pair of integers i, j, both lying between 0 and n. Lemma 8.9.6 tells us that $\mathbb{P}^n_i \cap \mathbb{P}^n_j = \mathbb{P}^n_{\{i,j\}}$. From Remark 8.9.10 we know that the affine scheme

$$\left(\mathbb{P}^n_i \cap \mathbb{P}^n_j, \mathcal{O}_{\mathbb{P}^n_i \cap \mathbb{P}^n_j}\right) = \left(\mathbb{P}^n_{\{i,j\}}, \mathcal{O}_{\mathbb{P}^n_{\{i,j\}}}\right)$$

identifies with $\left(\mathrm{Spec}(S^{G^{\mathrm{an}}}_{\{i,j\}}), \widetilde{S^{G^{\mathrm{an}}}_{\{i,j\}}}\right)$. Furthermore, Lemma 8.9.15 tells us that the open immersion

$$\left(\mathbb{P}^n_i \cap \mathbb{P}^n_j, \mathcal{O}_{\mathbb{P}^n_i \cap \mathbb{P}^n_j}\right) \longrightarrow \left(\mathbb{P}^n_i, \mathcal{O}_{\mathbb{P}^n_i}\right)$$

must be $\left(\mathrm{Spec}(\psi^{\{i\}}_{\{i,j\}}), \widetilde{\psi^{\{i\}}_{\{i,j\}}}\right)$, where $\psi^{\{i\}}_{\{i,j\}} : S^{G^{\mathrm{an}}}_{\{i\}} \longrightarrow S^{G^{\mathrm{an}}}_{\{i,j\}}$ is the inclusion. We remind the reader: the rings $S^{G^{\mathrm{an}}}_{\{i\}}$ and $S^{G^{\mathrm{an}}}_{\{i,j\}}$ both consist of homogeneous Laurent polynomials of degree zero. The difference is that in $S^{G^{\mathrm{an}}}_{\{i\}}$ only the variable x_i is allowed to have negative exponents, while in $S^{G^{\mathrm{an}}}_{\{i,j\}}$ the exponents of x_i, x_j are both unrestricted.

We know, quite explicitly, an open cover of $\mathbb{P}^n$ by open affine subsets $\left(\mathrm{Spec}(S^{G^{\mathrm{an}}}_{\{i\}}), \widetilde{S^{G^{\mathrm{an}}}_{\{i\}}}\right)$. Furthermore, we know how to glue them; we learned, in Lemma 8.9.15, how the intersection $\mathbb{P}^n_i \cap \mathbb{P}^n_j$ embeds in each of $\mathbb{P}^n_i$ and $\mathbb{P}^n_j$. The way to glue $\mathbb{P}^n_i$ and $\mathbb{P}^n_j$ is to identify the image of the open immersion

$$\left(\operatorname{Spec}(S^{G^{\mathrm{an}}}_{\{i,j\}}), \widetilde{S^{G^{\mathrm{an}}}_{\{i,j\}}}\right) \xrightarrow{\left(\operatorname{Spec}(\psi^{\{i\}}_{\{i,j\}}), \widetilde{\psi^{\{i\}}_{\{i,j\}}}\right)} \left(\operatorname{Spec}(S^{G^{\mathrm{an}}}_{\{i\}}), \widetilde{S^{G^{\mathrm{an}}}_{\{i\}}}\right)$$

with the image of the open immersion

$$\left(\operatorname{Spec}(S^{G^{\mathrm{an}}}_{\{i,j\}}), \widetilde{S^{G^{\mathrm{an}}}_{\{i,j\}}}\right) \xrightarrow{\left(\operatorname{Spec}(\psi^{\{j\}}_{\{i,j\}}), \widetilde{\psi^{\{j\}}_{\{i,j\}}}\right)} \left(\operatorname{Spec}(S^{G^{\mathrm{an}}}_{\{j\}}), \widetilde{S^{G^{\mathrm{an}}}_{\{j\}}}\right) .$$

Remark 8.9.17. The lowbrow way to define projective space is via the cover $\mathbb{P}^n = \cup_{i=0}^n \mathbb{P}^n_i$. We have an atlas of projective space, a cover by open affines $\left(\operatorname{Spec}(S^{G^{\mathrm{an}}}_{\{i\}}), \widetilde{S^{G^{\mathrm{an}}}_{\{i\}}}\right)$. We know how to glue the open affines in pairs. This determines the ringed space $\mathbb{P}^n$.

In this book we chose the highbrow approach. It is more abstract, but cleaner. In the lowbrow approach, there are grubby facts one has to check. For example, it needs to be verified that the gluing data are compatible. It also is not clear, from the construction, that it is independent of the choice of generators for $S = \mathbb{C}[x_0, x_1, \dots, x_n]$. If, instead of the generators $\{x_0, x_1, \dots, x_n\}$, we elected to look at the generators $\{x_0 + x_1, x_1, \dots, x_n\}$, it is unclear that the resulting space would be the same.

Remark 8.9.18. If Remark 8.9.17 was not lowbrow enough for you, you might wish to understand, even more concretely, what the rings $S^{G^{\mathrm{an}}}_J$ really are. We will do this forthwith.

Notation 8.9.19. We have been assuming, throughout this section, that the subset $J \subset \{0, 1, \dots, n\}$ is non-empty. It is convenient, in the next statement, to assume that $0 \in J$; it makes the statement slightly simpler. It goes without saying that we could replace 0 by any other element of J, and the analogous assertions will hold.

As we have said, to keep the notation simple we will assume $0 \in J$. We will write $J = \{0\} \cup J'$, where $J' \subset \{1, 2, \dots, n\}$ is any subset, possibly empty. We will consider two rings:

(i) $S = \mathbb{C}[x_0, x_1, \dots, x_n]$ is the polynomial ring in $(n+1)$ variables.
(ii) $S' = \mathbb{C}[z_1, z_2, \dots, z_n]$ is the polynomial ring in n variables.

Lemma 8.9.20. *In the terminology of Notation 8.9.19, there is an isomorphism*

$$\sigma : S'_{J'} \longrightarrow S^{G^{\mathrm{an}}}_J .$$

The isomorphism is given by the formula $\sigma(z_j) = x_j / x_0$.

Proof. We are assuming that $0 \in J$, and hence $x_0^{-1}x_j = x_j/x_0$ is certainly an element of S_J; negative powers of x_0 are allowed. Also, $x_0^{-1}x_j$ has degree zero; it is G^{an}–invariant. Thus the homomorphism $\sigma : S'_{J'} \longrightarrow S_J^{G^{\text{an}}}$, sending z_j to $x_0^{-1}x_j$, is certainly well defined. We need to show the σ is both surjective and injective.

To prove the assertion, both the injectivity and the surjectivity, it suffices to show that a basis for the vector space $S'_{J'}$ maps, bijectively, to a basis for the vector space $S_J^{G^{\text{an}}}$. The basis we will consider for $S'_{J'}$ is the collection of all monomials

$$\mathbf{z^b} \quad = \quad z_1^{b_1} z_2^{b_2} \cdots z_n^{b_n} \ ,$$

where $b_j \geq 0$ if $j \notin J'$, while if $j \in J'$ then $b_j \in \mathbb{Z}$ is unrestricted. The homomorphism σ takes the monomial $\mathbf{z^b}$ to

$$\sigma(\mathbf{z^b}) \quad = \quad x_0^{-\langle \mathbf{b} \rangle} x_1^{b_1} x_2^{b_2} \cdots x_n^{b_n} \ ,$$

where $\langle \mathbf{b} \rangle = b_1 + b_2 + \cdots + b_n$. As $\mathbf{z^b}$ runs over all the basis elements of $S'_{J'}$, the images $\sigma(\mathbf{z^b})$ are precisely the basis elements $\{\mathbf{x^a} \mid \langle \mathbf{a} \rangle = 0\}$ of $S_J^{G^{\text{an}}}$. □

Remark 8.9.21. If the lowbrow construction of projective space, given in Remark 8.9.16, was not hands-on enough for you, we will try to do better here.

We know that the affine scheme $\left(\mathbb{P}_0^n, \mathcal{O}_{\mathbb{P}_0^n}\right)$ is $\left(\operatorname{Spec}(S_{\{0\}}^{G^{\text{an}}}), \widetilde{S_{\{0\}}^{G^{\text{an}}}}\right)$. Lemma 8.9.20 gives an isomorphism of $S_{\{0\}}^{G^{\text{an}}}$ with $S' = S'_\emptyset$. That is, $S_{\{0\}}^{G^{\text{an}}}$ is isomorphic to the polynomial ring

$$S' \quad = \quad \mathbb{C}[z_1, z_2, \dots, z_n] \ .$$

Symmetry says that the affine scheme $\left(\mathbb{P}_i^n, \mathcal{O}_{\mathbb{P}_i^n}\right)$ is $\left(\operatorname{Spec}(S_{\{i\}}^{G^{\text{an}}}), \widetilde{S_{\{i\}}^{G^{\text{an}}}}\right)$, with $S_{\{i\}}^{G^{\text{an}}}$ also isomorphic to the polynomial ring S'. In other words, all the affine schemes $\mathbb{P}_i^n$ are isomorphic, with each being isomorphic to $\left(\operatorname{Spec}(S'), \widetilde{S'}\right)$. The only problem is to understand the gluing.

Lemma 8.9.20 explicitly told us the isomorphism $\sigma : S' \longrightarrow S_{\{0\}}^{G^{\text{an}}}$, but it is no great effort to interchange x_0 and x_i and produce an isomorphism $\sigma_i : S' \longrightarrow S_{\{i\}}^{G^{\text{an}}}$. The isomorphism is given by the formula

$$\sigma_i(z_j) \quad = \quad \begin{cases} x_i^{-1}x_{j-1} & \text{if } j \leq i, \\ x_i^{-1}x_j & \text{if } j > i. \end{cases}$$

Perhaps a more enlightening way to write this is that the isomorphism σ_i sends the variables $z_1, z_2, \dots, z_n$, in order, to the Laurent monomials

$$\frac{x_0}{x_i}\ , \ \frac{x_1}{x_i}\ , \ \cdots \frac{x_{i-1}}{x_i}\ , \ \frac{x_{i+1}}{x_i}\ , \ \cdots \frac{x_n}{x_i} \ .$$

The homomorphism σ_0 is simply σ, but for $i > 0$ we have produced more homomorphisms.

Lemma 8.9.20 also went on to identify for us the ring $S_{\{0,j\}}^{G^{\mathrm{an}}}$. It is isomorphic to $S'_{\{j\}}$, that is to the Laurent polynomial ring

$$S'_{\{j\}} \quad = \quad \mathbb{C}[z_1, z_2, \ldots, z_n, z_j^{-1}] \, .$$

If we identify $\left(\mathbb{P}_0^n, \mathcal{O}_{\mathbb{P}_0^n}\right)$ with $\left(\operatorname{Spec}(S'), \widetilde{S'}\right)$, via the map $\sigma = \sigma_0$ of Lemma 8.9.20, then $\mathbb{P}_0^n \cap \mathbb{P}_j^n \subset \mathbb{P}_0^n$ maps to the open set $\{\operatorname{Spec}(S')\}_{z_j} \subset \operatorname{Spec}(S')$. A more enlightening way to write this is to say that the ring of invariants $S_{\{0,j\}}^{G^{\mathrm{an}}}$, that is the ring

$$\mathbb{C}[x_0, x_1, \ldots, x_n, x_0^{-1}, x_j^{-1}]^{G^{\mathrm{an}}} \qquad ,$$

can also be written as

$$\mathbb{C}\left[\frac{x_1}{x_0}, \frac{x_2}{x_0}, \ldots, \frac{x_n}{x_0}, \left(\frac{x_j}{x_0}\right)^{-1}\right]$$

which is nothing other than $\mathbb{C}[z_1, z_2, \ldots, z_n, z_j^{-1}]$, via the isomorphism $\sigma = \sigma_0$. Interchanging the roles of 0 and j we obtain that the ring $S_{\{0,j\}}^{G^{\mathrm{an}}}$ can be written as

$$\mathbb{C}\left[\frac{x_0}{x_j}, \frac{x_1}{x_j}, \ldots, \frac{x_{j-1}}{x_j}, \frac{x_{j+1}}{x_j}, \ldots, \frac{x_n}{x_j}, \left(\frac{x_0}{x_j}\right)^{-1}\right] \qquad ,$$

and, via the isomorphism $\sigma_j : S' \longrightarrow S_{\{j\}}^{G^{\mathrm{an}}}$, this becomes identified with

$$S'_{\{1\}} \quad = \quad \mathbb{C}[z_1, z_2, \ldots, z_n, z_1^{-1}] \, .$$

The rule for gluing $\left(\mathbb{P}_0^n, \mathcal{O}_{\mathbb{P}_0^n}\right)$ with $\left(\mathbb{P}_j^n, \mathcal{O}_{\mathbb{P}_j^n}\right)$, both of which are isomorphic to $\left(\operatorname{Spec}(S'), \widetilde{S'}\right)$, becomes to use the isomorphisms

$$S'\left[\frac{1}{z_j}\right] \xrightarrow{\ \sigma_0\ } S_{\{0,j\}}^{G^{\mathrm{an}}} \xleftarrow{\ \sigma_j\ } S'\left[\frac{1}{z_1}\right]$$

above, to identify the open subset $\{\operatorname{Spec}(S')\}_{z_j} \subset \mathbb{P}_0^n$ with the open subset $\{\operatorname{Spec}(S')\}_{z_1} \subset \mathbb{P}_j^n$. If the reader prefers to see the isomorphism written in terms of the ring S', without ever mentioning a ring of invariants $S_J^{G^{\mathrm{an}}}$, that can also be done. The formula, which I find singularly unenlightening, is that the isomorphism $\alpha : S'[1/z_j] \longrightarrow S'[1/z_1]$ is given by the rule

$$\alpha(z_i) \quad = \quad \begin{cases} z_1^{-1} z_{i+1} & \text{if } i < j, \\ z_1^{-1} & \text{if } i = j, \\ z_1^{-1} z_i & \text{if } i > j. \end{cases}$$

If we take the case $i < j$, this opaque description means that $z_i = x_i/x_0$ maps to

$$\alpha(z_i) \quad = \quad \left(\frac{x_0}{x_j}\right)^{-1} \cdot \left(\frac{x_i}{x_j}\right) \quad ;$$

we leave the other cases to the reader. The reader can also, if she so desires, explicitly work out the gluing of $\mathbb{P}^n_i \cong \operatorname{Spec}(S')$ with $\mathbb{P}^n_j \cong \operatorname{Spec}(S')$, where we do not assume that $i = 0$. I have no authority to stop the reader from wasting her time.

Let us now leave the lowbrow description of projective space, given above by explicitly gluing the open affines $\mathbb{P}^n_i$. In passing we should mention a couple more easy corollaries of Lemma 8.9.20.

Corollary 8.9.22. *Let $J \subset \{0, 1, \ldots, n\}$ be a non-empty subset. Then the map $(\pi, \pi^*) : (X_J, \mathcal{O}_{X_J}) \longrightarrow (\mathbb{P}^n_J, \mathcal{O}_{\mathbb{P}^n_J})$ admits a splitting.*

Proof. Without loss of generality we may assume $0 \in J$. The map (π, π^*) is equal to $\left(\operatorname{Spec}(\theta), \widetilde{\theta}\right)$, for the inclusion map $\theta : S_J^{G^{\mathrm{an}}} \longrightarrow S_J$. Hence it suffices to show that θ admits a splitting. Even better, it suffices to show that the composite $\theta\sigma$ admits a splitting, where σ is the isomorphism $\sigma : S'_{J'} \longrightarrow S_J^{G^{\mathrm{an}}}$ of Lemma 8.9.20. The map $\theta\sigma$ takes $z_j \in S'_{J'}$ to $x_0^{-1}x_j \in S_J$. We define $\rho : S_J \longrightarrow S'_{J'}$ by the formula

$$\rho(x_j) \quad = \quad \begin{cases} z_j & \text{if } j \geq 1, \\ 1 & \text{if } j = 0, \end{cases}$$

and observe that

$$\rho\theta\sigma(z_j) \quad = \quad \rho(x_0^{-1}x_j) \quad = \quad \rho(x_0)^{-1}\rho(x_j) \quad = \quad z_j\ .$$

This implies that $\rho\theta\sigma = 1$. □

Remark 8.9.23. It follows that the map $\pi^{\mathrm{an}} : U^{\mathrm{an}} \longrightarrow \{\mathbb{P}^n\}^{\mathrm{an}}$ is surjective. The question is local in $\{\mathbb{P}^n\}^{\mathrm{an}}$, so it suffices to prove that the map $\pi^{\mathrm{an}} : X_J^{\mathrm{an}} \longrightarrow \{\mathbb{P}^n_J\}^{\mathrm{an}}$ is surjective. But Corollary 8.9.22 tells us even more; it guarantees the existence of a splitting, namely $\left\{\operatorname{Spec}(\rho)\right\}^{\mathrm{an}}$.

Corollary 8.9.24. *If $\mathbf{y}, \mathbf{y}'$ are points in $U^{\mathrm{an}} = \mathbb{C}^{n+1} - \{0\}$, with $\pi(\mathbf{y}) = \pi(\mathbf{y}')$, then $\mathbf{y}$ and $\mathbf{y}'$ lie in the same G^{an}–orbit.*

Proof. We know that $\mathbb{P}^n$ may be covered by open affine subsets $\mathbb{P}^n_i$. Hence $\pi(y) = \pi(y')$ lies in some $\mathbb{P}^n_i$; without loss of generality we may assume that $\pi(y) = \pi(y') \in \mathbb{P}^n_0 \subset \mathbb{P}^n$. Next we apply Lemma 8.9.20, with $J = \{0\}$; this makes $J' = J - \{0\}$ empty.

By Lemma 8.9.20, the map $\pi : X_{\{0\}} \longrightarrow \mathbb{P}^n_{\{0\}}$ is $\mathrm{Spec}(\theta\sigma)$, where $\theta\sigma : S' \longrightarrow S_{\{0\}}$ is the map given by the formula

$$\theta\sigma(z_j) \quad = \quad x_0^{-1} x_j \ .$$

The induced map $\pi^{\mathrm{an}} : X^{\mathrm{an}}_{\{0\}} \longrightarrow \{\mathbb{P}^n_{\{0\}}\}^{\mathrm{an}}$ is therefore the map taking the point $\mathbf{y} = (y_0, y_1, \ldots, y_n) \in \mathbb{C}^{n+1}_{x_0}$ to the point

$$\pi(\mathbf{y}) \quad = \quad \left(\frac{y_1}{y_0}, \frac{y_2}{y_0}, \ldots, \frac{y_n}{y_0}\right) \quad \in \quad \mathbb{C}^n \ .$$

If $\pi(\mathbf{y}) = \pi(\mathbf{y}')$, then

$$\left(\frac{y_1}{y_0}, \frac{y_2}{y_0}, \ldots, \frac{y_n}{y_0}\right) \quad = \quad \left(\frac{y'_1}{y'_0}, \frac{y'_2}{y'_0}, \ldots, \frac{y'_n}{y'_0}\right) ,$$

from which we easily deduce that

$$y'_0 \cdot (y_0, y_1, \ldots, y_n) \quad = \quad y_0 \cdot (y'_0, y'_1, \ldots, y'_n) \ ;$$

that is, $\mathbf{y}$ and $\mathbf{y}'$ lie in the same G^{an}–orbit. □

Remark 8.9.25. Taken together, Remark 8.9.23 and Corollary 8.9.24 tell us that the closed points in $\mathbb{P}^n$ are in bijection with orbits of closed points in U. Remark 8.9.23 says that $\pi^{\mathrm{an}} : U^{\mathrm{an}} \longrightarrow \{\mathbb{P}^n\}^{\mathrm{an}}$ is surjective, and Corollary 8.9.24 identifies the fibers as G^{an}–orbits. This makes it plausible that we should view $\mathbb{P}^n$ as U/G; at the very least the closed points are right. At the level of sets, we have that $\{\mathbb{P}^n\}^{\mathrm{an}}$ is the orbit set $U^{\mathrm{an}}/G^{\mathrm{an}}$.

Given all the facts we have been proving, using the explicit, local description of $\mathbb{P}^n$ as the union of $\mathbb{P}^n = \cup\mathbb{P}^n_i$, it seems only right to show the reader one example of a statement that is easier to prove with the coordinate-free description of the space $\mathbb{P}^n$.

Lemma 8.9.26. *The space $\{\mathbb{P}^n\}^{\mathrm{an}}$ is Hausdorff.*

Proof. We need to show that we can separate any two points in $\{\mathbb{P}^n\}^{\mathrm{an}}$. By Remark 8.9.23 any closed point in $\mathbb{P}^n$ is the image of a closed point in U. We therefore need to show that if b and b' are closed points in U, with distinct images in $\mathbb{P}^n$, then $\pi(b)$ and $\pi(b')$ can be separated.

Therefore let b and b' be two points in $U^{\mathrm{an}} = \mathbb{C}^{n+1} - \{0\}$ with distinct images in $\mathbb{P}^n$. There is always a linear function which does not vanish at either b or b'; in other words there is a homogeneous polynomial $f \in \mathbb{C}[x_0, x_1, \ldots, x_n]$, of degree 1, with $f(b) \neq 0 \neq f(b')$. The open set $X_f \subset X$ is one of the open subsets satisfying Hypothesis 8.8.1(i), and b, b' both lie in X_f. It therefore suffices to separate $\pi(b)$ from $\pi(b')$ in $\{X_f/G\}^{\mathrm{an}} \subset$

$\{\mathbb{P}^n\}^{\text{an}}$. But Refinement 8.9.1 tells us that X_f/G is affine, meaning that $\{X_f/G\}^{\text{an}} = \{\operatorname{Spec}(T)\}^{\text{an}}$ for some finitely generated $\mathbb{C}$–algebra T. Definition 4.5.1 tells us that the complex topology of $\{\operatorname{Spec}(T)\}^{\text{an}}$ is that of some subspace of $\mathbb{C}^m$. Since $\mathbb{C}^m$ is Hausdorff we can separate $\pi(b)$ from $\pi(b')$. □

An easy consequence is:

Lemma 8.9.27. *The space $\{\mathbb{P}^n\}^{\text{an}}$ is compact and Hausdorff.*

Proof. Lemma 8.9.26 tells us that $\{\mathbb{P}^n\}^{\text{an}}$ is Hausdorff; we need to prove it compact. Remark 8.9.23 say that the map $\pi_U^{\text{an}} : U^{\text{an}} \longrightarrow \{\mathbb{P}^n\}^{\text{an}}$ is surjective. Corollary 8.9.24 establishes that the fibers of the map are precisely the orbits for $\mathbb{C}^* = G^{\text{an}}$; in particular, $\pi(b) = \pi(\lambda b)$ for all $\lambda \in \mathbb{C}^*$. Let $\|b\|$ stand for the Euclidean norm of $b \in \mathbb{C}^{n+1} - \{0\}$. Any point of $\{\mathbb{P}^n\}^{\text{an}}$ can be written as $\pi(b)$ with $b \in \mathbb{C}^{n+1} - \{0\}$, and we have the identity

$$\pi(b) = \pi\big(b/\|b\|\big), \qquad \text{with } \big\|b/\|b\|\big\| = 1\,.$$

Hence every point in $\{\mathbb{P}^n\}^{\text{an}}$ can be expressed as $\pi(c)$ with $\|c\| = 1$, that is c lies on the unit sphere $S^{2n+1} \subset \mathbb{C}^{n+1} - \{0\}$. This means that the map $\pi : S^{2n+1} \longrightarrow \{\mathbb{P}^n\}^{\text{an}}$ is surjective. The sphere is compact; hence its image $\{\mathbb{P}^n\}^{\text{an}}$ must be compact. □

Before we end this section we should see what we can learn about the sheaf of analytic functions on projective space. Until now we have focused on the rings $\Gamma(\mathbb{P}^n_J, \mathcal{O})$; how about the Fréchet rings $\Gamma(\{\mathbb{P}^n_J\}^{\text{an}}, \mathcal{O}^{\text{an}})$?

We know that all the open sets $\mathbb{P}^n_J$ are affine; each can be written as $\operatorname{Spec}(T)$ for some finitely generated $\mathbb{C}$–algebra T. This means that there is a ring homomorphism $\mu : T \longrightarrow \Gamma(\{\mathbb{P}^n_J\}^{\text{an}}, \mathcal{O}^{\text{an}})$, with a dense image. The ring $\Gamma(\{\mathbb{P}^n_J\}^{\text{an}}, \mathcal{O}^{\text{an}})$ may be viewed as the completion of T with respect to a pre-Fréchet topology $[T, V]$, where V is the polydisc $V = \{\operatorname{Spec}(T)\}^{\text{an}} = \{\mathbb{P}^n_J\}^{\text{an}}$.

We have, in the course of this section, considered several descriptions of the ring T. These various descriptions must, of course, agree up to isomorphism. We have written T as $\Gamma(\mathbb{P}^n_J, \mathcal{O})$, as $S_J^{G^{\text{an}}}$, as $S'_{J'}$ where $J = \{0\} \cup J'$, and even as $S'_{J''}$ where $J = \{i\} \cup J''$. The topology $[T, V]$, where $V = \{\operatorname{Spec}(T)\}^{\text{an}} = \{\mathbb{P}^n_J\}^{\text{an}}$, is intrinsic; it is independent of the particular description we choose for the ring T. In the remainder of this section we want to understand the topology.

Remark 8.9.28. For any finitely generated $\mathbb{C}$–algebra T we will, in this section, only consider the polydisc $V = \{\operatorname{Spec}(T)\}^{\text{an}}$. Hence we will speak of "the" pre-Fréchet topology of T, meaning the topology $\left[T, \{\operatorname{Spec}(T)\}^{\mathrm{an}}\right]$.

Remark 8.9.29. Let T and T' be finitely generated $\mathbb{C}$–algebras, and let $\varphi : T \longrightarrow T'$ be any isomorphism of $\mathbb{C}$–algebras. Then, by Lemma 5.6.7, both φ and φ^{-1} must be continuous. If we can somehow compute the pre-Fréchet topology on T', then any isomorphism computes for us the pre-Fréchet topology on $T \cong T'$. In the case where $T = T'$ we note that all automorphisms of T are homeomorphisms.

We are interested in the special case $T = \Gamma(\mathbb{P}^n_J, \mathcal{O})$. The description of T which is most helpful, for computing the topology, is $T \cong S'_{J'}$. Let us therefore start by assuming that $0 \in J \subset \{0, 1, \ldots, n\}$. As in Notation 8.9.19 we will write $J = \{0\} \cup J'$, with $J' \subset \{1, 2, \ldots, n\}$. We know, from Remark 8.9.10 and Lemma 8.9.20, that there is an (explicit) isomorphism of schemes

$$\left(\mathbb{P}^n_J, \mathcal{O}_{\mathbb{P}^n_J}\right) \quad \cong \quad \left(\operatorname{Spec}(S'_{J'}), \widetilde{S'_{J'}}\right) .$$

Recall that $S'_{J'}$ is the ring of Laurent polynomials

$$S'_{J'} \quad = \quad \mathbb{C}[z_1, z_2, \ldots, z_n, z_{j_1}^{-1}, z_{j_2}^{-1}, \ldots, z_{j_\ell}^{-1}] \quad ,$$

with $J' = \{j_1, j_2, \ldots, j_\ell\}$.

Remark 5.8.28 applies. It tell us explicitly the pre-Fréchet topology of the ring $S'_{J'}$, when viewed as $\left[S'_{J'}, \{\operatorname{Spec}(S'_{J'})\}^{\mathrm{an}}\right]$. This topology is determined by the sequence of seminorms $\{q_\ell\}$ of Construction 5.8.23(iii). Let us remind the reader.

Reminder 8.9.30. The elements of the ring $S'_{J'}$ are the finite sums $\sum_{\mathbf{z}^{\mathbf{b}} \in S'_{J'}} \lambda_{\mathbf{b}} \mathbf{z}^{\mathbf{b}}$. The seminorm q_ℓ, of Remark 5.8.28, was given by the formula

$$q_\ell \left(\sum_{\mathbf{z}^{\mathbf{b}} \in S'_{J'}} \lambda_{\mathbf{b}} \mathbf{z}^{\mathbf{b}} \right) \quad = \quad \max_{\mathbf{b}} \left(|\lambda_{\mathbf{b}}| \cdot \ell^{|\mathbf{b}|} \right) ,$$

where $|\mathbf{b}| = |b_1| + |b_2| + \cdots + |b_n|$. The seminorms $\{q_\ell \mid \ell \geq 1\}$, taken together, give the ring $S'_{J'}$ its pre-Fréchet topology.

Remark 8.9.31. Next we want to see what is the topology on the ring $S_J^{G^{\mathrm{an}}}$. The isomorphism $\sigma : S'_{J'} \longrightarrow S_J^{G^{\mathrm{an}}}$, given in Lemma 8.9.20, is a homeomorphism by Remark 8.9.29. But σ is explicit; it sends the monomial

$$\mathbf{z}^{\mathbf{b}} \quad = \quad z_1^{b_1} z_2^{b_2} \cdots z_n^{b_n}$$

to the monomial

$$\sigma(\mathbf{z}^{\mathbf{b}}) \quad = \quad x_0^{-\langle \mathbf{b} \rangle} x_1^{b_1} x_2^{b_2} \cdots x_n^{b_n} \ .$$

The seminorm q_ℓ, on the ring $S'_{J'}$, pushes forward via the homeomorphism σ to a seminorm, which we will also call q_ℓ, on the ring $S_J^{G^{\mathrm{an}}}$. It is straightforward to compute the seminorm q_ℓ on $S_J^{G^{\mathrm{an}}}$; it is given by the formula

$$\begin{aligned} q_\ell\left(\sum_{\mathbf{x}^{\mathbf{a}} \in S_J^{G^{\mathrm{an}}}} \lambda_{\mathbf{a}} \mathbf{x}^{\mathbf{a}} \right) &= \max_{\mathbf{a}} \left(|\lambda_{\mathbf{a}}| \cdot \ell^{|a_1|+|a_2|+\cdots+|a_n|} \right) \\ &= \max_{\mathbf{a}} \left(|\lambda_{\mathbf{a}}| \cdot \ell^{|\mathbf{a}|-|a_0|} \right) , \end{aligned}$$

where $|\mathbf{a}|$ is, as expected, $|\mathbf{a}| = |a_0| + |a_1| + |a_2| + \cdots + |a_n|$, and hence

$$|a_1| + |a_2| + \cdots + |a_n| \quad = \quad |\mathbf{a}| - |a_0| \ .$$

What we do not like about the formula is that it is not symmetric in the variables $\{x_0, x_1, \ldots, x_n\}$. Suppose $\{0,1\} \subset J$; that is J contains both 0 and 1. We could, instead of starting out by writing $J = \{0\} \cup J'$, begin by setting $J = \{1\} \cup J''$. Then we could follow the entire argument above, interchanging everywhere the roles of x_0 and x_1. We would end up with seminorms q'_ℓ on $S_J^{G^{\mathrm{an}}}$, given by the formula

$$q'_\ell\left(\sum_{\mathbf{x}^{\mathbf{a}} \in S_J^{G^{\mathrm{an}}}} \lambda_{\mathbf{a}} \mathbf{x}^{\mathbf{a}} \right) \quad = \quad \max_{\mathbf{a}} \left(|\lambda_{\mathbf{a}}| \cdot \ell^{|\mathbf{a}|-|a_1|} \right) .$$

We know that the sequences of seminorms $\{q_\ell\}$ and $\{q'_\ell\}$ must induce the same topology on the ring $S_J^{G^{\mathrm{an}}}$; the topology is intrinsic, it is the topology $\left[S_J^{G^{\mathrm{an}}}, \left\{ \mathrm{Spec}(S_J^{G^{\mathrm{an}}})\right\}^{\mathrm{an}}\right]$. What we want next is to see this explicitly by comparing the seminorms. It turns out to be better to compare both seminorms to a third, to a seminorm that is clearly symmetric in the variables.

Definition 8.9.32. *The seminorm r_ℓ, on the ring $S_J^{G^{\mathrm{an}}}$, is given by the formula*

$$\begin{aligned} r_\ell\left(\sum_{\mathbf{x}^{\mathbf{a}} \in S_J^{G^{\mathrm{an}}}} \lambda_{\mathbf{a}} \mathbf{x}^{\mathbf{a}} \right) &= \max_{\mathbf{a}} \left(|\lambda_{\mathbf{a}}| \cdot \ell^{|\mathbf{a}|} \right) \\ &= \max_{\mathbf{a}} \left(|\lambda_{\mathbf{a}}| \cdot \ell^{|a_0|+|a_1|+\cdots+|a_n|} \right) . \end{aligned}$$

Lemma 8.9.33. *For any ℓ, and any Laurent polynomial $f \in S_J^{G^{\mathrm{an}}}$, we have the inequalities*

$$q_\ell(f) \le r_\ell(f) \le q_{\ell^2}(f) .$$

Proof. The Laurent polynomial $f \in S_J^{G^{\mathrm{an}}}$ is a linear combination of monomials $\mathbf{x}^{\mathbf{a}}$, where the degree $\langle \mathbf{a} \rangle = 0$. That is,

$$a_0 \quad = \quad -a_1 - a_2 - \cdots - a_n .$$

This makes

$$\begin{aligned} |a_0| &\le |a_1| + |a_2| + \cdots + |a_n| \\ &= |\mathbf{a}| - |a_0| . \end{aligned}$$

Hence

$$|\mathbf{a}| \quad = \quad \big(|\mathbf{a}| - |a_0|\big) + |a_0| \quad \le \quad 2\big(|\mathbf{a}| - |a_0|\big) .$$

Now consider the following inequalities, the first of which is obvious

$$|\mathbf{a}| - |a_0| \quad \le \quad |\mathbf{a}| \quad \le \quad 2\big(|\mathbf{a}| - |a_0|\big) .$$

They induce inequalities

$$|\lambda_{\mathbf{a}}| \cdot \ell^{|\mathbf{a}| - |a_0|} \quad \le \quad |\lambda_{\mathbf{a}}| \cdot \ell^{|\mathbf{a}|} \quad \le \quad |\lambda_{\mathbf{a}}| \cdot \ell^{2\left(|\mathbf{a}| - |a_0|\right)} .$$

Taking the maximum over $\mathbf{a}$, we obtain the inequality

$$q_\ell(f) \le r_\ell(f) \le q_{\ell^2}(f) .$$

□

Remark 8.9.34. This means that the sequence of seminorms $\{q_\ell\}$ is equivalent to the sequence of seminorms $\{r_\ell\}$. Either one determines the pre-Fréchet topology $\left[S_J^{G^{\mathrm{an}}}, \left\{\operatorname{Spec}(S_J^{G^{\mathrm{an}}})\right\}^{\mathrm{an}}\right]$. An element in $\Gamma\big(\{\mathbb{P}_J^n\}^{\mathrm{an}}, \mathcal{O}^{\mathrm{an}}\big)$, which may be viewed as an element in the completion of $S_J^{G^{\mathrm{an}}}$ in this topology, can be written as a Laurent series $\sum_{\mathbf{x}^{\mathbf{a}} \in S_J^{G^{\mathrm{an}}}} \lambda_{\mathbf{a}} \mathbf{x}^{\mathbf{a}}$, where

$$r_\ell \left(\sum_{\mathbf{x}^{\mathbf{a}} \in S_J^{G^{\mathrm{an}}}} \lambda_{\mathbf{a}} \mathbf{x}^{\mathbf{a}} \right) \quad = \quad \sup_{\mathbf{a}} \left(|\lambda_{\mathbf{a}}| \cdot \ell^{|\mathbf{a}|} \right)$$

is finite, for every $\ell \ge 1$.

Remark 8.9.35. Suppose we are given two non-empty subsets $J \subset J' \subset \{0, 1, \ldots, n\}$. The formulas, for the seminorms r_ℓ, tell us that the inclusion homomorphism $S_J^{G^{\mathrm{an}}} \longrightarrow S_{J'}^{G^{\mathrm{an}}}$ is an isometry. The fact that the inclusion is continuous is not remarkable; continuity comes from

Lemma 5.6.7. What is not formal, and is special to the seminorms r_ℓ, is that they are independent of J; the map $S_J^{G^{\mathrm{an}}} \longrightarrow S_{J'}^{G^{\mathrm{an}}}$ is not merely continuous, it also embeds $S_J^{G^{\mathrm{an}}}$ isometrically as a subspace of $S_{J'}^{G^{\mathrm{an}}}$.

9

Projective space: the proofs

Chapter 8 developed geometric invariant theory. More accurately it developed the simplified, bare-bones version which suffices for us. In Section 8.8, especially in Theorem 8.8.6, we stated the main results of this simplified version of geometric invariant theory. Now we have to prove these claims. This chapter is devoted to the proofs.

In previous chapters the opening paragraphs consisted of a summary of the main results. In this chapter such a summary would be out of place, for the simple reason that it already occurred in Theorem 8.8.6. As we learned, in Theorem 8.8.6, the main results of this chapter tell us that certain open subsets of the ringed space $(X/G, \mathcal{O}_{X/G})$ are schemes of finite type over $\mathbb{C}$. The results then go on to describe the coherent sheaves on the quotient, and the fact that exact sequences of coherent G–sheaves on X yield exact sequences of coherent sheaves on the open subset $V \subset X/G$.

Since there is little point in repeating the statement of Theorem 8.8.6 in detail, we will spend the remainder of these opening paragraphs telling the reader what else can be found in the chapter. In the course of the proof we do introduce certain ideas which might be worth summarizing.

We spend Sections 9.2 and 9.3 discussing the linear representations of affine group schemes $(G, \mathcal{O}_G)$. Once again this is a subject of tremendous independent interest. Many wonderful books have been written in the field, and what we will say is very minimal. The interested reader is advised to go to the more advanced books for further information. Any book with the words "linear algebraic groups" in the title will deal with this field, and many of these books are truly excellent. What we give in Sections 9.2 and 9.3 is very sketchy, barely covering what is absolutely necessary for our proofs. In Section 9.2 we explain what a linear representation is, and explain why all linear representations are unions of finite dimensional subrepresentations. This part of the theory is true for arbitrary linear algebraic groups. In Section 9.3 we go into much greater detail with the particular algebraic group $(G, \mathcal{O}_G)$ that interests us most, the group whose analytification is $G^{\mathrm{an}} = \mathbb{C}^*$. The representation theory of this group is straightforward. All representations are direct sums of 1–dimensional subrepresentations, and the 1–dimensional subrepresentations are very simple. The only algebraic actions of $\mathbb{C}^*$ on a 1–dimensional vector space V take a pair (λ, x), with $\lambda \in \mathbb{C}^*$ and $x \in V$, to the element $\lambda^n v \in V$. It is not difficult to deduce that, if V is any linear representation, then the invariant subspace $V^{G^{\mathrm{an}}}$, the subspace on which the group $G^{\mathrm{an}} = \mathbb{C}^*$ acts trivially, is naturally a direct summand of V. The projection to this direct summand is called the *Reynolds operator,* and is denoted $E : V \longrightarrow V^{G^{\mathrm{an}}}$.

This ends our treatment of representation theory. We inform the reader, without proof, that the Reynolds operator $E : V \longrightarrow V^{G^{\mathrm{an}}}$ exists for certain other linear algebraic groups, the so-called *reductive groups.* For us this is the definition of reductive groups; see Definition 9.3.11. Making this the definition has the disadvantage that it is not at all clear, *a priori,* that there are any non-trivial examples other than the one we exhibited. In the remainder of the chapter we use the Reynolds operator extensively. The statements of the theorems all hold either for arbitrary reductive groups, or occasionally we need to assume the groups connected. We state the theorems in this generality but, to the extent that our proofs rely on the existence of the Reynolds operator (which they do, very heavily), it is not clear that they apply to any examples beyond the special $(G, \mathcal{O}_G)$ we really care about, the trivial group and the group with $G^{\mathrm{an}} = \mathbb{C}^*$.

The chapter is about proving Theorem 8.8.6, and Theorem 8.8.6 asserts that certain open subsets of X/G are schemes of finite type over $\mathbb{C}$. This means that some open subsets of X/G are of the form $\left(\mathrm{Spec}(T), \widetilde{T}\right)$, where T is a finitely generated $\mathbb{C}$–algebra. At some point in this chapter we will have to construct finitely generated $\mathbb{C}$–algebras. This point comes in Section 9.4. In Section 9.4 we prove a celebrated theorem of Hilbert's, which says that, if S is a finitely generated $\mathbb{C}$–algebra and $(G, \mathcal{O}_G)$ is a reductive group acting on S, then the subring of invariants $S^{G^{\mathrm{an}}} \subset S$ is also a finitely generated $\mathbb{C}$–algebra. The proof is simple and beautiful, and we include it in Section 9.4. Of course to the extent that the proof relies on the Reynolds operator the reader has a choice: either she specializes to the case $G^{\mathrm{an}} = \mathbb{C}^*$, where we proved the existence, or else she believes us that there are other interesting examples.

The following Sections, 9.5 and 9.6, use the Reynolds operator to prove Theorem 8.8.6(i), (ii) and (iii). These are the straightforward, local parts. Once we have proved these facts we know that the open set $V \subset X/G$ is a scheme of finite type over $\mathbb{C}$, that the sheaves $\{\pi_*\widetilde{M}\}^{G^{\mathrm{an}}}\big|_V$ are coherent on V, and that exact sequences of coherent G–sheaves on X descend to exact sequences of coherent sheaves on V. The only remaining statement is Theorem 8.8.6(iv), which asserts that every coherent sheaf on $V \subset X/G$ is of the form $\{\pi_*\widetilde{M}\}^{G^{\mathrm{an}}}\big|_V$.

Then come the technical Sections 9.7 and 9.8, which are devoted to the proof of Theorem 8.8.6(iv). In a first reading of the book the reader might do well to skip Sections 9.7 and 9.8.

The final Section 9.9 discusses the case where the group $(G, \mathcal{O}_G)$ is trivial. Parts (i), (ii) and (iii) of Theorem 8.8.6 are content-free when $(G, \mathcal{O}_G)$ is the trivial group, but part (iv) is not. It gives us valuable information about exact sequences of coherent sheaves, and about extending coherent sheaves from open subsets of a scheme.

9.1 A reminder of symmetric powers

Let V be a finite dimensional vector space over $\mathbb{C}$. We can form the tensor power

$$V^{\otimes r} \quad = \quad \underbrace{V \otimes_{\mathbb{C}} V \otimes_{\mathbb{C}} \cdots \otimes_{\mathbb{C}} V}_{r \text{ times}},$$

that is the tensor product of r copies of V. The symmetric group Σ_r acts on $V^{\otimes r}$, by permuting the factors. We recall that the symmetric power $\mathrm{Sym}^r V$ is defined to be the quotient $V^{\otimes r}/\Sigma_r$. That is, $\mathrm{Sym}^r V$ is the quotient vector space $V^{\otimes r}/W$, where W is the subspace of $V^{\otimes r}$ spanned by all vectors $y - g(y)$, with $y \in V^{\otimes r}$ and $g \in \Sigma_r$.

It might help to do a simple example. Assume V is 2–dimensional, and let x, y be a basis of V. Then one basis for $V^{\otimes 3}$ consists of the vectors

$$x{\otimes}x{\otimes}x,\ x{\otimes}x{\otimes}y,\ x{\otimes}y{\otimes}x,\ x{\otimes}y{\otimes}y,\ y{\otimes}x{\otimes}x,\ y{\otimes}x{\otimes}y,\ y{\otimes}y{\otimes}x,\ y{\otimes}y{\otimes}y.$$

Dividing by the action of the symmetric group Σ_3 has the effect of identifying the three elements

$$x \otimes x \otimes y, \quad x \otimes y \otimes x \quad \text{and} \quad y \otimes x \otimes x,$$

and it also identifies

$$x \otimes y \otimes y, \quad y \otimes x \otimes y \quad \text{and} \quad y \otimes y \otimes x.$$

The usual way to write the basis for $\mathrm{Sym}^3 V$ is as

$$x^3, \quad x^2y, \quad xy^2 \quad \text{and} \quad y^3 .$$

The vector space $\mathrm{Sym}^3 V$ is simply the vector space of homogeneous polynomials in x and y of degree 3. More generally, $\mathrm{Sym}^r V$ is the vector space of homogeneous polynomials in x and y of degree r. If we sum over all degrees, the vector space

$$\mathbb{C}[V] \quad = \quad \sum_{i=0}^{\infty} \mathrm{Sym}^i V$$

is the polynomial ring in the two variables x and y. The point of writing it as $\mathbb{C}[V]$ is that there is no choice of basis involved; the construction is basis free.

In the above we discussed the special case where the dimension of V is 2, but the argument easily generalizes to all dimensions.

9.2 Generators

We want to prove Theorem 8.8.6, and in Remark 8.8.7 we reduced it to a local statement. In the past we have often used generators to prove local statements. Given a finitely generated $\mathbb{C}$–algebra S we have frequently chosen generators for S as an algebra over $\mathbb{C}$. If we choose generators $\{b_1, b_2, \ldots, b_k\} \subset S$ they give a surjective $\mathbb{C}$–algebra homomorphism from the polynomial ring $A = \mathbb{C}[x_1, x_2, \ldots, x_k]$ to S, this induces an embedding $\left(\mathrm{Spec}(S), \widetilde{S}\right) \longrightarrow \left(\mathrm{Spec}(A), \widetilde{A}\right)$, and we have often used this embedding. The point has always been that $\left(\mathrm{Spec}(A), \widetilde{A}\right)$ is a very simple, explicit ringed space. The embedding permits us to prove facts about $\left(\mathrm{Spec}(S), \widetilde{S}\right)$ by reducing them to statements about $\left(\mathrm{Spec}(A), \widetilde{A}\right)$. Now that we are permitting an affine group scheme $(G, \mathcal{O}_G)$ to act, it helps to work out what the right notion of generators ought to be. In this section we will do this. The first observation is that it is possible to make the construction of the homomorphism $\theta : A \longrightarrow S$ slightly more coordinate-free as follows.

Observation 9.2.1. Let S be a finitely generated $\mathbb{C}$–algebra. Given any finite dimensional $\mathbb{C}$–vector space V, as well as a linear map $f : V \longrightarrow S$, there is a unique way to extend f to a $\mathbb{C}$–algebra homomorphism $\varphi : \mathbb{C}[V] \longrightarrow S$. Here $\mathbb{C}[V]$ stands for the symmetric algebra

$$\mathbb{C}[V] \quad = \quad \sum_{i=0}^{\infty} \mathrm{Sym}^i V \ ,$$

where $\mathrm{Sym}^i V$ is the ith symmetric power of V, as in Section 9.1.

Note that this is nothing other than the old map from the polynomial ring to S. If we choose a basis, $\{x_1, x_2, \ldots, x_k\} \subset V$, then the linear map $f : V \longrightarrow S$ is entirely determined by the elements $f(x_1), f(x_2), \ldots, f(x_k) \in S$. And of course there is a unique ring homomorphism $\varphi : \mathbb{C}[x_1, x_2, \ldots, x_k] \longrightarrow S$ with $\varphi(x_i) = f(x_i)$. The ring $\mathbb{C}[V]$ is simply a basis-free way of writing the polynomial ring $\mathbb{C}[x_1, x_2, \ldots, x_k]$.

The next question is what does it mean for the polynomial ring $\mathbb{C}[V]$ to be acted on by an affine group scheme $(G, \mathcal{O}_G)$ of finite type over $\mathbb{C}$. There are many actions, but the simplest are the ones that arise from linear actions on V. In case the reader has not seen this elsewhere, we recall what is meant by a linear representation of the group $(G, \mathcal{O}_G)$:

Reminder 9.2.2. Let $(G, \mathcal{O}_G)$ be an affine group scheme of finite type over $\mathbb{C}$. More explicitly suppose $(G, \mathcal{O}_G) = \left(\mathrm{Spec}(R), \widetilde{R}\right)$. A *linear representation* of $(G, \mathcal{O}_G)$ is a $\mathbb{C}$–vector space V, and a map

$$a' : V \longrightarrow R \otimes_{\mathbb{C}} V ,$$

for which

(i) The following square commutes

$$\begin{array}{ccc} V & \xrightarrow{\ a'\ } & R \otimes_{\mathbb{C}} V \\ {\scriptstyle a'}\big\downarrow & & \big\downarrow{\scriptstyle 1 \otimes a'} \\ R \otimes_{\mathbb{C}} V & \xrightarrow[\mu' \otimes 1]{} & R \otimes_{\mathbb{C}} R \otimes_{\mathbb{C}} V \end{array} .$$

(ii) The following triangle commutes

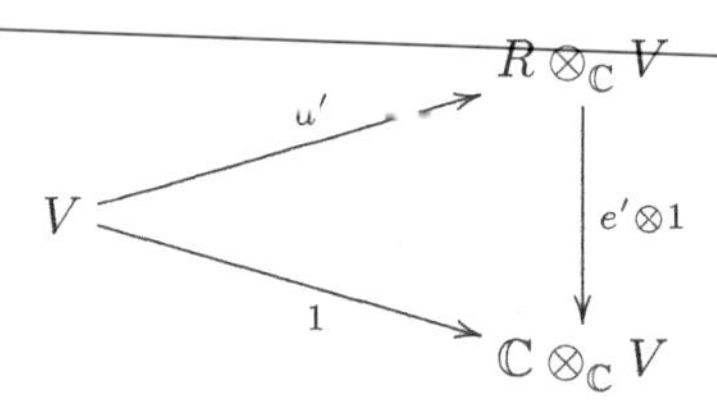

A G–homomorphism $\varphi : V \longrightarrow W$ of linear representations of $(G, \mathcal{O}_G)$ is a linear map $\varphi : V \longrightarrow W$, for which the following square commutes

$$\begin{array}{ccc} V & \xrightarrow{\ a'_V\ } & R \otimes_{\mathbb{C}} V \\ {\scriptstyle \varphi}\big\downarrow & & \big\downarrow{\scriptstyle 1 \otimes \varphi} \\ W & \xrightarrow[a'_W]{} & R \otimes_{\mathbb{C}} W \end{array} .$$

Example 9.2.3. If S is a G–ring, or if M is a G–module over S, then, among other things, they are linear representations of $(G, \mathcal{O}_G)$; see Definition 8.3.1 and Remark 8.3.16.

Remark 9.2.4. If V is a linear representation of $(G, \mathcal{O}_G)$ then the group G^{an} acts on V. The element $g \in G^{\mathrm{an}}$ acts by the linear map

$$V \xrightarrow{\ a'\ } R \otimes_{\mathbb{C}} V \xrightarrow{\ \varphi_{g^{-1}} \otimes 1\ } \mathbb{C} \otimes_{\mathbb{C}} V \quad = \quad V\,.$$

We leave it to the reader to check that this defines a homomorphism from the group G^{an} to the automorphism group $\mathrm{Aut}(V)$. Furthermore if V and W are linear representations, and $\varphi : V \longrightarrow W$ is a G–homomorphism, then φ intertwines the G^{an}–action.

Remark 9.2.5. Note that Remark 9.2.4 is compatible with the previous definitions of actions of G^{an}. Let S be a G-ring; then, for any G–invariant open set $U \subset \mathrm{Spec}(S)$, there is an action of G^{an} on $\Gamma(U, \widetilde{S})$. In the special case, where $U = \mathrm{Spec}(S)$, this gives an action of the group G^{an} on S, and in Remark 8.4.6 we worked out explicitly what it is. It agrees with the formula of Remark 9.2.4. Similarly, if M is a G–module over the G–ring S, and U is a G–invariant open subset of $\mathrm{Spec}(S)$, then G^{an} acts on $\Gamma(U, \widetilde{M})$. Again, the special case when $U = \mathrm{Spec}(S)$ gives an action of G^{an} on M. In Remark 8.4.11 we worked out this version, and the action is also the same as Remark 9.2.4.

Remark 9.2.6. Let V be a finite dimensional linear representation of $(G, \mathcal{O}_G)$. We assert that the polynomial ring $\mathbb{C}[V]$ has an obvious G–ring structure. This structure comes about as follows. We have a linear map

$$V \xrightarrow{\ a'_V\ } R \otimes_{\mathbb{C}} V \xrightarrow{\ 1 \otimes i\ } R \otimes_{\mathbb{C}} \mathbb{C}[V]\,,$$

where $i : V \longrightarrow \mathbb{C}[V]$ is the natural inclusion. By Observation 9.2.1 the linear map $V \longrightarrow R \otimes_{\mathbb{C}} \mathbb{C}[V]$, from V to the $\mathbb{C}$–algebra $R \otimes_{\mathbb{C}} \mathbb{C}[V]$, has a unique factorization through a $\mathbb{C}$–algebra homomorphism $a'_{\mathbb{C}[V]} : \mathbb{C}[V] \longrightarrow R \otimes_{\mathbb{C}} \mathbb{C}[V]$. In other words there is a unique $\mathbb{C}$–algebra homomorphism $a'_{\mathbb{C}[V]}$ rendering commutative the square

$$\begin{array}{ccc} V & \xrightarrow{\ a'_V\ } & R \otimes_{\mathbb{C}} V \\ {\scriptstyle i}\downarrow & & \downarrow{\scriptstyle 1 \otimes i} \\ \mathbb{C}[V] & \xrightarrow[\ a'_{\mathbb{C}[V]}\ ]{} & R \otimes_{\mathbb{C}} \mathbb{C}[V] \end{array}\,.$$

Exercise 9.2.7. Let the notation be as in Remark 9.2.6. We leave it to the reader to check that the two diagrams of $\mathbb{C}$–algebra homomorphisms, of Definition 8.3.1(i) and (ii), both commute; hence $\mathbb{C}[V]$ really is a G–ring.

Next we note an easy fact.

Lemma 9.2.8. *Suppose S is a G–ring, V is a finite dimensional linear representation of $(G, \mathcal{O}_G)$ and $f : V \longrightarrow S$ is a homomorphism of linear representations. Then the unique $\mathbb{C}$–algebra homomorphism $\varphi : \mathbb{C}[V] \longrightarrow S$ of Observation 9.2.1, through which $f : V \longrightarrow S$ factors, is a G–homomorphism of G–rings.*

Proof. Consider the diagram

$$\begin{array}{ccccc} V & \xrightarrow{\ i\ } & \mathbb{C}[V] & \xrightarrow{a'_{\mathbb{C}[V]}} & R \otimes_{\mathbb{C}} \mathbb{C}[V] \\ & & \varphi \big\downarrow & & \big\downarrow 1\otimes\varphi \\ & & S & \xrightarrow[a'_S]{} & R \otimes_{\mathbb{C}} S \end{array} .$$

The two composites $V \longrightarrow R \otimes_{\mathbb{C}} S$ agree, and the uniqueness, of the factorization through $\mathbb{C}[V]$, tells us that the square of $\mathbb{C}$–algebra homomorphisms must commute. □

Remark 9.2.9. Given any finitely generated $\mathbb{C}$–algebra T, as well as elements $\{b_1, b_2, \ldots, b_k\} \subset T$, there is an induced $\mathbb{C}$–algebra homomorphism $\theta : \mathbb{C}[x_1, x_2, \ldots, x_k] \longrightarrow T$. We have just learned the G–ring analogue: given any homomorphism of linear representations $f : V \longrightarrow S$, where V is a finite dimensional linear representation of $(G, \mathcal{O}_G)$ and S is a G–ring, Lemma 9.2.8 establishes that there is an induced G–homomorphism of G–rings $\varphi : \mathbb{C}[V] \longrightarrow S$. When there is no group acting we know how to make the homomorphism $\theta : \mathbb{C}[x_1, x_2, \ldots, x_k] \longrightarrow T$ surjective; all we need to do is make sure that $\{b_1, b_2, \ldots, b_k\} \subset T$ generate T. The main result of this section will be that, in the G–ring context, it is also possible to guarantee the surjectivity of $\varphi : \mathbb{C}[V] \longrightarrow S$.

Lemma 9.2.10. *Let $(G, \mathcal{O}_G)$ be an affine group scheme of finite type over $\mathbb{C}$ and let W be a linear representation of $(G, \mathcal{O}_G)$. We assert that any finite dimensional $\mathbb{C}$–vector subspace $V \subset W$ is contained in a finite dimensional G^{an}–invariant subspace $V' \subset W$.*

Proof. Let $(G, \mathcal{O}_G) = \left(\operatorname{Spec}(R), \widetilde{R}\right)$. The fact that W is a linear representation of $(G, \mathcal{O}_G)$ means that there is a homomorphism $a' : W \longrightarrow R \otimes_{\mathbb{C}} W$ satisfying the conditions of Reminder 9.2.2(i) and (ii). The way a $g \in G^{\mathrm{an}}$ acts on the representation W was given in Remark 9.2.4; g acts by the linear map

$$W \xrightarrow{a'} R \otimes_{\mathbb{C}} W \xrightarrow{\varphi_{g^{-1}} \otimes 1} \mathbb{C} \otimes_{\mathbb{C}} W = W .$$

Now let V be a finite dimensional $\mathbb{C}$–vector subspace of W, and let $v_1, v_2, \ldots, v_k$ be a basis for V. We can choose elements $w_1, w_2, \ldots, w_\ell$ in W so that there are formulas

$$a'(v_i) = \sum_{j=1}^{\ell} r_{ij} \otimes w_j ,$$

with $r_{ij} \in R$. Let $V'' \subset W$ be the vector space spanned by $w_1, w_2, \ldots, w_\ell$. For each $g \in G^{\mathrm{an}}$ the action of g on v_i is given, using the above, by the formula

$$gv_i = \sum_{j=1}^{\ell} \varphi_{g^{-1}}(r_{ij}) \cdot w_j .$$

Clearly this means that gv_i is a linear combination of $w_1, w_2, \ldots, w_\ell$. We conclude that $gv_i \in V''$ for all $g \in G^{\mathrm{an}}$ and $1 \le i \le k$.

Now let V' be the vector subspace of V'' spanned by all gv_i, with $g \in G^{\mathrm{an}}$ and $1 \le i \le k$. It is clear that V' is invariant under the action of G^{an}, that it contains V, and that it is finite dimensional (being a subspace of V''). □

Next we observe the little lemma:

Lemma 9.2.11. *Let $(G, \mathcal{O}_G) = \left(\mathrm{Spec}(R), \widetilde{R}\right)$ be an affine group scheme of finite type over $\mathbb{C}$. Let V be a vector space, let y be an element of $R \otimes_{\mathbb{C}} V$, and assume that y lies in the kernel of the map*

$$\varphi_g \otimes 1 : R \otimes_{\mathbb{C}} V \longrightarrow \mathbb{C} \otimes_{\mathbb{C}} V$$

for every $g \in G^{\mathrm{an}}$. Then $y = 0$.

Proof. This is another argument that is much simpler in the two cases $R = \mathbb{C}$ or $R = \mathbb{C}[t, t^{-1}]$ than in general. In the case $R = \mathbb{C}$ the statement is trivial and is left to the reader. The way we will proceed for non-trivial G is as follows. We give the beginning of the proof, which is no simpler in the special case. Then we will explain what needs to be done in general, and show that it is very easy for $R = \mathbb{C}[t, t^{-1}]$. And finally, in an aside to the experts, we will explain how to treat the general case.

The element y can be expressed as $y = \sum r_i \otimes v_i$, with $r_i \in R$ and v_i linearly independent elements of V. We know that y is killed by $\varphi_g \otimes 1$; this means that $\sum \varphi_g(r_i) \cdot v_i$ vanishes in the vector space V. Since the

v_i are linearly independent it follows that $\varphi_g(r_i) = 0$ for all $g \in G^{\text{an}}$ and all r_i. What we are therefore reduced to is showing that, if r is an element of R and $\varphi_g(r) = 0$ for all $g \in G^{\text{an}}$, then $r = 0$.

In the special case where $R = \mathbb{C}[t, t^{-1}]$ this is obvious. The element r is a Laurent polynomial $r = r(t, t^{-1})$ vanishing at all points $g \in \mathbb{C}^*$. It clearly must vanish. □

Aside for the Experts 9.2.12. For a general affine group scheme $(G, \mathcal{O}_G) = \left(\text{Spec}(R), \widetilde{R}\right)$, of finite type over $\mathbb{C}$, we note that the ring R must be reduced by [5, page 101, Theorem]. If we have an element $r \in R$, with $\varphi_g(r) = 0$ for all $g \in G^{\text{an}}$, then r must lie in all the maximal ideals of R. The strong Nullstellensatz (see Reminder 3.8.1(ii)) tells us that r must vanish.

Next we give a couple of corollaries of Lemma 9.2.11.

Corollary 9.2.13. *Let $(G, \mathcal{O}_G) = \left(\text{Spec}(R), \widetilde{R}\right)$ be an affine group scheme of finite type over $\mathbb{C}$. Let V be a linear representation of $(G, \mathcal{O}_G)$. An element $v \in V$ belongs to $V^{G^{\text{an}}} \subset V$ if and only if $a'(v) = 1 \otimes v$. That is, v is fixed by the action of G^{an} if and only if $a'(v) = 1 \otimes v$.*

Proof. Consider the element $w = 1 \otimes v - a'(v) \in R \otimes_{\mathbb{C}} V$. The vector v is fixed by the group G^{an} if and only if, for every $g \in G^{\text{an}}$, the map

$$\varphi_{g^{-1}} \otimes 1 : R \otimes_{\mathbb{C}} V \longrightarrow \mathbb{C} \otimes_{\mathbb{C}} V$$

takes w to zero. By Lemma 9.2.11 this is equivalent to the vanishing of $w = 1 \otimes v - a'(v)$. □

A slightly less immediate corollary of Lemma 9.2.11 asserts

Corollary 9.2.14. *Let $(G, \mathcal{O}_G) = \left(\text{Spec}(R), \widetilde{R}\right)$ be an affine group scheme of finite type over $\mathbb{C}$. Let M be a linear representation of $(G, \mathcal{O}_G)$, and let $V \subset M$ be a subspace invariant under the action of G^{an} on M. Let $i : V \longrightarrow M$ be the inclusion. Then the composite*

$$V \xrightarrow{\ i\ } M \xrightarrow{\ a'_M\ } R \otimes_{\mathbb{C}} M$$

takes V to $R \otimes_{\mathbb{C}} V \subset R \otimes_{\mathbb{C}} M$; that is, there is a unique factorization

$$\begin{array}{ccc} V & \xrightarrow{\ a'_V\ } & R \otimes_{\mathbb{C}} V \\ {\scriptstyle i}\downarrow & & \downarrow{\scriptstyle 1 \otimes i} \\ M & \xrightarrow[\ a'_M\ ]{} & R \otimes_{\mathbb{C}} M \end{array} .$$

Furthermore, this map $a'_V : V \longrightarrow R \otimes_{\mathbb{C}} V$ gives V the structure of a linear representation of $(G, \mathcal{O}_G)$, making the map i a homomorphism of linear representations.

Proof. Let g be a closed point in G; that is $g \in G^{\mathrm{an}}$. Let $\varphi_g : R \longrightarrow \mathbb{C}$ be the corresponding homomorphism. Let $p : M \longrightarrow M/V$ be the projection to the quotient vector space. Consider the commutative diagram

$$\begin{array}{ccccccc} V & \xrightarrow{i} & M & \xrightarrow{a'} & R \otimes_{\mathbb{C}} M & \xrightarrow{1\otimes p} & R \otimes_{\mathbb{C}} \{M/V\} \\ & & & & \Big\downarrow{\scriptstyle \varphi_g\otimes 1} & & \Big\downarrow{\scriptstyle \varphi_g\otimes 1} \\ & & & & \mathbb{C} \otimes_{\mathbb{C}} M & \xrightarrow[1\otimes p]{} & \mathbb{C} \otimes_{\mathbb{C}} \{M/V\} \end{array} .$$

The composite

$$\begin{array}{ccccc} V & \xrightarrow{i} & M & \xrightarrow{a'} & R \otimes_{\mathbb{C}} M \\ & & & & \Big\downarrow{\scriptstyle \varphi_g\otimes 1} \\ & & & & \mathbb{C} \otimes_{\mathbb{C}} M \end{array}$$

takes $v \in V$ to $g^{-1}v \in M$; by the hypothesis of the corollary this lies in $V \subset M$. Hence the composite

$$\begin{array}{ccccccc} V & \longrightarrow & M & \xrightarrow{a'} & R \otimes_{\mathbb{C}} M & & \\ & & & & \Big\downarrow{\scriptstyle \varphi_g\otimes 1} & & \\ & & & & \mathbb{C} \otimes_{\mathbb{C}} M & \xrightarrow[1\otimes p]{} & \mathbb{C} \otimes_{\mathbb{C}} \{M/V\} \end{array}$$

must vanish. It follows that the composite

$$V \xrightarrow{i} M \xrightarrow{a'} R \otimes_{\mathbb{C}} M \xrightarrow{1\otimes p} R \otimes_{\mathbb{C}} \{M/V\}$$

takes the element $v \in V$ to the element $\{1 \otimes p\}a'i(v) \in R \otimes_{\mathbb{C}} \{M/V\}$, which is annihilated by all maps $\varphi_g : R \longrightarrow \mathbb{C}$. By Lemma 9.2.11 we conclude that $\{1 \otimes p\}a'i(v)$ vanishes. Hence the composite

$$V \xrightarrow{i} M \xrightarrow{a'} R \otimes_{\mathbb{C}} M$$

factors through $R \otimes_{\mathbb{C}} V \subset R \otimes_{\mathbb{C}} M$. We deduce a commutative diagram

$$\begin{array}{ccc} V & \xrightarrow{a'_V} & R \otimes_{\mathbb{C}} V \\ {\scriptstyle i}\Big\downarrow & & \Big\downarrow{\scriptstyle 1\otimes i} \\ M & \xrightarrow[a'_M]{} & R \otimes_{\mathbb{C}} M \end{array} .$$

The fact that the map $a'_V : V \longrightarrow R \otimes_{\mathbb{C}} V$ makes V into a linear representation, of the group $(G, \mathcal{O}_G)$, is the assertion that the two diagrams of Reminder 9.2.2(i) and (ii) commute. This follows easily from the fact that M is a representation; we leave it to the reader. The fact that $i : V \longrightarrow M$ is a G–homomorphism of linear representations is immediate from the commutativity of the square above. □

Now we state

Proposition 9.2.15. *Let $(G, \mathcal{O}_G)$ be an affine group scheme of finite type over $\mathbb{C}$, and let W be a linear representation of $(G, \mathcal{O}_G)$. Let V be a finite dimensional vector subspace $V \subset W$. There exists a finite dimensional subspace $V' \subset W$, containing V, and such that*

(i) *V' is a linear representation of $(G, \mathcal{O}_G)$.*
(ii) *The inclusion $i : V' \longrightarrow W$ is a G–homomorphism of linear representations.*

Proof. By Lemma 9.2.10 we know that V is contained in a finite dimensional subspace V' invariant under the action of G^{an}. Corollary 9.2.14 permits us to deduce that the inclusion $i : V' \longrightarrow W$ is an inclusion of linear representations of $(G, \mathcal{O}_G)$. □

One of the important consequences is

Corollary 9.2.16. *Let $(G, \mathcal{O}_G)$ be an affine group scheme, of finite type over $\mathbb{C}$, and let S be a G–ring. There exists a finite dimensional linear representation V of $(G, \mathcal{O}_G)$, and a surjective G–homomorphism of G–rings $\varphi : \mathbb{C}[V] \longrightarrow S$.*

Proof. The ring S is a finitely generated $\mathbb{C}$–algebra (all G–rings are assumed to be). Choose generators, and let $V \subset S$ be the vector space spanned by them. By Proposition 9.2.15 the vector space V is contained in a larger, finite dimensional vector space $V' \subset S$, so that the inclusion $V' \longrightarrow S$ is a homomorphism of linear representations of $(G, \mathcal{O}_G)$. That is, the following square commutes

$$\begin{array}{ccc} V' & \xrightarrow{\ a'_{V'}\ } & R \otimes_{\mathbb{C}} V' \\ \downarrow & & \downarrow \\ S & \xrightarrow[\ a'_S\ ]{} & R \otimes_{\mathbb{C}} S \end{array} .$$

Lemma 9.2.8 tells us that the natural homomorphism $\varphi : \mathbb{C}[V'] \longrightarrow S$ is a G–homomorphism of G–rings. Since V' contains V, and V generates S, we conclude that φ must be surjective. □

Remark 9.2.17. We will make heavy use of Corollary 9.2.16. But this is not the only way in which Proposition 9.2.15 will enter our argument. Perhaps equally important is the following observation. Every vector space is the union of its finite dimensional subspaces. In particular every linear representation W of $(G, \mathcal{O}_G)$ is the union of its finite dimensional subspaces. Proposition 9.2.15 tells us that all finite dimensional subspaces of W are contained in finite dimensional $(G, \mathcal{O}_G)$ subrepresentations. Therefore W is the union of its finite dimensional subrepresentations.

So far in this section we have not mentioned modules at all. We observe

Corollary 9.2.18. *Let $(G, \mathcal{O}_G)$ be an affine group scheme, of finite type over $\mathbb{C}$, and let S be a G–ring. Suppose M is a G–module for the G–ring S. Then there exists a finite dimensional linear representation V of $(G, \mathcal{O}_G)$, and a homomorphism of $(G, \mathcal{O}_G)$–representations $V \longrightarrow M$ whose image generates M as an S–module.*

Proof. The ring $S \oplus M$ is a G–ring, and by Corollary 9.2.16 it admits a surjective G–homomorphism of G–rings $\varphi : \mathbb{C}[V] \longrightarrow S \oplus M$. We can consider the composite

$$V \longrightarrow \mathbb{C}[V] \longrightarrow S \oplus M \longrightarrow M ;$$

it is a G–homomorphism of linear representations. The reader should check:

Exercise 9.2.19. The image of V, under the composite $V \longrightarrow M$ above, generates M as an S–module. □

Exercise 9.2.20. Let $(G, \mathcal{O}_G)$ be an affine group scheme, of finite type over $\mathbb{C}$. In this exercise we will speak about modules not necessarily finite over S; the reader is referred to Remark 8.3.16 for a discussion.

The reader should prove the following:

(i) Let V be a linear representation of $(G, \mathcal{O}_G)$, and let S be a G–ring. Consider the S–module $S \otimes_{\mathbb{C}} V$. We have a linear map

$$\begin{array}{ccc} S \otimes_{\mathbb{C}} V \xrightarrow{a'_S \otimes a'_V} \{R \otimes_{\mathbb{C}} S\} \otimes_{\mathbb{C}} \{R \otimes_{\mathbb{C}} V\} & & \\ \| & & \\ \{R \otimes_{\mathbb{C}} R\} \otimes_{\mathbb{C}} \{S \otimes_{\mathbb{C}} V\} & \xrightarrow[m \otimes 1]{} & R \otimes_{\mathbb{C}} \{S \otimes_{\mathbb{C}} V\} , \end{array}$$

where $m : R \otimes_{\mathbb{C}} R \longrightarrow R$ is the multiplication in the ring R. Prove that this map, which we will call

$$a'_{S\otimes_{\mathbb{C}}V} : S \otimes_{\mathbb{C}} V \longrightarrow R \otimes_{\mathbb{C}} \{S \otimes_{\mathbb{C}} V\} ,$$

gives $S \otimes_{\mathbb{C}} V$ the structure of a large G–module over the G–ring S.

(ii) Given any large G–module M, over the G–ring S, and a G–homomorphism of G–representations $\eta : V \longrightarrow M$, there is always the natural map of S–modules $S \otimes_{\mathbb{C}} V \longrightarrow M$. Prove that it is a G–homomorphism of large G–modules, where $S \otimes_{\mathbb{C}} V$ is made into a large G–module as in (i).

Remark 9.2.21. Combining Corollary 9.2.18 and Exercise 9.2.20 we learn the following. Let M be a G–module for the G–ring S. This time we mean a G–module, not a large G–module; M is finitely generated over S. There exists a finite dimensional representation V and a surjective G–homomorphism, of G–modules over the G–ring S,

$$S \otimes_{\mathbb{C}} V \longrightarrow M .$$

9.3 Finite dimensional representations of $\mathbb{C}^*$

In Remark 9.2.17 we learned that every linear representation of $(G, \mathcal{O}_G)$ is a union of finite dimensional linear subrepresentations. In particular this is true for all G–rings and all G–modules. It is therefore interesting to understand more closely the finite dimensional linear representations of $(G, \mathcal{O}_G)$. In this section we will study them in the special case where $(G, \mathcal{O}_G) = \left(\mathrm{Spec}(R), \widetilde{R}\right)$ for our favorite Hopf algebra $R = \mathbb{C}[t, t^{-1}]$. Here the theory is particularly simple. We begin with

Lemma 9.3.1. *Let V be a finite dimensional representation of the group $\mathbb{C}^*$. Consider the subgroup $H \subset \mathbb{C}^*$ of all roots of unity; that is*

$$H = \{\zeta \in \mathbb{C}^* \mid \exists n \in \mathbb{N} \text{ with } \zeta^n = 1\} .$$

Then there is a basis $\{v_1, v_2, \dots, v_n\} \subset V$ of H–eigenvectors. That is, for all $h \in H$ and all $1 \le i \le n$ we have that hv_i is a multiple of v_i.

Proof. Let $\rho : H \longrightarrow \mathrm{Aut}(V)$ be the homomorphism from the group H to the automorphism group of V. Any $h \in H$ is a root of unity; it satisfies $h^n = 1$ for some $n \in \mathbb{N}$. Hence $\rho(h)$ also satisfies $\rho(h)^n = 1$. Thus $\rho(h)$ is an $n \times n$ matrix satisfying the equation $x^n - 1 = 0$, and the polynomial $x^n - 1$ is square-free. Linear algebra tells us that the matrix $\rho(h)$ is diagonalizable.

For every $h \in H$ we have a diagonalizable matrix $\rho(h)$, and these matrices all commute. So what the lemma comes down to is the old result that any set of commuting, diagonalizable matrices, acting on a finite dimensional vector space, is simultaneously diagonalizable.

Let us briefly remind the reader of the proof. We prove this by induction on the dimension of V. If the dimension is one there is nothing to prove. Assume we know the result for vector spaces V of dimension $< n$, and we have a vector space V of dimension n, and on it a set H of commuting, diagonalizable matrices. If all the matrices in H are multiples of the identity there is nothing to prove. Assume therefore that there is an $h \in H$ which is not a multiple of the identity. Let $\{\lambda_1, \lambda_2, \ldots, \lambda_k\}$ be the eigenvalues of h. Then $V = V_1 \oplus V_2 \oplus \cdots \oplus V_k$, where h acts on V_i with the eigenvalue λ_i. But the matrices in H commute and hence, for any $h' \in H$ and any $v \in V_i$, we have

$$hh'v = h'hv = h'\lambda_i v = \lambda_i h'v,$$

that is $h'V_i \subset V_i$. By induction on the dimension we know that, on each of the spaces V_i, the matrices are simultaneously diagonalizable. □

Lemma 9.3.2. *Let V be a finite dimensional linear representation of the group $(G, \mathcal{O}_G) = \left(\text{Spec}(R), \widetilde{R}\right)$, with $R = \mathbb{C}[t, t^{-1}]$. Then V has a basis $\{v_1, v_2, \ldots, v_n\} \subset V$ so that the map $a' : V \longrightarrow R \otimes_{\mathbb{C}} V$ takes each v_i to $a'(v_i) = r_i \otimes v_i$, with $r_i \in R$.*

Proof. By Lemma 9.3.1 we can choose a basis $\{v_1, v_2, \ldots, v_n\} \subset V$ of H–eigenvectors, where $H \subset \mathbb{C}^*$ is the group of roots of unity. Now for each v_i in the basis we have a formula

$$a'(v_i) \quad = \quad \sum_{j=1}^{n} r_{ij} \otimes v_j \; .$$

The fact that v_i is an h–eigenvector, for any $h \in H$, says that

$$hv_i \quad = \quad \sum_{j=1}^{n} \varphi_{h^{-1}}(r_{ij}) \cdot v_j$$

is a multiple of v_i. In other words $\varphi_{h^{-1}}(r_{ij}) = 0$ unless $i = j$. Whenever $i \neq j$, the Laurent polynomial $r_{ij} = r_{ij}(t, t^{-1}) \in \mathbb{C}[t, t^{-1}]$ vanishes at the infinitely many points $h \in H$, and hence must vanish. That is, the formula for $a'(v_i)$ reduces to $a'(v_i) = r_{ii} \otimes v_i$. □

Remark 9.3.3. Lemma 9.3.2 establishes that any finite dimensional linear representation of $(G, \mathcal{O}_G)$, for the special case where $R = \mathbb{C}[t, t^{-1}]$, is a direct sum of 1–dimensional subrepresentations. Next we observe what the 1–dimensional representations are.

Lemma 9.3.4. *Suppose V is a 1–dimensional linear representation of $(G, \mathcal{O}_G)$, with $R = \mathbb{C}[t,t^{-1}]$ as above. Then there exists an integer $n \in \mathbb{Z}$ so that, for all $v \in V$, the map $a' : V \longrightarrow R \otimes_{\mathbb{C}} V$ takes v to*

$$a'(v) = t^n \otimes v\,.$$

Proof. Choose a non-zero $v \in V$. Because V is 1–dimensional we know that $a'(v) = r \otimes v$ for some $r \in R = \mathbb{C}[t,t^{-1}]$. But we know further that, for each element $g \in \mathbb{C}^*$, $g^{-1}v = \varphi_g(r) \cdot v$ is non-zero; the group $\mathbb{C}^*$ acts by automorphisms. Hence the polynomial $r = r(t,t^{-1})$ does not vanish at any point in $\mathbb{C}^*$. Now write

$$r(t,t^{-1}) \quad = \quad t^n(a_0 + a_1 t + \cdots + a_m t^m)\,,$$

with $a_0 \neq 0$. If $m > 0$ the polynomial $a_0 + a_1 t + \cdots + a_m t^m$ will have to have a non-zero root. Therefore we must have $m = 0$; it follows that $r(t,t^{-1}) = at^n$. But now in the commutative square

$$\begin{array}{ccc} V & \xrightarrow{a'} & R \otimes_{\mathbb{C}} V \\ {\scriptstyle a'}\downarrow & & \downarrow{\scriptstyle 1\otimes a'} \\ R \otimes_{\mathbb{C}} V & \xrightarrow[\mu' \otimes 1]{} & R \otimes_{\mathbb{C}} R \otimes_{\mathbb{C}} V \end{array}$$

the map

$$\begin{array}{ccc} V & \xrightarrow{a'} & R \otimes_{\mathbb{C}} V \\ & & \downarrow{\scriptstyle 1\otimes a'} \\ & & R \otimes_{\mathbb{C}} R \otimes_{\mathbb{C}} V \end{array}$$

takes v to $a^2 \cdot t^n \otimes t^n \otimes v$, while the composite

$$\begin{array}{ccc} V & & \\ {\scriptstyle a'}\downarrow & & \\ R \otimes_{\mathbb{C}} V & \xrightarrow[\mu' \otimes 1]{} & R \otimes_{\mathbb{C}} R \otimes_{\mathbb{C}} V \end{array}$$

takes v to $a \cdot t^n \otimes t^n \otimes v$. The equality means $a = 1$. Linearity tells us that $a'(\lambda v) = t^n \otimes \{\lambda v\}$, and every vector in the 1–dimensional space V is of the form λv. □

Definition 9.3.5. *Let V be a linear representation, of the affine group* *scheme* $(G, \mathcal{O}_G) = \left(\operatorname{Spec}(R), \widetilde{R}\right)$, *with* $R = \mathbb{C}[t, t^{-1}]$. *The* nth weight space $V_n \subset V$ *is the vector subspace*

$$V_n \quad = \quad \{v \in V \mid a'(v) = t^{-n} \otimes v\} \, .$$

If V is any linear representation of an affine group $(G, \mathcal{O}_G)$, Remark 9.2.4 told us that the group G^{an} acts on V. If we specialize to the $(G, \mathcal{O}_G)$ of Definition 9.3.5, this yields an action of $G^{\mathrm{an}} = \mathbb{C}^*$ on the vector space V. We observe:

Remark 9.3.6. In the special case where $n = 0$, Corollary 9.2.13 identifies V_0 as the subset $V^{G^{\mathrm{an}}} \subset V$, consisting of all vectors fixed by the action of $G^{\mathrm{an}} = \mathbb{C}^*$.

Remark 9.3.7. More generally, if $v \in V_n$ is an element of the nth weight space, then the formula for the action is

$$g(v) = g^n v \, ,$$

where the left hand side should be read as the group element $g \in G^{\mathrm{an}}$ acting on the vector v, while on the right we mean the complex number g^n multiplied by v.

The proof of the assertion is to work through the definitions. Since $v \in V_n$, we have $a'(v) = t^{-n} \otimes v$. The action of the element $g \in G^{\mathrm{an}}$, given in Remark 9.2.4, was to apply the map $\varphi_{g^{-1}} \otimes 1$ to $a'(v) = t^{-n} \otimes v$. This means we should evaluate at $t = g^{-1}$, obtaining $g(v) = g^n v$.

Proposition 9.3.8. *Let* $(G, \mathcal{O}_G) = \left(\operatorname{Spec}(R), \widetilde{R}\right)$, *with* $R = \mathbb{C}[t, t^{-1}]$. *Let* W *be a linear representation of* $(G, \mathcal{O}_G)$. *Then* W *is the direct sum of its weight spaces* W_i.

Proof. Proposition 9.2.15 tells us that any finite dimensional subspace $V \subset W$ is contained in a finite dimensional subspace $V' \subset W$ which is a $(G, \mathcal{O}_G)$–subrepresentation of W. In particular any element $w \in W$ generates a 1–dimensional subspace $V = \mathbb{C}w$, which is contained in a finite dimensional subrepresentation $V' \subset W$. Lemmas 9.3.2 and 9.3.4 together establish that any vector $v \in V'$ can be written as $v = \lambda_1 v_1 + \lambda_2 v_2 + \cdots + \lambda_n v_n$, with $a'(v_i) = t^{n_i} \otimes v_i$; that is the weight spaces span V'. Hence the weight spaces span W.

We need to show that the sum is direct. Suppose therefore that

$$0 \quad = \quad \sum_{i \in \mathbb{Z}} w_i$$

with $w_i \in W_i$, the ith weight space. Applying $a' : W \longrightarrow R \otimes_{\mathbb{C}} W$ to this identity we have

$$0 \quad = \quad \sum_{i\in\mathbb{Z}} a'(w_i) \quad = \quad \sum_{i\in\mathbb{Z}} t^{-i}\otimes w_i \ ,$$

but, as the elements $t^{-i}\in\mathbb{C}[t,t^{-1}]$ are linearly independent, this forces $w_i=0$. □

In the remainder of the section we will explore several immediate consequences of Proposition 9.3.8.

Example 9.3.9. Recall our favorite G–ring of Example 8.3.5. The ring S is the polynomial ring $S=\mathbb{C}[x_0,x_1,\ldots,x_n]$, and the action is by $a'(x_i)=t^{-1}\otimes x_i$. The ith weight space S_i is precisely the set of homogeneous polynomials of degree i.

What about G–modules M for the G–ring S? Any such module will be the direct sum of its weight spaces M_i, $i\in\mathbb{Z}$. And the rule of Remark 8.3.16(iii) says that the following diagram commutes

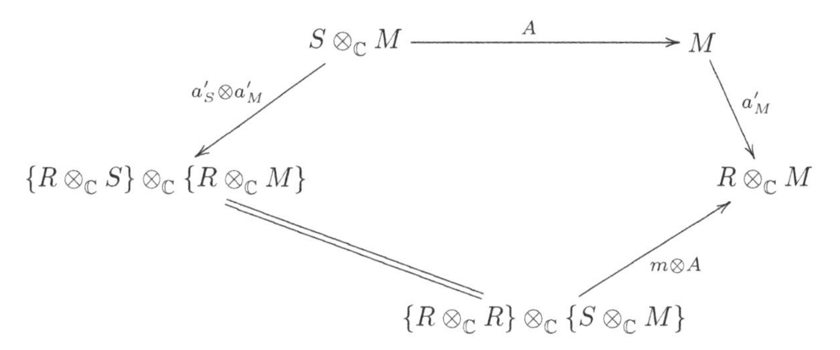

Taking an $s\in S_i$ and $m\in M_j$ and applying the two composites to the element $s\otimes m\in S\otimes_{\mathbb{C}} M$, we get the identity

$$a'_M(sm)=t^{-i}t^{-j}\otimes sm\,;$$

in other words $sm\in M_{i+j}$. Put concretely, a G–module for the G–ring S is nothing more nor less than a finitely generated graded module for the polynomial ring $S=\mathbb{C}[x_0,x_1,\ldots,x_n]$. We leave it to the reader to check that G–homomorphisms of G–modules are precisely graded homomorphisms of graded S–modules.

Remark 9.3.10. Any G–homomorphism $\varphi:V\longrightarrow W$, of linear representations of $(G,\mathcal{O}_G)$, takes the weight space $V_i\subset V$ to $W_i\subset W$. In particular any G–homomorphism of G–rings $S\longrightarrow S'$ will send S_i to S'_i, and any G–homomorphism of G–modules $M\longrightarrow N$ will send $M_i\subset M$ to $N_i\subset N$.

Definition 9.3.11. *A* reductive group *will be an affine group scheme* $(G, \mathcal{O}_G)$*, of finite type over* $\mathbb{C}$*, so that*

(i) *Every linear representation* V *decomposes, uniquely, as a direct sum of linear representations* $V = V^{G^{\mathrm{an}}} \oplus V'$.

(ii) *Any homomorphism* $V \longrightarrow W$ *of linear representations respects the decomposition in (i).*

Example 9.3.12. If $(G, \mathcal{O}_G)$ is trivial then, for any linear representation V, we have $V = V^{G^{\mathrm{an}}}$. We take $V' = 0$, and (i) and (ii) are obvious.

If $(G, \mathcal{O}_G) = \left(\operatorname{Spec}(R), \widetilde{R}\right)$ with $R = \mathbb{C}[t, t^{-1}]$, and V is a linear representation of $(G, \mathcal{O}_G)$, we let

$$V' \quad = \quad \sum_{i \neq 0} V_i \, ,$$

that is V' is the sum of the non-zero weight spaces. By Proposition 9.3.8 we know that $V = V_0 \oplus V'$, while Remark 9.3.6 tells us that $V_0 = V^{G^{\mathrm{an}}}$. Remark 9.3.10 assures us that any G–homomorphism $\varphi : V \longrightarrow W$ satisfies $\varphi\left(V^{G^{\mathrm{an}}}\right) \subset W^{G^{\mathrm{an}}}$ and $\varphi(V') \subset W'$.

It remains to prove the uniqueness of the decomposition. Let $V = V^{G^{\mathrm{an}}} \oplus V'$, and we will compute what V' must be. Now V' is a linear representation of G, and by Proposition 9.3.8 it can be decomposed as a direct sum of its weight spaces

$$V' \quad = \quad \bigoplus_{i \in \mathbb{Z}} V'_i \, .$$

This makes

$$V \quad = \quad V^{G^{\mathrm{an}}} \oplus V' \quad = \quad V_0 \oplus \left(\bigoplus_{i \in \mathbb{Z}} V'_i \right) .$$

It follows that V'_0 must be $\{0\}$ and, for $i \neq 0$, we must have $V'_i = V_i$.

Remark 9.3.13. In Example 9.3.12 we saw that the trivial group, as well as our favorite group $(G, \mathcal{O}_G)$ with $G^{\mathrm{an}} = \mathbb{C}^*$, are both reductive. There are other reductive groups, but in this book we do not care.

Lemma 9.3.14. *Suppose* $0 \longrightarrow U \longrightarrow V \longrightarrow W \longrightarrow 0$ *is a short exact sequence of linear representations of a reductive group* $(G, \mathcal{O}_G)$*. Then the sequence*

$$0 \longrightarrow U^{G^{\mathrm{an}}} \longrightarrow V^{G^{\mathrm{an}}} \longrightarrow W^{G^{\mathrm{an}}} \longrightarrow 0$$

is also exact.

Proof. By Definition 9.3.11 the exact sequence $0 \longrightarrow U \longrightarrow V \longrightarrow W \longrightarrow 0$ decomposes as a direct sum of two sequences

$$\begin{array}{ccccccccc} 0 & \longrightarrow & U^{G^{\mathrm{an}}} & \longrightarrow & V^{G^{\mathrm{an}}} & \longrightarrow & W^{G^{\mathrm{an}}} & \longrightarrow & 0 \\ 0 & \longrightarrow & U' & \longrightarrow & V' & \longrightarrow & W' & \longrightarrow & 0 \end{array} ,$$

both of which must therefore be exact. □

Definition 9.3.15. *Let $(G, \mathcal{O}_G)$ be a reductive group, and let V be a linear representation. Then the* Reynolds operator *is defined to be the projection $E : V \longrightarrow V^{G^{\mathrm{an}}}$, that is the projection to the direct summand*

$$V = V^{G^{\mathrm{an}}} \oplus V' \longrightarrow V^{G^{\mathrm{an}}} .$$

Much of what we do in the rest of the chapter will hinge on playing around with the Reynolds operator. Let us begin with a general observation.

Remark 9.3.16. Because any G–homomorphism $\varphi : V \longrightarrow W$ respects the decomposition of Definition 9.3.11, we see that φ commutes with the Reynolds operator. That is, the following square always commutes

$$\begin{array}{ccc} V & \xrightarrow{\ \varphi\ } & W \\ {\scriptstyle E}\downarrow & & \downarrow{\scriptstyle E} \\ V^{G^{\mathrm{an}}} & \xrightarrow[\ \varphi\ ]{} & W^{G^{\mathrm{an}}} \end{array} .$$

The G–homomorphisms, to which we will be applying the Reynolds operator, fall into three classes:

(i) The G–homomorphisms of G–rings $\varphi : S \longrightarrow S'$.
(ii) The inclusions of G^{an}–invariant ideals $i : I \longrightarrow S$, where S is a G–ring.
(iii) The maps $f : S \longrightarrow S$ of the form

$$f(s) = bs$$

where $b \in S_0 = S^{G^{\mathrm{an}}}$ is G^{an}–invariant.

Perhaps we should elaborate a little on (ii) and (iii) above. Let us state the results as two lemmas.

Lemma 9.3.17. *Let $(G, \mathcal{O}_G)$ be a reductive group, and let S be a G–ring. Let I be an ideal of S and assume $gI = I$ for all $g \in G^{\mathrm{an}}$. Then the Reynolds operator $E : S \longrightarrow S$ takes I to itself.*

Proof. Corollary 9.2.14 tells us that $I \subset S$ is a subrepresentation of $(G, \mathcal{O}_G)$; it is a representation and the inclusion $i : I \longrightarrow S$ is a G–homomorphism of representations. But then Remark 9.3.16 asserts that the Reynolds operator commutes with the inclusion $i : I \longrightarrow S$. Given an element $a \in I$ we conclude $i\big(E(a)\big) = E\big(i(a)\big)$. But $i\big(E(a)\big)$ is clearly an element of I, and $E\big(i(a)\big)$ is the map $E : S \longrightarrow S$ applied to $i(a) \in S$. □

Lemma 9.3.18. *Let $(G, \mathcal{O}_G)$ be a reductive group, and let S be a G–ring. The Reynolds operator $E : S \longrightarrow S^{G^{\mathrm{an}}}$ is a $\mathbb{C}$–linear map, with the property that $E(bs) = bE(s)$ for any $b \in S^{G^{\mathrm{an}}} = S_0$ and any $s \in S$.*

Proof. The $\mathbb{C}$–linearity is obvious. As in Remark 9.3.16 we note that the map $f : S \longrightarrow S$, taking $s \in S$ to $f(s) = bs$, is a G–homomorphism of linear representations of $(G, \mathcal{O}_G)$; after all

$$a'(bs) = a'(b)a'(s) = (1 \otimes b)a'(s)\,;$$

that is the square

$$\begin{array}{ccc} S & \xrightarrow{\ a'\ } & R \otimes_{\mathbb{C}} S \\ {\scriptstyle f}\big\downarrow & & \big\downarrow{\scriptstyle 1\otimes f} \\ S & \xrightarrow[\ a'\ ]{} & R \otimes_{\mathbb{C}} S \end{array}$$

commutes. Remark 9.3.10 therefore implies that $E\big(f(s)\big) = f\big(E(s)\big)$, which means $E(bs) = bE(s)$. □

9.4 The finite generation of the ring of invariants

It is now time to start proving Theorem 8.8.6. We remind the reader.

Reminder 9.4.1. In Remark 8.8.7 we reduced parts (i), (ii) and (iii) of Theorem 8.8.6 to the local case. We have an affine G–scheme $(X, \mathcal{O}_X) = \Big(\mathrm{Spec}(S), \widetilde{S}\Big)$. In Sections 8.6 and 8.7 we constructed a map of ringed spaces over $\mathbb{C}$

$$(\pi, \pi^*) : (X, \mathcal{O}_X) \longrightarrow (X/G, \mathcal{O}_{X/G})\,,$$

where the sheaf of rings $\mathcal{O}_{X/G}$ was defined by the formula $\mathcal{O}_{X/G} = \{\pi_*\mathcal{O}_X\}^{G^{\mathrm{an}}}$. Furthermore for every G–module M, over the G–ring S, we constructed a sheaf of $\mathcal{O}_{X/G}$–modules $\{\pi_*\widetilde{M}\}^{G^{\mathrm{an}}}$ on X/G. The local case of Theorem 8.8.6, which we have to prove, asserts that, if the G^{an}–orbit of every closed point in X is Zariski closed, then the following holds:

(i) $(X/G, \mathcal{O}_{X/G})$ is a scheme of finite type over $\mathbb{C}$.

(ii) The sheaves $\{\pi_* \widetilde{M}\}^{G^{\mathrm{an}}}$ are coherent.

(iii) Any exact sequence $0 \longrightarrow M' \longrightarrow M \longrightarrow M'' \longrightarrow 0$, of G–modules over the G–ring S, gives rise over $(X/G, \mathcal{O}_{X/G})$ to an exact sequence of coherent sheaves

$$0 \longrightarrow \{\pi_* \widetilde{M'}\}^{G^{\mathrm{an}}} \longrightarrow \{\pi_* \widetilde{M}\}^{G^{\mathrm{an}}} \longrightarrow \{\pi_* \widetilde{M''}\}^{G^{\mathrm{an}}} \longrightarrow 0\,.$$

Remark 9.4.2. Suppose the group $(G, \mathcal{O}_G)$ is trivial. Example 8.6.11 tells us that the map $\pi : X \longrightarrow X/G$ is the identity, and Example 8.7.10 further informs us that $\{\pi_* \widetilde{M}\}^{G^{\mathrm{an}}} = \widetilde{M}$. Assertions (i), (ii) and (iii) of Theorem 8.8.6 are free. We will return to the trivial group when we prove Theorem 8.8.6(iv).

Remark 9.4.3. In Reminder 9.4.1 you find what we promised the reader. We will actually prove more; see Refinement 8.9.1. We will show that the scheme $(X/G, \mathcal{O}_{X/G})$ is affine, and that the sheaf $\{\pi_* \widetilde{M}\}^{G^{\mathrm{an}}}$ is of the form $\widetilde{N}$. The way the proof will proceed will be to produce a $\mathbb{C}$–algebra homomorphism $\theta : T \longrightarrow S$, where T is a finitely generated $\mathbb{C}$–algebra, and show that the natural map

$$\left(\mathrm{Spec}(\theta), \widetilde{\theta}\right) : \left(\mathrm{Spec}(S), \widetilde{S}\right) \longrightarrow \left(\mathrm{Spec}(T), \widetilde{T}\right)$$

can be identified with the map $(\pi, \pi^*) : (X, \mathcal{O}_X) \longrightarrow (X/G, \mathcal{O}_{X/G})$. The ring T will be the subring $S^{G^{\mathrm{an}}} \subset S$ of all G^{an}–invariant elements of S, and the homomorphism $\theta : T \longrightarrow S$ will be nothing more than the inclusion $\theta : S^{G^{\mathrm{an}}} \longrightarrow S$. The first step is therefore to prove that the ring $S^{G^{\mathrm{an}}}$ is a finitely generated $\mathbb{C}$–algebra. This is a celebrated 1892 theorem due to Hilbert:

Theorem 9.4.4. *Let $(G, \mathcal{O}_G)$ be a reductive group over $\mathbb{C}$. Let S be a G–ring. Then the subring $S^{G^{\mathrm{an}}}$ is a finitely generated $\mathbb{C}$–algebra.*

Proof. By Corollary 9.2.16 there exists a finite dimensional linear representation V of the group $(G, \mathcal{O}_G)$, as well as an onto G–homomorphism of G–rings $\theta : \mathbb{C}[V] \longrightarrow S$. Lemma 9.3.14 establishes that the map $\theta^{G^{\mathrm{an}}} : \mathbb{C}[V]^{G^{\mathrm{an}}} \longrightarrow S^{G^{\mathrm{an}}}$ is surjective. To prove that $S^{G^{\mathrm{an}}}$ is finitely generated it clearly suffices to show the finite generation of $\mathbb{C}[V]^{G^{\mathrm{an}}}$. We may therefore assume $S = \mathbb{C}[V]$. Thus

$$S \quad = \quad \mathbb{C}[V] \quad = \quad \sum_{i=0}^{\infty} \mathrm{Sym}^i V$$

is a graded ring, and the action of the group G^{an} preserves the grading.

Furthermore, in degree zero we have the constants $\mathbb{C}$, on which G^{an} acts trivially.

Consider the ideal $I \subset S$, spanned by all the G^{an}–invariant elements with a zero constant term. That is, we take the intersection of $S^{G^{\text{an}}}$ with $\sum_{i=1}^{\infty} \operatorname{Sym}^i V$, and form the ideal this generates in S. By the Hilbert Basis Theorem the ideal I is finitely generated; see Theorem 5.2.1. We can choose a finite number of generators for I. Each of the generators lies in the ideal I, and hence can be written as

$$a_1 b_1 + a_2 b_2 + \cdots a_\ell b_\ell$$

with $a_j \in S$, and with $b_j \in S^{G^{\text{an}}}$ having zero constant term. Replacing each generator by the set $\{b_1, b_2, \ldots, b_\ell\}$ above, we may assume that the generators of the ideal I lie in $S^{G^{\text{an}}}$. Now each element of $S^{G^{\text{an}}} \cap I$, that is each element of $S^{G^{\text{an}}}$ with a zero constant term, can be written as a sum

$$b = b_1 + b_2 + \cdots b_m \ ,$$

with each b_j homogeneous of degree > 0. Because the action of G^{an} respects the grading on S, each of the homogeneous components b_j will lie in the subring $S^{G^{\text{an}}} \subset S$. Replacing each b by its homogeneous components $\{b_1, b_2, \ldots, b_m\}$, we may choose in $S^{G^{\text{an}}}$ finitely many homogeneous elements $b_1, b_2, \ldots, b_n$, of degree > 0, which generate the ideal I. We assert that $b_1, b_2, \ldots, b_n$ generate $S^{G^{\text{an}}}$ as an algebra over $\mathbb{C}$.

We prove this by contradiction. Suppose they do not generate the $\mathbb{C}$–algebra $S^{G^{\text{an}}}$. Then there must be some homogeneous element of $S^{G^{\text{an}}}$ not in the subalgebra generated by the $b_1, b_2, \ldots, b_n$. Choose such an element s of minimal degree d, and we will prove a contradiction. Note that d must be greater than 0, since in degree 0 we have only the constants, which lie in the subalgebra generated by $b_1, b_2, \ldots, b_n$.

Because $s \in S^{G^{\text{an}}}$ is homogeneous, of degree $d > 0$, it must lie in the ideal I, and the ideal is generated by $b_1, b_2, \ldots, b_n$. Hence there is an identity

$$s \quad = \quad \sum_{i=1}^{n} b_i s_i$$

with $s_i \in S$. Applying the Reynolds operator we have

$$s \quad = \quad E(s) \quad = \quad \sum_{i=1}^{n} E(b_i s_i) \quad = \quad \sum_{i=1}^{n} b_i E(s_i) \ ,$$

where $E(s) = s$ because s is G^{an}–invariant, and $E(b_i s_i) = b_i E(s_i)$ by Lemma 9.3.18 and because b_i is G^{an}–invariant. Now recall that b_i is homogeneous, of degree > 0; let its degree be d_i. Write each $E(s_i)$ as a sum of its homogeneous components

$$E(s_i) = {}^0y_i + {}^1y_i + \cdots + {}^\ell y_i \ ;$$

that is, ${}^jy_i$ is homogeneous of degree j. Consider the identity

$$s \quad = \quad \sum_{i=1}^{n} b_i E(s_i) \ .$$

The degree–d homogeneous component of the left hand side must equal the degree–d homogeneous component on the right. Therefore we have

$$s \quad = \quad \sum_{i=1}^{n} b_i \cdot \left({}^{d-d_i}y_i \right) \ .$$

We know that $E(s_i)$ is G^{an}–invariant, hence so are all of its homogeneous components ${}^jy_i$. In particular ${}^{d-d_i}y_i$ is in $S^{G^{\mathrm{an}}}$. Since its degree is $d - d_i < d$, the minimality of d forces ${}^{d-d_i}y_i$ to belong to the subalgebra generated by $b_1, b_2, \ldots, b_n$. This being true for every ${}^{d-d_i}y_i$, we have proved a contradiction. □

Before we end the section let us observe an easy consequence of Theorem 9.4.4, which we will need later.

Corollary 9.4.5. *Let $(G, \mathcal{O}_G)$ be a reductive group over $\mathbb{C}$ and let S be a G–ring. Let M be a G–module for the G–ring S. Then $M^{G^{\mathrm{an}}}$ is a finite module over the ring $S^{G^{\mathrm{an}}}$.*

Proof. The ring $S \oplus M$ is a G–ring, and Theorem 9.4.4 says that the ring $\{S \oplus M\}^{G^{\mathrm{an}}} = S^{G^{\mathrm{an}}} \oplus M^{G^{\mathrm{an}}}$ is a finitely generated $\mathbb{C}$–algebra. Choose a finite set $\{s_1 \oplus m_1, s_2 \oplus m_2, \ldots, s_n \oplus m_n\}$ of $\mathbb{C}$–algebra generators. Then certainly the set

$$\{s_1, s_2, \ldots, s_n, m_1, m_2, \ldots, m_n\}$$

also generates. But then the elements $m_1, m_2, \ldots, m_n$ must generate $M^{G^{\mathrm{an}}}$ as a module over $S^{G^{\mathrm{an}}}$. □

9.5 The topological facts about $\pi : X \longrightarrow X/G$

So far we have shown that the subring $S^{G^{\mathrm{an}}} \subset S$ is a finitely generated $\mathbb{C}$–algebra. The natural inclusion $\theta : S^{G^{\mathrm{an}}} \longrightarrow S$ induces a map of ringed spaces over $\mathbb{C}$

$$\left(\mathrm{Spec}(\theta), \widetilde{\theta}\right) : \left(\mathrm{Spec}(S), \widetilde{S}\right) \longrightarrow \left(\mathrm{Spec}(S^{G^{\mathrm{an}}}), \widetilde{S^{G^{\mathrm{an}}}}\right) \ .$$

The program we set forth, in Remark 9.4.3, was to study the map $\left(\mathrm{Spec}(\theta),\widetilde{\theta}\right)$, and to identify it with the morphism of ringed spaces $(\pi,\pi^*) : (X,\mathcal{O}_X) \longrightarrow (X/G,\mathcal{O}_{X/G})$. We now begin this program. In this section we will concern ourselves only with identifying the continuous maps $\pi : X \longrightarrow X/G$ and $\mathrm{Spec}(\theta) : X \longrightarrow \mathrm{Spec}(S^{G^{\mathrm{an}}})$. We will study the sheaves of rings in Section 9.6.

Lemma 9.5.1. *Let $(G,\mathcal{O}_G)$ be a reductive group over $\mathbb{C}$. Let S be a G–ring, and let $\theta : S^{G^{\mathrm{an}}} \longrightarrow S$ be the inclusion of the invariant subring. Then the continuous map*

$$\mathrm{Spec}(\theta) : \mathrm{Spec}(S) \longrightarrow \mathrm{Spec}(S^{G^{\mathrm{an}}})$$

is surjective.

Proof. Let $\mathfrak{p}$ be a prime ideal $\mathfrak{p} \subset S^{G^{\mathrm{an}}}$; we need to exhibit a prime ideal $\mathfrak{q} \subset S$ with $\theta^{-1}\mathfrak{q} = \mathfrak{p}$. Consider therefore the ideal $S\mathfrak{p}$, that is the ideal in S generated by $\mathfrak{p}$. I assert first that $S^{G^{\mathrm{an}}} \cap S\mathfrak{p} = \mathfrak{p}$. The inclusion $\mathfrak{p} \subset S^{G^{\mathrm{an}}} \cap S\mathfrak{p}$ is obvious; we have to prove the reverse inclusion.

Suppose therefore that a lies in $S^{G^{\mathrm{an}}} \cap S\mathfrak{p}$; we need to prove that $a \in \mathfrak{p}$. We have

$$a = \sum_{i=1}^{n} s_i p_i$$

with $s_i \in S$ and $p_i \in \mathfrak{p}$. Applying the Reynolds operator we have

$$a = E(a) = \sum_{i=1}^{n} E(s_i p_i) = \sum_{i=1}^{n} p_i E(s_i) ,$$

where the third equality is by Lemma 9.3.18 and because $p_i \in \mathfrak{p} \subset S^{G^{\mathrm{an}}}$. But the right hand side is a linear combination of the elements $p_i \in \mathfrak{p}$, with coefficients $E(s_i) \in S^{G^{\mathrm{an}}}$, and hence it lies in $\mathfrak{p}$. We deduce that $S^{G^{\mathrm{an}}} \cap S\mathfrak{p} \subset \mathfrak{p}$.

We therefore have an injective homomorphism $S^{G^{\mathrm{an}}}/\mathfrak{p} \longrightarrow S/S\mathfrak{p}$. The integral domain $S^{G^{\mathrm{an}}}/\mathfrak{p}$ embeds in the ring $S/S\mathfrak{p}$. The set Σ, of non-zero elements in the integral domain $S^{G^{\mathrm{an}}}/\mathfrak{p}$, is a multiplicatively closed set. If we localize with respect to Σ we get an inclusion

$$\Sigma^{-1}\{S^{G^{\mathrm{an}}}/\mathfrak{p}\} \longrightarrow \Sigma^{-1}\{S/S\mathfrak{p}\} .$$

But $\Sigma^{-1}\{S^{G^{\mathrm{an}}}/\mathfrak{p}\} = k(\mathfrak{p})$ is the quotient field of the integral domain $S^{G^{\mathrm{an}}}/\mathfrak{p}$, and the above tells us that $k(\mathfrak{p})$ embeds in the ring $\Sigma^{-1}\{S/S\mathfrak{p}\}$. Choose any maximal ideal $\mathfrak{m} \subset \Sigma^{-1}\{S/S\mathfrak{p}\}$; we have a commutative diagram of ring homomorphisms

$$\begin{array}{ccccccc} & & S^{G^{\rm an}} & \xrightarrow{\theta} & S & & \\ & & \downarrow & & \downarrow & & \\ k(\mathfrak{p}) & = & \Sigma^{-1}\{S^{G^{\rm an}}/\mathfrak{p}\} & \longrightarrow & \Sigma^{-1}\{S/S\mathfrak{p}\} & \longrightarrow & k(\mathfrak{m}) \end{array} .$$

The bottom row composes to give a ring homomorphism $k(\mathfrak{p}) \longrightarrow k(\mathfrak{m})$, where $k(\mathfrak{p})$ and $k(\mathfrak{m})$are fields. The map must be injective. The kernel of the composite

$$\begin{array}{ccccccc} & & S^{G^{\rm an}} & & & & \\ & & \downarrow & & & & \\ k(\mathfrak{p}) & = & \Sigma^{-1}\{S^{G^{\rm an}}/\mathfrak{p}\} & \longrightarrow & \Sigma^{-1}\{S/S\mathfrak{p}\} & \longrightarrow & k(\mathfrak{m}) \end{array}$$

therefore agrees with the kernel of $S^{G^{\rm an}} \longrightarrow k(\mathfrak{p})$, which is $\mathfrak{p}$. This means that the kernel of

$$\begin{array}{ccc} S & & \\ \downarrow & & \\ \Sigma^{-1}\{S/S\mathfrak{p}\} & \longrightarrow & k(\mathfrak{m}) \end{array}$$

is a prime ideal $\mathfrak{q} \subset S$ with $\theta^{-1}\mathfrak{q} = \mathfrak{p}$. □

So far we know that the map $\mathrm{Spec}(\theta) : X = \mathrm{Spec}(S) \longrightarrow \mathrm{Spec}(S^{G^{\rm an}})$ is surjective. Next we show that it factors through the surjective map $\pi : X \longrightarrow X/G$. We prove

Lemma 9.5.2. *Let $(G, \mathcal{O}_G)$ be a connected affine group scheme of finite type over $\mathbb{C}$. Let S be a G–ring, put $X = \mathrm{Spec}(S)$, and let $\pi : X \longrightarrow X/G$ be the map of Definition 8.6.7. If x and y are points of X, with $\pi(x) = \pi(y)$, then the map $\mathrm{Spec}(\theta) : X \longrightarrow \mathrm{Spec}(S^{G^{\rm an}})$ also satisfies $\{\mathrm{Spec}(\theta)\}(x) = \{\mathrm{Spec}(\theta)\}(y)$.*

Proof. Recall that X/G comes with an embedding $i : X/G \longrightarrow X$; the subspace $X/G \subset X$ was defined in Definition 8.6.2. For any point $x \in X$ the point $\pi(x)$ can be thought of as belonging to X, and therefore we can apply the map $\mathrm{Spec}(\theta) : X \longrightarrow \mathrm{Spec}(S^{G^{\rm an}})$ to the point $\pi(x) \in X/G \subset X$. We will show that, for any $x \in X$, we have $\{\mathrm{Spec}(\theta)\}(x) = \{\mathrm{Spec}(\theta)\}\big(\pi(x)\big)$. If $\pi(x) = \pi(y)$ it then follows

$$\{\mathrm{Spec}(\theta)\}(x) = \{\mathrm{Spec}(\theta)\}\big(\pi(x)\big) = \{\mathrm{Spec}(\theta)\}\big(\pi(y)\big) = \{\mathrm{Spec}(\theta)\}(y) .$$

Therefore let x be a point in $X = \mathrm{Spec}(S)$, that is a prime ideal $x \subset S$. The point $\pi(x)$, which is also a prime ideal in S, was defined as the intersection over all $g \in G^{\mathrm{an}}$ of gx. We have an inclusion of prime ideals $\pi(x) \subset x$. If $\theta : S^{G^{\mathrm{an}}} \longrightarrow S$ is the natural embedding then the inclusion $\theta^{-1}\pi(x) \subset \theta^{-1}x$ is obvious. We have to prove the reverse inclusion.

Let a be any element of $\theta^{-1}x$. That is $a \in S^{G^{\mathrm{an}}}$ is a G^{an}–invariant element of the ring S, and a lies in the prime ideal x. Since a is G^{an}–invariant it must lie in gx for all $g \in G^{\mathrm{an}}$. It follows that it lies in the intersection of gx over all $g \in G^{\mathrm{an}}$. That is $a \in \pi(x)$. □

Until now we have proved that, at least on the level of sets, the map $\mathrm{Spec}(\theta) : X \longrightarrow \mathrm{Spec}(S^{G^{\mathrm{an}}})$ is surjective and factors through $\pi : X \longrightarrow X/G$. This has been completely painless in the sense that, so far, we have not had to make any assumption on $X = \mathrm{Spec}(S)$. We have had to assume that $(G, \mathcal{O}_G)$ is reductive and connected, but nothing much about X. Now we are about to prove that distinct points of X/G map to distinct points of $\mathrm{Spec}(S^{G^{\mathrm{an}}})$, and that the topologies agree. For this we will need our more restrictive hypotheses. In addition to the assumptions on G, we will appeal to the hypothesis that the G^{an}–orbit of every $x \in X^{\mathrm{an}}$ is Zariski closed. Next we prove a little technical lemma which will do most of the work.

Lemma 9.5.3. *Let $(G, \mathcal{O}_G)$ be a reductive group over $\mathbb{C}$, and let S be a G–ring. Let $X = \mathrm{Spec}(S)$, and suppose every G^{an}–orbit in X^{an} is Zariski closed. Suppose $I \subset S$ is an ideal such that $gI = I$ for all $g \in G^{\mathrm{an}}$. If $\mathfrak{m} \subset S$ is a maximal ideal, not containing I, then there is an element in $S^{G^{\mathrm{an}}}$ which lies in $I - \mathfrak{m}$.*

Proof. The maximal ideal $\mathfrak{m}$ is a closed point in X, that is a point in X^{an}. We are assuming its G^{an}–orbit to be Zariski closed; there is an ideal $J \subset S$ with $V(J) \cap X^{\mathrm{an}}$ being the orbit $G^{\mathrm{an}}\mathfrak{m}$ of the point $\mathfrak{m}$. The fact that $g\mathfrak{m} \in V(J)$ means we have an inclusion of ideals $J \subset g\mathfrak{m}$ for every $g \in G^{\mathrm{an}}$. Hence the ideal $J' = \cap_g g\mathfrak{m}$ has the property that $J \subset J' \subset \mathfrak{m}$ and J' is G^{an}–invariant. Because $J \subset J'$ we have $V(J') \subset V(J)$, but because J' is G^{an}–invariant and $J' \subset \mathfrak{m}$ it follows that $V(J')$ contains the orbit of $\mathfrak{m}$. We conclude that $V(J') \cap X^{\mathrm{an}}$ is precisely the orbit of $\mathfrak{m}$. Replacing J by J' we may assume the ideal $J \subset S$ is invariant under the action of G^{an}.

We now have two G^{an}–invariant ideals of S: the ideal J we constructed above and the ideal I given to us in the hypotheses of the lemma. The ideal I is assumed G^{an}–invariant and assumed not to be contained in $\mathfrak{m}$; hence it cannot be contained in any point $g\mathfrak{m}$ with $g \in G^{\mathrm{an}}$. The property of $V(J)$ is the opposite: the only maximal ideals containing it are the $g\mathfrak{m}$, $g \in G^{\mathrm{an}}$. We conclude that $V(I) \cap V(J) \cap X^{\mathrm{an}}$ is empty.

Since $V(I) \cap V(J) = V(I+J)$ (see Lemma 3.1.9) we conclude that $V(I+J) \cap X^{\mathrm{an}} = \emptyset$. But then $I+J$ is an ideal of S not contained in any maximal ideal; this can only happen if $I+J=S$.

Therefore we can write $1 = i + j$ with $i \in I$ and $j \in J$. This means $1 = E(1) = E(i) + E(j)$. The ideals I and J are both invariant under the action of G^{an}; Lemma 9.3.17 allows us to conclude that $E(i) \in I$ and $E(j) \in J$. The fact that $E(j) = 1 - E(i)$ lies in $J \subset \mathfrak{m}$ means that $E(i) \notin \mathfrak{m}$. We have produced an element $E(i) \in S^{G^{\mathrm{an}}}$, with $E(i) \in I - \mathfrak{m}$. □

Now we are ready to prove that, at least on the level of sets, the maps $\pi : X \longrightarrow X/G$ and $\mathrm{Spec}(\theta) : X \longrightarrow \mathrm{Spec}(S^{G^{\mathrm{an}}})$ agree. We already know that $\mathrm{Spec}(\theta)$ is surjective and factors through π. If $i : X/G \longrightarrow X$ is the inclusion of Definition 8.6.2 then the composite

$$X/G \xrightarrow{\ i\ } X \xrightarrow{\ \pi\ } X/G$$

is the identity by the proof of Lemma 8.6.9. It therefore suffices to show that the composite

$$X/G \xrightarrow{\ i\ } X \xrightarrow{\mathrm{Spec}(\theta)} \mathrm{Spec}(S^{G^{\mathrm{an}}})$$

is injective.

Lemma 9.5.4. *Let $(G, \mathcal{O}_G)$ be a connected, reductive group over $\mathbb{C}$ and let S be a G–ring. Let $X = \mathrm{Spec}(S)$, and suppose every G^{an}–orbit in X^{an} is Zariski closed. If $\mathfrak{p}$ and $\mathfrak{q}$ are points in $X/G \subset X$, and $\mathfrak{p} \neq \mathfrak{q}$, then the ideals $\mathfrak{p} \cap S^{G^{\mathrm{an}}}$ and $\mathfrak{q} \cap S^{G^{\mathrm{an}}}$ are also different. That is, the map $\mathrm{Spec}(\theta) : X \longrightarrow \mathrm{Spec}(S^{G^{\mathrm{an}}})$ takes $\mathfrak{p}, \mathfrak{q} \in X/G \subset X$ to distinct points in $\mathrm{Spec}(S^{G^{\mathrm{an}}})$.*

Proof. We are given $\mathfrak{p} \neq \mathfrak{q}$. Therefore either $\mathfrak{p} \not\subset \mathfrak{q}$ or $\mathfrak{q} \not\subset \mathfrak{p}$. Without loss of generality we may assume $\mathfrak{p} \not\subset \mathfrak{q}$. We will prove that $\mathfrak{p} \cap S^{G^{\mathrm{an}}}$ is not contained in $\mathfrak{q} \cap S^{G^{\mathrm{an}}}$. The ideal $\mathfrak{q} \subset S$ is a prime ideal in the finitely generated $\mathbb{C}$–algebra S, and the strong Nullstellensatz tells us that

$$\mathfrak{q} \quad = \bigcap_{\substack{\mathfrak{m} \supset \mathfrak{q} \\ \mathfrak{m} \text{ a maximal ideal of} \\ \text{the ring } S}} \mathfrak{m} \quad ;$$

see Reminder 3.8.1(ii). We also know that $\mathfrak{p} \not\subset \mathfrak{q}$, that is

$$\mathfrak{p} \quad \not\subset \quad \bigcap_{\substack{\mathfrak{m} \supset \mathfrak{q} \\ \mathfrak{m} \text{ a maximal ideal of} \\ \text{the ring } S}} \quad \mathfrak{m} \quad .$$

This means "it is not true that $\mathfrak{p} \subset \mathfrak{m}$, for every maximal $\mathfrak{m} \supset \mathfrak{q}$". Hence there exists a maximal ideal $\mathfrak{m}$ containing $\mathfrak{q}$ but not $\mathfrak{p}$.

Now $\mathfrak{p}$ is a point in $X/G \subset X$, hence $\mathfrak{p}$ is a prime ideal of S such that $g\mathfrak{p} = \mathfrak{p}$ for all $g \in G^{\text{an}}$. The maximal ideal $\mathfrak{m}$ does not contain $\mathfrak{p}$; Lemma 9.5.3 tells us that there exists an element $a \in \{\mathfrak{p} \cap S^{G^{\text{an}}}\} - \mathfrak{m}$. Since $a \notin \mathfrak{m}$ and $\mathfrak{q} \subset \mathfrak{m}$ we have $a \notin \mathfrak{q}$; we have found an element $a \in \mathfrak{p} \cap S^{G^{\text{an}}}$ which does not lie in $\mathfrak{q}$. □

So far we have shown that the maps $\pi : X \longrightarrow X/G$ and $\text{Spec}(\theta) : X \longrightarrow \text{Spec}(S^{G^{\text{an}}})$ agree as maps of sets. It is time to worry about the topology. We need to show that the open sets $V \subset X/G$ agree with the open sets $W \subset \text{Spec}(S^{G^{\text{an}}})$. Since both $\pi : X \longrightarrow X/G$ and $\text{Spec}(\theta) : X \longrightarrow \text{Spec}(S^{G^{\text{an}}})$ are surjective it is enough to show that open sets $\pi^{-1}V \subset X$ and the open sets $\{\text{Spec}(\theta)\}^{-1}W \subset X$ coincide. Lemma 8.6.9 tells us that the sets $\pi^{-1}V \subset X$, where $V \subset X/G$ is open, are precisely all the G–invariant open subsets $U \subset X$. To prove that the topologies on X/G and $\text{Spec}(S^{G^{\text{an}}})$ agree it therefore suffices to establish that the sets $\{\text{Spec}(\theta)\}^{-1}W \subset X$ are also all the G–invariant open subsets. If $W \subset \text{Spec}(S^{G^{\text{an}}})$ is open then $\{\text{Spec}(\theta)\}^{-1}W$ is certainly G–invariant and open. What we must show is that every G–invariant open set can be expressed as $\{\text{Spec}(\theta)\}^{-1}W$. Or in terms of the complements, we must show

Lemma 9.5.5. *Let $(G, \mathcal{O}_G)$ be a connected, reductive group over $\mathbb{C}$ and let S be a G–ring. Let $X = \text{Spec}(S)$, and suppose every G^{an}–orbit in X^{an} is Zariski closed. Then every G–invariant closed subset $Z \subset X$ is the inverse image of a closed subset in* $\text{Spec}(S^{G^{\text{an}}})$.

Proof. Let Z be a G–invariant closed subset of X. Being closed Z must be of the form $Z = V(I)$ for some ideal $I \subset S$. Since Z is G–invariant we have that $Z = gZ = V(gI)$ for every $g \in G^{\text{an}}$. But then

$$Z \quad = \quad \bigcap_{g \in G^{\text{an}}} V(gI) \quad = \quad V\left(\sum_{g \in G^{\text{an}}} gI\right)$$

expresses Z as $V(I')$, for an ideal $I' \subset S$ invariant under the action of

G^{an}. Let $J = I' \cap S^{G^{\text{an}}}$. We assert that the inverse image of the closed set $V(J) \subset \text{Spec}(S^{G^{\text{an}}})$ is exactly $Z = V(I) = V(I') \subset X$.

The map $\text{Spec}(\theta) : X \longrightarrow \text{Spec}(S^{G^{\text{an}}})$ takes a point $\mathfrak{p} \in X$, that is a prime ideal $\mathfrak{p} \subset S$, to its inverse image under the inclusion $\theta : S^{G^{\text{an}}} \longrightarrow S$. That is it takes $\mathfrak{p}$ to $\mathfrak{p} \cap S^{G^{\text{an}}}$. The inverse image of $V(J) \subset \text{Spec}(S^{G^{\text{an}}})$ is the set of prime ideals $\mathfrak{p} \subset S$ such that $J \subset \mathfrak{p} \cap S^{G^{\text{an}}}$. Since J was defined as $I' \cap S^{G^{\text{an}}}$, it follows that $I' \subset \mathfrak{p}$ implies $J = I' \cap S^{G^{\text{an}}} \subset \mathfrak{p} \cap S^{G^{\text{an}}}$. That is, if $\mathfrak{p} \in V(I') = Z$ it follows that $\{\text{Spec}(\theta)\}(\mathfrak{p}) \in V(J)$, or, expressed slightly differently, $\mathfrak{p} \in \{\text{Spec}(\theta)\}^{-1} V(J)$. In other words we have proved the easy inclusion

$$Z \quad \subset \quad \{\text{Spec}(\theta)\}^{-1} V(J) \ ;$$

we need to prove the reverse inclusion.

Let us first prove that, if $\mathfrak{m} \subset S$ is a maximal ideal not in $Z = V(I')$, then $\mathfrak{m} \notin \{\text{Spec}(\theta)\}^{-1} V(J)$. We are given a maximal ideal $\mathfrak{m}$ not containing the ideal $I' \subset S$, and I' is invariant under the action of G^{an}. Lemma 9.5.3 tells us that there exists an element $a \in I' \cap S^{G^{\text{an}}} - \mathfrak{m}$, that is $a \in J - \mathfrak{m}$. It follows that $\mathfrak{m} \cap S^{G^{\text{an}}}$ does not contain J, that is $\{\text{Spec}(\theta)\}(\mathfrak{m}) \notin V(J)$.

So far we have two Zariski closed sets in $X = \text{Spec}(S)$, namely $Z = V(I')$ and $\{\text{Spec}(\theta)\}^{-1} V(J)$, and we have proved that their intersections with $X^{\text{an}} \subset X$ agree; they have precisely the same subsets of closed points. Our lemma therefore follows from the following general fact:

9.5.5.1. *Let S be a finitely generated $\mathbb{C}$–algebra, let $X = \text{Spec}(S)$ and let $Z, Z' \subset X$ be two Zariski closed subsets of X. If $Z \cap X^{\text{an}} = Z' \cap X^{\text{an}}$ then $Z = Z'$.*

It remains to prove 9.5.5.1. Put $Z = V(I)$ and $Z' = V(I')$, for I, I' ideals in S. We want to show that $V(I) = V(I')$. In other words we want to show that, if $\mathfrak{p} \subset S$ is a prime ideal, then

$$\{I \subset \mathfrak{p}\} \quad \Longleftrightarrow \quad \{I' \subset \mathfrak{p}\} \ .$$

What we are given is that the intersection of $V(I), V(I')$ with X^{an} agree; this means that, for maximal ideals $\mathfrak{m}$, we have

$$\{I \subset \mathfrak{m}\} \quad \Longleftrightarrow \quad \{I' \subset \mathfrak{m}\} \ .$$

Let $\mathfrak{p}$ be a prime ideal in S. Recalling the strong Nullstellensatz (Reminder 3.8.1(ii)), we know

$$\mathfrak{p} \quad = \quad \bigcap_{\substack{\mathfrak{m} \supset \mathfrak{p} \\ \mathfrak{m} \text{ a maximal ideal of} \\ \text{the ring } S}} \mathfrak{m} \,.$$

Therefore

$$\begin{aligned} \{I \subset \mathfrak{p}\} \quad &\Longleftrightarrow \quad \{I \subset \mathfrak{m}, \text{ for all maximal } \mathfrak{m} \supset \mathfrak{p}\} \\ &\Longleftrightarrow \quad \{I' \subset \mathfrak{m}, \text{ for all maximal } \mathfrak{m} \supset \mathfrak{p}\} \\ &\Longleftrightarrow \quad \{I' \subset \mathfrak{p}\} \,. \end{aligned}$$

□

9.6 The sheaves on X/G

So far we have identified the continuous map $\pi : X \longrightarrow X/G$ with the map $\text{Spec}(\theta) : X \longrightarrow \text{Spec}(S^{G^{\text{an}}})$. In this section we will prove parts (i), (ii) and (iii) of Theorem 8.8.6. All three parts are sheaf theoretic statements about $X/G \simeq \text{Spec}(S^{G^{\text{an}}})$. We begin with Theorem 8.8.6(i).

The map of ringed spaces

$$\left(\text{Spec}(\theta), \widetilde{\theta}\right) : \left(\text{Spec}(S), \widetilde{S}\right) \longrightarrow \left(\text{Spec}(S^{G^{\text{an}}}), \widetilde{S^{G^{\text{an}}}}\right)$$

can now be identified as

$$(\pi, \widetilde{\theta}) : (X, \widetilde{S}) \longrightarrow \left(X/G, \widetilde{S^{G^{\text{an}}}}\right) \,.$$

To prove Theorem 8.8.6(i) it suffices to establish that the induced map of sheaves of rings on X/G

$$\widetilde{\theta}^* : \widetilde{S^{G^{\text{an}}}} \longrightarrow \pi_* \widetilde{S}$$

identifies $\widetilde{S^{G^{\text{an}}}}$ with the subsheaf $\{\pi_* \widetilde{S}\}^{G^{\text{an}}} \subset \pi_* \widetilde{S}$. This is because the sheaf of rings $\mathcal{O}_{X/G}$ was given as $\{\pi_* \mathcal{O}_X\}^{G^{\text{an}}} = \{\pi_* \widetilde{S}\}^{G^{\text{an}}}$; see Definition 8.7.6.

Put very concretely, we need to show that the map of sheaves on X/G

$$\widetilde{\theta}^* : \widetilde{S^{G^{\text{an}}}} \longrightarrow \pi_* \widetilde{S}$$

is injective, and that the image is precisely the subsheaf of G^{an}–invariant elements of $\pi_* \widetilde{S}$. In other words we need to show that the sheaf $\ker(\widetilde{\theta}^*)$ vanishes, and then that the subsheaf $\widetilde{S^{G^{\text{an}}}} \subset \pi_* \widetilde{S}$ is precisely the invariants. Both are statements that can be checked on a basis for the topology of $X/G = \operatorname{Spec}(S^{G^{\mathrm{an}}})$. Our favorite basis has always been the basic open sets. Choose therefore an element $f \in S^{G^{\mathrm{an}}}$. If we put $Y = \operatorname{Spec}(S^{G^{\mathrm{an}}})$ and Y_f the basic open set corresponding to f, Remark 3.6.4 tells us that the inverse image of $Y_f \subset Y$ by the map $\operatorname{Spec}(\theta) : X \longrightarrow Y$ is $X_{\theta(f)} \subset X$. In our case $\theta : S^{G^{\mathrm{an}}} \longrightarrow S$ is just the obvious inclusion. We will identify $f \in S^{G^{\mathrm{an}}}$ with $\theta(f) \in S$; hence the inverse image of Y_f is X_f. What we have to prove is that the ring homomorphism

$$\widetilde{\theta}^*_{Y_f} : \Gamma(Y_f, \widetilde{S^{G^{\mathrm{an}}}}) \longrightarrow \Gamma(Y_f, \pi_*\widetilde{S}) \quad = \quad \Gamma(X_f, \widetilde{S})$$

is injective, and that the image is precisely the G^{an}–invariant elements in $\Gamma(X_f, \widetilde{S})$. Put more concretely, we need to show that the map from the ring $S^{G^{\mathrm{an}}}[1/f] = \Gamma(Y_f, \widetilde{S^{G^{\mathrm{an}}}})$ to the ring $S[1/f] = \Gamma(X_f, \widetilde{S})$ is injective, and identifies $S^{G^{\mathrm{an}}}[1/f]$ as the G^{an}–invariant elements. We need to prove:

Lemma 9.6.1. *Let $(G, \mathcal{O}_G)$ be a reductive group over $\mathbb{C}$ and let S be a G–ring. Let f be an element of $S^{G^{\mathrm{an}}} \subset S$. Then the natural ring homomorphism*

$$S^{G^{\mathrm{an}}}[1/f] \longrightarrow S[1/f]$$

is injective, and its image is precisely $\{S[1/f]\}^{G^{\mathrm{an}}} \subset S[1/f]$.

Proof. The map $\theta : S^{G^{\mathrm{an}}} \longrightarrow S$ is clearly injective, and the injectivity $S^{G^{\mathrm{an}}}[1/f] \longrightarrow S[1/f]$ follows from Reminder 7.3.1. We need to prove that the image is precisely $\{S[1/f]\}^{G^{\mathrm{an}}}$. That is, we must show that any G^{an}–invariant $x/f^n \in S[1/f]$ is the image of some element in $S^{G^{\mathrm{an}}}[1/f]$.

The element $f \in S^{G^{\mathrm{an}}} \subset S$ is invariant under the action of G^{an} on S. Corollary 9.2.13 tells us that the map $a' : S \longrightarrow R \otimes_{\mathbb{C}} S$ takes $f \in S$ to $1 \otimes f \in R \otimes_{\mathbb{C}} S$. We are in the situation of Exercise 8.3.12: the ring $S[1/f]$ is a G ring and $\alpha_f : S \longrightarrow S[1/f]$ is a G–homomorphism of G–rings. Remark 9.2.4 tells us that G^{an} acts on S and on $S[1/f]$, and that the map α_f intertwines the actions. Of course we already have actions of G^{an} on $S = \Gamma(X, \widetilde{S})$ and on $S[1/f] = \Gamma(X_f, \widetilde{S})$, which we obtained in Lemma 8.4.4, and, if we are going to be pedantic, the lemma we are proving now is about the old actions of Section 8.4. But the old actions agree with the new; for the ring S this was noted in Remark 9.2.5, while in Example 8.4.7 we observed that there is only one possible action of G^{an} on $S[1/f]$ for which $\alpha_f : S \longrightarrow S[1/f]$ is an intertwining operator.

It follows that $S[1/f]$ has a Reynolds operator $E : S[1/f] \longrightarrow S[1/f]$, and that the homomorphism $\alpha_f : S \longrightarrow S[1/f]$ commutes with the Reynolds operators: see Remark 9.3.16(i). In our notation we have often identified $x \in S$ with its image $\alpha_f(x) \in S[1/f]$. The fact that the Reynolds operator commutes with α_f says this will not lead to any confusion. If $x \in S$ and we write $E(x) \in S[1/f]$ we could either mean $E(\alpha_f(x))$ or $\alpha_f(E(x))$, and the fact that $E\alpha_f = \alpha_f E$ says these are equal.

The element f is G^{an} invariant, and hence so is $1/f^n$ for any integer n. Let $x/f^n \in S[1/f]$ be G^{an}–invariant. We compute

$$\frac{x}{f^n} = E\left(\frac{x}{f^n}\right) = \left(\frac{1}{f^n}\right)E(x)$$

where the second equality is by Lemma 9.3.18. This exhibits x/f^n as $E(x)/f^n$ with $E(x)$ in $S^{G^{\mathrm{an}}}$. Thus x/f^n is in the image of $S^{G^{\mathrm{an}}}[1/f]$. □

We have now completed the proof that the morphism of ringed spaces $(\pi, \pi^*) : (X, \mathcal{O}_X) \longrightarrow (X/G, \mathcal{O}_{X/G})$ identifies with the map

$$\left(\mathrm{Spec}(\theta), \widetilde{\theta}\right) : \left(\mathrm{Spec}(S), \widetilde{S}\right) \longrightarrow \left(\mathrm{Spec}(S^{G^{\mathrm{an}}}), \widetilde{S^{G^{\mathrm{an}}}}\right) .$$

That is, we have completed the proof of Theorem 8.8.6(i). In this section we still want to prove Theorem 8.8.6(ii) and (iii). That is, we want the statements about G–modules M for the G–ring S. Both are statements about the sheaves $\{\pi_* \widetilde{M}\}^{G^{\mathrm{an}}}$. The idea is that by now we understand very well what are the sheaves $\{\pi_* \widetilde{S}\}^{G^{\mathrm{an}}} = \{\pi_* \mathcal{O}_X\}^{G^{\mathrm{an}}}$, and Nagata's trick will allow us to reduce ourselves to this case. To use Nagata's trick, together with our understanding of the sheaf $\{\pi_* \mathcal{O}_X\}^{G^{\mathrm{an}}}$, we must be willing to replace $\mathcal{O}_X = \widetilde{S}$ by some sheaf of rings $\widetilde{S \oplus M}$ with a square-zero sheaf of ideals. We begin by untangling our definition of the sheaf $\{\pi_* \widetilde{M}\}^{G^{\mathrm{an}}}$.

Recall that Remark 8.3.14 tells us that the ring homomorphisms

$$S \xrightarrow{\ \psi\ } S \oplus M \xrightarrow{\ \varphi\ } S$$

are G–homomorphisms of G–rings. On the space X they induce homomorphisms of sheaves of rings

$$\widetilde{S} \xrightarrow{\ \widetilde{\psi}\ } \widetilde{S \oplus M} \xrightarrow{\ \widetilde{\varphi}\ } \widetilde{S} \,,$$

and, for any G–invariant open set $U \subset X$, the homomorphisms

$$\Gamma(U, \widetilde{S}) \xrightarrow{\Gamma(U, \widetilde{\psi})} \Gamma(U, \widetilde{S \oplus M}) \xrightarrow{\Gamma(U, \widetilde{\varphi})} \Gamma(U, \widetilde{S})$$

commute with the action of G^{an}. The action of G^{an} on $\Gamma(U, \widetilde{M}) = \ker\left(\Gamma(U, \widetilde{\varphi})\right)$ was defined as the action on the submodule of $\Gamma(U, \widetilde{S \oplus M})$; see Remark 8.4.9 and Corollary 8.4.10. If we let $U \subset X$ vary over all the open sets $U = \pi^{-1}V$, with $V \subset X/G$ an open subset, the ring homomorphisms above give, on X/G, morphisms of G^{an}–sheaves of rings

$$\pi_*\widetilde{S} \xrightarrow{\pi_*\widetilde{\psi}} \pi_*(\widetilde{S \oplus M}) \xrightarrow{\pi_*\widetilde{\varphi}} \pi_*\widetilde{S} .$$

The sheaf $\pi_*\widetilde{M}$ is obviously the kernel of the map $\pi_*\widetilde{\varphi}$, and the identification is compatible with the action of G^{an}. Hence the sheaf $\{\pi_*\widetilde{M}\}^{G^{\text{an}}}$ is the kernel of the map $\{\pi_*\widetilde{\varphi}\}^{G^{\text{an}}}$ below

$$\{\pi_*\widetilde{S}\}^{G^{\text{an}}} \xrightarrow{\{\pi_*\widetilde{\psi}\}^{G^{\text{an}}}} \{\pi_*(\widetilde{S \oplus M})\}^{G^{\text{an}}} \xrightarrow{\{\pi_*\widetilde{\varphi}\}^{G^{\text{an}}}} \{\pi_*\widetilde{S}\}^{G^{\text{an}}} .$$

But in Lemma 9.6.1 we completed the identification of these sheaves of rings. On the space $X/G = \operatorname{Spec}(S^{G^{\text{an}}}) = \operatorname{Spec}(\{S \oplus M\}^{G^{\text{an}}})$ these become

$$\widetilde{S^{G^{\text{an}}}} \xrightarrow{\widetilde{\psi^{G^{\text{an}}}}} \widetilde{\{S \oplus M\}^{G^{\text{an}}}} \xrightarrow{\widetilde{\varphi^{G^{\text{an}}}}} \widetilde{S^{G^{\text{an}}}} .$$

Since the action of G^{an} on the ring $S \oplus M$ respects the direct sum decomposition, we have $\{S \oplus M\}^{G^{\text{an}}} = S^{G^{\text{an}}} \oplus M^{G^{\text{an}}}$. This makes $\widetilde{\{S \oplus M\}^{G^{\text{an}}}} = \widetilde{S^{G^{\text{an}}} \oplus M^{G^{\text{an}}}}$. The sequence of morphisms of sheaves of rings on X/G can be rewritten ever so slightly as

$$\widetilde{S^{G^{\text{an}}}} \xrightarrow{\widetilde{\psi^{G^{\text{an}}}}} \widetilde{S^{G^{\text{an}}} \oplus M^{G^{\text{an}}}} \xrightarrow{\widetilde{\varphi^{G^{\text{an}}}}} \widetilde{S^{G^{\text{an}}}} .$$

Theorem 9.4.4 establishes that $S^{G^{\text{an}}}$ is a finitely generated $\mathbb{C}$–algebra, while Corollary 9.4.5 guarantees that $M^{G^{\text{an}}}$ is a finite module over $S^{G^{\text{an}}}$. The rewritten version of the sheaves of rings, on $X/G \cong \operatorname{Spec}(S^{G^{\text{an}}})$, is easily recognized from Definition 7.2.6. The kernel of the map $\widetilde{\varphi^{G^{\text{an}}}}$ is, by definition, the sheaf $\widetilde{M^{G^{\text{an}}}}$. Therefore, with all the identifications above, we have rewritten $\{\pi_*\widetilde{M}\}^{G^{\text{an}}}$ as $\widetilde{M^{G^{\text{an}}}}$. We note this for future reference:

Remark 9.6.2. The sheaf $\{\pi_*\widetilde{M}\}^{G^{\text{an}}}$ is isomorphic, as a sheaf of modules over $\{\pi_*\widetilde{S}\}^{G^{\text{an}}} = \widetilde{S^{G^{\text{an}}}}$, to the sheaf $\widetilde{M^{G^{\text{an}}}}$. In particular it is coherent.

This proves Theorem 8.8.6(ii); for this section only Theorem 8.8.6(iii) remains. We remind the reader: Theorem 8.8.6(iii) asserts that, given any exact sequence of G–modules $0 \longrightarrow M' \longrightarrow M \longrightarrow M'' \longrightarrow 0$ over the G–ring S, then the induced sequence of coherent sheaves on X/G, that is

$$0 \longrightarrow \{\pi_*\widetilde{M'}\}^{G^{\text{an}}} \longrightarrow \{\pi_*\widetilde{M}\}^{G^{\text{an}}} \longrightarrow \{\pi_*\widetilde{M''}\}^{G^{\text{an}}} \longrightarrow 0 ,$$

is exact. Now Remark 9.6.2 tells us that this sequence of coherent sheaves identifies with

$$0 \longrightarrow \widetilde{\{M'\}^{G^{\mathrm{an}}}} \longrightarrow \widetilde{M^{G^{\mathrm{an}}}} \longrightarrow \widetilde{\{M'\}^{G^{\mathrm{an}}}} \longrightarrow 0\,.$$

We need to prove this exact, and by Lemma 7.6.14 it suffices to show the exactness of the sequence of modules

$$0 \longrightarrow \{M'\}^{G^{\mathrm{an}}} \longrightarrow M^{G^{\mathrm{an}}} \longrightarrow \{M'\}^{G^{\mathrm{an}}} \longrightarrow 0\,.$$

But this we know, by Lemma 9.3.14. Thus the proof of Theorem 8.8.6(iii) is also complete.

9.7 Two technical lemmas

Consider the following. If $(X, \mathcal{O}) = \left(\mathrm{Spec}(S), \widetilde{S}\right)$ is an affine scheme of finite type over $\mathbb{C}$, and if M is a finite S–module, we have a sheaf $\widetilde{M}$ on X. Suppose f is an element of S and X_f is the corresponding basic open subset. Lemma 7.4.1 tells us that $\Gamma(X_f, \widetilde{M}) = M[1/f]$. In the case $f = 1$ this gives $\Gamma(X, \widetilde{M}) = M$. If we are willing to make our notation more cumbersome we could combine the two to obtain

$$\Gamma(X_f, \widetilde{M}) \quad = \quad \Gamma(X, \widetilde{M})[1/f]\,.$$

So far we have only reformulated facts we know. The technical improvement we want is the following:

Lemma 9.7.1. *As above, let $(X, \mathcal{O}) = \left(\mathrm{Spec}(S), \widetilde{S}\right)$ be an affine scheme of finite type over $\mathbb{C}$, and let M be a finite S–module. Suppose we are given an element $f \in S$, as well as an open subset U so that $X_f \subset U \subset X$. Then*

$$\Gamma(X_f, \widetilde{M}) \quad = \quad \Gamma(U, \widetilde{M})[1/f]\,.$$

In words rather than symbols, provided U contains the open set X_f, we can recover the S–module $\Gamma(X_f, \widetilde{M})$ from $\Gamma(U, \widetilde{M})$ by just inverting $f \in S$. For technical reasons, that will become apparent in the next section, we want to prove a more general version of this. In reading the remainder of the section it might help to bear in mind the simple example where $Y = X = \mathrm{Spec}(S)$ and $\mathscr{S} = \widetilde{M}$.

Definition 9.7.2. *A topological space Y is called* noetherian *if every open set is quasicompact. That is, every open cover of every open set has a finite subcover.*

Remark 9.7.3. If Y is a scheme of finite type over $\mathbb{C}$ then Corollary 5.2.3 tells us that Y is a noetherian topological space. Noetherian topological spaces have an especially simple sheaf theory; in this section we will prove a couple of facts about them.

Notation 9.7.4. Let us now set up the framework in which we want to state the next lemma. Let S be a ring, let Y be a noetherian topological space and let $\mathscr{S}$ be a sheaf of S–modules over Y. Suppose we are given some multiplicative subset $\Sigma \subset S$, and, for any $f \in \Sigma$, an open set $Y_f \subset Y$. We want to assume further:

(i) For any $f, g \in \Sigma$ we have $Y_f \cap Y_g = Y_{fg}$.
(ii) The open sets $Y_f, f \in \Sigma$ form a basis for the topology of Y.
(iii) For each $f \in \Sigma$ the element $f \in S$ acts invertibly on $\Gamma(Y_f, \mathscr{S})$; in other words $\Gamma(Y_f, \mathscr{S})$ is an $S[1/f]$ module.

Based on just the above assumptions we note

9.7.4.1. Let $f \in \Sigma$, and let $U \subset Y$ be an open set containing Y_f. Since $\mathscr{S}$ is a sheaf of S–modules the restriction homomorphism $\mathrm{res}^U_{Y_f}$: $\Gamma(U, \mathscr{S}) \longrightarrow \Gamma(Y_f, \mathscr{S})$ is a homomorphism of S–modules. By (iii) the element $f \in S$ acts invertibly on $\Gamma(Y_f, \mathscr{S})$. Reminder 7.3.1 tells us there is a unique factorization

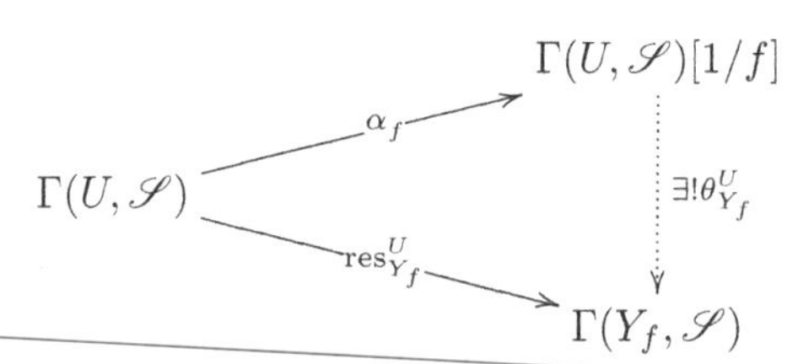

The map $\theta^U_{Y_f}$ exists for every pair $Y_f \subset U$. One special case is the inclusion $Y_{fg} \subset Y_f$, where $f, g \in \Sigma$. We prove

Lemma 9.7.5. *Let the conventions be as in Notation 9.7.4. Suppose that, for any $f, g \in \Sigma$, the map $\theta^{Y_f}_{Y_{fg}}$ of 9.7.4.1 is an isomorphism. Then it follows that, for any open sets $Y_f \subset U$, the map*

$$\theta^U_{Y_f} : \Gamma(U, \mathscr{S})[1/f] \longrightarrow \Gamma(Y_f, \mathscr{S})$$

must also be an isomorphism.

Proof. By Notation 9.7.4(ii) the open sets Y_g, $g \in \Sigma$ form a basis for the topology of Y. It is therefore possible to cover $U \subset Y$ by open sets Y_g. Because Y is noetherian it is possible to find a finite cover. Let us choose and fix such a finite cover; for the rest of the proof $U = \cup_{i=1}^{n} Y_{g_i}$, with $g_i \in \Sigma$.

We want to prove the map $\theta_{Y_f}^U$ to be an isomorphism. We need to prove it injective and surjective. We start with the injectivity. Suppose $s/f^m \in \Gamma(U, \mathscr{S})[1/f]$ lies in the kernel of $\theta_{Y_f}^U$. From 9.7.4.1 we have a commutative triangle

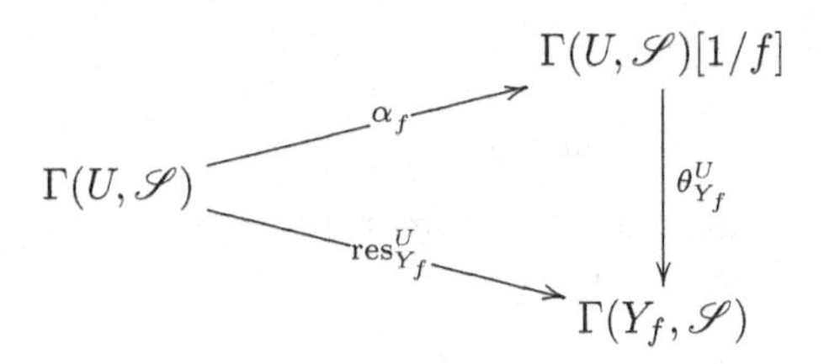

and by chasing the image $s \in \Gamma(U, \mathscr{S})$ by the two maps we conclude that s must be annihilated by the map $\mathrm{res}_{Y_f}^U : \Gamma(U, \mathscr{S}) \longrightarrow \Gamma(Y_f, \mathscr{S})$. We want to show that $f^N s = 0$ for some sufficient large integer N. Now U is a union $U = \cup_{i=1}^{n} Y_{g_i}$, and the sheaf axiom (see Definition 2.2.7(i)) tells us it suffices to show that all the restrictions $\mathrm{res}_{Y_{g_i}}^U (f^N s) = f^N \,\mathrm{res}_{Y_{g_i}}^U (s)$ vanish. Because the union is finite it suffices to show that for each $1 \le i \le n$ there exists an integer N_i, possibly depending on i, so that $f^{N_i} \,\mathrm{res}_{Y_{g_i}}^U (s) = 0$. But now consider the commutative square

$$\begin{array}{ccc} \Gamma(U,\mathscr{S}) & \xrightarrow{\mathrm{res}_{Y_f}^U} & \Gamma(Y_f,\mathscr{S}) \\ {\scriptstyle \mathrm{res}_{Y_{g_i}}^U}\downarrow & & \downarrow{\scriptstyle \mathrm{res}_{Y_{fg_i}}^{Y_f}} \\ \Gamma(Y_{g_i},\mathscr{S}) & \xrightarrow[\mathrm{res}_{Y_{fg_i}}^{Y_{g_i}}]{} & \Gamma(Y_{fg_i},\mathscr{S}) \end{array}$$

The composite

$$\begin{array}{ccc} \Gamma(U,\mathscr{S}) & \xrightarrow{\mathrm{res}_{Y_f}^U} & \Gamma(Y_f,\mathscr{S}) \\ & & \downarrow{\scriptstyle \mathrm{res}_{Y_{fg_i}}^{Y_f}} \\ & & \Gamma(Y_{fg_i},\mathscr{S}) \end{array}$$

certainly takes s to zero. Hence so does the composite from top left to

bottom right in the diagram

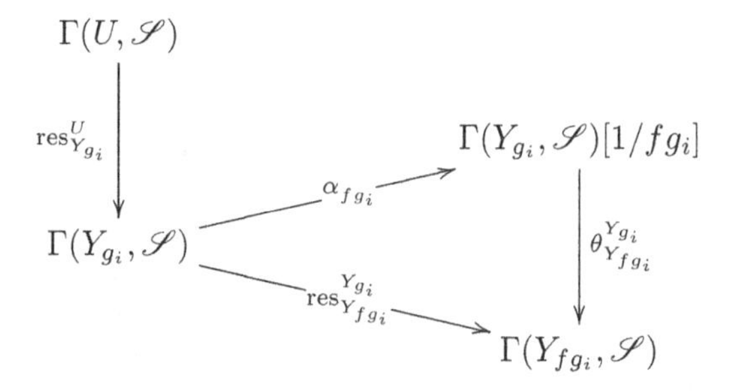

The map $\theta^{Y_{g_i}}_{Y_{fg_i}}$ is assumed an isomorphism, and hence the image of $\mathrm{res}^U_{Y_{g_i}}(s) \in \Gamma(Y_{g_i},\mathscr{S})$ by the map

$$\alpha_{fg_i} : \Gamma(Y_{g_i},\mathscr{S}) \longrightarrow \Gamma(Y_{g_i},\mathscr{S})[1/fg_i]$$

must vanish. But this means that, for some integer $N_i > 0$, we have the identity $(fg_i)^{N_i} \mathrm{res}^U_{Y_{g_i}}(s) = 0$. Since g_i acts invertibly on $\Gamma(Y_{g_i},\mathscr{S})$ we conclude that $f^{N_i} \mathrm{res}^U_{Y_{g_i}}(s) = 0$. This proves the injectivity.

Next we must prove the surjectivity. Let $s \in \Gamma(Y_f,\mathscr{S})$; we must exhibit s as $\theta^U_{Y_f}(r/f^m)$ for some $r/f^m \in \Gamma(U,\mathscr{S})[1/f]$. Put differently we must show that there exists some $m > 0$ so that $f^m s$ is the image of some $r \in \Gamma(U,\mathscr{S})$ by the composite

$$\Gamma(U,\mathscr{S}) \xrightarrow{\ \alpha_f\ } \Gamma(U,\mathscr{S})[1/f] \xrightarrow{\ \theta^U_{Y_f}\ } \Gamma(Y_f,\mathscr{S}) .$$

By 9.7.4.1 this composite is the map $\mathrm{res}^U_{Y_f} : \Gamma(U,\mathscr{S}) \longrightarrow \Gamma(Y_f,\mathscr{S})$.

For each $1 \le i \le n$ let $s_i \in \Gamma(Y_{fg_i},\mathscr{S})$ be $s_i = \mathrm{res}^{Y_f}_{Y_{fg_i}}(s)$. In the commutative triangle

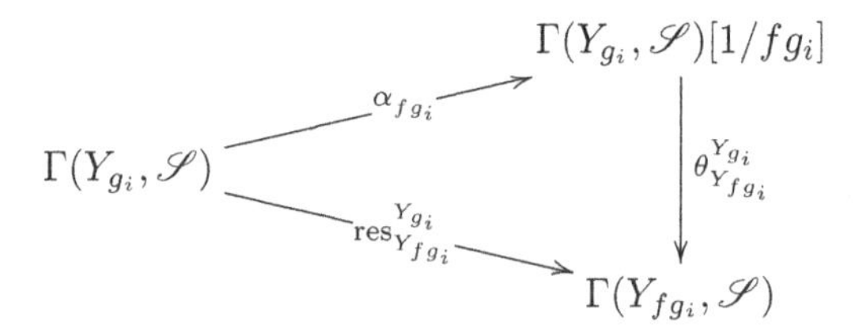

the map $\theta^{Y_{g_i}}_{Y_{fg_i}}$ is an isomorphism, in particular it is surjective. Hence we may choose an $\widetilde{s}_i \in \Gamma(Y_{g_i},\mathscr{S})[1/fg_i]$ with $\theta^{Y_{g_i}}_{Y_{fg_i}}(\widetilde{s}_i) = s_i$. And the map α_{fg_i} always has the property that, for any element $\widetilde{s}_i \in \Gamma(Y_{g_i},\mathscr{S})[1/fg_i]$, there exists an integer $m_i > 0$ so that $(fg_i)^{m_i}\widetilde{s}_i$ is in the image of α_{fg_i}. We can choose an element $\widetilde{r}_i \in \Gamma(Y_{g_i},\mathscr{S})$ with $(fg_i)^{m_i}\widetilde{s}_i = \alpha_{fg_i}(\widetilde{r}_i)$.

Since g_i acts invertibly on $\Gamma(Y_{g_i}, \mathscr{S})$ we can define $r_i = g_i^{-m_i}\widetilde{r}_i \in \Gamma(Y_{g_i}, \mathscr{S})$. Our formulas combine to give

$$\begin{aligned} \mathrm{res}^{Y_{g_i}}_{Y_{fg_i}}(r_i) &= \theta^{Y_{g_i}}_{Y_{fg_i}} \alpha_{fg_i}(g_i^{-m_i}\widetilde{r}_i) &= g_i^{-m_i}\theta^{Y_{g_i}}_{Y_{fg_i}}((fg_i)^{m_i}\widetilde{s}_i) \\ & &= f^{m_i}s_i\,. \end{aligned}$$

If $m = \max(m_i)$ for the finitely many integers $m_i > 0$, then we can replace r_i by $f^{m-m_i}r_i$ and obtain the simpler identities $f^m s_i = \mathrm{res}^{Y_{g_i}}_{Y_{fg_i}}(r_i)$.

Now consider the diagram where all the homomorphisms are restrictions

$$\begin{array}{ccccc} r_i \in \Gamma(Y_{g_i}, \mathscr{S}) & \longrightarrow & \Gamma(Y_{fg_i}, \mathscr{S}) & \longleftarrow & \Gamma(Y_f, \mathscr{S}) \ni f^m s \\ \downarrow & & \downarrow & & \\ \Gamma(Y_{g_ig_j}, \mathscr{S}) & \longrightarrow & \Gamma(Y_{fg_ig_j}, \mathscr{S}) & & \end{array}$$

The paragraph above establishes that the images in $\Gamma(Y_{fg_ig_j}, \mathscr{S})$ of r_i and $f^m s$ agree. The commutativity of the square means that the image of $\mathrm{res}^{Y_{g_i}}_{Y_{g_ig_j}}(r_i)$ by the map

$$\mathrm{res}^{Y_{g_ig_j}}_{Y_{fg_ig_j}} : \Gamma(Y_{g_ig_j}, \mathscr{S}) \longrightarrow \Gamma(Y_{fg_ig_j}, \mathscr{S})$$

is equal to $\mathrm{res}^{Y_f}_{Y_{fg_ig_j}}(f^m s)$, which is clearly independent of the order of g_i, g_j. In other words

$$\mathrm{res}^{Y_{g_ig_j}}_{Y_{fg_ig_j}} \mathrm{res}^{Y_{g_i}}_{Y_{g_ig_j}}(r_i) = \mathrm{res}^{Y_f}_{Y_{fg_ig_j}}(f^m s) = \mathrm{res}^{Y_{g_ig_j}}_{Y_{fg_ig_j}} \mathrm{res}^{Y_{g_j}}_{Y_{g_ig_j}}(r_j)\,.$$

On the open set $Y_{g_ig_j}$ we have two elements $\mathrm{res}^{Y_{g_i}}_{Y_{g_ig_j}}(r_i)$ and $\mathrm{res}^{Y_{g_j}}_{Y_{g_ig_j}}(r_j)$, and their restrictions to the subset $Y_{fg_ig_j}$ agree. Now recall the commutative triangle

$$\begin{array}{ccc} & & \Gamma(Y_{g_ig_j}, \mathscr{S})[1/fg_ig_j] \\ & \nearrow^{\alpha_{fg_ig_j}} & \downarrow \theta^{Y_{g_ig_j}}_{Y_{fg_ig_j}} \\ \Gamma(Y_{g_ig_j}, \mathscr{S}) & & \\ & \searrow_{\mathrm{res}^{Y_{g_ig_j}}_{Y_{fg_ig_j}}} & \\ & & \Gamma(Y_{fg_ig_j}, \mathscr{S}) \end{array}$$

We reasoned above that the two elements

$$\mathrm{res}^{Y_{g_i}}_{Y_{g_ig_j}}(r_i),\ \mathrm{res}^{Y_{g_j}}_{Y_{g_ig_j}}(r_j) \quad \in \quad \Gamma(Y_{g_ig_j}, \mathscr{S})$$

have the same image in $\Gamma(Y_{fg_ig_j}, \mathscr{S})$. The fact that $\theta^{Y_{g_ig_j}}_{Y_{fg_ig_j}}$ is an isomorphism means their images must agree already in $\Gamma(Y_{g_ig_j}, \mathscr{S})[1/fg_ig_j]$.

Hence there exists an integer N_{ij} with

$$(fg_ig_j)^{N_{ij}} \operatorname{res}^{Y_{g_i}}_{Y_{g_ig_j}}(r_i) \quad = \quad (fg_ig_j)^{N_{ij}} \operatorname{res}^{Y_{g_j}}_{Y_{g_ig_j}}(r_j) \,.$$

Since g_ig_j acts invertibly on $\Gamma(Y_{g_ig_j}, \mathscr{S})$ this identity also means

$$f^{N_{ij}} \operatorname{res}^{Y_{g_i}}_{Y_{g_ig_j}}(r_i) \quad = \quad f^{N_{ij}} \operatorname{res}^{Y_{g_j}}_{Y_{g_ig_j}}(r_j) \,.$$

Now there are finitely many integers N_{ij}. Replacing by the maximum $N = \max(N_{ij})$ we have that

$$\operatorname{res}^{Y_{g_i}}_{Y_{g_ig_j}}(f^N r_i) \quad = \quad \operatorname{res}^{Y_{g_j}}_{Y_{g_ig_j}}(f^N r_j) \,.$$

By the sheaf axiom of Definition 2.2.7(ii) the elements $f^N r_i \in \Gamma(Y_{g_i}, \mathscr{S})$ glue to a global section $r \in \Gamma(U, \mathscr{S})$, with $U = \cup_{i=1}^n Y_{g_i}$ as above.

Now the composite

$$\Gamma(U, \mathscr{S}) \xrightarrow{\operatorname{res}^U_{Y_{g_i}}} \Gamma(Y_{g_i}, \mathscr{S}) \xrightarrow{\operatorname{res}^{Y_{g_i}}_{Y_{fg_i}}} \Gamma(Y_{fg_i}, \mathscr{S})$$

takes $r \in \Gamma(U, \mathscr{S})$ to

$$\operatorname{res}^{Y_{g_i}}_{Y_{fg_i}} \operatorname{res}^U_{Y_{g_i}}(r) \quad = \quad \operatorname{res}^{Y_{g_i}}_{Y_{fg_i}}(f^N r_i) \quad = \quad f^N f^m s_i$$

But $s_i \in \Gamma(Y_{fg_i}, \mathscr{S})$ was defined as $s_i = \operatorname{res}^{Y_f}_{Y_{fg_i}}(s)$. This means that the two elements

$$f^{m+N} s, \quad \operatorname{res}^U_{Y_f}(r) \quad \in \quad \Gamma(Y_f, \mathscr{S})$$

have the same restrictions to each $Y_{fg_i} \subset Y_f$. Since $U = \cup_{i=1}^n Y_{g_i}$ it follows that $Y_f = \cup_{i=1}^n \{Y_f \cap Y_{g_i}\} = \cup_{i=1}^n Y_{fg_i}$. By the sheaf axiom of Definition 2.2.7(i) we conclude that $f^{m+N} s = \operatorname{res}^U_{Y_f}(r)$. □

One more lemma about noetherian topological spaces.

Lemma 9.7.6. *Let Y be a noetherian topological space, let $\mathscr{S}$ be a sheaf of $\mathbb{C}$–vector spaces on Y, and let R be some $\mathbb{C}$–vector space. For any open set $U \subset Y$, define $\Gamma(U, R \otimes_{\mathbb{C}} \mathscr{S})$ by the formula*

$$\Gamma(U, R \otimes_{\mathbb{C}} \mathscr{S}) \quad = \quad R \otimes_{\mathbb{C}} \Gamma(U, \mathscr{S}) \,.$$

With the obvious restriction maps $R \otimes_{\mathbb{C}} \mathscr{S}$ becomes a sheaf on Y.

Discussion 9.7.7. It is clear that $R \otimes_{\mathbb{C}} \mathscr{S}$ is a presheaf; we need to prove that the sheaf axiom holds. Let $V \subset Y$ be an open set, and let $V = \cup_{i \in I} U_i$ be an open cover. By Lemma 2.2.10 it suffices to prove that the sequence

$$0 \longrightarrow \Gamma(V, R \otimes_{\mathbb{C}} \mathscr{S}) \xrightarrow{\alpha} \prod_{i \in I} \Gamma(U_i, R \otimes_{\mathbb{C}} \mathscr{S}) \xrightarrow{\beta} \prod_{i,j \in I} \Gamma(U_i \cap U_j, R \otimes_{\mathbb{C}} \mathscr{S})$$

is exact. The proof proceeds via two lemmas; Lemma 9.7.8 assumes that the set I is finite, and Lemma 9.7.9 handles the general case.

Lemma 9.7.8. *For every open subset $V \subset Y$, and for any* finite *open cover $V = \cup_{i=1}^n U_i$, the sequence of Discussion 9.7.7 is exact.*

Proof. The sheaf $\mathscr{S}$ gives an exact sequence of vector spaces over $\mathbb{C}$

$$0 \longrightarrow \Gamma(V, \mathscr{S}) \longrightarrow \prod_{i=1}^n \Gamma(U_i, \mathscr{S}) \longrightarrow \prod_{i,j=1}^n \Gamma(U_i \cap U_j, \mathscr{S}) .$$

Tensoring with R we obtain an exact sequence

$$0 \longrightarrow R \otimes_{\mathbb{C}} \Gamma(V, \mathscr{S}) \longrightarrow \prod_{i=1}^n R \otimes_{\mathbb{C}} \Gamma(U_i, \mathscr{S}) \longrightarrow \prod_{i,j=1}^n R \otimes_{\mathbb{C}} \Gamma(U_i \cap U_j, \mathscr{S})$$

which identifies with

$$0 \longrightarrow \Gamma(V, R \otimes_{\mathbb{C}} \mathscr{S}) \xrightarrow{\alpha} \prod_{i=1}^n \Gamma(U_i, R \otimes_{\mathbb{C}} \mathscr{S}) \xrightarrow{\beta} \prod_{i,j=1}^n \Gamma(U_i \cap U_j, R \otimes_{\mathbb{C}} \mathscr{S})$$

Hence the lemma. □

Lemma 9.7.9. *Let $V \subset Y$ be an open subset, and let $V = \cup_{i \in I} U_i$ be any open cover. Then the sequence of Discussion 9.7.7 is exact.*

Proof. Because Y is noetherian every open set is quasicompact. The open cover for V has a finite subcover. Relabel the elements of the set I so that $U_1, U_2, \ldots, U_n$ cover U, and the other open sets in the cover are labeled U_t over the indices $t \in I - \{1, 2, \ldots, n\}$. We have a commutative diagram

$$\begin{array}{ccccccc}
0 \longrightarrow & \Gamma(V, R \otimes_{\mathbb{C}} \mathscr{S}) & \xrightarrow{\alpha} & \prod_{i \in I} \Gamma(U_i, R \otimes_{\mathbb{C}} \mathscr{S}) & \xrightarrow{\beta} & \prod_{i,j \in I} \Gamma(U_i \cap U_j, R \otimes_{\mathbb{C}} \mathscr{S}) \\
& \| & & \downarrow \pi & & \downarrow \pi' \\
0 \longrightarrow & \Gamma(V, R \otimes_{\mathbb{C}} \mathscr{S}) & \xrightarrow[\alpha']{} & \prod_{i=1}^n \Gamma(U_i, R \otimes_{\mathbb{C}} \mathscr{S}) & \xrightarrow[\beta']{} & \prod_{i,j=1}^n \Gamma(U_i \cap U_j, R \otimes_{\mathbb{C}} \mathscr{S})
\end{array}$$

where π and π' are the obvious projections. What we know so far, from Lemma 9.7.8 applied to the finite open cover $V = \cup_{i=1}^n U_i$, is that the bottom row in this diagram is exact. From the commutative square on the left we learn that $\alpha' = \pi\alpha$; the fact that α' is injective means that so is α.

Next choose an element $\prod_{i \in I} s_i$ in the kernel of β. The commutative square on the right informs us that the projection π takes $\prod_{i\in I} s_i$ to an element in the kernel of β', which equals the image of α'. It follows that $\prod_{i=1}^n s_i$ will lie in the image of α'. There is an element $s \in \Gamma(V, R\otimes_{\mathbb{C}}\mathscr{S})$ with $\mathrm{res}_{U_i}^V s = s_i$ for all $1 \le i \le n$. It clearly suffices to show that $\mathrm{res}_{U_t}^V(s) = s_t$ for all $t \in I - \{1,2,\dots,n\}$. Now take the two sections

$$s_t,\ \mathrm{res}_{U_t}^V(s) \quad\in\quad \Gamma(U_t, R\otimes_{\mathbb{C}}\mathscr{S})$$

and restrict to $U_i\cap U_t$, with $1\le i\le n$. Because $\prod_{i\in I} s_i$ lies in the kernel of β we have $\mathrm{res}_{U_i\cap U_t}^{U_t}(s_t) = \mathrm{res}_{U_i\cap U_t}^{U_i}(s_i)$. Therefore

$$\begin{aligned}\mathrm{res}_{U_i\cap U_t}^{U_t}\,\mathrm{res}_{U_t}^V(s) &= \mathrm{res}_{U_i\cap U_t}^{V}(s) &= \mathrm{res}_{U_i\cap U_t}^{U_i}\,\mathrm{res}_{U_i}^V(s)\\ &= \mathrm{res}_{U_i\cap U_t}^{U_i}(s_i) &= \mathrm{res}_{U_i\cap U_t}^{U_t}(s_t)\,.\end{aligned}$$

This means that $s_t - \mathrm{res}_{U_t}^V(s)$ maps to $0\in\Gamma(U_i\cap U_t, R\otimes_{\mathbb{C}}\mathscr{S})$ under the restriction maps $\mathrm{res}_{U_i\cap U_t}^{U_t}$. Lemma 9.7.8, applied to the finite open cover $U_t = \cup_{i=1}^n\{U_i\cap U_t\}$ of the open set $U_t\subset Y$, gives the exactness of the sequence

$$0 \longrightarrow \Gamma(U_t, R\otimes_{\mathbb{C}}\mathscr{S}) \xrightarrow{\ \alpha''\ } \prod_{i=1}^n \Gamma(U_i\cap U_t, R\otimes_{\mathbb{C}}\mathscr{S})\,.$$

We have shown that the element $s_t - \mathrm{res}_{U_t}^V(s)$ lies in the kernel of α'', and the injectivity of α'' allows us to conclude that $\mathrm{res}_{U_t}^V(s) = s_t$. □

9.8 The global statement about coherent sheaves

We have proved all the local statements in Theorem 8.8.6. The assertion that remains is Theorem 8.8.6(iv); this says that any coherent sheaf on $V\subset X/G$ is isomorphic to the restriction, to $V\subset X/G$, of the sheaf $\{\pi_*\widetilde{M}\}^{G^{\mathrm{an}}}$, for some G–module M over the G–ring S. If $\mathscr{S}$ is a coherent sheaf on V we now have to construct something global.

We return to the notation in Hypothesis 8.8.1. We are given an open set $U\subset X$, and we assume that it has an open cover by open sets X_f where the f's are somewhat special. Perhaps it might help to remind ourselves more precisely:

Reminder 9.8.1. We are given $(G, \mathcal{O}_G) = \left(\mathrm{Spec}(R), \widetilde{R}\right)$, which is a connected, reductive group. Also given is an affine scheme $(X, \mathcal{O}_X) = \left(\mathrm{Spec}(S), \widetilde{S}\right)$ acted on by $(G,\mathcal{O}_G)$. The action is determined by a ring homomorphism $a' : S \longrightarrow R\otimes_{\mathbb{C}} S$. The open subset $U\subset X$, also part of the data of Hypothesis 8.8.1, has an open cover by basic open sets $X_f \subset X$, where $f \in S$ satisfies $a'(f) = r \otimes f$.

We like the open sets described above so much that we will make a definition

Definition 9.8.2. *The set $\Sigma \subset S$ will be given by*

$$\Sigma \quad = \quad \left\{ f \in S \;\middle|\; \begin{array}{c} a'(f) = r \otimes f \\ \text{and } X_f \subset U \end{array} \right\}.$$

The basic open sets X_f, $f \in \Sigma$ will be called the good open sets.

Reminder 9.8.3. By Remark 8.4.14 the good open sets are all G–invariant.

Remark 9.8.4. If $(G, \mathcal{O}_G)$ is the trivial group, acting on the G–ring S as in Example 8.3.4, then any $f \in S$ satisfies $a'(f) = 1 \otimes f$. Thus Σ simplifies to

$$\Sigma \quad = \quad \{f \in S \mid X_f \subset U\}.$$

The next lemma is immediate for the trivial group, but needs a little argument in the general case.

Lemma 9.8.5. *The set $\Sigma \subset S$, and the good open subsets $X_f \subset U$, have the following two properties:*

(i) *If $f, g \in \Sigma$ then $fg \in \Sigma$ and $X_f \cap X_g = X_{fg}$.*
(ii) *Any G–invariant open subset of U admits an open cover by good open sets.*

Proof. To prove (i), suppose $f, g \in \Sigma$; that is $a'(f) = r \otimes f$, $a'(g) = r' \otimes g$. Then $a'(fg) = rr' \otimes fg$. The fact that $X_f \cap X_g = X_{fg}$ was proved in Lemma 3.2.5.

Next we need to prove (ii). Let W be a G–invariant open subset of U; we need to show that W can be covered by good open sets. Hypothesis 8.8.1(i) tells us that U can be covered by open sets X_f, where $X_f \subset X$ is good. It therefore suffices to show that $W \cap X_f$ can be covered by good open sets. Replacing W by $W \cap X_f$, if necessary, we may assume $W \subset X_f$, for some good open set X_f. For such a $W \subset X_f$, we must exhibit it as a union of good open sets.

The proof of Theorem 8.8.6(i) showed that the map

$$(\pi, \pi^*) : (X_f, \mathcal{O}_{X_f}) \longrightarrow (X_f/G, \mathcal{O}_{X_f/G})$$

can be identified with

$$\left(\operatorname{Spec}(S[1/f]), \widetilde{S[1/f]}\right) \xrightarrow{(\operatorname{Spec}(\varphi),\widetilde{\varphi})} \left(\operatorname{Spec}(S[1/f]^{G^{\mathrm{an}}}), \widetilde{S[1/f]^{G^{\mathrm{an}}}}\right) .$$

We are given a G–invariant open subset $W \subset X_f = \operatorname{Spec}(S[1/f])$, and Lemma 9.5.5 (or Lemma 8.6.9) tells us there is an open set $W' \subset X_f/G = \operatorname{Spec}(S[1/f]^{G^{\mathrm{an}}})$ with $W = \pi^{-1}W'$. The open set W' can be covered by basic open sets in the affine scheme X_f/G; it therefore suffices to show that the inverse image, in X_f, of a basic open set in X_f/G is a good open subset of U.

Any basic open subset of $X_f/G = \operatorname{Spec}(S[1/f]^{G^{\mathrm{an}}})$ is of the form $\{X_f/G\}_{g/f^n}$ with $g/f^n \in S[1/f]^{G^{\mathrm{an}}}$. The inverse image, by the map $\pi : X_f \longrightarrow X_f/G$, is the basic open set $\{X_f\}_{g/f^n}$ in $X_f = \operatorname{Spec}(S[1/f])$; see Remark 3.6.4. Since f is a unit in $S[1/f]$, Remark 3.2.3 tells us that $\{X_f\}_{1/f^n} = X_f$. Lemma 3.2.5 permits us to compute

$$\{X_f\}_{g/f^n} \quad = \quad \{X_f\}_g \cap \{X_f\}_{1/f^n} \quad = \quad \{X_f\}_g .$$

But now the inverse image of the basic open set $X_g \subset X = \operatorname{Spec}(S)$ by the open immersion $\operatorname{Spec}(\alpha_f) : \operatorname{Spec}(S[1/f]) \longrightarrow \operatorname{Spec}(S)$, which we identify with the inclusion $X_f \subset X$, is the open set $\{X_f\}_g$; see Remark 3.6.4. This makes $\{X_f\}_g = X_g \cap X_f = X_{fg}$. Now we have to study $a'(fg)$ to see if X_{fg} is good.

We know that $g/f^n \in S[1/f]^{G^{\mathrm{an}}}$ is invariant under the action of G^{an}. That is the map $a'_{S[1/f]} : S[1/f] \longrightarrow R \otimes_{\mathbb{C}} S[1/f]$ satisfies $a'_{S[1/f]}(g/f^n) = 1 \otimes g/f^n$. We are assuming $a'_S(f) = r \otimes f$. Hence

$$\begin{aligned} a'_{S[1/f]}(g) \;=\; a'_{S[1/f]}(g/f^n) \cdot a'_{S[1/f]}(f^n) \quad &= \quad \{1 \otimes g/f^n\} \cdot \{r^n \otimes f^n\} \\ &= \quad r^n \otimes g \, . \end{aligned}$$

This tells us that the element $a'_S(g) - r^n \otimes g \in R \otimes_{\mathbb{C}} S$ is in the kernel of the localization map $R \otimes_{\mathbb{C}} S \longrightarrow R \otimes_{\mathbb{C}} S[1/f]$. There exists an integer $m > 0$ with $(1 \otimes f^m)(a'_S(g) - r^n \otimes g) = 0$, the identity being in $R \otimes_{\mathbb{C}} S$. But then

$$a'_S(f^m g) \quad = \quad a'_S(f^m) a'_S(g) \quad = \quad (r^m \otimes f^m) a'_S(g) \quad = \quad r^{m+n} \otimes f^m g$$

and $X_{fg} = X_f \cap X_g = X_{f^m} \cap X_g = X_{f^m g}$. □

In the notation of Hypothesis 8.8.1 and Theorem 8.8.6 we had a continuous map $\pi : X \longrightarrow X/G$, an open set $V \subset X/G$ and its inverse image $U = \pi^{-1}V \subset X$. We define

Definition 9.8.6. *Let Σ be as in Definition 9.8.2. For each $f \in \Sigma$ define $V_f \subset V$ to be the unique open set with $\pi^{-1}V_f = X_f$.*

Remark 9.8.7. When $f \in \Sigma$ then the open set $X_f \subset U$ is G–invariant, and Lemma 8.6.9 tells us there is a unique $V_f \subset V$ with $\pi^{-1}V_f = X_f$. Definition 9.8.6 makes sense.

Lemma 9.8.8. *The sets V_f, $f \in \Sigma$ of Definition 9.8.6 satisfy the following properties:*

(i) *For any $f, g \in \Sigma$ we have $V_f \cap V_g = V_{fg}$.*
(ii) *The open sets V_f, $f \in \Sigma$ form a basis for the topology of V.*
(iii) *The scheme $(V_f, \{\mathcal{O}_{X/G}\}|_{V_f})$ of finite type over $\mathbb{C}$ is affine; more explicitly it is* $\left(\mathrm{Spec}(S[1/f]^{G^{\mathrm{an}}}), \widetilde{S[1/f]^{G^{\mathrm{an}}}}\right)$.

Remark 9.8.9. If $(G, \mathcal{O}_G)$ is the trivial group, acting on the G–ring S as in Example 8.3.4, then the map $(\pi, \pi^*) : (X, \mathcal{O}_X) \longrightarrow (X/G, \mathcal{O}_{X/G})$ is the identity and the lemma is immediate. Much of what follows simplifies in the case of the trivial group. We leave this to the reader; we note only that Theorem 8.8.6(iv) is interesting for the trivial group and we will apply it.

Proof. We begin with (i). The fact that $fg \in \Sigma$ is by Lemma 9.8.5(i). Next observe that

$$\pi^{-1}(V_f \cap V_g) \quad = \quad \pi^{-1}V_f \cap \pi^{-1}V_g \quad = \quad X_f \cap X_g \quad = \quad X_{fg}$$

Since V_{fg} is the unique open set in V with $\pi^{-1}V_{fg} = X_{fg}$ we conclude that $V_{fg} = V_f \cap V_g$.

To prove (ii) let W be any open subset of V. Then $\pi^{-1}W$ is a G–invariant open subset of U, and Lemma 9.8.5(ii) says there is an open cover of $\pi^{-1}W$ by open sets X_f, $f \in \Sigma$. Write X_f as $X_f = \pi^{-1}V_f$, and we have covered W by open sets V_f, $f \in \Sigma$.

Finally (iii) is by the proof of Theorem 8.8.6(i); see Remark 9.4.3. □

Construction 9.8.10. Now suppose we have a coherent sheaf $\mathscr{S}$ on V. We define $\Sigma' \subset \Sigma$ as follows

$$\Sigma' \quad = \quad \left\{ f \in \Sigma \;\middle|\; \begin{array}{c} \text{There exists a finite module } N \\ \text{for the ring } S[1/f]^{G^{\mathrm{an}}}, \text{ and an isomorphism} \\ \widetilde{N} \cong \mathscr{S}|_{V_f} \text{ of sheaves over } (V_f, \{\mathcal{O}_V\}|_{V_f}) \end{array} \right\}.$$

We know that $(V_f, \{\mathcal{O}_V\}|_{V_f}) = \left(\mathrm{Spec}(S[1/f]^{G^{\mathrm{an}}}), \widetilde{S[1/f]^{G^{\mathrm{an}}}}\right)$, from Lemma 9.8.8(iii); hence it makes sense to ask whether $\mathscr{S}|_{V_f} \cong \widetilde{N}$. In the remainder of the section we work with $\Sigma' \subset S$ and the corresponding open sets V_f, $f \in \Sigma'$.

Lemma 9.8.11. *The set $\Sigma' \subset S$ of Construction 9.8.10 satisfies the following two properties:*

(i) *For any $f, g \in \Sigma'$ we have $fg \in \Sigma'$ and $V_f \cap V_g = V_{fg}$.*
(ii) *The open sets V_f, $f \in \Sigma'$ form a basis for the topology of V.*

Proof. The proofs of both (i) and (ii) become easy once we remember the helpful Lemma 7.6.18(i). The lemma asserts that, if $(Y, \mathcal{O})$ is affine and $(W, \mathcal{O}|_W)$ is an open affine subset, then the restriction $\widetilde{M}|_W$ of an $\widetilde{M}$ on $(Y, \mathcal{O})$ is isomorphic to an $\widetilde{M'}$ on $(W, \mathcal{O}|_W)$.

For the proof of (i) observe that if $f, g \in \Sigma'$ then V_{fg} is an affine subset of V_f. Because $f \in \Sigma'$ the sheaf $\mathscr{S}|_{V_f}$ must be isomorphic to $\widetilde{M}$, and hence, on the open affine subset $V_{fg} \subset V_f$, we have an isomorphism $\mathscr{S}|_{V_{fg}} \cong \widetilde{M'}$.

To prove (ii) we need to show that every open subset $W \subset V$ can be covered by V_f, $f \in \Sigma'$. Now the coherence of $\mathscr{S}$ means that V can be covered by open affine sets W_i with $\mathscr{S}|_{W_i} \cong \widetilde{M_i}$. It suffices to show that each $W \cap W_i$ can be covered by open sets V_f, $f \in \Sigma'$. By Lemma 9.8.8 we know that $W \cap W_i$ has a cover by open sets V_f, $f \in \Sigma$. But the sheaf $\mathscr{S}|_{W_i}$ is isomorphic to $\widetilde{M_i}$, and for any open subset $V_f \subset W \cap W_i \subset W_i$ Lemma 7.6.18(i) tells us that $\mathscr{S}|_{V_f}$ must be isomorphic to $\widetilde{M_f}$. In other words if V_f, $f \in \Sigma$ is such that $V_f \subset W_i$, then $f \in \Sigma' \subset \Sigma$. □

Lemma 9.8.12. *For $f \in \Sigma'$ we give a formula*

$$\Gamma(V_f, \pi_*\pi^*\mathscr{S}) = S[1/f] \otimes_{S[1/f]^{G^{\mathrm{an}}}} \Gamma(V_f, \mathscr{S}) .$$

Let the restriction maps be the obvious choice. There is a unique extension of this to all open sets in V, which gives a sheaf $\pi_*\pi^*\mathscr{S}$ of S-modules on V. Furthermore, the sheaf $\pi_*\pi^*\mathscr{S}$ and the set $\Sigma' \subset S$ satisfy the hypotheses of Notation 9.7.4 and Lemma 9.7.5.

Proof. Perhaps we ought to note that $\Gamma(V_f, \mathscr{S})$ is a module over the ring $\Gamma(V_f, \mathcal{O}_V)$, and Lemma 9.8.8(iii) tells us that $\Gamma(V_f, \mathcal{O}_V) = S[1/f]^{G^{\mathrm{an}}}$. This is the meaning of the formula

$$\begin{aligned} \Gamma(V_f, \pi_*\pi^*\mathscr{S}) &= S[1/f] \otimes_{\Gamma(V_f, \mathcal{O}_V)} \Gamma(V_f, \mathscr{S}) \\ &= S[1/f] \otimes_{S[1/f]^{G^{\mathrm{an}}}} \Gamma(V_f, \mathscr{S}) \end{aligned}$$

defining $\Gamma(V_f, \pi_*\pi^*\mathscr{S})$. Let us first prove that, if the sheaf $\pi_*\pi^*\mathscr{S}$ exists, then it satisfies the hypotheses of Notation 9.7.4 and Lemma 9.7.5. Parts (i) and (ii) of Notation 9.7.4 are topological statements about the open sets V_f, $f \in \Sigma'$ and follow from Lemma 9.8.11. For part (iii) of Notation 9.7.4 notice the formula

$$\Gamma(V_f, \pi_*\pi^*\mathscr{S}) = S[1/f] \otimes_{S[1/f]^{G^{\mathrm{an}}}} \Gamma(V_f, \mathscr{S}) ,$$

which establishes that $\Gamma(V_f, \pi_*\pi^*\mathscr{S})$ is a module over $S[1/f]$. We must prove that the extra hypothesis of Lemma 9.7.5 is satisfied. To do so observe that, if $f, g \in \Sigma'$, then $V_{fg} \subset V_f$ are both affine; the helpful Lemma 7.6.18(ii) applies, and informs us that

$$\begin{aligned}\Gamma(V_{fg}, \mathscr{S}) &= \Gamma(V_{fg}, \mathcal{O}_V) \otimes_{\Gamma(V_f, \mathcal{O}_V)} \Gamma(V_f, \mathscr{S}) \\ &= S[1/fg]^{G^{\mathrm{an}}} \otimes_{S[1/f]^{G^{\mathrm{an}}}} \Gamma(V_f, \mathscr{S}) .\end{aligned}$$

This makes

$$\begin{aligned}\Gamma(V_{fg}, \pi_*\pi^*\mathscr{S}) &= S[1/fg] \otimes_{S[1/fg]^{G^{\mathrm{an}}}} \Gamma(V_{fg}, \mathscr{S}) \\ &= S[1/fg] \otimes_{S[1/fg]^{G^{\mathrm{an}}}} S[1/fg]^{G^{\mathrm{an}}} \otimes_{S[1/f]^{G^{\mathrm{an}}}} \Gamma(V_f, \mathscr{S}) \\ &= S[1/fg] \otimes_{S[1/f]^{G^{\mathrm{an}}}} \Gamma(V_f, \mathscr{S}) \\ &= S[1/fg] \otimes_{S[1/f]} S[1/f] \otimes_{S[1/f]^{G^{\mathrm{an}}}} \Gamma(V_f, \mathscr{S}) \\ &= \Gamma(V_f, \pi_*\pi^*\mathscr{S})[1/fg] ,\end{aligned}$$

and the hypothesis of Lemma 9.7.5 follows.

Now we have to show that the sheaf $\pi_*\pi^*\mathscr{S}$ exists. The formula specifies its values on a basis of the open sets; in this case, on the basis V_f, $f \in \Sigma'$. Lemma 3.4.10 tells us that, to check that a sheaf exists, we need only verify that, whenever we have an open cover of a basic open set V_f by basic open sets $V_f = \cup_{i \in I} V_{g_i}$, the sequence of Lemma 2.2.10 is exact. We may therefore confine attention to the open set V_f; it suffices to produce on V_f a sheaf extending, to all open sets in V_f, the formula for $\Gamma(V_g, \pi_*\pi^*\mathscr{S})$. To do this, let us consider that $S[1/f]$–module

$$M \quad = \quad \Gamma(V_f, \pi_*\pi^*\mathscr{S}) \quad = \quad S[1/f] \otimes_{S[1/f]^{G^{\mathrm{an}}}} \Gamma(V_f, \mathscr{S}) .$$

On $X_f = \mathrm{Spec}(S[1/f])$ there is a sheaf $\widetilde{M}$, and by Lemma 7.4.1 we have the equality $\Gamma(X_{fg}, \widetilde{M}) = M[1/fg]$. The sheaf $\pi_*\widetilde{M}$ is a sheaf on $V_f = X_f/G$ by Lemma 7.1.14, and

$$\Gamma(V_{fg}, \pi_*\widetilde{M}) \quad = \quad \Gamma(\pi^{-1}V_{fg}, \widetilde{M}) \quad = \quad \Gamma(X_{fg}, \widetilde{M}) \quad = \quad M[1/fg] ,$$

in other words

$$\Gamma(V_{fg}, \pi_*\widetilde{M}) \quad = \quad \Gamma(V_f, \pi_*\pi^*\mathscr{S})[1/fg] .$$

The first paragraph of the proof told us that this agrees with the S–module $\Gamma(V_{fg}, \pi_*\pi^*\mathscr{S})$. That is

$$\Gamma(V_{fg}, \pi_*\widetilde{M}) \quad = \quad \Gamma(V_{fg}, \pi_*\pi^*\mathscr{S}) .$$

If $f, g \in \Sigma'$ and $V_g \subset V_f$ then $V_{fg} = V_f \cap V_g = V_g$. Thus the formula

$$\Gamma(V_g, \pi_*\widetilde{M}) \quad = \quad \Gamma(V_g, \pi_*\pi^*\mathscr{S})$$

holds for any $g \in \Sigma'$ with $V_g \subset V_f$. $\square$

Aside for the Experts 9.8.13. We have a map $\pi : U \longrightarrow V$, and a coherent sheaf $\mathscr{S}$ on V. There is a coherent sheaf $\pi^*\mathscr{S}$ on U, and a quasicoherent $\pi_*\pi^*\mathscr{S}$ on V. Since we have not defined the sheaf $\pi^*\mathscr{S}$ and have not told the reader what a quasicoherent sheaf is, we constructed the sheaf directly in Lemma 9.8.12.

Remark 9.8.14. Let $(G, \mathcal{O}_G) = \left(\operatorname{Spec}(R), \widetilde{R}\right)$ be the affine group scheme acting on the space $(X, \mathcal{O}_X)$. The sheaf $\pi_*\pi^*\mathscr{S}$ was defined in Lemma 9.8.12, and Lemma 9.7.6 defined for us a sheaf $R \otimes_{\mathbb{C}} \pi_*\pi^*\mathscr{S}$. For each open set V_f, $f \in \Sigma'$ there is a homomorphism

$$S[1/f] \otimes_{S[1/f]^{G^{\mathrm{an}}}} \Gamma(V_f, \mathscr{S}) \xrightarrow{a' \otimes 1} R \otimes_{\mathbb{C}} S[1/f] \otimes_{S[1/f]^{G^{\mathrm{an}}}} \Gamma(V_f, \mathscr{S})$$

which, with our definitions of $\pi_*\pi^*\mathscr{S}$ and $R \otimes_{\mathbb{C}} \pi_*\pi^*\mathscr{S}$, comes to a homomorphism

$$a'_{V_f} : \Gamma(V_f, \pi_*\pi^*\mathscr{S}) \longrightarrow \Gamma(V_f, R \otimes_{\mathbb{C}} \pi_*\pi^*\mathscr{S}) \,.$$

These homomorphisms are clearly compatible with restrictions. Explicitly this means that, if $V_g \subset V_f$ with $f, g \in \Sigma'$, then the square below commutes

$$\begin{array}{ccc}
\Gamma(V_f, \pi_*\pi^*\mathscr{S}) & \xrightarrow{a'_{V_f}} & \Gamma(V_f, R \otimes_{\mathbb{C}} \pi_*\pi^*\mathscr{S}) \\
\Big\downarrow \mathrm{res}^{V_f}_{V_g} & & \Big\downarrow \mathrm{res}^{V_f}_{V_g} \\
\Gamma(V_g, \pi_*\pi^*\mathscr{S}) & \xrightarrow[a'_{V_g}]{} & \Gamma(V_g, R \otimes_{\mathbb{C}} \pi_*\pi^*\mathscr{S})
\end{array} \quad .$$

This already means that there is a morphism of sheaves $a' : \pi_*\pi^*\mathscr{S} \longrightarrow R \otimes_{\mathbb{C}} \pi_*\pi^*\mathscr{S}$ extending the above to all open sets $W \subset V$. The reader can check that this defines an action of the group $(G, \mathcal{O}_G)$ on the sheaf of S–modules $\pi_*\pi^*\mathscr{S}$. That is

(i) The following square commutes

$$\begin{array}{ccc}
\pi_*\pi^*\mathscr{S} & \xrightarrow{a'} & R \otimes_{\mathbb{C}} \pi_*\pi^*\mathscr{S} \\
\Big\downarrow a' & & \Big\downarrow 1 \otimes a' \\
R \otimes_{\mathbb{C}} \pi_*\pi^*\mathscr{S} & \xrightarrow[\mu' \otimes 1]{} & R \otimes_{\mathbb{C}} R \otimes_{\mathbb{C}} \pi_*\pi^*\mathscr{S}
\end{array} \quad .$$

(ii) The following triangle commutes

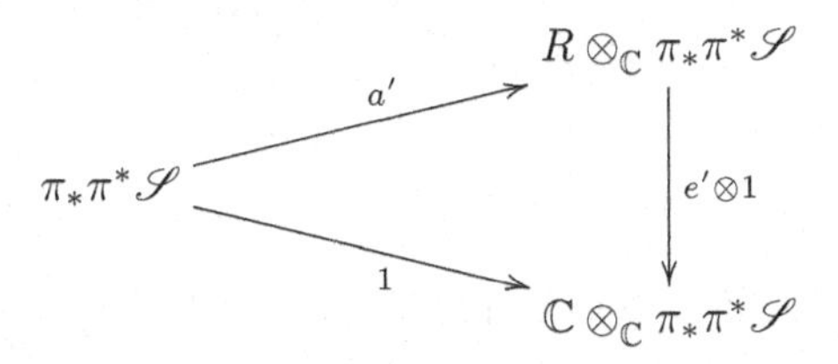

(iii) The following diagram commutes

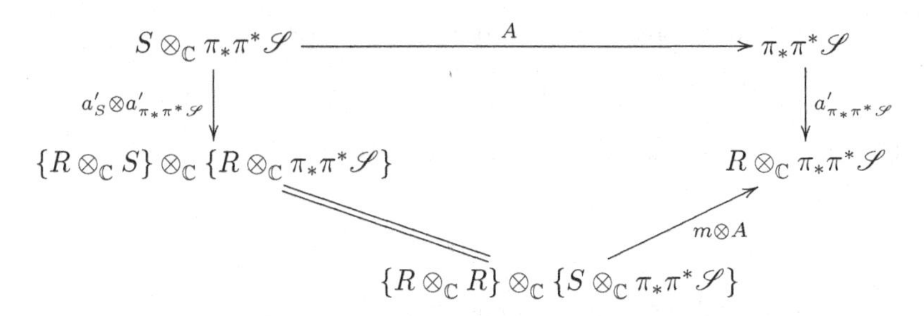

The hint is that the commutativity of the above diagrams can be checked on any basis of open sets, and it is easy on the open sets V_f, $f \in \Sigma'$.

Remark 9.8.15. We conclude that, for every open set $W \subset V$, the vector space $\Gamma(W, \pi_*\pi^*\mathscr{S})$ is a large G–module for the G–ring S. We have a map

$$\Gamma(W, \pi_*\pi^*\mathscr{S}) \xrightarrow{\Gamma(W,a')} \Gamma(W, R \otimes_{\mathbb{C}} \pi_*\pi^*\mathscr{S}) \quad = \quad R \otimes_{\mathbb{C}} \Gamma(W, \pi_*\pi^*\mathscr{S})$$

where the equality on the right is by Lemma 9.7.6. The three commutative diagrams of maps of sheaves in Remark 9.8.14(i), (ii) and (iii) mean that, on every open set $W \subset V$, the diagrams of Remark 8.3.16 commute. There is, however, no guarantee that the module $\Gamma(W, \pi_*\pi^*\mathscr{S})$ will be finite over S. In fact it almost never is. The reader can consider the case where $W = V_f$, $f \in \Sigma'$ as an example.

Remark 9.8.16. Among other things Remark 9.8.15 tells us that the group G^{an} acts on $\Gamma(W, \pi_*\pi^*\mathscr{S})$, for every open set $W \subset V$, and that the action is compatible with restriction maps. We are in the situation of Construction 8.7.2; we can form a sheaf $\{\pi_*\pi^*\mathscr{S}\}^{G^{\text{an}}}$ on V by the formula

$$\Gamma\left(W, \{\pi_*\pi^*\mathscr{S}\}^{G^{\text{an}}}\right) \quad = \quad \Gamma\left(W, \pi_*\pi^*\mathscr{S}\right)^{G^{\text{an}}}.$$

Lemma 9.8.17. *The sheaf* $\{\pi_*\pi^*\mathscr{S}\}^{G^{\text{an}}}$ *is isomorphic to* $\mathscr{S}$.

Proof. This can be checked on a basis of the open sets, and we check it on the V_f, $f \in \Sigma'$. What is being asserted is that $\Gamma(V_f, \mathscr{S})$ agrees with the G^{an}–invariant elements of $\Gamma\big(V_f, \pi_*\pi^*\mathscr{S}\big)$. The formula for $\Gamma\big(V_f, \pi_*\pi^*\mathscr{S}\big)$ was

$$\Gamma\big(V_f, \pi_*\pi^*\mathscr{S}\big) \quad = \quad S[1/f] \otimes_{S[1/f]^{G^{\mathrm{an}}}} \Gamma(V_f, \mathscr{S}) \ .$$

Simplifying the notation, by putting $T = S[1/f]$ and $M = \Gamma(V_f, \mathscr{S})$, we must prove that, if T is G–ring and M is a finite $T^{G^{\mathrm{an}}}$–module, then

$$M \quad = \quad \{T \otimes_{T^{G^{\mathrm{an}}}} M\}^{G^{\mathrm{an}}} \ .$$

Now M is finitely presented. Choose a presentation

$$\{T^{G^{\mathrm{an}}}\}^m \longrightarrow \{T^{G^{\mathrm{an}}}\}^n \longrightarrow M \longrightarrow 0 \ .$$

The tensor product is right exact; if we tensor with T we have an exact sequence

$$T^m \longrightarrow T^n \longrightarrow T \otimes_{T^{G^{\mathrm{an}}}} M \longrightarrow 0 \ .$$

Lemma 9.3.14 allows us to conclude that

$$\{T^{G^{\mathrm{an}}}\}^m \longrightarrow \{T^{G^{\mathrm{an}}}\}^n \longrightarrow \{T \otimes_{T^{G^{\mathrm{an}}}} M\}^{G^{\mathrm{an}}} \longrightarrow 0$$

is also exact, and hence we have an identification

$$M \quad = \quad \{T \otimes_{T^{G^{\mathrm{an}}}} M\}^{G^{\mathrm{an}}} \ .$$

□

Remark 9.8.18. We have now exhibited $\mathscr{S}$ as $\mathscr{T}^{G^{\mathrm{an}}}$, where $\mathscr{T} = \pi_*\pi^*\mathscr{S}$ is a G^{an}–sheaf on V. It remains to find a G–module M for the G–ring S, and an isomorphism of the G^{an}–sheaf $\pi_*\pi^*\mathscr{S}$ with $\pi_*\widetilde{M}|_V$. This is what we will do in the rest of this section.

Lemma 9.8.19. *With the notation as above the sheaf $R \otimes_{\mathbb{C}} \pi_*\pi^*\mathscr{S}$, viewed as a sheaf of S–modules over the space V, satisfies the hypotheses of Notation 9.7.4 and Lemma 9.7.5.*

Proof. Parts (i) and (ii) of Notation 9.7.4 are topological statements about the open sets V_f, $f \in \Sigma'$ and follow from Lemma 9.8.11. The formula

$$\Gamma(V_f, R \otimes_{\mathbb{C}} \pi_*\pi^*\mathscr{S}) \quad = \quad R \otimes_{\mathbb{C}} S[1/f] \otimes_{S[1/f]^{G^{\mathrm{an}}}} \Gamma(V_f, \mathscr{S})$$

establishes that $\Gamma(V_f, R \otimes_{\mathbb{C}} \pi_*\pi^*\mathscr{S})$ is a module over $S[1/f]$, which means that Notation 9.7.4(iii) holds. It remains to establish the hypothesis of Lemma 9.7.5.

Now Lemma 9.8.12 tells us that if $f, g \in \Sigma'$ then

$$\Gamma(V_{fg}, \pi_*\pi^*\mathscr{S}) \quad = \quad \Gamma(V_f, \pi_*\pi^*\mathscr{S})[1/fg]\ ,$$

which means that

$$R \otimes_{\mathbb{C}} \Gamma(V_{fg}, \pi_*\pi^*\mathscr{S}) \quad = \quad R \otimes_{\mathbb{C}} \Gamma(V_f, \pi_*\pi^*\mathscr{S})[1/fg]$$

or, in other words,

$$\Gamma(V_{fg}, R \otimes_{\mathbb{C}} \pi_*\pi^*\mathscr{S}) \quad = \quad \Gamma(V_f, R \otimes_{\mathbb{C}} \pi_*\pi^*\mathscr{S})[1/fg]\ .$$

This establishes our lemma. □

Finally we are ready for the proof of Theorem 8.8.6(iv). We prove

Lemma 9.8.20. *Theorem 8.8.6(iv) is true. That is every coherent sheaf $\mathscr{S}$ on V is isomorphic to $\{\pi_*\widetilde{M}\}^{G^{\mathrm{an}}}$, for some G–module M for the G–ring S.*

Proof. By Remark 9.8.18 we only need to show that our G^{an}–sheaf $\pi_*\pi^*\mathscr{S}$ is isomorphic to $\pi_*\widetilde{M}$, for some G–module M over the G–ring S. Now note that both the sheaves $\pi_*\pi^*\mathscr{S}$ and $R \otimes_{\mathbb{C}} \pi_*\pi^*\mathscr{S}$ satisfy the hypotheses of Lemma 9.7.5. For $\pi_*\pi^*\mathscr{S}$ we proved this in Lemma 9.8.12, while for $R \otimes_{\mathbb{C}} \pi_*\pi^*\mathscr{S}$ see Lemma 9.8.19. It follows that the conclusion of Lemma 9.7.5 holds for both sheaves. For any $f \in \Sigma'$ the maps

$$\begin{array}{rlcl} \theta^V_{V_f}: & \Gamma(V, \pi_*\pi^*\mathscr{S})[1/f] & \longrightarrow & \Gamma(V_f, \pi_*\pi^*\mathscr{S}) \\ \theta^V_{V_f}: & \Gamma(V, R \otimes_{\mathbb{C}} \pi_*\pi^*\mathscr{S})[1/f] & \longrightarrow & \Gamma(V_f, R \otimes_{\mathbb{C}} \pi_*\pi^*\mathscr{S}) \end{array}$$

are both isomorphisms. In the commutative square

$$\begin{array}{ccc} \Gamma(V, \pi_*\pi^*\mathscr{S}) & \xrightarrow{a'_V} & \Gamma(V, R \otimes_{\mathbb{C}} \pi_*\pi^*\mathscr{S}) \\ {\scriptstyle \mathrm{res}^V_{V_f}}\downarrow & & \downarrow{\scriptstyle \mathrm{res}^V_{V_f}} \\ \Gamma(V_f, \pi_*\pi^*\mathscr{S}) & \xrightarrow[a'_{V_f}]{} & \Gamma(V_f, R \otimes_{\mathbb{C}} \pi_*\pi^*\mathscr{S}) \end{array}$$

the bottom row can be identified, by making use of the isomorphisms $\theta^V_{V_f}$ of Lemma 9.7.5, with

$$a'_V[1/f]\ :\ \Gamma(V, \pi_*\pi^*\mathscr{S})[1/f] \longrightarrow \Gamma(V, R \otimes_{\mathbb{C}} \pi_*\pi^*\mathscr{S})[1/f]\ .$$

Since the maps are the same we will use the briefer a'_{V_f} for $a'_V[1/f]$. In any case we know the map a'_{V_f}; the map of sheaves $a' : \pi_*\pi^*\mathscr{S} \longrightarrow R \otimes_{\mathbb{C}} \pi_*\pi^*\mathscr{S}$ was defined in terms of what it does on the open sets V_f, $f \in \Sigma'$. See Remark 9.8.14. Let us simplify the notation by putting $N_f = \Gamma(V_f, \mathscr{S})$ and $M_f = \Gamma(V_f, \pi_*\pi^*\mathscr{S}) = S[1/f] \otimes_{S[1/f]^{G^{\mathrm{an}}}} N_f$. The most important fact is that N_f is a finite $S[1/f]^{G^{\mathrm{an}}}$–module and the formula above shows that M_f is a finite $S[1/f]$–module. We have a G–module M_f for the G–ring $S[1/f]$. By Corollary 9.2.18 there is a finite dimensional linear representation W_f of $(G, \mathcal{O}_G)$, and a map of representations $W_f \longrightarrow M_f = \Gamma(V, \pi_*\pi^*\mathscr{S})[1/f]$, so that the image of W_f generates M_f as a module over the ring $S[1/f]$.

Replacing W_f by its image we may assume the map $W_f \longrightarrow M_f$ is injective. This gives us a finite dimensional vector subspace $W_f \subset \Gamma(V, \pi_*\pi^*\mathscr{S})[1/f]$, and the action a'_{V_f} takes W_f to $R \otimes_{\mathbb{C}} W_f$. Choose a basis of W_f, that is choose elements $\{w_1/f^m, w_2/f^m, \ldots, w_n/f^m\} \subset \Gamma(V, \pi_*\pi^*\mathscr{S})[1/f]$ which form a basis for the subspace. The elements $\{w_1, w_2, \ldots, w_n\} \subset \Gamma(V, \pi_*\pi^*\mathscr{S})$ generate a vector subspace which we will call W'_f. It need not be true that $a'_V W'_f \subset R \otimes_{\mathbb{C}} W'_f$; however when we project via the map $\alpha_f : \Gamma(V, \pi_*\pi^*\mathscr{S}) \longrightarrow \Gamma(V, \pi_*\pi^*\mathscr{S})[1/f]$ we have an identity

$$a'_{V_f}\left(f^m \cdot \frac{w_i}{f^m}\right) = a'_{V_f}(f^m) \cdot a'_{V_f}\left(\frac{w_i}{f^m}\right) = (r^m \otimes f^m) \cdot a'_{V_f}\left(\frac{w_i}{f^m}\right) .$$

We also know that $a'_{V_f} W_f \subset R \otimes_{\mathbb{C}} W_f$; there must be identities

$$a'_{V_f}\left(\frac{w_i}{f^m}\right) = \sum_{j=1}^{n} r_{ij} \otimes \left(\frac{w_j}{f^m}\right) .$$

Combining these, we have the identity $a'_{V_f}(w_i) = \sum_{j=1}^{n} r^m r_{ij} \otimes w_j$. This is an identity in $\Gamma(V_f, R \otimes \pi_*\pi^*\mathscr{S}) = \Gamma(V, R \otimes \pi_*\pi^*\mathscr{S})[1/f]$, but we know that the restriction map $\mathrm{res}^V_{V_f}$ respects the action a'. In other words, the expressions

$$a'_V(w_i) - \sum_{j=1}^{n} r^m r_{ij} \otimes w_j$$

are elements of $\Gamma(V, R \otimes \pi_*\pi^*\mathscr{S})$, which are annihilated by $\mathrm{res}^V_{V_f}$. They must be killed by some sufficiently high power of f. There exists an integer N so that, for all i,

$$(1 \otimes f^N) \cdot a'_V(w_i) = (1 \otimes f^N) \cdot \sum_{j=1}^{n} r^m r_{ij} \otimes w_j .$$

Therefore

$$\begin{aligned} a'_V(f^N w_i) &= a'(f^N) \cdot a'_V(w_i) \\ &= (r^N \otimes f^N) \cdot \sum_{j=1}^{n} r^m r_{ij} \otimes w_j \\ &= \sum_{j=1}^{n} r^{m+N} r_{ij} \otimes f^N w_j \ , \end{aligned}$$

and we conclude that $a'_V(f^N W'_f) \subset R \otimes_{\mathbb{C}} (f^N W'_f)$.

The topological space V can be covered by open sets V_f, $f \in \Sigma'$. Since V is quasicompact there is a finite cover. Let us choose and fix such a cover; put $V = \cup_{i=1}^n V_{f_i}$. Let us compactify the notation by writing $V_i = V_{f_i}$. The beginning of the proof tells us that, for each $1 \le i \le n$, we may choose a finite dimensional linear subrepresentation $W'_i = W'_{f_i} \subset \Gamma(V, \pi_*\pi^*\mathscr{S})$, so that the image of W'_i in $\Gamma(V_i, \pi_*\pi^*\mathscr{S})$ generates $\Gamma(V_i, \pi_*\pi^*\mathscr{S})$ as a module over $S[1/f_i]$. Let $W \subset \Gamma(V, \pi_*\pi^*\mathscr{S})$ be the subspace generated by all the W'_i, $1 \le i \le n$. The vector subspace W is finite dimensional, it is a G–subrepresentation of $\Gamma(V, \pi_*\pi^*\mathscr{S})$, and its image in each $\Gamma(V_i, \pi_*\pi^*\mathscr{S})$ generates the module over $S[1/f_i]$.

We have constructed a G–homomorphism, from the finite dimensional representation W of $(G, \mathcal{O}_G)$ to the large G–module $\Gamma(V, \pi_*\pi^*\mathscr{S})$ for the G–ring S. Exercise 9.2.20 extends it to a G–homomorphism of large G–modules

$$\Phi : S \otimes_{\mathbb{C}} W \longrightarrow \Gamma(V, \pi_*\pi^*\mathscr{S}) \ .$$

Let M be the image of Φ. Then Φ factors as

$$S \otimes_{\mathbb{C}} W \xrightarrow{\ \alpha\ } M \xrightarrow{\ \beta\ } \Gamma(V, \pi_*\pi^*\mathscr{S}) \ .$$

with α surjective and β injective. Both α and β are G–homomorphisms of large G–modules. Since $S \otimes_{\mathbb{C}} W$ is finite over S and α is surjective we deduce that M is also finite; this makes M a G–module over the G–ring S. The proof of the lemma will be concluded once we show that

$$\pi_*\widetilde{M}|_V \ \cong \ \pi_*\pi^*\mathscr{S} \ ,$$

as G^{an}–sheaves on V.

For every V_f, $f \in \Sigma'$ we can consider the composite

$$M[1/f] \xrightarrow{\ \beta[1/f]\ } \Gamma(V, \pi_*\pi^*\mathscr{S})[1/f] \xrightarrow{\ \theta^V_{V_f}\ } \Gamma(V_f, \pi_*\pi^*\mathscr{S}) \ .$$

It defines a G^{an}–map from the $S[1/f]$–module $\Gamma(V_f, \pi_*\widetilde{M}) = M[1/f]$ to the $S[1/f]$–module $\Gamma(V_f, \pi_*\pi^*\mathscr{S})$. Let us denote this map

$$\beta_f : \Gamma(V_f, \pi_*\widetilde{M}) \longrightarrow \Gamma(V_f, \pi_*\pi^*\mathscr{S}) .$$

The maps β_f commute with restriction; the reader should check that, if $f, g \in \Sigma'$ and $V_f \subset V_g$, then the square

$$\begin{array}{ccc} \Gamma(V_g, \pi_*\widetilde{M}) & \xrightarrow{\beta_g} & \Gamma(V_g, \pi_*\pi^*\mathscr{S}) \\ {\scriptstyle \mathrm{res}^{V_g}_{V_f}}\downarrow & & \downarrow{\scriptstyle \mathrm{res}^{V_g}_{V_f}} \\ \Gamma(V_f, \pi_*\widetilde{M}) & \xrightarrow[\beta_f]{} & \Gamma(V_f, \pi_*\pi^*\mathscr{S}) \end{array}$$

commutes. It follows that there is an extension to a G^{an}–morphism of G^{an}–sheaves $\widetilde{\beta} : \pi_*\widetilde{M} \longrightarrow \pi_*\pi^*\mathscr{S}$. It suffices to prove that $\widetilde{\beta}$ is an isomorphism.

Now for the f_i, $1 \le i \le n$, which we used above to construct W, we know that the composite

$$\{S \otimes_{\mathbb{C}} W\}[1/f_i] \xrightarrow{\alpha[1/f_i]} M[1/f_i] \xrightarrow{\beta[1/f_i]} \Gamma(V, \pi_*\pi^*\mathscr{S})[1/f_i]$$

is surjective. Hence $\beta[1/f_i]$ is surjective. Furthermore β is injective, and Reminder 7.3.1 tells us that $\beta[1/f_i]$ must also be injective. Therefore $\beta[1/f_i]$ is an isomorphism for each $1 \le i \le n$. Inverting another element $g \in \Sigma'$ preserves isomorphisms; hence $\beta[1/f_i g]$ must be an isomorphism for every $g \in \Sigma'$ and every $1 \le i \le n$. In the composite

$$M[1/f_i g] \xrightarrow{\beta[1/f_i g]} \Gamma(V, \pi_*\pi^*\mathscr{S})[1/f_i g] \xrightarrow{\theta^V_{V_{f_i g}}} \Gamma(V_{f_i g}, \pi_*\pi^*\mathscr{S}) .$$

we have have just proved $\beta[1/f_i g]$ an isomorphism, and Lemma 9.7.5 tells us that the map $\theta^V_{V_{f_i g}}$ is also an isomorphism. Thus the composite is an isomorphism. That is

$$\beta_{f_i g} : \Gamma(V_{f_i g}, \pi_*\widetilde{M}) \longrightarrow \Gamma(V_{f_i g}, \pi_*\pi^*\mathscr{S})$$

is an isomorphism on every open set $V_{f_i g} = V_{f_i} \cap V_g$. In other words for any open set V_g, satisfying the two conditions

(i) The g is an element of Σ'

(ii) The open set V_g is contained in V_{f_i} for some $1 \le i \le n$,

the map $\Gamma(V_g, \widetilde{\beta}) : \Gamma(V_g, \pi_*\widetilde{M}) \longrightarrow \Gamma(V_g, \pi_*\pi^*\mathscr{S})$ is an isomorphism. Since the open sets V_g satisfying (i) and (ii) form a basis for the topology of V we conclude that $\widetilde{\beta} : \pi_*\widetilde{M} \longrightarrow \pi_*\pi^*\mathscr{S}$ is an isomorphism. □

9.9 The case of the trivial group

In Section 9.8 we proved Theorem 8.8.6(iv). Consider now the case where $(G, \mathcal{O}_G)$ is the trivial group of Example 8.2.7. Every finitely generated $\mathbb{C}$–algebra is a G–ring, as in Example 8.3.4, and every open set $U \subset X = \operatorname{Spec}(S)$ is G–invariant, by Example 8.4.15. By Example 8.8.4 the technical conditions of Hypothesis 8.8.1 are satisfied, for every open set $U \subset X$. Already in the special case, where $U = X$, Theorem 8.8.6(iv) teaches us something we did not know. In fact, it is something we promised the reader in Remark 7.6.4.

Corollary 9.9.1. *Suppose that $\mathscr{S}$ is a coherent sheaf over the affine scheme $(X, \mathcal{O}_X) = \left(\operatorname{Spec}(S), \widetilde{S}\right)$, where S is a finitely generated $\mathbb{C}$–algebra. Then there exists a finite S–module M with $\mathscr{S} \cong \widetilde{M}$.*

Proof. Theorem 8.8.6(iv) tells us that $\mathscr{S} = \{\pi_* \widetilde{M}\}^{G^{\mathrm{an}}}$, and the argument above says $\{\pi_* \widetilde{M}\}^{G^{\mathrm{an}}} = \widetilde{M}$. □

Remark 9.9.2. If we apply the theorem to a general open set $U \subset X$ we learn more. We learn that every coherent sheaf $\mathscr{S}$ on U is isomorphic to $\widetilde{M}|_U$, where M is a finite S–module. Every coherent sheaf on U is the restriction of some coherent sheaf on X. We will not use this fact in the book, but it has important implications which have been studied extensively elsewhere. The analytic analogue is quite false; coherent analytic sheaves need not extend from open subsets.

For us the important consequence of Theorem 8.8.6(iv), in the special case where the group G is trivial, is the following:

Lemma 9.9.3. *Let S be a finitely generated $\mathbb{C}$–algebra. Suppose that $0 \longrightarrow \mathscr{R} \longrightarrow \mathscr{S} \longrightarrow \mathscr{T} \longrightarrow 0$ is a short exact sequence of coherent sheaves over the affine scheme $(X, \mathcal{O}_X) = \left(\operatorname{Spec}(S), \widetilde{S}\right)$. Then the sequence*

$$0 \longrightarrow \Gamma(X, \mathscr{R}) \longrightarrow \Gamma(X, \mathscr{S}) \longrightarrow \Gamma(X, \mathscr{T}) \longrightarrow 0$$

is also exact.

Proof. Corollary 9.9.1 tells us that $\mathscr{R} \cong \widetilde{M'}$, $\mathscr{S} \cong \widetilde{M}$ and $\mathscr{T} \cong \widetilde{M''}$. The result now follows from Lemma 7.6.14. □

Remark 9.9.4. Lemma 9.9.3 generalizes to the analytic setting. If $(X, \mathcal{O}_X)$ is an affine scheme of finite type over $\mathbb{C}$, and if $0 \longrightarrow \mathscr{R}' \longrightarrow \mathscr{S}' \longrightarrow \mathscr{T}' \longrightarrow 0$ is an exact sequence of coherent analytic sheaves on

$(X^{\mathrm{an}}, \mathcal{O}_X^{\mathrm{an}})$, then

$$0 \longrightarrow \Gamma(X^{\mathrm{an}}, \mathscr{R}') \longrightarrow \Gamma(X^{\mathrm{an}}, \mathscr{S}') \longrightarrow \Gamma(X^{\mathrm{an}}, \mathscr{T}') \longrightarrow 0$$

is an exact sequence. This is true because X^{an} can be embedded, as a closed analytic subspace, in some $\mathbb{C}^n$, and the exact sequence of coherent analytic sheaves on X^{an} extends trivially (see [3, page 145, Theorem 8]) to an exact sequence of coherent analytic sheaves on $\mathbb{C}^n$. By [3, page 209, Example 1] $\mathbb{C}^n$ is Stein, and [3, page 243, Theorem 14] now says that the sequence

$$0 \longrightarrow \Gamma(X^{\mathrm{an}}, \mathscr{R}') \longrightarrow \Gamma(X^{\mathrm{an}}, \mathscr{S}') \longrightarrow \Gamma(X^{\mathrm{an}}, \mathscr{T}') \longrightarrow 0$$

must be exact.

10
The proof of GAGA

Every book must eventually come to the punchline, and we are rapidly approaching ours. This chapter will give the proof of GAGA. We will prove, for projective space $\mathbb{P}^n$, the two statements

(i) Let $\mathscr{S}$ and $\mathscr{T}$ be coherent algebraic sheaves and suppose that $f : \mathscr{S}^{\mathrm{an}} \longrightarrow \mathscr{T}^{\mathrm{an}}$ is an analytic map between them. Then there exists a unique algebraic map $\varphi : \mathscr{S} \longrightarrow \mathscr{T}$ so that $f = \varphi^{\mathrm{an}}$.
(ii) Every coherent analytic sheaf $\mathscr{S}'$ is isomorphic to $\mathscr{S}^{\mathrm{an}}$, for some coherent algebraic sheaf $\mathscr{S}$.

The category theorist would compactify the two statements above to

(iii) The functor $(-)^{\mathrm{an}}$, which takes a coherent algebraic sheaf to its analytification, is an equivalence of categories. It identifies the category of coherent algebraic sheaves on $\mathbb{P}^n$ with the category of coherent analytic sheaves.

Perhaps we should elaborate. Part (i) says that the functor $(-)^{\mathrm{an}}$ is fully faithful; the set of morphisms $\varphi : \mathscr{S} \longrightarrow \mathscr{T}$ maps bijectively to the set of morphisms $f : \mathscr{S}^{\mathrm{an}} \longrightarrow \mathscr{T}^{\mathrm{an}}$. Part (ii) is the assertion that every coherent analytic sheaf is isomorphic to a sheaf in the image of the functor $(-)^{\mathrm{an}}$.

For the purpose of comparing with earlier claims in the book: part (i) and part (ii) correspond, respectively, to parts (i) and (ii) of Theorem 7.9.1.

From a technical point view we are essentially ready for the proof of (i); we understand quite well the coherent algebraic sheaves $\mathscr{S}$ on $\mathbb{P}^n$. We have spent two chapters describing them all as $\{\pi_* \widetilde{M}\}^{G^{\mathrm{an}}}$, and it will not take much effort to move from this description to the proof of (i). We have so far said close to nothing about coherent analytic sheaves on $\{\mathbb{P}^n\}^{\text{an}}$, and before we can make much progress on (ii) we will need to develop the theory a little.

The way the chapter is structured is the following. We begin with four sections, Sections 10.1, 10.2, 10.3 and 10.4, which are devoted to the elementary sheaf theory of $\mathbb{P}^n$. We discuss the line bundles $\mathcal{O}(m)$, we compute the Fréchet topology on $\Gamma\big(\mathbb{P}^n_J, \mathcal{O}(m)\big)$, we produce some maps $\mathcal{O}(m) \longrightarrow \mathcal{O}(m')$, and we study the sheaves $\mathscr{H}om(\mathcal{O}(m)\,, \mathscr{S})$. We do this mostly in the analytic category because, as we have already explained, it is the analytic sheaves which we do not understand. For the proof of Theorem 7.9.1(ii) we need to know a little more about these sheaves.

After that comes Section 10.5, which tell us about sheaf cohomology. The entire point of the book is that chasing around long exact sequences can prove great theorems. There is no escaping the fact that the reader will need some elementary familiarity with homological algebra, to understand the proof of GAGA. We do not assume much, but knowing about long exact sequences, that come from short exact sequences of chain complexes, is a must. So is familiarity with the 5–lemma. We do, however, give an almost self-contained description of the particular chain complexes we need, of the chain maps, and of the short exact sequences of chain complexes which will play a role. In Section 10.5 the reader can find the definitions.

Next comes Section 10.6; in it we generalize Theorem 7.9.1(i) to a statement about higher sheaf cohomology. The statement only makes sense once we have defined the vector spaces $H^i(\mathscr{S})$; we will not give it here. We give a very general theorem, about all $H^i(\mathscr{S})$, and Theorem 7.9.1(i) becomes the special case where $i = 0$. The general case, strangely enough, is simpler to prove. After stating the more general result we then continue, still in Section 10.6, to prove it for the sheaves $\mathcal{O}(m)$. This turns out to be a computation we can do. Following that, Section 10.7 completes the proof of Theorem 7.9.1(i), by reducing the general statement to the computation of Section 10.6.

The next two sections, Sections 10.8 and 10.9, deal with skyscraper sheaves. Skyscraper sheaves are among the world's stupidest sheaves, but the proof of Theorem 7.9.1(ii) is by reducing the general case to the stupid one. In our discussion of skyscraper sheaves we are hamstrung by the fact that we never fully developed the local theory; we have treated sheaves on $\mathbb{P}^n$ globally, constructing them as $\{\pi_*\widetilde{M}\}^{G^{\text{an}}}$. This makes it a slight nuisance to derive the elementary properties, of skyscraper sheaves, which we need. In this book, the presentation we have given is aimed at overall economy and, in Sections 10.8 and 10.9, we are forced to pay a small price for not providing more information earlier.

Then Section 10.10 concludes the proof of GAGA.

10.1 The sheaves $\mathcal{O}(m)$

In this chapter we will prove Theorem 7.9.1 for the scheme $\mathbb{P}^n$. The theorem is a statement about coherent sheaves, algebraic and analytic, on projective space $\mathbb{P}^n$. Before we get seriously started we must learn a little about the space $\mathbb{P}^n$ and the sheaves on it.

Notation 10.1.1. For the next few sections the notation is as in Example 8.3.5. We fix the Hopf algebra $R = \mathbb{C}[t, t^{-1}]$, the group will be $(G, \mathcal{O}_G) = \left(\mathrm{Spec}(R), \widetilde{R}\right)$, the G–ring S will be the polynomial ring $S = \mathbb{C}[x_0, x_1, \dots, x_n]$ in $(n+1)$ variables, the action of $(G, \mathcal{O}_G)$ on S is by the map $a' : S \longrightarrow R \otimes_{\mathbb{C}} S$ with $a'(x_i) = t^{-1} \otimes x_i$, the scheme $(X, \mathcal{O}_X) = \left(\mathrm{Spec}(S), \widetilde{S}\right)$ admits an induced action of $(G, \mathcal{O}_G)$, the quotient map

$$(\pi, \pi^*) : (X, \mathcal{O}_X) \longrightarrow (X/G, \mathcal{O}_{X/G})$$

is as in Definitions 8.6.2 and 8.7.6, the G–invariant open subset we wish to consider is $U = X - \{0\}$, and $\mathbb{P}^n \subset X/G$ is the open subset whose inverse image $\pi^{-1}\mathbb{P}^n$ is the G–invariant open set $U \subset X$.

Definition 10.1.2. *The 1–dimensional vector space $\mathbb{C}e_m$ will be the space with basis e_m. We make it a linear representation for the group $(G, \mathcal{O}_G)$ by setting $a'(e_m) = t^m \otimes e_m$.*

Remark 10.1.3. The action $a' : \mathbb{C}e_m \longrightarrow R \otimes_{\mathbb{C}} \mathbb{C}e_m$, given in Definition 10.1.2, makes the 1–dimensional representation of G purely of weight $(-m)$, in the notation of Definition 9.3.5.

Definition 10.1.4. *The sheaf $\mathcal{O}(m)$ on projective space $\mathbb{P}^n$ is defined to be $\{\pi_* \widetilde{M}\}^{G^{\mathrm{an}}}\big|_{\mathbb{P}^n}$, where the G–module M for the G–ring S is $M = S \otimes_{\mathbb{C}} \mathbb{C}e_m$.*

Remark 10.1.5. We remind the reader that, for any finite dimensional linear representation V of the group $(G, \mathcal{O}_G)$, we saw, in Exercise 9.2.20(i), how to make $S \otimes_{\mathbb{C}} V$ into a G–module for the G–ring S. To form the sheaves $\mathcal{O}(m)$ we begin with the 1–dimensional linear representation $\mathbb{C}e_m$, form the G–module $M = S \otimes_{\mathbb{C}} \mathbb{C}e_m$, then apply to it the construction $\{\pi_* \widetilde{M}\}^{G^{\mathrm{an}}}$, which gives a sheaf of $\mathcal{O}_{X/G}$–modules on X/G, and finally restrict to the open subset $\mathbb{P}^n \subset X/G$ to obtain a coherent sheaf on $\mathbb{P}^n$. The coherence is by Theorem 8.8.6(ii).

In this section, we prove two important facts about the sheaves $\mathcal{O}(m)$.

Lemma 10.1.6. *Every coherent algebraic sheaf $\mathscr{T}$ on $\mathbb{P}^n$ admits a surjection $\mathscr{S} \longrightarrow \mathscr{T}$, where $\mathscr{S}$ is a finite direct sum of $\mathcal{O}(m)$'s.*

Proof. By Theorem 8.8.6(iv) we know that the coherent sheaf $\mathscr{T}$ is isomorphic to the restriction to the open set $\mathbb{P}^n \subset X/G$ of $\{\pi_*\widetilde{M}\}^{G^{\mathrm{an}}}$, for some G–module M for the G–ring S. Remark 9.2.21 tells us that there is a finite dimensional linear representation V of $(G, \mathcal{O}_G)$ and a surjection $S \otimes_{\mathbb{C}} V \longrightarrow M$. Theorem 8.8.6(iii) then informs us that the induced morphism of sheaves

$$\{\pi_*(\widetilde{S \otimes_{\mathbb{C}} V})\}^{G^{\mathrm{an}}} \longrightarrow \{\pi_*\widetilde{M}\}^{G^{\mathrm{an}}}$$

gives, on the open subset $\mathbb{P}^n \subset X/G$, a surjection of coherent sheaves. Let $\mathscr{S}$ be the restriction to $\mathbb{P}^n$ of the sheaf $\{\pi_*(\widetilde{S \otimes_{\mathbb{C}} V})\}^{G^{\mathrm{an}}}$. It remains to show that $\mathscr{S}$ is a finite direct sum of $\mathcal{O}(m)$'s.

But V is a finite dimensional linear representation of the group scheme $(G, \mathcal{O}_G)$, and Lemmas 9.3.2 and 9.3.4 tell us that V is a direct sum of 1–dimensional subrepresentations, each isomorphic to $\mathbb{C}e_m$ for some m. The passage from a module M to the sheaf $\{\pi_*\widetilde{M}\}^{G^{\mathrm{an}}}$ takes exact sequences to exact sequences and hence must respect direct sums. We conclude that $\mathscr{S}$ must be a finite direct sum of $\mathcal{O}(m)$'s. □

Lemma 10.1.7. *Let $y \in S$ be any homogeneous polynomial of degree* 1. *On the open set $\mathbb{P}^n_y = X_y/G$ the restriction of the sheaf $\mathcal{O}(m)$ is trivial. The word* "trivial" *means, in this context, that it is isomorphic to the restriction of the sheaf $\mathcal{O}$.*

Proof. Maybe we should remind the reader that, because y is homogeneous (the degree does not matter), we have $a'(y) = r \otimes y$; see Example 8.3.13. The basic open set $X_y = \mathrm{Spec}(S[1/y])$ is contained in $U = X - \{0\}$, and Remark 8.8.7 tells us that the open subset $\mathbb{P}^n_y \subset \mathbb{P}^n \subset X/G$, whose inverse image is X_y, is isomorphic to X_y/G, even as a ringed space. Furthermore the restriction to $\mathbb{P}^n_y \subset X/G$ of a sheaf $\{\pi_*\widetilde{M}\}^{G^{\mathrm{an}}}$ is isomorphic to $\{\pi_*\widetilde{M[1/y]}\}^{G^{\mathrm{an}}}$.

This means that, in order to show that the sheaves $\mathcal{O}$ and $\mathcal{O}(m)$ have isomorphic restrictions to $\mathbb{P}^n_y$, it suffices to establish that the G–modules $\{S \otimes_{\mathbb{C}} \mathbb{C}e_m\}[1/y]$ and $S[1/y]$ are isomorphic over the G–ring $S[1/y]$. The isomorphism is the map

$$\eta : S[1/y] \longrightarrow \{S \otimes_{\mathbb{C}} \mathbb{C}e_m\}[1/y]$$

with $\eta(1) = y^m \otimes e_m$. Note that this is certainly an isomorphism of $S[1/y]$–modules. We need only worry about the G–action. And we compute that $a'(y^m) = t^{-m} \otimes y^m$, $a'(e_m) = t^m \otimes e_m$, and hence

$$a'(y^m \otimes e_m) \quad = \quad 1 \otimes [y^m \otimes e_m] \ .$$

□

10.2 Another visit to the concrete world

In Section 10.1 we defined the sheaves $\mathcal{O}(m)$, and proved a couple of lemmas about them. Next we want to come down to earth and understand it all concretely. We first untangle the definition of $\mathcal{O}(m)$ a little.

Example 10.2.1. Definition 10.1.4 began with $M = S \otimes_{\mathbb{C}} \mathbb{C}e_m$, which is a G–module for the G–ring S. Let $W \subset X$ be a G–invariant open subset. Corollary 8.4.10 tells us that the group G^{an} acts on $\Gamma(W, \widetilde{M})$. Remark 8.4.11 computes this action, in the special case where the open set is $W = X$. The element $g \in G^{\mathrm{an}}$ acts on $M = \Gamma(X, \widetilde{M})$ via the composite

$$M \xrightarrow{\ a'\ } R \otimes_{\mathbb{C}} M \xrightarrow{\ \varphi_{g^{-1}} \otimes 1\ } \mathbb{C} \otimes_{\mathbb{C}} M \quad = \quad M \ .$$

Let us work out explicitly, on a basis for $M = S \otimes_{\mathbb{C}} \mathbb{C}e_m$, just exactly what $g \in \mathbb{C}^*$ does.

Our choice of basis for S has always been the monomials

$$\mathbf{x}^{\mathbf{a}} \quad = \quad x_0^{a_0} x_1^{a_1} \cdots x_n^{a_n} \ .$$

The basis we choose for $M = S \otimes_{\mathbb{C}} \mathbb{C}e_m$ consists of the elements $\mathbf{x}^{\mathbf{a}} \otimes e_m$. With the convention that $\langle \mathbf{a} \rangle = a_0 + a_1 + \cdots + a_n$, we have that $a'(\mathbf{x}^{\mathbf{a}}) = t^{-\langle \mathbf{a} \rangle} \otimes \mathbf{x}^{\mathbf{a}}$, while $a'(e_m) = t^m \otimes e_m$. In view of Exercise 9.2.20(i), this makes

$$a'(\mathbf{x}^{\mathbf{a}} \otimes e_m) \quad = \quad t^{m - \langle \mathbf{a} \rangle} \otimes [\mathbf{x}^{\mathbf{a}} \otimes e_m] \ .$$

Applying the map $\varphi_{g^{-1}} \otimes 1$ means evaluating at $t = g^{-1}$; the formula becomes

$$g(\mathbf{x}^{\mathbf{a}} \otimes e_m) \quad = \quad g^{\langle \mathbf{a} \rangle - m} \cdot \mathbf{x}^{\mathbf{a}} \otimes e_m \ .$$

The way we want to say this is the following. There is an action of G^{an} on $\Gamma(X, \mathcal{O}) = S$; in Example 8.5.1 we worked this out, the formula is

$g(\mathbf{x}^{\mathbf{a}}) = g^{\langle \mathbf{a} \rangle}\mathbf{x}^{\mathbf{a}}$. There is also an action of G^{an} on the linear representation $\mathbb{C}e_m$; we computed this in Remark 9.3.7, and the formula turns out to be $g(e_m) = g^{-m}e_m$. The way G^{an} acts on $\Gamma(X, \widetilde{M}) = M = S \otimes_{\mathbb{C}} \mathbb{C}e_m$ is best thought of as the diagonal action:

$$g(f \otimes e_m) \quad = \quad g(f) \otimes g(e_m) \ .$$

Discussion 10.2.2. Slightly more interesting is the following. Suppose $J \subset \{0, 1, \ldots, n\}$ is some subset. As always, let $f_J = \prod_{i \in J} x_i$. Then $X_J = X_{f_J}$ is a G–invariant open subset of X, and G^{an} must act on $\Gamma(X_J, \widetilde{M})$. What is this action?

Remark 8.4.12 comes to the rescue. Every element of $\Gamma(X_J, \widetilde{M}) = M[1/f_J]$ can be written as y/f_J^ℓ, for some $y \in M$ and some integer $\ell \geq 0$. Remark 8.4.12 informs us that the way $g \in \mathbb{C}^*$ acts is by the formula

$$g\left(\frac{y}{f_J^\ell}\right) \quad = \quad g\left(\frac{1}{f_J^\ell}\right) \cdot g(y) \ ,$$

where by $g(y)$ we mean the action of $g \in G^{\text{an}}$ on $M = \Gamma(X, \widetilde{M})$, computed in Example 10.2.1, while $g(1/f_J^\ell)$ comes from the action of g on $S[1/f_J]$.

Now express $y \in M = S \otimes_{\mathbb{C}} \mathbb{C}e_m$ in the form $y = h \otimes e_m$, with $h \in S$. The formula becomes

$$\begin{aligned} g\left(\left[\frac{h}{f_J^\ell}\right] \otimes e_m\right) \quad &= \quad g\left(\frac{h \otimes e_m}{f_J^\ell}\right) \\ &= \quad g\left(\frac{1}{f_J^\ell}\right) \cdot g(h \otimes e_m) \\ &= \quad g\left(\frac{1}{f_J^\ell}\right) \cdot g(h) \otimes g(e_m) \\ &= \quad g\left(\frac{h}{f_J^\ell}\right) \otimes g(e_m) \ . \end{aligned}$$

In other words, the action is still diagonal. If we identify $M[1/f_J]$ with $S[1/f_J] \otimes_{\mathbb{C}} \mathbb{C}e_m$, then the action of $g \in G^{\text{an}}$ is by the simple formula

$$g(h' \otimes e_m) \quad = \quad g(h') \otimes g(e_m) \ .$$

The reader should recall that we understand very well the ring $S[1/f_J]$ and the G^{an} action on it. In Exercise 3.5.4(i) we gave an isomorphism of $S[1/f_J]$ with the Laurent polynomial ring S_J, and in Remark 8.5.5 we figured out the G^{an}–action on the ring S_J. With our identifications, $\Gamma(X_J, \widetilde{M})$ becomes $S_J \otimes_{\mathbb{C}} \mathbb{C}e_m$, where G^{an} acts diagonally. Explicitly,

$$g(\mathbf{x}^{\mathbf{a}} \otimes e_m) \quad = \quad g(\mathbf{x}^{\mathbf{a}}) \otimes g(e_m) \quad = \quad g^{\langle \mathbf{a} \rangle - m}\mathbf{x}^{\mathbf{a}} \otimes e_m$$

for all Laurent monomials $\mathbf{x}^{\mathbf{a}} \in S_J$.

Remark 10.2.3. It follows that the element $\mathbf{x}^{\mathbf{a}} \otimes e_m \in S_J \otimes_{\mathbb{C}} \mathbb{C}e_m$ will be G^{an}–invariant if and only if $\langle \mathbf{a} \rangle = m$, that is if and only if $\mathbf{x}^{\mathbf{a}}$ is a Laurent monomial of degree m. All the finite linear combinations of such $\mathbf{x}^{\mathbf{a}} \otimes e_m$ are invariant, and these are the only invariants. That is, the vector space $\{S_J \otimes_{\mathbb{C}} \mathbb{C}e_m\}^{G^{\text{an}}}$ consists of elements $f \otimes e_m$, where f is a homogeneous Laurent polynomial of degree m.

This leads us to a definition:

Definition 10.2.4. *The vector space $S_J(m) \subset S_J$ is the subspace of all homogeneous Laurent polynomials of degree m.*

Remark 10.2.5. In the notation of Definition 10.2.4, the observation of Remark 10.2.3 can be rephrased: it says that the invariant subspace of $S_J \otimes_{\mathbb{C}} \mathbb{C}e_m$ is precisely $S_J(m) \otimes_{\mathbb{C}} \mathbb{C}e_m$. In symbols:

$$\{S_J \otimes_{\mathbb{C}} \mathbb{C}e_m\}^{G^{\text{an}}} \quad = \quad S_J(m) \otimes_{\mathbb{C}} \mathbb{C}e_m \;.$$

Remark 10.2.6. In Discussion 10.2.2 we produced a G^{an}–isomorphism of $S_J \otimes_{\mathbb{C}} \mathbb{C}e_m$ with $\Gamma(X_J, \widetilde{M})$. Given an inclusion $J \subset J' \subset \{0, 1, \ldots, n\}$, we can ask whether the square below commutes

$$\begin{array}{ccc} S_J \otimes_{\mathbb{C}} \mathbb{C}e_m & \xrightarrow{\text{inclusion}} & S_{J'} \otimes_{\mathbb{C}} \mathbb{C}e_m \\ \downarrow & & \downarrow \\ \Gamma(X_J, \widetilde{M}) & \xrightarrow[\text{res}^{X_J}_{X_{J'}}]{} & \Gamma(X_{J'}, \widetilde{M}) \end{array} \quad ,$$

where the vertical maps are the identifications of Discussion 10.2.2. The unsurprising answer is Yes, and we will now sketch the argument. Note that, in checking the assertion, we only need to worry whether the two composites are equal; we do not have to concern ourselves with the G^{an}–action. Each of the homomorphisms respects the G^{an}–action, and therefore so does any composite.

For any S–module M, we proved in Lemma 7.4.1 that the square below commutes

$$\begin{array}{ccc} M[1/f_J] & \xrightarrow{\alpha^{f_J}_{f_{J'}}} & M[1/f_{J'}] \\ \downarrow & & \downarrow \\ \Gamma(X_J, \widetilde{M}) & \xrightarrow[\text{res}^{X_J}_{X_{J'}}]{} & \Gamma(X_{J'}, \widetilde{M}) \end{array} \quad .$$

In our case the module is $M = S \otimes_{\mathbb{C}} \mathbb{C}e_m$, and there is a dumb isomorphism $S[1/f_J] \otimes_{\mathbb{C}} \mathbb{C}e_m \cong M[1/f_J]$. This is the isomorphism sending $(h/f_J^\ell) \otimes e_m$ to $(h \otimes e_m)/f_J^\ell$. Clearly the square

$$\begin{array}{ccc} S[1/f_J] \otimes_{\mathbb{C}} \mathbb{C}e_m & \xrightarrow{\alpha_{f_{J'}}^{f_J} \otimes 1} & S[1/f_{J'}] \otimes_{\mathbb{C}} \mathbb{C}e_m \\ \downarrow & & \downarrow \\ M[1/f_J] & \xrightarrow[\alpha_{f_{J'}}^{f_J}]{} & M[1/f_{J'}] \end{array}$$

commutes. It only remains to check the commutativity of the square

$$\begin{array}{ccc} S[1/f_J] \otimes_{\mathbb{C}} \mathbb{C}e_m & \xrightarrow{\alpha_{f_{J'}}^{f_J} \otimes 1} & S[1/f_{J'}] \otimes_{\mathbb{C}} \mathbb{C}e_m \\ \downarrow & & \downarrow \\ S_J \otimes_{\mathbb{C}} \mathbb{C}e_m & \xrightarrow[\text{inclusion}]{} & S_{J'} \otimes_{\mathbb{C}} \mathbb{C}e_m \end{array} ,$$

which is obtained by taking the commutative square of Exercise 3.5.4(iii) and tensoring with the 1–dimensional vector space $\mathbb{C}e_m$.

Example 10.2.7. In Definition 10.1.4 we declared that $\mathcal{O}(m)$ is the sheaf $\{\pi_*\widetilde{M}\}^{G^{\text{an}}}\Big|_{\mathbb{P}^n}$, where $M = S \otimes_{\mathbb{C}} \mathbb{C}e_m$. Take any non-empty subset $J \subset \{0, 1, \ldots, n\}$, and let $\mathbb{P}^n_J \subset \mathbb{P}^n$ be the open set of Notation 8.9.4(ii); that is, $\pi^{-1}\mathbb{P}^n_J = X_J$. The definition of the sheaf $\mathcal{O}(m)$ means that

$$\Gamma\big(\mathbb{P}^n_J, \mathcal{O}(m)\big) \quad = \quad \Gamma(X_J, \widetilde{M})^{G^{\text{an}}} .$$

In Discussion 10.2.2 we produced an identification, compatible with the G^{an}–action, of $\Gamma(X_J, \widetilde{M})$ with $S_J \otimes_{\mathbb{C}} \mathbb{C}e_m$. Combining our identifications, we have

$$\begin{array}{rclcl} \Gamma\big(\mathbb{P}^n_J, \mathcal{O}(m)\big) & = & \Gamma(X_J, \widetilde{M})^{G^{\text{an}}} & \cong & \{S_J \otimes_{\mathbb{C}} \mathbb{C}e_m\}^{G^{\text{an}}} \\ & & & = & S_J(m) \otimes_{\mathbb{C}} \mathbb{C}e_m , \end{array}$$

where the equality on the right is by Remark 10.2.5. That is the sections of the sheaf $\mathcal{O}(m)$, on the open subset $\mathbb{P}^n_J \subset \mathbb{P}^n$, can be naturally identified with tensors $f \otimes e_m$, where $f \in S_J$ is a homogeneous Laurent polynomial of degree m.

Remark 10.2.8. If $J \subset J' \subset \{0, 1, \ldots, n\}$ are two non-empty subsets,

Remark 10.2.6 tells us that the diagram

$$\begin{array}{ccc} S_J \otimes_{\mathbb{C}} \mathbb{C}e_m & \xrightarrow{\text{inclusion}} & S_{J'} \otimes_{\mathbb{C}} \mathbb{C}e_m \\ \downarrow & & \downarrow \\ \Gamma(X_J, \widetilde{M}) & \xrightarrow[\text{res}^{X_J}_{X_{J'}}]{} & \Gamma(X_{J'}, \widetilde{M}) \end{array} ,$$

commutes, where the vertical maps are the G^{an}–isomorphisms of Discussion 10.2.2. Taking G^{an}–invariants, we learn that the square

$$\begin{array}{ccc} S_J(m) \otimes_{\mathbb{C}} \mathbb{C}e_m & \xrightarrow{\text{inclusion}} & S_{J'}(m) \otimes_{\mathbb{C}} \mathbb{C}e_m \\ \downarrow & & \downarrow \\ \Gamma\big(\mathbb{P}^n_J, \mathcal{O}(m)\big) & \xrightarrow[\text{res}^{\mathbb{P}^n_J}_{\mathbb{P}^n_{J'}}]{} & \Gamma\big(\mathbb{P}^n_{J'}, \mathcal{O}(m)\big) \end{array}$$

also commutes, where the vertical maps are the identifications of Example 10.2.7.

Discussion 10.2.9. So far all we have used, from Section 10.1, is Definition 10.1.4; we untangled the definition to compute the vector spaces $\Gamma\big(\mathbb{P}^n_J, \mathcal{O}(m)\big)$. Next we want to work out the Fréchet topology on these vector spaces. For this purpose it helps to use Lemma 10.1.7, or more precisely it helps to recall the proof of the lemma.

In the proof of Lemma 10.1.7, put $y = x_0$. The lemma tells us that the sheaves $\mathcal{O}$ and $\mathcal{O}(m)$ have isomorphic restrictions to the open set $\mathbb{P}^n_{x_0} \subset \mathbb{P}^n$, and the proof is a little more precise: it informs us that the map $\eta : S[1/x_0] \longrightarrow \{S \otimes_{\mathbb{C}} \mathbb{C}e_m\}[1/x_0]$ induces an isomorphism. It goes without saying that η induces an isomorphism on any smaller open set. For example, if $J \subset \{0, 1, \dots, n\}$ is a set with $0 \in J$, then η induces an isomorphism when restricted to $\mathbb{P}^n_J \subset \mathbb{P}^n_{x_0}$.

The map η induces an isomorphism of sheaves on $\mathbb{P}^n_J$. In an abuse of notation, we will use the name η also for this induced map

$$\eta : \mathcal{O}|_{\mathbb{P}^n_J} \longrightarrow \mathcal{O}(m)|_{\mathbb{P}^n_J} .$$

It follows that the analytification of η, that is the map η^{an}, is an isomorphism of coherent analytic sheaves. If $V \subset \{\mathbb{P}^n_J\}^{\text{an}}$ is any polydisc, we obtain a commutative square

$$\begin{array}{ccc} \Gamma(\mathbb{P}^n_J,\mathcal{O}) & \xrightarrow{\Gamma(\mathbb{P}^n_J,\eta)} & \Gamma\big(\mathbb{P}^n_J,\mathcal{O}(m)\big) \\ {\scriptstyle\lambda^*}\downarrow & & \downarrow{\scriptstyle\lambda^*} \\ \Gamma\big(V,\mathcal{O}^{\text{an}}\big) & \xrightarrow[\Gamma(V,\eta^{\text{an}})]{} & \Gamma\big(V,\mathcal{O}(m)^{\text{an}}\big) \end{array} .$$

In this commutative square the top horizontal map, $\Gamma(\mathbb{P}^n_J,\eta)$, is an isomorphism. The bottom horizontal map is more: it is a homeomorphism in the Fréchet topology. We deduce that the map of pre-Fréchet spaces

$$\Gamma(\mathbb{P}^n_J,\eta)\ :\ \big[\Gamma(\mathbb{P}^n_J,\mathcal{O})\ ,\ V\big] \longrightarrow \big[\Gamma\big(\mathbb{P}^n_J,\mathcal{O}(m)\big)\ ,\ V\big]$$

is a homeomorphism. We should perhaps remind the reader of the notation; the symbol $\big[\Gamma\big(\mathbb{P}^n_J,\mathcal{O}(m)\big)\ ,\ V\big]$ means that we take the vector space $\Gamma\big(\mathbb{P}^n_J,\mathcal{O}(m)\big)$, and endow it with the topology it inherits from the map to $\Gamma\big(V,\mathcal{O}(m)^{\text{an}}\big)$. Fact 7.8.9(iii)(a) tells us that $\Gamma\big(V,\mathcal{O}(m)^{\text{an}}\big)$ is the completion of the pre-Fréchet space $\big[\Gamma\big(\mathbb{P}^n_J,\mathcal{O}(m)\big)\ ,\ V\big]$.

10.2.9.1. We could, in principle, consider any polydisc $V \subset \{\mathbb{P}^n_J\}^{\text{an}}$. However, we will not. As in Remark 8.9.28, the only polydisc we care about is $V = \{\mathbb{P}^n_J\}^{\text{an}}$. Therefore, in the rest of this section, when we speak of "the" topology of $\Gamma\big(\mathbb{P}^n_J,\mathcal{O}(m)\big)$ we will mean the topology $\big[\Gamma\big(\mathbb{P}^n_J,\mathcal{O}(m)\big)\ ,\ \{\mathbb{P}^n_J\}^{\text{an}}\big]$.

Now let us figure out what the map η is. On the open set X_{x_0} we began with an isomorphism of G–modules $\eta : S[1/x_0] \longrightarrow S[1/x_0]\otimes_{\mathbb{C}}\mathbb{C}e_m$. On the open subset $X_J \subset X_{x_0}$ this restricts to an isomorphism $\eta[1/f_J]$: $S[1/f_J] \longrightarrow S[1/f_J]\otimes_{\mathbb{C}}\mathbb{C}e_m$. With the identification $S[1/f_J] = S_J$, the homomorphism becomes $\eta' : S_J \longrightarrow S_J\otimes_{\mathbb{C}}\mathbb{C}e_m$; with all this rewriting it remains the simple map given by $\eta'(f) = x_0^m f\otimes e_m$. The induced map

$$\Gamma(X_J,\widetilde{S_J}) \longrightarrow \Gamma\big(X_J,\widetilde{S_J\otimes_{\mathbb{C}}\mathbb{C}e_m}\big)$$

is nothing other than $\eta' : S_J \longrightarrow S_J\otimes_{\mathbb{C}}\mathbb{C}e_m$, and

$$\Gamma(\mathbb{P}^n_J,\eta) : \Gamma(\mathbb{P}^n_J,\mathcal{O}) \longrightarrow \Gamma\big(\mathbb{P}^n_J,\mathcal{O}(m)\big)$$

is, by definition,

$$\Gamma(X_J,\widetilde{S_J})^{G^{\text{an}}} \longrightarrow \Gamma\big(X_J,\widetilde{S_J\otimes_{\mathbb{C}}\mathbb{C}e_m}\big)^{G^{\text{an}}} .$$

In Lemma 8.9.12 and Remark 10.2.5 we worked out the subspaces $S_J^{G^{\text{an}}} \subset$

S_J and $\{S_J \otimes_{\mathbb{C}} \mathbb{C}e_m\}^{G^{\text{an}}} \subset S_J \otimes_{\mathbb{C}} \mathbb{C}e_m$; in the notation of Definition 10.2.4 they are identified as

$$S_J^{G^{\text{an}}} \quad = \quad S_J(0)\,, \qquad\qquad \{S_J \otimes_{\mathbb{C}} \mathbb{C}e_m\}^{G^{\text{an}}} \quad = \quad S_J(m) \otimes_{\mathbb{C}} \mathbb{C}e_m\,.$$

The homomorphism $\Gamma(\mathbb{P}^n_J, \eta)$ comes down to the map taking $f \in S_J(0)$, that is a homogeneous Laurent polynomial f of degree 0, to the element $x_0^m f \otimes e_m \in S_J(m) \otimes_{\mathbb{C}} \mathbb{C}e_m$, where $x_0^m f$ is a homogeneous Laurent polynomial of degree m.

We know, as a result of the preceding paragraphs, that this map is a homeomorphism in the pre-Fréchet topology. We also know the topology on $S_J^{G^{\text{an}}}$; see Definition 8.9.32, Lemma 8.9.33 and Remark 8.9.34. Surely we ought to be able to come up with a hands-on, explicit understanding of the topology on $\Gamma\big(\mathbb{P}^n_J, \mathcal{O}(m)\big) = S_J(m) \otimes_{\mathbb{C}} \mathbb{C}e_m$.

Definition 10.2.10. *As in Definition 10.2.4, let $S_J(m)$ be the vector space consisting of homogeneous Laurent polynomials of degree m. On the space $S_J(m)$ we will define a sequence of seminorms s_ℓ, by the formula*

$$s_\ell\left(\sum_{\mathbf{x}^{\mathbf{a}} \in S_J} \lambda_{\mathbf{a}} \mathbf{x}^{\mathbf{a}}\right) \quad = \quad \max_{\mathbf{a}} \left(|\lambda_{\mathbf{a}}| \cdot \ell^{|\mathbf{a}|}\right)\,.$$

In this formula $|\mathbf{a}|$ means, as expected, $|\mathbf{a}| = |a_0| + |a_1| + \cdots + |a_n|$.

Remark 10.2.11. In the case $m = 0$ we know that $S_J(0)$ agrees with $S_J^{G^{\text{an}}}$. We already have a seminorm r_ℓ on $S_J^{G^{\text{an}}}$, from Definition 8.9.32. Note that the seminorms r_ℓ and s_ℓ are identical, in the special case $m = 0$ where both are defined.

Lemma 10.2.12. *As in Discussion 10.2.9, let $J \subset \{0, 1, \ldots, n\}$ be a subset containing 0. The map $\eta' : S_J(0) \longrightarrow S_J(m) \otimes_{\mathbb{C}} \mathbb{C}e_m$ is a homeomorphism, where $S_J(0) = S_J^{G^{\text{an}}}$ has the seminorms $r_\ell = s_\ell$ of Definitions 8.9.32 or 10.2.10, while the seminorms s_ℓ, on $S_J(m)$, are as in Definition 10.2.10.*

Proof. We note that $\eta'(f) = x_0^m f \otimes e_m$. This means that

$$\eta'(x_0^{a_0} x_1^{a_1} \cdots x_n^{a_n}) \quad = \quad x_0^{a_0+m} x_1^{a_1} \cdots x_n^{a_n} \otimes e_m\,.$$

Now observe the inequality

$$|a_0| - |m| \quad \le \quad |a_0 + m| \quad \le \quad |a_0| + |m|\,.$$

Adding to this $|a_1| + |a_2| + \cdots + |a_n|$, the inequality becomes

$$|\mathbf{a}| - |m| \quad \le \quad |a_0 + m| + |a_1| + |a_2| + \cdots + |a_n| \quad \le \quad |\mathbf{a}| + |m|\,.$$

Hence

$$|\lambda_{\mathbf{a}}|\cdot \ell^{|\mathbf{a}|-|m|} \quad \leq \quad |\lambda_{\mathbf{a}}|\cdot \ell^{|a_0+m|+|a_1|+|a_2|+\cdots+|a_n|} \quad \leq \quad |\lambda_{\mathbf{a}}|\cdot \ell^{|\mathbf{a}|+|m|} ,$$

and taking the maximum over $\mathbf{a}$, we discover the inequality

$$\ell^{-|m|}r_\ell(f) \quad \leq \quad s_\ell\big(\eta'(f)\big) \quad \leq \quad \ell^{|m|}r_\ell(f) .$$

Therefore both $\eta' : S_J(0) \longrightarrow S_J(m) \otimes_{\mathbb{C}} \mathbb{C}e_m$ and its inverse are continuous. □

Corollary 10.2.13. *Suppose $J \subset \{0, 1, \ldots, n\}$ is a non-empty subset. Let us give the vector space $S_J(m) \otimes_{\mathbb{C}} \mathbb{C}e_m$ its pre-Fréchet topology determined by the seminorms s_ℓ. Then the isomorphism $S_J(m) \otimes_{\mathbb{C}} \mathbb{C}e_m \cong \Gamma\big(\mathbb{P}^n_J, \mathcal{O}(m)\big)$, of Example 10.2.7, is a homeomorphism.*

Proof. Assume $0 \in J$. We have a commutative square

$$\begin{array}{ccc} S_J^{G^{\mathrm{an}}} & \xrightarrow{\ \eta'\ } & S_J(m) \otimes_{\mathbb{C}} \mathbb{C}e_m \\ {\scriptstyle i_0}\big\downarrow & & \big\downarrow{\scriptstyle i_m} \\ \Gamma(\mathbb{P}^n_J, \mathcal{O}) & \xrightarrow[\Gamma(\mathbb{P}^n_J, \eta)]{} & \Gamma\big(\mathbb{P}^n_J, \mathcal{O}(m)\big) \end{array} .$$

We want to show that i_m is a homeomorphism. Now Discussion 10.2.9 showed that $\Gamma(\mathbb{P}^n_J, \eta)$ is a homeomorphism, Lemma 10.2.12 proved that so is η', and in Remark 8.9.34 we saw that i_0 also is. Hence the result, in the case $0 \in J$.

Since the assertion is symmetric in the variables, it must be true for every non-empty J. □

Remark 10.2.14. We learn that the completion of $S_J(m) \otimes_{\mathbb{C}} \mathbb{C}e_m$, with respect to the sequence of seminorms $\{s_\ell\}$, is naturally identified with $\Gamma\big(\{\mathbb{P}^n_J\}^{\mathrm{an}}, \mathcal{O}(m)^{\mathrm{an}}\big)$. Therefore elements of $\Gamma\big(\{\mathbb{P}^n_J\}^{\mathrm{an}}, \mathcal{O}(m)^{\mathrm{an}}\big)$ can be thought of as Laurent series

$$\sum_{\mathbf{x}^{\mathbf{a}} \in S_J,\ \langle \mathbf{a} \rangle = m} \lambda_{\mathbf{a}} \mathbf{x}^{\mathbf{a}}$$

where the coefficients are rapidly decreasing. This means precisely that the quantity

$$\sup_{\mathbf{a}} \big(|\lambda_{\mathbf{a}}| \cdot \ell^{|\mathbf{a}|}\big)$$

must be finite, for every $\ell \geq 1$.

Remark 10.2.15. Suppose $J \subset J' \subset \{0, 1, \ldots, n\}$ are non-empty subsets. The reader should note that, with our seminorms s_ℓ, the inclusion map $S_J(m) \longrightarrow S_{J'}(m)$ is an isometry. The case $m = 0$ was already observed in Remark 8.9.35.

Remark 10.2.16. The information we have, about the sheaf theory of $\mathbb{P}^n$, is already enough to prove half of GAGA. More precisely we know enough for the proof of Theorem 7.9.8 for projective space; as we said in Remark 7.9.3 we will only prove GAGA for projective space $\mathbb{P}^n$. We do not need to know much, about the sheaf theory of $\mathbb{P}^n$, to prove half of GAGA. We remind the reader: Theorem 7.9.8 asserts that, for any coherent algebraic sheaf $\mathscr{S}$ on $\mathbb{P}^n$, the natural map

$$\Gamma(\mathbb{P}^n, \lambda^*) \;:\; \Gamma(\mathbb{P}^n, \mathscr{S}) \longrightarrow \Gamma(\mathbb{P}^n, \lambda_* \mathscr{S}^{\mathrm{an}}) \quad = \quad \Gamma\big(\{\mathbb{P}^n\}^{\mathrm{an}}, \mathscr{S}^{\mathrm{an}}\big)$$

is an isomorphism. Recall also that this is equivalent to Theorem 7.9.1(i), which says that, if $\mathscr{S}$ and $\mathscr{T}$ are coherent algebraic sheaves on $\mathbb{P}^n$ and $f : \mathscr{S}^{\mathrm{an}} \longrightarrow \mathscr{T}^{\mathrm{an}}$ is a morphism of sheaves of $\mathcal{O}_{\mathbb{P}^n}^{\mathrm{an}}$–modules on $\{\mathbb{P}^n\}^{\mathrm{an}}$, then there exists a unique map $\varphi : \mathscr{S} \longrightarrow \mathscr{T}$, of sheaves of $\mathcal{O}_{\mathbb{P}^n}$–modules, for which $f = \varphi^{\mathrm{an}}$. The second half of GAGA asserts that every coherent analytic sheaf is isomorphic to $\mathscr{S}^{\mathrm{an}}$, for some coherent algebraic sheaf $\mathscr{S}$.

10.3 Maps between the sheaves $\mathcal{O}(m)$

In Section 10.2 we gave one application of Lemma 10.1.7, using it to compute the pre-Fréchet topology on $S_J(m) \otimes_{\mathbb{C}} \mathbb{C}e_m \cong \Gamma\big(\mathbb{P}^n_J, \mathcal{O}(m)\big)$. In this section we will apply the lemma in completely different directions.

Remark 10.3.1. The $(n+1)$ elements $x_0, x_1, \ldots, x_n \in S$ are all homogeneous and of degree 1; therefore Lemma 10.1.7 applies to them. Lemma 10.1.7 teaches us that the sheaves $\mathcal{O}(m)$ are trivial on $\mathbb{P}^n_i = \mathbb{P}^n_{x_i}$. From Lemma 8.9.8 we know that the open sets $\mathbb{P}^n_i = \mathbb{P}^n_{x_i}$ cover $\mathbb{P}^n$. Taken together, the two lemmas assert that the sheaves $\mathcal{O}(m)$ are locally trivial on $\mathbb{P}^n$. Recall that, in the terminology of Lemma 10.1.7, this means that there is an open cover, $\mathbb{P}^n = \cup V_j$, so that the restrictions to V_j of $\mathcal{O}(m)$ and $\mathcal{O}$ are isomorphic. A locally trivial sheaf, that is a sheaf locally isomorphic to $\mathcal{O}$, is called a *line bundle.* The term *invertible sheaf* is also used; it is a synonym.

Remark 10.3.2. As stated, Lemma 8.9.8 was not coordinate free. Perhaps we should remind the reader of Observation 9.2.1. The polynomial ring $S = \mathbb{C}[x_0, x_1, \ldots, x_n]$ can be thought of, in a more coordinate-free fashion, by taking a vector space V of dimension $(n+1)$ and forming

$$S \quad = \quad \mathbb{C}[V] \quad = \quad \sum_{i=0}^{\infty} \mathrm{Sym}^i V \ ,$$

where $\mathrm{Sym}^i V$ is the ith symmetric power of V. If we choose a basis $\{x_0, x_1, \ldots, x_n\}$ for the vector space V then, with respect to this basis, $S = \mathbb{C}[V]$ is just the usual polynomial ring. The notation $\mathbb{C}[V]$ is meant to be more basis-free.

Of course Lemma 8.9.8 works for any choice of basis for V. Given another basis $\{y_0, y_1, \ldots, y_n\}$, we still have that the open sets $\mathbb{P}^n_{y_i} \subset \mathbb{P}^n$ cover $\mathbb{P}^n$. And Lemma 10.1.7 still tells us that the restrictions of the sheaves $\mathcal{O}(m)$, to the open affines $\mathbb{P}^n_{y_i} \subset \mathbb{P}^n$, are all trivial.

Remark 10.3.3. Our terminology is not completely standard at this point; the use of the word "trivial" in Remarks 10.3.1 and 10.3.2 is slightly unusual. In the literature one often says that a sheaf is locally trivial if it is locally isomorphic to $\mathcal{O}^\ell$, that is to the free module of rank ℓ over $\mathcal{O}$. A sheaf that is locally isomorphic to $\mathcal{O}^\ell$ is called a *vector bundle of rank* ℓ. A line bundle is nothing but a vector bundle of rank 1.

Construction 10.3.4. For each homogeneous polynomial $f \in S$ of degree $j > 0$, and for any integer m, the element $f \otimes e_m$ can be viewed as belonging to $S \otimes_{\mathbb{C}} \mathbb{C}e_m$. The action of G on it is given by

$$\begin{aligned} a'(f \otimes_{\mathbb{C}} e_m) \ = \ a'(f) \otimes_R a'(e_m) \quad &= \quad \{t^{-j} \otimes f\} \otimes_R \{t^m \otimes e_m\} \\ &= \quad t^{m-j} \otimes \{f \otimes e_m\} \ . \end{aligned}$$

That is $f \otimes e_m$ lies in the $(j-m)$ weight space of $S \otimes_{\mathbb{C}} \mathbb{C}e_m$. We could rephrase this to say that we have found a map

$$\mathbb{C}e_{m-j} \longrightarrow S \otimes_{\mathbb{C}} \mathbb{C}e_m \ ,$$

which takes the basis vector $e_{m-j} \in \mathbb{C}e_{m-j}$ to $(f \otimes e_m) \in S \otimes_{\mathbb{C}} \mathbb{C}e_m$. This map is a G–homomorphism of linear representations. By Exercise 9.2.20 it extends to a G–homomorphism of G–modules

$$f \ : \ S \otimes_{\mathbb{C}} \mathbb{C}e_{m-j} \longrightarrow S \otimes_{\mathbb{C}} \mathbb{C}e_m \ .$$

Remark 10.3.5. On the level of maps of S–modules this is the homomorphism $f : S \longrightarrow S$ which multiplies $g \in S$ by f. The whole exercise above was about keeping track of the way the group G acts, in other words keeping track of the homogeneous degrees. See Example 9.3.9.

Remark 10.3.6. If $f, g \in S$ are two homogeneous elements of degree j then each induces a map as in Construction 10.3.4. We have two G–homomorphisms of G–modules

$$f, g \ : \ S \otimes_{\mathbb{C}} \mathbb{C}e_{m-j} \longrightarrow S \otimes_{\mathbb{C}} \mathbb{C}e_m \ .$$

If λ and μ are any two complex numbers then $\lambda f + \mu g \in S$ is also a homogeneous element of degree j. It induces a map

$$\lambda f + \mu g \ : \ S \otimes_{\mathbb{C}} \mathbb{C}e_{m-j} \longrightarrow S \otimes_{\mathbb{C}} \mathbb{C}e_m \ .$$

We assert that the obvious identity holds: the map induced by $\lambda f + \mu g$ is

$$\lambda\{\text{map induced by } f\} \ + \ \mu\{\text{map induced by } g\} \ .$$

Definition 10.3.7. *For any homogeneous element $f \in S$, of degree $j > 0$, and for any integer $m \in \mathbb{Z}$, we define the map $f : \mathcal{O}(m-j) \longrightarrow \mathcal{O}(m)$ to be the map of sheaves on $\mathbb{P}^n$ induced by the homomorphism of G–modules*

$$f \ : \ S \otimes_{\mathbb{C}} \mathbb{C}e_{m-j} \longrightarrow S \otimes_{\mathbb{C}} \mathbb{C}e_m$$

of Construction 10.3.4.

Lemma 10.3.8. *If $f, g \in S$ are two homogeneous elements of degree j, and if $\lambda, \mu \in \mathbb{C}$ are complex numbers, then the map $\mathcal{O}(m-j) \longrightarrow \mathcal{O}(m)$ induced by $\lambda f + \mu g$ is*

$$\lambda\{\textit{map induced by } f\} \ + \ \mu\{\textit{map induced by } g\} \ .$$

Proof. Immediate from Remark 10.3.6. □

Lemma 10.3.9. *The map $f : \mathcal{O}(m-j) \longrightarrow \mathcal{O}(m)$ is an isomorphism on the open set $\mathbb{P}^n_f \subset \mathbb{P}^n$.*

Proof. The map

$$f \ : \ \{S \otimes_{\mathbb{C}} \mathbb{C}e_{m-j}\}[1/f] \longrightarrow \{S \otimes_{\mathbb{C}} \mathbb{C}e_m\}[1/f]$$

is an isomorphism of G–modules over the G–ring $S[1/f]$, and hence induces an isomorphism of the sheaves on $\mathbb{P}^n_f$. □

Lemma 10.3.10. *If $f \in S$ is non-zero and homogeneous of degree j, then the morphism of sheaves $f : \mathcal{O}(m-j) \longrightarrow \mathcal{O}(m)$ of Definition 10.3.7 is injective.*

Proof. Forget the G–action for a second. The homomorphism

$$f \ : \ S \otimes_{\mathbb{C}} \mathbb{C}e_{m-j} \longrightarrow S \otimes_{\mathbb{C}} \mathbb{C}e_m$$

is multiplication by $f \neq 0$ in the integral domain S, and is therefore injective. The functor taking a G–module M to the sheaf $\{\pi_* \widetilde{M}\}^{G^{\text{an}}}$ on $\mathbb{P}^n$ is exact by Theorem 8.8.6(iii), and hence $f : \mathcal{O}(m-j) \longrightarrow \mathcal{O}(m)$ is injective. □

Remark 10.3.11. Next we want to formalize the fact that, if we fix the homogeneous element $f \in S$ and vary the integer $m \in \mathbb{Z}$, the maps $f : \mathcal{O}(m-j) \longrightarrow \mathcal{O}(m)$ are all the same. Of course this should not be taken too literally, since they are maps between different sheaves. Still, if we invert x_i then the maps do identify with each other. Let us make this precise. Choose a subset $\mathbb{P}^n_{x_i} \subset \mathbb{P}^n$ on which the sheaves $\mathcal{O}(m)$ all trivialize, as in Lemmas 10.1.7 and 8.9.8. On the open set $\mathbb{P}^n_{x_i}$ choose trivializations of all the sheaves $\mathcal{O}(m)$; they all become isomorphic to $\mathcal{O}$. On the open set $\mathbb{P}^n_{x_i} \subset \mathbb{P}^n$ we risk becoming confused which map f we are talking about. We will let $f_m : \mathcal{O}|_{\mathbb{P}^n_{x_i}} \longrightarrow \mathcal{O}|_{\mathbb{P}^n_{x_i}}$ be the restriction of the map $f : \mathcal{O}(m-j) \longrightarrow \mathcal{O}(m)$, in our chosen trivializations. It might also help to remember Exercise 7.1.16(iii): a map $f_m : \mathcal{O}|_{\mathbb{P}^n_{x_i}} \longrightarrow \mathcal{O}|_{\mathbb{P}^n_{x_i}}$ must take $s \in \Gamma(U, \mathcal{O})$ to $a_m s \in \Gamma(U, \mathcal{O})$, for some element $a_m \in \Gamma(\mathbb{P}^n_{x_i}, \mathcal{O})$.

Lemma 10.3.12. *Let the notation be as in Remark 10.3.11. The maps* $f_m : \mathcal{O}|_{\mathbb{P}^n_{x_i}} \longrightarrow \mathcal{O}|_{\mathbb{P}^n_{x_i}}$ *agree up to unit. That is* $f_m = \varepsilon f_{m'}$ *where* $\varepsilon \in \Gamma(\mathbb{P}^n_{x_i}, \mathcal{O})$ *is invertible.*

Proof. In proving the lemma we can choose our favorite trivialization. By Exercise 7.1.16(iv) any isomorphism $\mathcal{O}|_{\mathbb{P}^n_{x_i}} \longrightarrow \mathcal{O}|_{\mathbb{P}^n_{x_i}}$ is multiplication by a unit in $\Gamma(\mathbb{P}^n_{x_i}, \mathcal{O})$. Since the statement of the lemma is only up to units, it makes no difference which isomorphism of $\mathcal{O}|_{\mathbb{P}^n_{x_i}}$ with $\mathcal{O}(m)|_{\mathbb{P}^n_{x_i}}$ we happen to choose. Now observe that, for the ring $S[1/x_i]$, we have a commutative diagram of G–homomorphisms of G–modules

$$\begin{array}{ccc} \{S \otimes_{\mathbb{C}} \mathbb{C}e_{m-j}\}[1/x_i] & \xrightarrow{\ f\ } & \{S \otimes_{\mathbb{C}} \mathbb{C}e_m\}[1/x_i] \\ {\scriptstyle x_i^{m'-m}}\downarrow & & \downarrow{\scriptstyle x_i^{m'-m}} \\ \{S \otimes_{\mathbb{C}} \mathbb{C}e_{m'-j}\}[1/x_i] & \xrightarrow[\ f\ ]{} & \{S \otimes_{\mathbb{C}} \mathbb{C}e_{m'}\}[1/x_i] \end{array}$$

where the vertical maps are isomorphisms. Descending to the induced diagram of coherent sheaves on $\mathbb{P}^n$ we have a commutative square

$$\begin{array}{ccc} \mathcal{O}(m-j)|_{\mathbb{P}^n_{x_i}} & \xrightarrow{\ f\ } & \mathcal{O}(m)|_{\mathbb{P}^n_{x_i}} \\ {\scriptstyle x_i^{m'-m}}\downarrow & & \downarrow{\scriptstyle x_i^{m'-m}} \\ \mathcal{O}(m'-j)|_{\mathbb{P}^n_{x_i}} & \xrightarrow[\ f\ ]{} & \mathcal{O}(m')|_{\mathbb{P}^n_{x_i}} \end{array}$$

where the vertical maps are again isomorphisms. The lemma is now immediate. □

Reminder 10.3.13. It is now time to remind the reader of Exercise 7.1.16. For every ringed space $(X, \mathcal{O})$ and any sheaf $\mathscr{S}$ of $\mathcal{O}$–modules on X we learned, in Exercise 7.1.16, about the sheaves $\mathscr{H}om(\mathcal{O}, \mathscr{S})$; we gave an explicit isomorphism $\mathscr{S} \longrightarrow \mathscr{H}om(\mathcal{O}, \mathscr{S})$. The sheaves $\mathcal{O}(m)$ are locally isomorphic to $\mathcal{O}$ by Remark 10.3.1. When we make local statements about $\mathscr{H}om(\mathcal{O}(m), \mathscr{S})$ we may, in the proof, restrict to an open cover over which $\mathcal{O}(m) \cong \mathcal{O}$ and use the isomorphism $\mathscr{S} = \mathscr{H}om(\mathcal{O}, \mathscr{S}) \cong \mathscr{H}om(\mathcal{O}(m), \mathscr{S})$.

Corollary 10.3.14. *Let m and ℓ be integers. Then there is an isomorphism of the sheaf $\mathscr{H}om(\mathcal{O}(\ell), \mathcal{O}(m))$ with $\mathcal{O}(m - \ell)$.*

Proof. The assertion is clear when $\ell = 0$; for any sheaf $\mathscr{S}$, Exercise 7.1.16(v) gives a canonical isomorphism

$$\mathscr{S} \quad = \quad \mathscr{H}om(\mathcal{O}, \mathscr{S})\,.$$

Next we prove the assertion when $\ell > 0$. Choose any non-zero homogeneous element $f \in S$ of degree ℓ. Lemma 10.3.12 says that, if we trivialize on $\mathbb{P}^n_{x_i}$, then the two maps

$$f_m : \mathcal{O}(m - \ell) \longrightarrow \mathcal{O}(m)\,, \qquad\qquad f_\ell : \mathcal{O} \longrightarrow \mathcal{O}(\ell)$$

agree up to unit. This means the following. On the open set $\mathbb{P}^n_{x_i}$ we choose isomorphisms $\mathcal{O} \cong \mathcal{O}(m - \ell)$, $\mathcal{O} \cong \mathcal{O}(m)$ and $\mathcal{O} \cong \mathcal{O}(\ell)$. The two maps above become

$$f_m : \mathcal{O} \longrightarrow \mathcal{O}\,, \qquad\qquad f_\ell : \mathcal{O} \longrightarrow \mathcal{O}\,.$$

Lemma 10.3.12 says there is a unit $\varepsilon \in \Gamma(\mathbb{P}^n_{x_i}, \mathcal{O})$ with $f_m = \varepsilon f_\ell$. Now Exercise 7.1.16(iii) tells us that the map $f_\ell : \mathcal{O} \longrightarrow \mathcal{O}$ is given by multiplication by some element $a \in \Gamma(\mathbb{P}^n_{x_i}, \mathcal{O})$. For any open set $U \subset \mathbb{P}^n_{x_i}$ the map f_ℓ takes $s \in \Gamma(U, \mathcal{O})$ to $as \in \Gamma(U, \mathcal{O})$. Therefore the map $f_m = \varepsilon f_\ell$ takes $s \in \Gamma(U, \mathcal{O})$ to the element $\varepsilon as \in \Gamma(U, \mathcal{O})$. Now consider the two maps

$$\begin{aligned} f_m : \mathcal{O}(m - \ell) \quad &\longrightarrow \quad \mathcal{O}(m)\,, \\ \mathscr{H}om(f_\ell, 1) \;:\; \mathscr{H}om(\mathcal{O}(\ell), \mathcal{O}(m)) \quad &\longrightarrow \quad \mathscr{H}om(\mathcal{O}, \mathcal{O}(m)) \;=\; \mathcal{O}(m)\,. \end{aligned}$$

On the open set $\mathbb{P}^n_{x_i}$ and after trivializing, the top map is multiplication by the section $\varepsilon a \in \Gamma(\mathbb{P}^n_{x_i}, \mathcal{O})$, while the bottom map is multiplication by $a \in \Gamma(\mathbb{P}^n_{x_i}, \mathcal{O})$. Locally we see that both maps are the same up to unit, and have the same image in $\mathcal{O}(m)$. By Lemma 10.3.10 we know that the top map is injective, and since the maps are the same locally they must both be injective. Each map identifies the domain as the same subsheaf of $\mathcal{O}(m)$; hence

$$\mathcal{O}(m-\ell) \quad \cong \quad \mathscr{H}om\big(\mathcal{O}(\ell), \mathcal{O}(m)\big) \ .$$

Next we need to show that the statement is also true for $-\ell$ with $\ell > 0$. Observe that, for any three sheaves $\mathscr{R}$, $\mathscr{S}$ and $\mathscr{T}$, there is a natural map of sheaves

$$\mathscr{H}om(\mathscr{S}, \mathscr{T}) \longrightarrow \mathscr{H}om\Big(\mathscr{H}om(\mathscr{R}, \mathscr{S}) \ , \ \mathscr{H}om(\mathscr{R}, \mathscr{T})\Big) \ .$$

On the open set $U \subset \mathbb{P}^n$, this map takes $\eta : \mathscr{S}|_U \longrightarrow \mathscr{T}|_U$ to the map taking $\theta : \mathscr{R}|_U \longrightarrow \mathscr{S}|_U$ to the composite $\eta\theta : \mathscr{R}|_U \longrightarrow \mathscr{T}|_U$. In the special case, where $\mathscr{R} = \mathcal{O}(\ell)$, $\mathscr{S} = \mathcal{O}$ and $\mathscr{T} = \mathcal{O}(\ell+m)$, all the sheaves are locally trivial, and locally this natural map is an isomorphism. Thus

$$\mathscr{H}om\big(\mathcal{O}, \mathcal{O}(\ell+m)\big) \cong \mathscr{H}om\Big(\mathscr{H}om\big(\mathcal{O}(\ell), \mathcal{O}\big) \ , \ \mathscr{H}om\big(\mathcal{O}(\ell), \mathcal{O}(\ell+m)\big)\Big).$$

Taking $\ell > 0$, and using the fact that we already know the Corollary for $\ell \geq 0$, we conclude

$$\mathcal{O}(\ell+m) \quad \cong \quad \mathscr{H}om\big(\mathcal{O}(-\ell), \mathcal{O}(m)\big) \ .$$

□

Remark 10.3.15. In Corollary 10.3.14 we establish the existence of an isomorphism

$$\mathcal{O}(m-\ell) \quad \cong \quad \mathscr{H}om\big(\mathcal{O}(\ell), \mathcal{O}(m)\big) \ .$$

We do not construct a canonical isomorphism. With a little more effort it is possible to give one, but we do not need to know this.

10.4 The coherent analytic version

In Sections 10.1, 10.2 and 10.3 we studied the coherent algebraic sheaves $\mathcal{O}(m)$. As we have already said, in Remark 10.2.16, what we learned in Sections 10.1 and 10.2 is enough for the proof of Theorem 7.9.8, which is equivalent to Theorem 7.9.1(i). We already know enough about the sheaf theory of projective space to prove that, given coherent algebraic sheaves $\mathscr{S}$ and $\mathscr{T}$ and an analytic map $f : \mathscr{S}^{\mathrm{an}} \longrightarrow \mathscr{T}^{\mathrm{an}}$, there is a unique algebraic map $\varphi : \mathscr{S} \longrightarrow \mathscr{T}$ with $f = \varphi^{\mathrm{an}}$. In the proof of this statement the sheaves involved are all algebraic, they can all be described as $\{\pi_* \widetilde{M}\}^{G^{\mathrm{an}}}$, and in Sections 10.1 and 10.2 we extracted the little information we will need in the proof.

Theorem 7.9.1(ii) is different. It asks us to show that any coherent analytic sheaf $\mathscr{S}'$ is isomorphic to $\mathscr{S}^{\text{an}}$, for some coherent algebraic sheaf $\mathscr{S}$. In the proof we will inevitably have to deal with coherent analytic sheaves $\mathscr{S}'$ and make some constructions on them. What we will do is elementary enough, and in this section we develop the machinery. We start with two definitions, the first algebraic and the second analytic.

Definition 10.4.1. *For any coherent algebraic sheaf $\mathscr{S}$ on $\mathbb{P}^n$, and for any integer $m \in \mathbb{Z}$, we define*

$$\mathscr{S}(m) \quad = \quad \mathscr{H}om\big(\mathcal{O}(-m), \mathscr{S}\big) \; .$$

Remark 10.4.2. Corollary 10.3.14 tells us that in the case $\mathscr{S} = \mathcal{O}$ we do not get confused. In that case we already have a definition of $\mathcal{O}(m)$, but Corollary 10.3.14 says that

$$\mathcal{O}(m) \quad \cong \quad \mathscr{H}om\big(\mathcal{O}(-m), \mathcal{O}\big) \; .$$

Now for the analytic version. In the analytic case we want to define not only the sheaves, but also some morphisms among them. The morphisms can also be defined for coherent algebraic sheaves, but we will not need those.

Definition 10.4.3. *Let $\mathscr{S}'$ be any coherent analytic sheaf on $\{\mathbb{P}^n\}^{\text{an}}$. Then*

(i) *For any integer m, the sheaf $\mathscr{S}'(m)$ will be $\mathscr{H}om\big(\mathcal{O}(-m)^{\text{an}}, \mathscr{S}'\big)$.*
(ii) *Given a homogeneous element f in the ring S, of degree $j > 0$, then the map $f : \mathscr{S}'(m-j) \longrightarrow \mathscr{S}'(m)$ will be*

$$\mathscr{H}om\big(\mathcal{O}(j-m)^{\text{an}}, \mathscr{S}'\big) \xrightarrow{\mathscr{H}om(f^{\text{an}},1)} \mathscr{H}om\big(\mathcal{O}(-m)^{\text{an}}, \mathscr{S}'\big)$$

where $f : \mathcal{O}(-m) \longrightarrow \mathcal{O}(j-m)$ is as in Definition 10.3.7.

Lemma 10.4.4. *If $\mathscr{S}$ is a coherent sheaf, algebraic or analytic, then $\mathscr{S}(0) = \mathscr{S}$.*

Proof. On any ringed space Exercise 7.1.16 tells us that $\mathscr{H}om(\mathcal{O}, \mathscr{S})$ is canonically isomorphic to $\mathscr{S}$. □

Lemma 10.4.5. *If $\mathscr{S}$ is a coherent algebraic sheaf, then there is a canonical isomorphism $\{\mathscr{S}(m)\}^{\text{an}} \cong \mathscr{S}^{\text{an}}(m)$.*

Proof. For any coherent algebraic sheaves $\mathscr{R}$ and $\mathscr{S}$, Fact 7.8.9(vi) gives a canonical isomorphism

$$\{\mathscr{H}om(\mathscr{R}, \mathscr{S})\}^{\text{an}} \quad \cong \quad \mathscr{H}om(\mathscr{R}^{\text{an}}, \mathscr{S}^{\text{an}}) \; .$$

To prove the lemma just put $\mathscr{R} = \mathcal{O}(-m)$. □

Remark 10.4.6. On the open sets $\mathbb{P}^n_{x_i} \subset \mathbb{P}^n$ the sheaves $\mathcal{O}(m)$ can all be trivialized. Any choice of trivializations gives isomorphisms $\mathcal{O} \cong \mathcal{O}(k)$ for every k. Let $\mathscr{S}$ be a coherent algebraic sheaf. The above means that the restriction to $\mathbb{P}^n_{x_i} \subset \mathbb{P}^n$ of

$$\mathscr{S} = \mathscr{H}om(\mathcal{O}, \mathscr{S}) \qquad \text{and} \qquad \mathscr{S}(m) = \mathscr{H}om(\mathcal{O}(-m), \mathscr{S})$$

are isomorphic. Similarly for the coherent analytic case. Let $\mathscr{S}'$ be a coherent analytic sheaf. The restriction to $\{\mathbb{P}^n_{x_i}\}^{\mathrm{an}} \subset \{\mathbb{P}^n\}^{\mathrm{an}}$ of the sheaves

$$\mathscr{S}' = \mathscr{H}om(\mathcal{O}^{\mathrm{an}}, \mathscr{S}') \qquad \text{and} \qquad \mathscr{S}'(m) = \mathscr{H}om(\mathcal{O}(-m)^{\mathrm{an}}, \mathscr{S}')$$

are isomorphic. Furthermore, under the identification above, of the coherent sheaves $\mathscr{S}'(j)$ with each other, the maps $f : \mathscr{S}'(m-j) \longrightarrow \mathscr{S}'(m)$ agree up to units. This follows from Lemma 10.3.12.

Lemma 10.4.7. *Given an exact sequence $\mathscr{R}' \longrightarrow \mathscr{S}' \longrightarrow \mathscr{T}'$ of coherent sheaves, algebraic or analytic, and an integer $m \in \mathbb{Z}$, then the sequence of sheaves $\mathscr{R}'(m) \longrightarrow \mathscr{S}'(m) \longrightarrow \mathscr{T}'(m)$, induced when we apply $\mathscr{H}om(\mathcal{O}(-m)\,,-)$ in the algebraic case, $\mathscr{H}om(\mathcal{O}(-m)^{\mathrm{an}}\,,-)$ in the analytic case, is also exact.*

Proof. We treat the analytic case and leave to the reader the algebraic. Consider the induced sequence

$$\mathscr{H}om(\mathcal{O}(-m)^{\mathrm{an}}, \mathscr{R}') \longrightarrow \mathscr{H}om(\mathcal{O}(-m)^{\mathrm{an}}, \mathscr{S}') \longrightarrow \mathscr{H}om(\mathcal{O}(-m)^{\mathrm{an}}, \mathscr{T}')\,.$$

Checking for exactness is a local problem. On the open sets $\{\mathbb{P}^n_{x_i}\}^{\mathrm{an}}$ the sheaf $\mathcal{O}(-m)^{\mathrm{an}}$ is isomorphic to $\mathcal{O}^{\mathrm{an}}$, and the restriction to $\{\mathbb{P}^n_{x_i}\}^{\mathrm{an}}$ of the sequences $\mathscr{R}'(m) \longrightarrow \mathscr{S}'(m) \longrightarrow \mathscr{T}'(m)$ and $\mathscr{R}' \longrightarrow \mathscr{S}' \longrightarrow \mathscr{T}'$ are isomorphic. □

Lemma 10.4.8. *The sheaf $\{\mathscr{S}'(m)\}(m')$ is isomorphic to $\mathscr{S}'(m+m')$.*

Proof. Corollary 10.3.14 informs us that

$$\mathcal{O}(-m') \cong \mathscr{H}om(\mathcal{O}(-m)\,,\, \mathcal{O}(-m-m'))\,.$$

Fact 7.8.9(vi) gives an isomorphism

$$\mathcal{O}(-m')^{\mathrm{an}} \cong \mathscr{H}om(\mathcal{O}(-m)^{\mathrm{an}}\,,\, \mathcal{O}(-m-m')^{\mathrm{an}})\,.$$

In general, for any sheaves $\mathscr{K}$, $\mathscr{L}$ and $\mathscr{M}$ on any space at all, there is a natural map

$$\mathscr{H}om(\mathscr{L}, \mathscr{M}) \longrightarrow \mathscr{H}om(\mathscr{H}om(\mathscr{K}, \mathscr{L})\,,\, \mathscr{H}om(\mathscr{K}, \mathscr{M}))\,.$$

If the space happens to be $\{\mathbb{P}^n\}^{\mathrm{an}}$, and the sheaves we choose are $\mathscr{K} = \mathcal{O}(-m)^{\mathrm{an}}$, $\mathscr{L} = \mathcal{O}(-m-m')^{\mathrm{an}}$ and $\mathscr{M} = \mathscr{S}'$, this becomes a map

$$\mathscr{S}'(m+m') \longrightarrow \mathscr{H}om\Big(\mathcal{O}(-m')^{\mathrm{an}}, \mathscr{S}'(m)\Big) \quad = \quad \{\mathscr{S}'(m)\}(m') .$$

Showing that this map is an isomorphism is a local problem, and on the open sets $\{\mathbb{P}^n_{x_i}\}^{\mathrm{an}}$ the sheaves $\mathscr{K} = \mathcal{O}(-m)^{\mathrm{an}}$ and $\mathscr{L} = \mathcal{O}(-m-m')^{\mathrm{an}}$ are trivial. □

We are particularly interested in homogeneous polynomials of degree 1, that is linear polynomials. Let $V \subset S$ be the $(n+1)$–dimensional vector space of homogeneous polynomials of degree 1.

Definition 10.4.9. *Let $\mathscr{S}'$ be a coherent analytic sheaf on $\{\mathbb{P}^n\}^{\mathrm{an}}$. Its* annihilator, *denoted* $\mathrm{ann}(\mathscr{S}')$, *will be the subset of V given by*

$$\mathrm{ann}(\mathscr{S}') \quad = \quad \{y \in V \mid y : \mathscr{S}' \longrightarrow \mathscr{S}'(1) \text{ is the zero map}\} .$$

Remark 10.4.10. The annihilator of any coherent analytic sheaf $\mathscr{S}'$ is a vector subspace of V; if $y, y' \in \mathrm{ann}(\mathscr{S}')$ and $\lambda, \lambda' \in \mathbb{C}$ then $\lambda y + \lambda' y' \in \mathrm{ann}(\mathscr{S}')$. This follows immediately from Lemma 10.3.8.

Lemma 10.4.11. *Let $\mathscr{S}'$ be a coherent analytic sheaf on $\{\mathbb{P}^n\}^{\mathrm{an}}$. If $y \in \mathrm{ann}(\mathscr{S}')$ then the maps $y : \mathscr{S}'(m-1) \longrightarrow \mathscr{S}'(m)$, given in Definition 10.4.3, vanish for every $m \in \mathbb{Z}$.*

Proof. We are given that $y \in \mathrm{ann}(\mathscr{S}')$, which means that the morphism $y : \mathscr{S}(0) \longrightarrow \mathscr{S}(1)$ vanishes. We need to show that so does the morphism of sheaves $y : \mathscr{S}'(m-1) \longrightarrow \mathscr{S}'(m)$. Being a morphism of sheaves, the map will vanish if and only if there is a cover by open sets so that the restriction to the open sets vanishes. But the open sets $\{\mathbb{P}^n_{x_i}\}^{\mathrm{an}}$ cover $\{\mathbb{P}^n\}^{\mathrm{an}}$, and Remark 10.4.6 says that on these open sets the restriction of the map $y : \mathscr{S}'(m-1) \longrightarrow \mathscr{S}'(m)$ agrees with the vanishing $y : \mathscr{S}'(0) \longrightarrow \mathscr{S}'(1)$, up to a unit. □

Lemma 10.4.12. *For any integer m the annihilators* $\mathrm{ann}\big(\mathscr{S}'(m)\big)$ *and* $\mathrm{ann}(\mathscr{S}')$ *agree.*

Proof. We need to compare the maps

$$y : \mathscr{S}' \longrightarrow \mathscr{S}'(1) \qquad \text{and} \qquad y : \mathscr{S}'(m) \longrightarrow \big\{\mathscr{S}'(m)\big\}(1) .$$

More precisely we need to know that if one of these maps is zero then so is the other. As in the proof of Lemma 10.4.11 the question is local and, by Remark 10.4.6, on the open sets $\{\mathbb{P}^n_{x_i}\}^{\mathrm{an}}$ the sheaves $\mathscr{S}'$ and $\mathscr{S}'(m)$ are isomorphic. □

Lemma 10.4.13. *If $y \in \mathrm{ann}(\mathscr{S}')$ then $\mathscr{S}'|_{\{\mathbb{P}^n_y\}^{\mathrm{an}}} = 0$.*

Proof. This is by Lemma 10.3.9: from the lemma we know that on $\mathbb{P}^n_y$ the map $y : \mathcal{O}(-1) \longrightarrow \mathcal{O}$ is an isomorphism. It follows that on $\{\mathbb{P}^n_y\}^{\mathrm{an}}$ the map $y^{\mathrm{an}} : \mathcal{O}(-1)^{\mathrm{an}} \longrightarrow \mathcal{O}^{\mathrm{an}}$ is an isomorphism. Applying $\mathscr{H}om(-, \mathscr{S}')$ to this isomorphism we conclude that $y : \mathscr{S}' \longrightarrow \mathscr{S}'(1)$ is an isomorphism on the open set $\{\mathbb{P}^n_y\}^{\mathrm{an}} \subset \{\mathbb{P}^n\}^{\mathrm{an}}$. We are assuming that $y \in \mathrm{ann}(\mathscr{S}')$, which means $y : \mathscr{S}' \longrightarrow \mathscr{S}'(1)$ is the zero map. On the open set $\{\mathbb{P}^n_y\}^{\mathrm{an}}$ the zero map is an isomorphism, forcing the restriction of $\mathscr{S}'$ to $\{\mathbb{P}^n_y\}^{\mathrm{an}}$ to vanish. □

Lemma 10.4.14. *If $\mathscr{S}'$ is a coherent analytic sheaf, and if $\mathscr{R}'$ is either a subsheaf or a quotient sheaf, then $\mathrm{ann}(\mathscr{S}') \subset \mathrm{ann}(\mathscr{R}')$.*

Proof. Equivalently we show that if $0 \longrightarrow \mathscr{R}' \longrightarrow \mathscr{S}' \longrightarrow \mathscr{T}' \longrightarrow 0$ is a short exact sequence of coherent analytic sheaves then $\mathrm{ann}(\mathscr{S}') \subset \mathrm{ann}(\mathscr{R}')$ and $\mathrm{ann}(\mathscr{S}') \subset \mathrm{ann}(\mathscr{T}')$. The point is that the morphism $y^{\mathrm{an}} : \mathcal{O}(-1)^{\mathrm{an}} \longrightarrow \mathcal{O}^{\mathrm{an}}$ can be mapped into the exact sequence above, giving a commutative diagram

$$\begin{array}{ccccccccc} 0 & \longrightarrow & \mathscr{R}' & \longrightarrow & \mathscr{S}' & \longrightarrow & \mathscr{T}' & \longrightarrow & 0 \\ & & {\scriptstyle y}\big\downarrow & & {\scriptstyle y}\big\downarrow & & {\scriptstyle y}\big\downarrow & & \\ 0 & \longrightarrow & \mathscr{R}'(1) & \longrightarrow & \mathscr{S}'(1) & \longrightarrow & \mathscr{T}'(1) & \longrightarrow & 0 \end{array} .$$

The top row is exact by hypothesis, the bottom row by Lemma 10.4.7. We immediately conclude that if $y : \mathscr{S}' \longrightarrow \mathscr{S}'(1)$ vanishes then so do $y : \mathscr{R}' \longrightarrow \mathscr{R}'(1)$ and $y : \mathscr{T}' \longrightarrow \mathscr{T}'(1)$. □

Construction 10.4.15. Let $\mathscr{T}'$ be a coherent analytic sheaf on $\{\mathbb{P}^n\}^{\mathrm{an}}$, and let $y \in V$ be any element. That is y is any homogeneous polynomial in S of degree 1. Definition 10.4.3(ii) provides a map $y : \mathscr{T}'(-1) \longrightarrow \mathscr{T}'$. Complete to an exact sequence of coherent analytic sheaves

$$0 \longrightarrow \mathscr{K}' \longrightarrow \mathscr{T}'(-1) \xrightarrow{\ y\ } \mathscr{T}' \longrightarrow \mathscr{Q}' \longrightarrow 0 .$$

Lemma 10.4.16. *If $\mathscr{T}'$ is isomorphic to $\mathscr{T}^{\mathrm{an}}$ for some coherent algebraic sheaf $\mathscr{T}$, then the complex of Construction 10.4.15 is isomorphic to the analytification of a complex*

$$0 \longrightarrow \mathscr{K} \longrightarrow \mathscr{T}(-1) \xrightarrow{\ y\ } \mathscr{T} \longrightarrow \mathscr{Q} \longrightarrow 0$$

of coherent algebraic sheaves.

Proof. The map $y : \mathcal{O} \longrightarrow \mathcal{O}(1)$ induces a map of coherent algebraic sheaves

$$\mathscr{H}om(y, 1)\ ,\ \mathscr{H}om(\mathcal{O}(1), \mathscr{T}) \longrightarrow \mathscr{H}om(\mathcal{O}, \mathscr{T})\ ,$$

which we denote $y : \mathscr{T}(-1) \longrightarrow \mathscr{T}$. Fact 7.8.9(vi) identifies the analytification of this map with $y : \mathscr{T}'(-1) \longrightarrow \mathscr{T}'$. If we extend $y : \mathscr{T}(-1) \longrightarrow \mathscr{T}$ to an exact sequence of coherent algebraic sheaves

$$0 \longrightarrow \mathscr{K} \longrightarrow \mathscr{T}(-1) \xrightarrow{y} \mathscr{T} \longrightarrow \mathscr{Q} \longrightarrow 0$$

then Fact 7.8.9(iv), which says that analytification is exact, permits us to identify the analytification with the complex of Construction 10.4.15.

□

Lemma 10.4.17. *Let $\mathscr{T}'$ be a coherent analytic sheaf, and let the exact sequence*

$$0 \longrightarrow \mathscr{K}' \longrightarrow \mathscr{T}'(-1) \xrightarrow{y} \mathscr{T}' \longrightarrow \mathscr{Q}' \longrightarrow 0$$

be as in Construction 10.4.15. The annihilators $\mathrm{ann}(\mathscr{K}')$ *and* $\mathrm{ann}(\mathscr{Q}')$ *contain* $\mathrm{ann}(\mathscr{T}')$ *as well as* $y \in V$.

Proof. By Lemma 10.4.12 the annihilators of $\mathscr{T}'$ and $\mathscr{T}'(-1)$ agree. Now $\mathscr{K}'$ is a subsheaf of $\mathscr{T}'(-1)$ and $\mathscr{Q}'$ is a quotient of $\mathscr{T}'$, and hence Lemma 10.4.14 says that both $\mathrm{ann}(\mathscr{K}')$ and $\mathrm{ann}(\mathscr{Q}')$ must contain $\mathrm{ann}(\mathscr{T}')$. It remains to prove that y lies in $\mathrm{ann}(\mathscr{K}')$ and $\mathrm{ann}(\mathscr{Q}')$.

In any case the sheaves $\mathscr{K}'$ and $\mathscr{Q}'$ can be thought of as the homology of the complex

$$0 \longrightarrow \mathscr{T}'(-1) \xrightarrow{y} \mathscr{T}' \longrightarrow 0\,.$$

We have a map of chain complexes

$$\begin{array}{ccccccc} 0 & \longrightarrow & \mathscr{T}'(-1) & \xrightarrow{y} & \mathscr{T}' & \longrightarrow & 0 \\ & & {\scriptstyle y}\big\downarrow & & \big\downarrow{\scriptstyle y} & & \\ 0 & \longrightarrow & \{\mathscr{T}'(-1)\}(1) & \xrightarrow[y]{} & \mathscr{T}'(1) & \longrightarrow & 0 \end{array}$$

and the question is whether it induces the zero map in homology. The question is local, and we will show that on the open sets $\{\mathbb{P}^n_{x_i}\}^{\mathrm{an}} \subset \{\mathbb{P}^n\}^{\mathrm{an}}$ the chain map above is null homotopic.

Let us therefore work on the open set $\mathbb{P}^n_{x_i}$. We have two maps, $y : \mathcal{O} \longrightarrow \mathcal{O}(1)$, and $y : \mathcal{O}(-1) \longrightarrow \mathcal{O}$. After trivializing, all the sheaves become $\mathcal{O}$, but the maps are slightly different. In the notation of Lemma 10.3.12 we are dealing with $y_1 : \mathcal{O} \longrightarrow \mathcal{O}$ and $y_0 : \mathcal{O} \longrightarrow \mathcal{O}$. They differ by a unit. Put $y_0 = \varepsilon y_1$ with $\varepsilon \in \Gamma(\mathbb{P}^n_{x_i}, \mathcal{O})$ a unit. In the commutative diagram above the rows were obtained by Homming the map $y_1 : \mathcal{O} \longrightarrow \mathcal{O}$ into the sheaf $\mathscr{T}'$, while the columns arise by Homming the map $y_0 : \mathcal{O} \longrightarrow \mathcal{O}$ into $\mathscr{T}'$. The commutative diagram above becomes, on the open set $\{\mathbb{P}^n_{x_i}\}^{\mathrm{an}} \subset \{\mathbb{P}^n\}^{\mathrm{an}}$,

$$\begin{array}{ccccccccc} 0 & \longrightarrow & \mathscr{T}' & \xrightarrow{y_1} & \mathscr{T}' & \longrightarrow & 0 \\ & & \downarrow{\scriptstyle \varepsilon y_1} & & \downarrow{\scriptstyle \varepsilon y_1} & & \\ 0 & \longrightarrow & \mathscr{T}' & \xrightarrow[y_1]{} & \mathscr{T}' & \longrightarrow & 0 \end{array} .$$

But this chain map is clearly homotopic to the null map. □

Combining the last two lemmas we prove

Lemma 10.4.18. *Let $\mathscr{R}$ be a coherent algebraic sheaf and let $\mathscr{T}'$ be a coherent analytic sheaf. Any map $\varphi : \mathscr{R}^{\mathrm{an}} \longrightarrow \mathscr{T}'$ factors as*

$$\mathscr{R}^{\mathrm{an}} \xrightarrow{\psi^{\mathrm{an}}} \mathscr{S}^{\mathrm{an}} \longrightarrow \mathscr{T}'$$

where $\psi : \mathscr{R} \longrightarrow \mathscr{S}$ is a morphism of coherent algebraic sheaves and $\mathrm{ann}(\mathscr{T}') \subset \mathrm{ann}(\mathscr{S}^{\mathrm{an}})$.

Proof. It clearly suffices to prove that, given any element $y \in \mathrm{ann}\mathscr{T}'$, there is a factorization of φ as

$$\mathscr{R}^{\mathrm{an}} \xrightarrow{\psi^{\mathrm{an}}} \mathscr{S}^{\mathrm{an}} \longrightarrow \mathscr{T}'$$

with $\mathrm{ann}(\mathscr{S}^{\mathrm{an}})$ containing both y and $\mathrm{ann}(\mathscr{R}^{\mathrm{an}})$. To do this take the map $y^{\mathrm{an}} : \mathscr{O}^{\mathrm{an}} \longrightarrow \mathscr{O}(1)^{\mathrm{an}}$ and Hom it into the map $\varphi : \mathscr{R}^{\mathrm{an}} \longrightarrow \mathscr{T}'$. We find a commutative square

$$\begin{array}{ccc} \mathscr{R}(-1)^{\mathrm{an}} & \xrightarrow{y} & \mathscr{R}^{\mathrm{an}} \\ \downarrow{\scriptstyle \varphi(-1)} & & \downarrow{\scriptstyle \varphi} \\ \mathscr{T}'(-1) & \xrightarrow[y]{} & \mathscr{T}' \end{array} .$$

We may complete this to a morphism of exact sequences

$$\begin{array}{ccccccccccc} 0 & \longrightarrow & \mathscr{K}'_{\mathscr{R}} & \longrightarrow & \mathscr{R}(-1)^{\mathrm{an}} & \xrightarrow{y} & \mathscr{R}^{\mathrm{an}} & \longrightarrow & \mathscr{Q}'_{\mathscr{R}} & \longrightarrow & 0 \\ & & \downarrow & & \downarrow{\scriptstyle \varphi(-1)} & & \downarrow{\scriptstyle \varphi} & & \downarrow{\scriptstyle \beta} & & \\ 0 & \longrightarrow & \mathscr{K}'_{\mathscr{T}'} & \longrightarrow & \mathscr{T}'(-1) & \xrightarrow[y]{} & \mathscr{T}' & \xrightarrow[\alpha]{} & \mathscr{Q}'_{\mathscr{T}'} & \longrightarrow & 0 \end{array} .$$

By Lemma 10.4.16 the map $\mathscr{R}^{\mathrm{an}} \longrightarrow \mathscr{Q}'_{\mathscr{R}}$ is the analytification of some morphism $\psi : \mathscr{R} \longrightarrow \mathscr{S}$ of coherent algebraic sheaves. Lemma 10.4.17 tells us that the annihilator of $\mathscr{S}^{\mathrm{an}} = \mathscr{Q}'_{\mathscr{R}}$ contains both y and $\mathrm{ann}(\mathscr{R}^{\mathrm{an}})$. Since $y \in \mathrm{ann}(\mathscr{T}')$, Lemma 10.4.11 says that the map $y : \mathscr{T}'(-1) \longrightarrow$

$\mathscr{T}'$ must vanish; exactness of the bottom row now forces $\alpha : \mathscr{T}' \longrightarrow \mathscr{Q}'_{\mathscr{T}'}$ to be an isomorphism. Hence the composite

$$\mathscr{R}^{\text{an}} \xrightarrow{\psi^{\text{an}}} \mathscr{S}^{\text{an}} \xrightarrow{\beta} \mathscr{Q}'_{\mathscr{T}'} \xrightarrow{\alpha^{-1}} \mathscr{T}'$$

does the job. □

10.5 Sheaf cohomology

If $(X, \mathcal{O}_X)$ is an affine scheme of finite type over $\mathbb{C}$ we showed, in Lemma 9.9.3, that any exact sequence $0 \longrightarrow \mathscr{R} \longrightarrow \mathscr{S} \longrightarrow \mathscr{T} \longrightarrow 0$ of coherent algebraic sheaves on X gives an exact sequence

$$0 \longrightarrow \Gamma(X, \mathscr{R}) \longrightarrow \Gamma(X, \mathscr{S}) \longrightarrow \Gamma(X, \mathscr{T}) \longrightarrow 0\,.$$

The exactness is very special to affine schemes X, and now we want to understand how to handle projective space. The idea is to deal with open affine subsets of $\mathbb{P}^n$. We already know some; we remind the reader.

Reminder 10.5.1. Recall the conventions of Section 8.9. In Notation 8.9.4(iii) we set $\mathbb{P}^n_i = \mathbb{P}^n_{x_i}$. For any non-empty open subset $J \subset \{0, 1, \ldots, n\}$ we defined, in Notation 8.9.4(ii), an affine open subset $\mathbb{P}^n_J \subset \mathbb{P}^n$, and, in Lemma 8.9.6, we noted that

$$\mathbb{P}^n_J = \bigcap_{i \in J} \mathbb{P}^n_i\,.$$

Finally, Lemma 8.9.8 says that the open sets $\mathbb{P}^n_i$ cover $\mathbb{P}^n$. The open sets $\mathbb{P}^n_i$ form a cover of $\mathbb{P}^n$ by open affines, in such a way that all the non-empty intersections $\mathbb{P}^n_J$ are also affine.

Notation 10.5.2. In Notation 8.9.4(ii) we defined, for any non-empty subset $J \subset \{0, 1, \ldots, n\}$, an affine open subset $\mathbb{P}^n_J$. For any sheaf of $\mathbb{C}$–vector spaces $\mathscr{S}$ on $\mathbb{P}^n$ we write

$$\mathscr{S}_J = \Gamma(\mathbb{P}^n_J, \mathscr{S})\,.$$

Remark 10.5.3. Suppose we have two non-empty subsets $J \subset J' \subset \{0, 1, \ldots, n\}$. Then $\mathbb{P}^n_{J'} \subset \mathbb{P}^n_J$, and the restriction homomorphisms

$$\mathscr{S}_J = \Gamma(\mathbb{P}^n_J, \mathscr{S}) \xrightarrow{\text{res}^{\mathbb{P}^n_J}_{\mathbb{P}^n_{J'}}} \Gamma(\mathbb{P}^n_{J'}, \mathscr{S}) = \mathscr{S}_{J'}$$

give us maps we will denote by the symbol $r^J_{J'}$. We will usually prefer the more compact notation $r^J_{J'} : \mathscr{S}_J \longrightarrow \mathscr{S}_{J'}$, but the reader should remember which restriction maps they are.

Construction 10.5.4. For each integer i in the range $0 \leq i \leq n$, and for each sheaf of $\mathbb{C}$–vector spaces $\mathscr{S}$ on $\mathbb{P}^n$, we define

$$\mathscr{S}_{[i]} \quad = \quad \prod_{|J|=i+1} \mathscr{S}_J \, .$$

This means we take the product of all the $\mathscr{S}_J$ where the cardinality of the non-empty subset $J \subset \{0, 1, \ldots, n\}$ is precisely $(i+1)$.

Next we want to define a map $\partial_i : \mathscr{S}_{[i]} \longrightarrow \mathscr{S}_{[i+1]}$. The two $\mathbb{C}$–vector spaces $\mathscr{S}_{[i]}$, $\mathscr{S}_{[i+1]}$ are finite direct sums of vector spaces $\mathscr{S}_J$, so the map will be a matrix of homomorphisms between the summands. We need to define, for the summands $\mathscr{S}_J \subset \mathscr{S}_{[i]}$ and $\mathscr{S}_{J'} \subset \mathscr{S}_{[i+1]}$, the matrix entry of ∂_i between them. If we denote this matrix entry by $\partial_i^{JJ'}$, then the rule is simple:

$$\partial_i^{JJ'} \quad = \quad \begin{cases} \varepsilon \cdot r_{J'}^J & \text{if } J \subset J' \\ 0 & \text{otherwise,} \end{cases}$$

where $\varepsilon = \pm 1$ is determined by the following recipe. We are assuming $J \subset J'$, and the cardinalities are $|J| = i+1$, $|J'| = i+2$. There is exactly one element in $J' - J$. Put $J' - J = \{j\}$. Consider the set

$$S_{JJ'} \quad = \quad \{s \in J' \mid s < j\} \ .$$

Then, if $|S_{JJ'}|$ stands for the cardinality of the finite set $S_{JJ'}$,

$$\varepsilon \quad = \quad \varepsilon(J, J') \quad = \quad (-1)^{|S_{JJ'}|} \, .$$

Exercise 10.5.5. The reader should check that, with these formulas, the composite

$$\mathscr{S}_{[i]} \xrightarrow{\ \partial_i\ } \mathscr{S}_{[i+1]} \xrightarrow{\ \partial_{i+1}\ } \mathscr{S}_{[i+2]}$$

vanishes.

Lemma 10.5.6. *Let $\mathscr{S}$ be a sheaf of $\mathbb{C}$–vector spaces on $\mathbb{P}^n$. The kernel of the map $\partial_0 : \mathscr{S}_{[0]} \longrightarrow \mathscr{S}_{[1]}$ is precisely $\Gamma(\mathbb{P}^n, \mathscr{S})$.*

Proof. Recall that $\mathscr{S}_{[0]}$ is defined to be the product of all $\mathscr{S}_J$ where the cardinality of J is exactly one. This makes

$$\mathscr{S}_{[0]} \quad = \quad \prod_{i=0}^{n} \Gamma(\mathbb{P}^n_{x_i}, \mathscr{S}) \ ,$$

while

$$\mathscr{S}_{[1]} \quad = \quad \prod_{i<j} \Gamma(\mathbb{P}^n_{\{i,j\}}, \mathscr{S}) \quad = \quad \prod_{i<j} \Gamma(\mathbb{P}^n_{x_i} \cap \mathbb{P}^n_{x_j}, \mathscr{S}) \ .$$

The map

$$\partial_0 \;:\; \prod_{i=0}^{n} \Gamma(\mathbb{P}^n_{x_i}, \mathscr{S}) \longrightarrow \prod_{i<j} \Gamma(\mathbb{P}^n_{x_i} \cap \mathbb{P}^n_{x_j}, \mathscr{S})$$

takes a string of $s_i \in \Gamma(\mathbb{P}^n_{x_i}, \mathscr{S})$ to the string of

$$\mathrm{res}^{\mathbb{P}^n_{x_j}}_{\mathbb{P}^n_{\{i,j\}}}(s_j) - \mathrm{res}^{\mathbb{P}^n_{x_i}}_{\mathbb{P}^n_{\{i,j\}}}(s_i) \quad \in \quad \Gamma(\mathbb{P}^n_{x_i} \cap \mathbb{P}^n_{x_j}, \mathscr{S}).$$

The fact that the open sets $\mathbb{P}^n_{x_i}$ cover $\mathbb{P}^n$ guarantees that the kernel is $\Gamma(\mathbb{P}^n, \mathscr{S})$. □

Summary 10.5.7. So far we have learnt the following. Starting with a sheaf $\mathscr{S}$, of $\mathbb{C}$–vector spaces on $\mathbb{P}^n$, we have produced a cochain complex

$$0 \longrightarrow \mathscr{S}_{[0]} \xrightarrow{\partial_0} \mathscr{S}_{[1]} \xrightarrow{\partial_1} \cdots \xrightarrow{\partial_{n-2}} \mathscr{S}_{[n-1]} \xrightarrow{\partial_{n-1}} \mathscr{S}_{[n]} \longrightarrow 0\,.$$

We will denote this cochain complex $C^*(\mathscr{S})$. The zeroth cohomology of this cochain complex, that is the kernel of the map ∂_0 above, was computed in Lemma 10.5.6 to be $\Gamma(\mathbb{P}^n, \mathscr{S})$.

Lemma 10.5.8. *A map of sheaves $\eta : \mathscr{S} \longrightarrow \mathscr{T}$ gives a cochain map of cochain complexes*

$$\begin{array}{ccccccccccccc}
0 & \longrightarrow & \mathscr{S}_{[0]} & \longrightarrow & \mathscr{S}_{[1]} & \longrightarrow \cdots \longrightarrow & \mathscr{S}_{[n-1]} & \longrightarrow & \mathscr{S}_{[n]} & \longrightarrow & 0 \\
 & & \downarrow & & \downarrow & & \downarrow & & \downarrow & & \\
0 & \longrightarrow & \mathscr{T}_{[0]} & \longrightarrow & \mathscr{T}_{[1]} & \longrightarrow \cdots \longrightarrow & \mathscr{T}_{[n-1]} & \longrightarrow & \mathscr{T}_{[n]} & \longrightarrow & 0
\end{array}$$

We will denote this map $C^*(\eta) : C^*(\mathscr{S}) \longrightarrow C^*(\mathscr{T})$.

Proof. All the vector spaces $\mathscr{S}_{[i]}$ are finite direct sums of $\Gamma(\mathbb{P}^n_J, \mathscr{S})$, and all components of the maps $\partial_i : \mathscr{S}_{[i]} \longrightarrow \mathscr{S}_{[i+1]}$ are (up to sign) restriction maps. Since $\eta : \mathscr{S} \longrightarrow \mathscr{T}$ is a morphism of sheaves the maps

$$\Gamma(\mathbb{P}^n_J, \eta) : \Gamma(\mathbb{P}^n_J, \mathscr{S}) \longrightarrow \Gamma(\mathbb{P}^n_J, \mathscr{T})$$

commute with restrictions. □

Lemma 10.5.9. *Given maps of sheaves of $\mathbb{C}$–vector spaces on $\mathbb{P}^n$*

$$\mathscr{R} \xrightarrow{\psi} \mathscr{S} \xrightarrow{\varphi} \mathscr{T}$$

then $C^*(\varphi\psi) = C^*(\varphi)C^*(\psi)$.

Proof. Obvious. □

Now we begin to specialize to the case we actually care about, namely coherent algebraic and analytic sheaves. For coherent algebraic sheaves life is simple. Any coherent algebraic sheaf $\mathscr{S}$ on $\mathbb{P}^n$ is, among other things, a sheaf of vector spaces. We have no problem forming the complex $C^*(\mathscr{S})$. What do we do with coherent analytic sheaves? They are defined on the space $\{\mathbb{P}^n\}^{\mathrm{an}}$, not $\mathbb{P}^n$, so we need to say a little bit. It is time for a little reminder.

For any scheme Y, locally of finite type over $\mathbb{C}$, there is a map $\lambda_Y : Y^{\mathrm{an}} \longrightarrow Y$, which is the inclusion of the closed points in all points. It is a continuous map if Y has the Zariski topology and Y^{an} the complex topology. Since, in most of this chapter, $Y = \mathbb{P}^n$ will be fixed, we drop it from the notation. We simply write the map as $\lambda : \{\mathbb{P}^n\}^{\mathrm{an}} \longrightarrow \mathbb{P}^n$.

Given any sheaf $\mathscr{T}$ of vector spaces on Y^{an} we can form the sheaf $\lambda_*\mathscr{T}$ on Y, as in Definition 7.1.13. The general rule is

$$\Gamma(U, \lambda_*\mathscr{T}) \quad = \quad \Gamma(\lambda^{-1}U, \mathscr{T}) \ ,$$

and, in the case of λ being the inclusion of $Y^{\mathrm{an}} \subset Y$, this simplifies because $\lambda^{-1}U = U^{\mathrm{an}}$. We define

Definition 10.5.10. *Let $\mathscr{T}$ be any sheaf of $\mathbb{C}$–vector spaces on $\{\mathbb{P}^n\}^{\mathrm{an}}$. The cochain complex $C^*(\lambda_*\mathscr{T})$ will be denoted $C^*(\mathscr{T})$.*

Facts 10.5.11. We assemble together the easy facts we will need:

(i) Any short exact sequence $0 \longrightarrow \mathscr{R}' \longrightarrow \mathscr{S}' \longrightarrow \mathscr{T}' \longrightarrow 0$, of coherent analytic sheaves on $\{\mathbb{P}^n\}^{\mathrm{an}}$, gives a short exact sequence $0 \longrightarrow C^*(\mathscr{R}') \longrightarrow C^*(\mathscr{S}') \longrightarrow C^*(\mathscr{T}') \longrightarrow 0$ of cochain complexes.

(ii) For any coherent algebraic sheaf $\mathscr{S}$ on $\mathbb{P}^n$ there is a map of cochain complexes $C^*(\lambda^*) : C^*(\mathscr{S}) \longrightarrow C^*(\mathscr{S}^{\mathrm{an}})$.

(iii) Given any morphism $\eta : \mathscr{S} \longrightarrow \mathscr{T}$, of coherent algebraic sheaves on $\mathbb{P}^n$, the following square of maps of cochain complexes must commute:

$$\begin{array}{ccc} C^*(\mathscr{S}) & \xrightarrow{C^*(\eta)} & C^*(\mathscr{T}) \\ {\scriptstyle C^*(\lambda^*)}\downarrow & & \downarrow{\scriptstyle C^*(\lambda^*)} \\ C^*(\mathscr{S}^{\mathrm{an}}) & \xrightarrow[C^*(\eta^{\mathrm{an}})]{} & C^*(\mathscr{T}^{\mathrm{an}}) \end{array} \ .$$

(iv) Any short exact sequence $0 \longrightarrow \mathscr{R} \longrightarrow \mathscr{S} \longrightarrow \mathscr{T} \longrightarrow 0$, of coherent algebraic sheaves on $\mathbb{P}^n$, gives a commutative diagram with exact rows

$$\begin{array}{ccccccccc}
0 & \longrightarrow & C^*(\mathscr{R}) & \longrightarrow & C^*(\mathscr{S}) & \longrightarrow & C^*(\mathscr{T}) & \longrightarrow & 0 \\
 & & \downarrow{\scriptstyle C^*(\lambda^*)} & & \downarrow{\scriptstyle C^*(\lambda^*)} & & \downarrow{\scriptstyle C^*(\lambda^*)} & & \\
0 & \longrightarrow & C^*(\mathscr{R}^{\mathrm{an}}) & \longrightarrow & C^*(\mathscr{S}^{\mathrm{an}}) & \longrightarrow & C^*(\mathscr{T}^{\mathrm{an}}) & \longrightarrow & 0 \ .
\end{array}$$

Proof. To prove (i) we need to show the exactness of the sequence of maps of vector spaces $0 \longrightarrow \mathscr{R}'_{[i]} \longrightarrow \mathscr{S}'_{[i]} \longrightarrow \mathscr{T}'_{[i]} \longrightarrow 0$ for each i. But this sequence is a product, over the non-empty subsets $J \subset \{0, 1, \ldots, n\}$ of cardinality $(i+1)$, of sequences $0 \longrightarrow \mathscr{R}'_J \longrightarrow \mathscr{S}'_J \longrightarrow \mathscr{T}'_J \longrightarrow 0$. These are

$$0 \longrightarrow \Gamma\big(\{\mathbb{P}^n_J\}^{\mathrm{an}}, \mathscr{R}'\big) \longrightarrow \Gamma\big(\{\mathbb{P}^n_J\}^{\mathrm{an}}, \mathscr{S}'\big) \longrightarrow \Gamma\big(\{\mathbb{P}^n_J\}^{\mathrm{an}}, \mathscr{T}'\big) \longrightarrow 0 \ ,$$

and are exact by Remark 9.9.4 and because $\mathbb{P}^n_J$ is affine.

To prove (ii) note that Remark 7.8.4 produces, for every coherent algebraic sheaf $\mathscr{S}$ on $\mathbb{P}^n$, a morphism $\lambda^* : \mathscr{S} \longrightarrow \lambda_* \mathscr{S}^{\mathrm{an}}$ of sheaves on $\mathbb{P}^n$. We simply apply $C^*(-)$ to this map.

The proof of (iii) comes from the commutative square of sheaves on $\mathbb{P}^n$

$$\begin{array}{ccc}
\mathscr{S} & \xrightarrow{\ \eta\ } & \mathscr{T} \\
{\scriptstyle \lambda^*}\downarrow & & \downarrow{\scriptstyle \lambda^*} \\
\lambda_* \mathscr{S}^{\mathrm{an}} & \xrightarrow[\lambda_* \eta^{\mathrm{an}}]{} & \lambda_* \mathscr{T}^{\mathrm{an}}
\end{array}$$

of Exercise 7.8.8(ii).

In (iv) the commutativity is clear by (iii); we only need to prove the exactness of the top and bottom rows. For the bottom row the exactness is because the exact sequence $0 \longrightarrow \mathscr{R} \longrightarrow \mathscr{S} \longrightarrow \mathscr{T} \longrightarrow 0$ on $\mathbb{P}^n$ gives, by Fact 7.8.9(iv), an exact sequence $0 \longrightarrow \mathscr{R}^{\mathrm{an}} \longrightarrow \mathscr{S}^{\mathrm{an}} \longrightarrow \mathscr{T}^{\mathrm{an}} \longrightarrow 0$ on $\{\mathbb{P}^n\}^{\mathrm{an}}$, and then (i) tells us that the induced sequence of cochain complexes is exact. For the top row observe, as in the proof of (i), that the sequence $0 \longrightarrow \mathscr{R}_{[i]} \longrightarrow \mathscr{S}_{[i]} \longrightarrow \mathscr{T}_{[i]} \longrightarrow 0$ is a product of sequences

$$0 \longrightarrow \Gamma(\mathbb{P}^n_J, \mathscr{R}) \longrightarrow \Gamma(\mathbb{P}^n_J, \mathscr{S}) \longrightarrow \Gamma(\mathbb{P}^n_J, \mathscr{T}) \longrightarrow 0 \ ,$$

and these are exact by Lemma 9.9.3 and because $\mathbb{P}^n_J$ is affine. □

Now we do some homological algebra.

Definition 10.5.12. *For any sheaf $\mathscr{S}$ of $\mathbb{C}$–vector spaces on $\mathbb{P}^n$, define the vector spaces $H^i(\mathscr{S})$ to be the* i*th cohomology of the cochain complex $C^*(\mathscr{S})$. If $\mathscr{S}'$ is a sheaf of $\mathbb{C}$–vector spaces on $\{\mathbb{P}^n\}^{\mathrm{an}}$, then Definition 10.5.10 adopted the notation $C^*(\mathscr{S}') = C^*(\lambda_* \mathscr{S}')$, and $H^i(\mathscr{S}')$ is now defined to be the* i*th cohomology of the cochain complex $C^*(\mathscr{S}') = C^*(\lambda_* \mathscr{S}')$.*

Reminder 10.5.13. More concretely, the vector spaces $H^i(\mathscr{S})$ are defined to be

$$H^i(\mathscr{S}) \quad = \quad \frac{\ker\big(\partial_i : \mathscr{S}_{[i]} \longrightarrow \mathscr{S}_{[i+1]}\big)}{\operatorname{im}\big(\partial_{i-1} : \mathscr{S}_{[i-1]} \longrightarrow \mathscr{S}_{[i]}\big)} \; .$$

Facts 10.5.14. What we need to note is:

(i) If $\mathscr{S}$ is a sheaf on $\mathbb{P}^n$ then $H^0(\mathscr{S}) = \Gamma(\mathbb{P}^n, \mathscr{S})$, while if $\mathscr{S}'$ is a sheaf on $\{\mathbb{P}^n\}^{\text{an}}$ then $H^0(\mathscr{S}') = \Gamma\big(\{\mathbb{P}^n\}^{\text{an}}, \mathscr{S}'\big)$.

(ii) The vector spaces $H^i(\mathscr{S})$ vanish if $i < 0$ or $i > n$.

(iii) There are two sequences of functors, both denoted $\{H^i,\ i \in \mathbb{Z}\}$,

$$\begin{array}{rcl} H^i : \quad \{\text{Sheaves on } \mathbb{P}^n\} & \longrightarrow & \{\mathbb{C}\text{–vector spaces}\} \\ H^i : \big\{\text{Sheaves on } \{\mathbb{P}^n\}^{\text{an}}\big\} & \longrightarrow & \{\mathbb{C}\text{–vector spaces}\} \; . \end{array}$$

(iv) If we restrict to coherent sheaves, meaning either coherent algebraic sheaves on $\mathbb{P}^n$ or coherent analytic sheaves on $\{\mathbb{P}^n\}^{\text{an}}$, then H^i are δ–functors. This means they take a short exact sequence $0 \longrightarrow \mathscr{R} \longrightarrow \mathscr{S} \longrightarrow \mathscr{T} \longrightarrow 0$ of coherent sheaves, algebraic or analytic, to the long exact sequence

$$H^{i-1}(\mathscr{T}) \xrightarrow{\delta_{i-1}} H^i(\mathscr{R}) \longrightarrow H^i(\mathscr{S}) \longrightarrow H^i(\mathscr{T}) \xrightarrow{\delta_i} H^{i+1}(\mathscr{R}) \; .$$

(v) Given any coherent algebraic sheaf $\mathscr{S}$ on $\mathbb{P}^n$, there is a natural map $H^i(\lambda^*) : H^i(\mathscr{S}) \longrightarrow H^i(\mathscr{S}^{\text{an}})$.

(vi) Given any morphism of coherent algebraic sheaves $\eta : \mathscr{S} \longrightarrow \mathscr{T}$, then the following square commutes

$$\begin{array}{ccc} H^i(\mathscr{S}) & \xrightarrow{H^i(\eta)} & H^i(\mathscr{T}) \\ {\scriptstyle H^i(\lambda^*)}\big\downarrow & & \big\downarrow{\scriptstyle H^i(\lambda^*)} \\ H^i(\mathscr{S}^{\text{an}}) & \xrightarrow[H^i(\eta^{\text{an}})]{} & H^i(\mathscr{T}^{\text{an}}) \end{array} \; .$$

(vii) Given a short exact sequence $0 \longrightarrow \mathscr{R} \longrightarrow \mathscr{S} \longrightarrow \mathscr{T} \longrightarrow 0$, of coherent algebraic sheaves, then the following square also commutes

$$\begin{array}{ccc} H^i(\mathscr{T}) & \xrightarrow{\delta_i} & H^{i+1}(\mathscr{R}) \\ {\scriptstyle H^i(\lambda^*)}\big\downarrow & & \big\downarrow{\scriptstyle H^{i+1}(\lambda^*)} \\ H^i(\mathscr{T}^{\text{an}}) & \xrightarrow[\delta_i]{} & H^{i+1}(\mathscr{R}^{\text{an}}) \end{array} \; .$$

Taken together with (vi), this says that the maps λ^* induce a chain map of long exact sequences

$$\begin{array}{ccccccccc}
H^{i-1}(\mathscr{T}) & \xrightarrow{\delta_{i-1}} & H^i(\mathscr{R}) & \longrightarrow & H^i(\mathscr{S}) & \longrightarrow & H^i(\mathscr{T}) & \xrightarrow{\delta_i} & H^{i+1}(\mathscr{R}) \\
\downarrow & & \downarrow & & \downarrow & & \downarrow & & \downarrow \\
H^{i-1}(\mathscr{T}^{\mathrm{an}}) & \xrightarrow[\delta_{i-1}]{} & H^i(\mathscr{R}^{\mathrm{an}}) & \longrightarrow & H^i(\mathscr{S}^{\mathrm{an}}) & \longrightarrow & H^i(\mathscr{T}^{\mathrm{an}}) & \xrightarrow[\delta_i]{} & H^{i+1}(\mathscr{R}^{\mathrm{an}})
\end{array}$$

Proof. Fact (i) follows from Lemma 10.5.6; in the case where $\mathscr{S}$ is a sheaf on $\mathbb{P}^n$ it just reiterates the statement of the lemma, while for a sheaf $\mathscr{S}'$ on $\{\mathbb{P}^n\}^{\mathrm{an}}$ the relevant observation is that

$$H^0(\mathscr{S}') \quad = \quad \Gamma(\mathbb{P}^n, \lambda_*\mathscr{S}') \quad = \quad \Gamma(\{\mathbb{P}^n\}^{\mathrm{an}}, \mathscr{S}') \,.$$

Fact (ii) is because the complex $C^*(\mathscr{S})$ vanishes outside the interval $0 \le i \le n$.

Fact (iii) follows from Lemma 10.5.9.

Fact (iv) is derived from Facts 10.5.11 (i) and (iv), by taking the long exact sequences in cohomology induced by the short exact sequences of cochain complexes.

Fact (v) follows from Fact 10.5.11(ii).

Fact (vi) follows from Fact 10.5.11(iii).

Fact (vii) is once again obtained from Fact 10.5.11(iv), by taking the long exact sequences in cohomology corresponding to the short exact sequences of cochain complexes. □

Remark 10.5.15. The construction of $H^i(\mathscr{S})$ is not basis free, in the sense of Remark 10.3.2. If we replace our cover $\mathbb{P}^n = \cup\mathbb{P}^n_{x_i}$ by another cover $\mathbb{P}^n = \cup\mathbb{P}^n_{y_i}$ it is not clear from the construction that the cohomology groups $H^i(\mathscr{S})$ will be unaffected. It is possible to prove this fact, but such a proof goes way beyond the scope of this book.

10.6 GAGA in terms of cohomology

Now that we have defined the vector spaces $H^i(\mathscr{S})$ we can state the generalization of Theorem 7.9.8 which we plan to prove.

Theorem 10.6.1. *Let $\mathscr{S}$ be any coherent algebraic sheaf on $\mathbb{P}^n$. For any integer i the natural map*

$$H^i(\lambda^*) : H^i(\mathscr{S}) \longrightarrow H^i(\mathscr{S}^{\mathrm{an}})$$

is an isomorphism.

Remark 10.6.2. We deem that the case where $n = 0$ is trivial. In this case $\mathbb{P}^0 = \mathrm{Spec}(\mathbb{C})$ is the one-point space, and the theorem is virtually content-free. We will therefore assume, below, that $n \ge 1$.

Remark 10.6.3. When $i = 0$, Theorem 10.6.1 comes down precisely to Theorem 7.9.8. Strangely enough it is easier to prove the more general statement, which permits i to be arbitrary. Perhaps we should explain this a little. There are a few sheaves $\mathscr{S}$ for which we know Theorem 10.6.1, simply because we can compute both sides. The idea of the proof is to reduce the general case to the few special cases we know, by chasing around short exact sequences of sheaves. This method can only work if we look at all the maps $H^i(\lambda^*) : H^i(\mathscr{S}) \longrightarrow H^i(\mathscr{S}^{\text{an}})$, for every possible value of the integer i. Even if we only care about H^0, the short exact sequences of sheaves, which play a key role in the proof, give rise to long exact sequences in cohomology. And these long exact sequences involve every H^i.

As Remark 10.6.3 explained, the proof of Theorem 10.6.1 has two parts. First there is a computation, which explicitly shows that the linear map $H^i(\lambda^*) : H^i(\mathscr{S}) \longrightarrow H^i(\mathscr{S}^{\text{an}})$ is an isomorphism for certain, carefully chosen $\mathscr{S}$'s. Then the general case is deduced from this. In this section we do the computational part. We will prove:

Lemma 10.6.4. *Assume $n \geq 1$. On $\mathbb{P}^n$, let us consider the sheaves $\mathcal{O}(m)$ of Definition 10.1.4. Then the following three statements are true:*

(i) *For every pair of integers $i, m \in \mathbb{Z}$, the natural map*

$$H^i(\lambda^*) : H^i\big(\mathcal{O}(m)\big) \longrightarrow H^i\big(\mathcal{O}(m)^{\text{an}}\big)$$

is an isomorphism.

(ii) *The vector spaces $H^i\big(\mathcal{O}(m)\big) \cong H^i\big(\mathcal{O}(m)^{\text{an}}\big)$ are all finite dimensional.*

(iii) *If $i \geq 1$ and $m \geq -n$, then we have*

$$H^i\big(\mathcal{O}(m)\big) \quad = \quad 0 \quad = \quad H^i\big(\mathcal{O}(m)^{\text{an}}\big) .$$

Remark 10.6.5. Part (i) of Lemma 10.6.4 is, as promised, the special case of Theorem 10.6.1 for which we will give an explicit, computational proof. Part (iii) of Lemma 10.6.4 is a technical statement; we will need it, somewhat later, when we prove the second half of GAGA. Part (ii) of Lemma 10.6.4 is an observation we include just for fun.

The rest of this section is devoted to the proof of Lemma 10.6.4. Let us begin by reminding the reader of the tensor product of cochain complexes. We do it in a special case, the case that interests us here.

Remark 10.6.6. The reader is advised, in reading what comes next, to compare with Notation 3.5.3, especially with parts (iv) and (vi).

Reminder 10.6.7. Suppose that, for every $0 \leq i \leq n$, we are given a cochain complex

$$\cdots \longrightarrow 0 \longrightarrow A_i \xrightarrow{\varphi_i} B_i \longrightarrow 0 \longrightarrow \cdots$$

of $\mathbb{C}$–vector spaces. That is, the ith cochain complex has at most two non-vanishing vector spaces, and these two non-vanishing spaces, and the map φ_i between them, are as indicated. Then the tensor product

$$C \quad = \quad \bigotimes_{i=0}^{n}{}_{\mathbb{C}} \{A_i \longrightarrow B_i\}$$

is a cochain complex. To define it we first establish the notation that, for every subset $J \subset \{0, 1, \ldots, n\}$, the vector space C_J is given by the formula

$$C_J \quad = \quad \bigotimes_{i=0}^{n}{}_{\mathbb{C}} C_i^J \ ,$$

where

$$C_i^J \quad = \quad \begin{cases} A_i & \text{if } i \notin J, \\ B_i & \text{if } i \in J. \end{cases}$$

Note that all the tensor products are over $\mathbb{C}$.

Next we define, for each integer i in the range $-1 \leq i \leq n$, a vector space

$$C_{[i]} \quad = \quad \prod_{|J|=i+1} C_J \ .$$

This means that $C_{[i]}$ is the product of all the C_J, where the cardinality of the subset $J \subset \{0, 1, \ldots, n\}$ is precisely $(i+1)$. The reader should compare this with Construction 10.5.4; note that, unlike Construction 10.5.4, here we allow J to be empty, and hence i lies in the range $-1 \leq i \leq n$.

Next we want to define a map $\partial_i : C_{[i]} \longrightarrow C_{[i+1]}$, much as in Construction 10.5.4. The two $\mathbb{C}$–vector spaces $C_{[i]}$, $C_{[i+1]}$ are finite direct sums of vector spaces C_J, so the map ∂_i will be a matrix of homomorphisms between the summands. We need to define, for the summands $C_J \subset C_{[i]}$ and $C_{J'} \subset C_{[i+1]}$, the matrix entry of ∂_i between them. If we denote this matrix entry by $\partial_i^{JJ'}$ then the rule is simple. It is

(i) If $J \not\subset J'$, then $\partial_i^{JJ'} = 0$.

(ii) Suppose $J \subset J'$. We are assuming that the cardinalities are $|J| = i+1$, $|J'| = i+2$. There is one element in $J' - J$. Put $J' - J = \{j\}$. Then we define

$$\partial_i^{JJ'} \quad = \quad \varepsilon \cdot \psi^{JJ'} \ ,$$

where $\varepsilon = \pm 1$ is a sign defined by the same rule as in Construction 10.5.4, while $\psi^{JJ'}$ follows the recipe of Notation 3.5.3(vi). We remind the reader of both:

(a) The map $\psi^{JJ'} : C_J \longrightarrow C_{J'}$ is the tensor product

$$\bigotimes_{i=0}^{n} \psi_i^{JJ'} \; : \; \bigotimes_{i=0}^{n} C_i^J \longrightarrow \bigotimes_{i=0}^{n} C_i^{J'} ,$$

where $\psi_i^{JJ'} : C_i^J \longrightarrow C_i^{J'}$ is given by the formula

$$\psi_i^{JJ'} = \begin{cases} 1 & \text{if } i \notin J' - J = \{j\}, \\ \varphi_i & \text{if } i \in J' - J. \end{cases}$$

In this formula, $\varphi_i : A_i \longrightarrow B_i$ is the given map.

(b) The rule determining the sign $\varepsilon = \pm 1$ goes as follows. Consider the set

$$S_{JJ'} = \{s \in J' \mid s < j\} .$$

Then, if $|S_{JJ'}|$ stands for the cardinality of the finite set $S_{JJ'}$,

$$\varepsilon = \varepsilon(J, J') = (-1)^{|S_{JJ'}|} .$$

It is easy to verify that, with the formulas as above, $\partial_{i+1}\partial_i = 0$. We extend C to an infinite cochain complex by setting $C_{[i]} = 0$ if either $i < -1$ or $i > n$.

Remark 10.6.8. Our choice of degrees is a little unorthodox; it would be more standard to put the complex

$$C = \bigotimes_{i=0}^{n} {}_{\mathbb{C}} \{A_i \longrightarrow B_i\}$$

in degrees $0 \le i \le n+1$. The reason for our convention is to be consistent with Construction 10.5.4.

There are a couple of easy little lemmas we will use.

Lemma 10.6.9. *The tensor product of complexes is distributive; that is, if $\varphi_j : A_j \longrightarrow B_j$ is isomorphic to the direct sum*

$$A_j' \oplus A_j'' \xrightarrow{\varphi_j' \oplus \varphi_j''} B_j' \oplus B_j'' ,$$

then the tensor product

$$\bigotimes_{i=0}^{n} {}_{\mathbb{C}} \{A_i \longrightarrow B_i\}$$

is isomorphic to the direct sum of

$$\left[\bigotimes_{i=0}^{j-1}{}_{\mathbb{C}}\left\{A_i \longrightarrow B_i\right\}\right] \otimes_{\mathbb{C}} \left\{A'_j \longrightarrow B'_j\right\} \otimes_{\mathbb{C}} \left[\bigotimes_{i=j+1}^{n}{}_{\mathbb{C}}\left\{A_i \longrightarrow B_i\right\}\right]$$

and

$$\left[\bigotimes_{i=0}^{j-1}{}_{\mathbb{C}}\left\{A_i \longrightarrow B_i\right\}\right] \otimes_{\mathbb{C}} \left\{A''_j \longrightarrow B''_j\right\} \otimes_{\mathbb{C}} \left[\bigotimes_{i=j+1}^{n}{}_{\mathbb{C}}\left\{A_i \longrightarrow B_i\right\}\right] .$$

Proof. Obvious from the definitions, and from the fact that the tensor product of vector spaces is distributive. □

Lemma 10.6.10. *If, for some j with $0 \leq j \leq n$, the map $\varphi_j : A_j \longrightarrow B_j$ is an isomorphism, then the cochain complex*

$$C \quad = \quad \bigotimes_{i=0}^{n}{}_{\mathbb{C}}\left\{A_i \longrightarrow B_i\right\}$$

is contractible.

Proof. Choose and fix a j so that $\varphi_j : A_j \longrightarrow B_j$ is an isomorphism. We need to give the formula for the contracting homotopy. That is, for each i with $0 \leq i \leq n$ we need to give a map $\Theta_i : C_{[i]} \longrightarrow C_{[i-1]}$. Since $C_{[i]}$ and $C_{[i-1]}$ are finite direct sums of C_J, it is enough to describe the components of the map $\Theta_i : C_{[i]} \longrightarrow C_{[i-1]}$. That is, for all J' with $|J'| = i+1$ and all J with $|J| = i$, we must write down a map $\Theta_i^{J'J} : C_{J'} \longrightarrow C_J$.

The formula is simple. It is

(i) Unless $J' = J \cup \{j\}$, for our fixed j with $\varphi_j : A_j \longrightarrow B_j$ an isomorphism, the map $\Theta_i^{J'J} : C_{J'} \longrightarrow C_J$ is zero.

(ii) If $J' = J \cup \{j\}$, then $C_{J'} = X \otimes B_j \otimes Y$ and $C_J = X \otimes A_j \otimes Y$. In that case, the map $\Theta_i^{J'J} : C_{J'} \longrightarrow C_J$ is

$$\varepsilon(J, J') \cdot (1 \otimes \varphi_j^{-1} \otimes 1) ,$$

where

$$X \otimes B_j \otimes Y \xrightarrow{1 \otimes \varphi_j^{-1} \otimes 1} X \otimes A_j \otimes Y$$

is the tensor product of $1 : X \longrightarrow X$ with $\varphi_j^{-1} : B_j \longrightarrow A_j$ with $1 : Y \longrightarrow Y$, and $\varepsilon(J, J') = \pm 1$ is the sign of Reminder 10.6.7(ii).

We leave it to the reader to check that, with the formulas for ∂_i and Θ_i as given, we have the identity

$$1 = \Theta_{i+1}\partial_i + \partial_{i-1}\Theta_i \ ;$$

that is, Θ is a homotopy of the identity with the null map. □

Construction 10.6.11. The case that interests us most is the following. Let $\{x_0, x_1, \dots, x_n\}$ be indeterminates. Let $\mathbb{C}[x_i]$ be the polynomial ring in the variable x_i; as a vector space over $\mathbb{C}$ it has the basis $\{x_i^\ell,\ \ell \geq 0\}$. Let $\mathbb{C}[x_i, x_i^{-1}]$ be the ring of Laurent polynomials; as a vector space over $\mathbb{C}$ it has the basis $\{x_i^\ell,\ \ell \in \mathbb{Z}\}$. Let $\varphi_i : \mathbb{C}[x_i] \longrightarrow \mathbb{C}[x_i, x_i^{-1}]$ be the natural inclusion. We care particularly about the cochain complex

$$C \quad = \quad \bigotimes_{i=0}^{n}{}_{\mathbb{C}} \left\{ \mathbb{C}[x_i] \longrightarrow \mathbb{C}[x_i, x_i^{-1}] \right\} .$$

Remark 10.6.12. Recall the Hopf algebra $R = \mathbb{C}[t, t^{-1}]$ of Example 8.2.8. In Example 8.3.5 we defined an action on $\mathbb{C}[x_i]$, by the rule

$$a'(x_i^\ell) = t^{-\ell} \otimes x_i^\ell \ .$$

The same formula defines an action on $\mathbb{C}[x_i, x_i^{-1}]$; the only difference is that in $\mathbb{C}[x_i, x_i^{-1}]$ the integer ℓ is allowed to be negative. With these actions we have that the inclusion $\varphi_i : \mathbb{C}[x_i] \longrightarrow \mathbb{C}[x_i, x_i^{-1}]$ is a homomorphism of representations of $(G, \mathcal{O}_G) = \left(\operatorname{Spec}(R), \widetilde{R}\right)$. The group G^{an} acts on both $\mathbb{C}[x_i]$ and $\mathbb{C}[x_i, x_i^{-1}]$, and the map φ_i intertwines the action. In Remark 9.3.7 we saw that this action is given by the rule $g(x_i^\ell) = g^\ell x_i^\ell$.

Next observe that

$$\mathbb{C}[x_i, x_i^{-1}] \quad = \quad \mathbb{C}[x_i] \oplus x_i^{-1}\mathbb{C}[x_i^{-1}] \ ;$$

in terms of the bases this means that, for every x_i^ℓ, either $\ell \geq 0$ or $\ell < 0$. The basis for $\mathbb{C}[x_i]$ is the set $\{x_i^\ell,\ \ell \geq 0\}$, while the basis for $x_i^{-1}\mathbb{C}[x_i^{-1}]$ is the set $\{x_i^\ell,\ \ell < 0\}$. The direct sum decomposition

$$\mathbb{C}[x_i, x_i^{-1}] \quad = \quad \mathbb{C}[x_i] \oplus x_i^{-1}\mathbb{C}[x_i^{-1}]$$

respects the G^{an}–action.

The key computation is

Proposition 10.6.13. *The complex C decomposes as a direct sum of complexes $C(J)$, over $J \subset \{0, 1, \dots, n\}$. The complex $C(J)$ is a tensor product*

$$C(J) \quad = \quad \bigotimes_{i=0}^{n}{}_{\mathbb{C}} \left\{ A_i^J \longrightarrow B_i^J \right\} ,$$

where $A_i^J \longrightarrow B_i^J$ is given by the rule

(i) *If $i \notin J$, then $A_i^J \longrightarrow B_i^J$ is equal to $1 : \mathbb{C}[x_i] \longrightarrow \mathbb{C}[x_i]$.*

(ii) *If $i \in J$, then $A_i^J \longrightarrow B_i^J$ is equal to $0 \longrightarrow x_i^{-1}\mathbb{C}[x_i^{-1}]$.*

The decomposition of C, as a direct sum of $C(J)$'s, respects the diagonal action of G^{an}.

Proof. The complex

$$\cdots \longrightarrow 0 \longrightarrow \mathbb{C}[x_i] \xrightarrow{\varphi_i} \mathbb{C}[x_i, x_i^{-1}] \longrightarrow 0 \longrightarrow \cdots$$

is the direct sum of the complexes

$$\cdots \longrightarrow 0 \longrightarrow \mathbb{C}[x_i] \xrightarrow{1} \mathbb{C}[x_i] \longrightarrow 0 \longrightarrow \cdots$$

$$\cdots \longrightarrow 0 \longrightarrow 0 \longrightarrow x_i^{-1}\mathbb{C}[x_i^{-1}] \longrightarrow 0 \longrightarrow \cdots$$

and this decomposition is compatible with the action of G^{an}. The result now follows from the distributivity of Lemma 10.6.9. □

Corollary 10.6.14. *The complex C decomposes as a direct sum $C = C_1 \oplus C_2$, where C_1 is contractible while C_2 is the cochain complex*

$$\cdots \longrightarrow 0 \longrightarrow 0 \longrightarrow \bigotimes_{i=0}^{n} x_i^{-1}\mathbb{C}[x_i^{-1}] \longrightarrow 0 \longrightarrow 0 \longrightarrow \cdots$$

In the cochain complex C_2, the only non-zero term is in degree n. Furthermore, the decomposition $C = C_1 \oplus C_2$ respects the action of G^{an}.

Proof. Proposition 10.6.13 informs us that C decomposes, compatibly with the action of G^{an}, as the direct sum of the complexes $C(J)$. If J is strictly contained in $\{0, 1, \dots, n\}$ then there exists some $i \notin J$, and hence $C(J)$ is the tensor product of $A_i^J \longrightarrow B_i^J$, at least one of which is $1 : \mathbb{C}[x_i] \longrightarrow \mathbb{C}[x_i]$. Lemma 10.6.10 tells us that $C(J)$ must be contractible.

Therefore the only summand, which might not be contractible, is the summand $C(J)$ with $J = \{0, 1, \dots, n\}$. This summand is easily computed to be C_2 above. □

Discussion 10.6.15. In this section we want to compute $H^i(\mathcal{O}(m))$ and $H^i(\mathcal{O}(m)^{\mathrm{an}})$. It remains to explain what the complex C might have to do with these computations.

Now recall that the C of Construction 10.6.11 is a cochain complex

$$C_{[-1]} \longrightarrow C_{[0]} \longrightarrow \cdots \longrightarrow C_{[n-1]} \longrightarrow C_{[n]} ,$$

where each $C_{[i]}$ is defined as a sum of C_J, and where each C_J is given by the rule

$$C_J = \bigotimes_{i=0}^{n} C_i^J ,$$

and C_i^J was defined as

$$C_i^J \quad = \quad \begin{cases} \mathbb{C}[x_i] & \text{if } i \notin J, \\ \mathbb{C}[x_i, x_i^{-1}] & \text{if } i \in J. \end{cases}$$

If the reader has not yet done so, at this point she should take a close look at Notation 3.5.3(iv). The definition is not just parallel; it is identical. The C_J above are precisely the C_J of Notation 3.5.3(iv).

In Notation 3.5.3(v) we defined a map $\tau : C_J \longrightarrow S_J$, and in Exercise 3.5.4(ii) we proved that τ is an isomorphism. Furthermore, in Exercise 3.5.4(iii) we showed that the square

$$\begin{array}{ccc} C_J & \xrightarrow{\psi^{JJ'}} & C_{J'} \\ \tau\Big\downarrow & & \Big\downarrow\tau \\ S_J & \xrightarrow[\text{inclusion}]{} & S_{J'} \end{array}$$

commutes. The complex C is therefore isomorphic to a complex S'†. The complex S' can be written

$$S'_{[-1]} \longrightarrow S'_{[0]} \longrightarrow \cdots \longrightarrow S'_{[n-1]} \longrightarrow S'_{[n]} ,$$

where the definition is parallel to that of C, with C_J replaced by S_J everywhere, and $\psi^{JJ'}$ replaced by the inclusions $S_J \longrightarrow S_{J'}$. We leave the details to the reader.

Now each S_J is the direct sum of its homogeneous components. In the notation of Definition 10.2.4 we have

$$S_J \quad = \quad \bigoplus_{m \in \mathbb{Z}} S_J(m) ,$$

where $S_J(m)$ is the set of homogeneous Laurent polynomials of degree m. Furthermore, all the differentials in the complex are, up to signs, inclusions $S_J \longrightarrow S_{J'}$; they preserve homogeneous degrees. It follows that the entire complex decomposes as the direct sum over m of its homogeneous components. And finally we note that, if J is non-empty, then Example 10.2.7 produces an isomorphism $S_J(m) \longrightarrow \Gamma\big(\mathbb{P}^n_J, \mathcal{O}(m)\big)$, while Remark 10.2.8 establishes the commutativity of the square

$$\begin{array}{ccc} S_J(m) & \xrightarrow{\text{inclusion}} & S_{J'}(m) \\ \Big\downarrow & & \Big\downarrow \\ \Gamma\big(\mathbb{P}^n_J, \mathcal{O}(m)\big) & \xrightarrow[r^J_{J'}]{} & \Gamma\big(\mathbb{P}^n_{J'}, \mathcal{O}(m)\big) \end{array} .$$

† I would write the complex as S, except that the letter S is already taken for the polynomial ring $S = \mathbb{C}[x_0, x_1, \ldots, x_n]$.

From this it follows that the degree–m part of the cochain complex

$$S'_{[0]} \longrightarrow S'_{[1]} \longrightarrow \cdots \longrightarrow S'_{[n-1]} \longrightarrow S'_{[n]}$$

identifies with the complex

$$\mathcal{O}(m)_{[0]} \longrightarrow \mathcal{O}(m)_{[1]} \longrightarrow \cdots \longrightarrow \mathcal{O}(m)_{[n-1]} \longrightarrow \mathcal{O}(m)_{[n]}$$

that one obtains by applying Construction 10.5.4 to the sheaf $\mathcal{O}(m)$. Note that we must leave out $S'[-1]$ since, in Construction 10.5.4, we insisted on J being non-empty.

In view of all the identifications of Discussion 10.6.15, we can now state a lemma.

Lemma 10.6.16. *The vector space $H^i\big(\mathcal{O}(m)\big)$ is isomorphic, via the identifications of Discussion 10.6.15, with the degree–m part of the vector space $H^i(C')$, where C' is the truncated complex*

$$0 \longrightarrow C_{[0]} \longrightarrow \cdots \longrightarrow C_{[n-1]} \longrightarrow C_{[n]} \longrightarrow 0\,.$$

That is, C' is obtained from C by leaving out $C_{[-1]}$.

Proof. The identification of the complex

$$0 \longrightarrow C_{[0]} \longrightarrow \cdots \longrightarrow C_{[n-1]} \longrightarrow C_{[n]} \longrightarrow 0$$

with

$$0 \longrightarrow S'_{[0]} \longrightarrow \cdots \longrightarrow S'_{[n-1]} \longrightarrow S'_{[n]} \longrightarrow 0$$

is via the map $\tau : C_J \longrightarrow S_J$, given by the formula

$$\tau(x_0^{a_0} \otimes x_1^{a_1} \otimes \cdots \otimes x_n^{a_n}) \quad = \quad x_0^{a_0} x_1^{a_1} \cdots x_n^{a_n}\,,$$

and it is clear that this map preserves homogeneous degrees. The degree–m part of the cochain complex

$$0 \longrightarrow S'_{[0]} \longrightarrow \cdots \longrightarrow S'_{[n-1]} \longrightarrow S'_{[n]} \longrightarrow 0$$

was identified, in Discussion 10.6.15, with the complex

$$0 \longrightarrow \mathcal{O}(m)_{[0]} \longrightarrow \mathcal{O}(m)_{[1]} \longrightarrow \cdots \longrightarrow \mathcal{O}(m)_{[n-1]} \longrightarrow \mathcal{O}(m)_{[n]} \longrightarrow 0\,.$$

Definition 10.5.12 tells us that $H^i\big(\mathcal{O}(m)\big)$ is precisely the ith cohomology of this complex. □

Consequences 10.6.17. Let us assemble what we have learned up to now, and show how much of Lemma 10.6.4 is an immediate consequence. What we know so far is the following:

(i) The complex C, that is

$$0 \longrightarrow C_{[-1]} \longrightarrow C_{[0]} \longrightarrow C_{[1]} \longrightarrow \cdots \longrightarrow C_{[n-1]} \longrightarrow C_{[n]} \longrightarrow 0$$

decomposes as the direct sum of a contractible complex, and of the vector space

$$\bigotimes_{i=0}^{n} x_i^{-1}\mathbb{C}[x_i^{-1}]$$

concentrated in degree n. This decomposition respects the action of G^{an}; it respects homogeneous degrees.

(ii) If C' is the truncated complex

$$0 \longrightarrow C_{[0]} \longrightarrow C_{[1]} \longrightarrow \cdots \longrightarrow C_{[n-1]} \longrightarrow C_{[n]} \longrightarrow 0\ ,$$

then the degree–m part of $H^i(C')$ is isomorphic to $H^i\big(\mathcal{O}(m)\big)$.

The proof of (i) may be found in Corollary 10.6.14, while (ii) follows from Lemma 10.6.16.

The truncation does not change H^i, if $i \geq 1$. That is, if $i \geq 1$ then $H^i(C') = H^i(C)$. Now, recalling our assumption that $n \geq 1$, this gives

(iii) $H^i(C') = 0$ if $0 < i < n$.

(iv) $H^n(C')$ is isomorphic, with an isomorphism respecting homogeneous degrees, to

$$\bigotimes_{i=0}^{n} x_i^{-1}\mathbb{C}[x_i^{-1}]\ .$$

(v) For $i = 0$ we argue slightly differently. We know that $n > 0$, and hence $H^0(C) = 0$. This means that the kernel of the map $\partial_0 : C_{[0]} \longrightarrow C_{[1]}$ must be precisely $C_{[-1]}$. That is, $H^0(C') = C_{[-1]}$. Furthermore, the formulas tell us that

$$C_{[-1]} \quad = \quad \bigotimes_{i=0}^{n}{}_{\mathbb{C}}\, \mathbb{C}[x_i] \quad \cong \quad S \quad = \quad \mathbb{C}[x_0, x_1, \ldots, x_n]\ .$$

Now, if we focus on the degree–m part, we deduce

(vi) $H^i\big(\mathcal{O}(m)\big) = 0$ if $i \neq 0$ and $i \neq n$.

(vii) $H^0\big(\mathcal{O}(m)\big)$ is the vector space of homogeneous polynomials $f \in \mathbb{C}[x_0, x_1, \ldots, x_n]$ of degree m. A basis for it is given by the monomials

$$x_0^{b_0} x_1^{b_1} \cdots x_n^{b_n}\ ,$$

with $b_i \geq 0$ and $b_0 + b_1 + \cdots + b_n = m$. There are finitely many basis vectors, and hence $H^0\big(\mathcal{O}(m)\big)$ is finite dimensional.

(viii) $H^n\big(\mathcal{O}(m)\big)$ is the subspace of

$$\bigotimes_{i=0}^{n} x_i^{-1}\mathbb{C}[x_i^{-1}] ,$$

consisting of homogeneous Laurent polynomials of degree m. A basis is given by

$$x_0^{b_0} x_1^{b_1} \cdots x_n^{b_n} ,$$

with $b_i < 0$ and $b_0 + b_1 + \cdots + b_n = m$. There are finitely many basis vectors, and hence $H^n\big(\mathcal{O}(m)\big)$ is finite dimensional. Furthermore, the least negative, that the integer m can possibly be, is when the monomial is

$$x_0^{-1} x_1^{-1} \cdots x_n^{-1} ,$$

which is of degree $-(n+1)$. Hence $H^n\big(\mathcal{O}(m)\big) = 0$ if $m \geq -n$.

Summary 10.6.18. If we look at parts (vi), (vii) and (viii) of Consequences 10.6.17, they contain a couple of the assertions we made in Lemma 10.6.4. More precisely: we have proved the two assertions about $H^i\big(\mathcal{O}(m)\big)$. We know that it is always finite dimensional and that, for $i > 0$ and $m \geq -n$, we have $H^i\big(\mathcal{O}(m)\big) = 0$. The only assertion that remains to be proved is the statement that the map $H^i(\lambda^*) : H^i\big(\mathcal{O}(m)\big) \longrightarrow H^i\big(\mathcal{O}(m)^{\text{an}}\big)$ is an isomorphism.

Discussion 10.6.19. The idea of the proof is simple enough. The complex C, together with its truncation C', computed for us the cohomology vector spaces $H^i\big(\mathcal{O}(m)\big)$. In order to compute $H^i\big(\mathcal{O}(m)^{\text{an}}\big)$ we will simply complete the complex C with respect to its pre-Fréchet topology.

In Discussion 10.6.15 we identified the complex C with an isomorphic complex S', that is with a cochain complex

$$S'_{[-1]} \longrightarrow S'_{[0]} \longrightarrow \cdots \longrightarrow S'_{[n-1]} \longrightarrow S'_{[n]} .$$

Each $S'_{[i]}$ was a product, over subsets $J \subset \{0, 1, \ldots, n\}$ of cardinality $(i+1)$, of Laurent polynomial rings S_J. The degree–m part of this complex is identified with a complex of $S_J(m)$'s, and each $S_J(m)$ was identified, via the isomorphism of Example 10.2.7, with $\Gamma\big(\mathbb{P}^n_J, \mathcal{O}(m)\big)$. If we want the analytic version of the complex, we need only complete each $\Gamma\big(\mathbb{P}^n_J, \mathcal{O}(m)\big)$ with respect to its pre-Fréchet topology, obtaining $\Gamma\big(\{\mathbb{P}^n_J\}^{\text{an}}, \mathcal{O}(m)^{\text{an}}\big)$. Equivalently, we could complete each $S_J(m)$ with respect to its pre-Fréchet topology; we computed this topology in Corollary 10.2.13, and learned that the topology is the one determined by the seminorms s_ℓ of Definition 10.2.10.

What all of this means, very concretely, is the following. In the cochain complex C, that is the complex

$$C_{[-1]} \longrightarrow C_{[0]} \longrightarrow \cdots \longrightarrow C_{[n-1]} \longrightarrow C_{[n]} ,$$

the elements of the vector spaces $C_{[i]}$ are finite linear combinations of certain Laurent monomials. To complete we allow ourselves to form, separately in each degree m, certain infinite sums of the same monomials. The condition, imposed by the seminorms s_ℓ, is that the coefficients decay rapidly; for each m and for each $\ell \geq 1$, the real numbers

$$|\lambda_{\mathbf{a}}| \cdot \ell^{|\mathbf{a}|}, \quad \langle \mathbf{a} \rangle = m$$

must be bounded, independent of $\mathbf{a}$. We obtain a completed complex $\widehat{C}$, that is

$$\widehat{C}_{[-1]} \longrightarrow \widehat{C}_{[0]} \longrightarrow \cdots \longrightarrow \widehat{C}_{[n-1]} \longrightarrow \widehat{C}_{[n]} .$$

If we truncate the complex $\widehat{C}$ by leaving out $\widehat{C}_{[-1]}$ we obtain a complex we will call $\widehat{C}'$. And the degree–m part of $H^i(\widehat{C}')$ is precisely $H^i\big(\mathcal{O}(m)^{\mathrm{an}}\big)$.

What we have to prove comes down to the assertion that the map of cochain complexes $C' \longrightarrow \widehat{C}'$, that is the cochain map

$$\begin{array}{ccccccccccc}
0 & \longrightarrow & C_{[0]} & \longrightarrow & \cdots & \longrightarrow & C_{[n-1]} & \longrightarrow & C_{[n]} & \longrightarrow & 0 \\
 & & \downarrow & & & & \downarrow & & \downarrow & & \\
0 & \longrightarrow & \widehat{C}_{[0]} & \longrightarrow & \cdots & \longrightarrow & \widehat{C}_{[n-1]} & \longrightarrow & \widehat{C}_{[n]} & \longrightarrow & 0 \ ,
\end{array}$$

induces an isomorphism in cohomology. The first reduction we want comes from the short exact sequence of cochain complexes

$$\begin{array}{ccccccccccc}
0 & \longrightarrow & 0 & \longrightarrow & C_{[0]} & \longrightarrow & \cdots & \longrightarrow & C_{[n]} & \longrightarrow & 0 \\
 & & \downarrow & & \downarrow & & & & \downarrow & & \\
0 & \longrightarrow & C_{[-1]} & \longrightarrow & C_{[0]} & \longrightarrow & \cdots & \longrightarrow & C_{[n]} & \longrightarrow & 0 \\
 & & \downarrow & & \downarrow & & & & \downarrow & & \\
0 & \longrightarrow & C_{[-1]} & \longrightarrow & 0 & \longrightarrow & \cdots & \longrightarrow & 0 & \longrightarrow & 0 \ ,
\end{array}$$

which we will write as $0 \longrightarrow C' \longrightarrow C \longrightarrow C_{[-1]} \longrightarrow 0$. The completion of this short exact sequence is also a short exact sequence of cochain complexes, and the map to the completion gives a homomorphism of short exact sequences of cochain complexes

$$\begin{array}{ccccccccc} 0 & \longrightarrow & C' & \longrightarrow & C & \longrightarrow & C_{[-1]} & \longrightarrow & 0 \\ & & \downarrow{\scriptstyle i'} & & \downarrow{\scriptstyle i} & & \downarrow{\scriptstyle i_{[-1]}} & & \\ 0 & \longrightarrow & \widehat{C}' & \longrightarrow & \widehat{C} & \longrightarrow & \widehat{C}_{[-1]} & \longrightarrow & 0 \end{array} .$$

We want to know that the map i' induces an isomorphism in cohomology; the long exact sequence in cohomology tells us that it suffices to show that i and $i_{[-1]}$ do.

For $i_{[-1]}$ there is not a lot to do; we have already identified $C_{[-1]}$, in Consequence 10.6.17(v), with the polynomial ring $S = \mathbb{C}[x_0, x_1, \ldots, x_n]$. The subspace $S(m) \subset S$ consists of the homogeneous polynomials of degree m. The space $S(m)$ is finite dimensional; taking (possibly) infinite sums of its basis elements, with rapidly decaying coefficients, does not change $S(m)$. We conclude that $C_{[-1]} = \widehat{C}_{[-1]}$. It remains therefore only to show that the map $i : C \longrightarrow \widehat{C}$ induces an isomorphism in cohomology.

Now we recall Proposition 10.6.13; it told us that the complex C decomposes as a direct sum of complexes $C(J)$, most of which are contractible. The next observation is

Exercise 10.6.20. Completion respects this decomposition, that is $\widehat{C}$ is the direct sum of the $\widehat{C(J)}$. [Hint: each $C_{[i]}$ decomposes, by definition, as a finite direct sum of $C_{J'}$'s, which decompose, in turn, as finite direct sums of $C(J)_{J'}$'s. We need to show that completion respects this direct sum decomposition. The reader might wish to look at Remark 5.8.30.]

It remains therefore only to check that, for each $J \subset \{0, 1, \ldots, n\}$, the inclusion $C(J) \longrightarrow \widehat{C(J)}$ gives a homology isomorphism. For $J = \{0, 1, \ldots, n\}$ there is little to check: in that case $C(J)$ is the complex

$$\cdots \longrightarrow 0 \longrightarrow 0 \longrightarrow \bigotimes_{i=0}^{n} x_i^{-1}\mathbb{C}[x_i^{-1}] \longrightarrow 0 \longrightarrow 0 \longrightarrow \cdots$$

of Corollary 10.6.14, and it is finite dimensional in each degree m. It is equal to its completion. All that remains to prove is therefore

Lemma 10.6.21. *Suppose $J \subset \{0, 1, \ldots, n\}$ is a proper subset. Then the complex $\widehat{C(J)}$ is contractible.*

Proof. Fix a proper subset $J \subset \{0, 1, \ldots, n\}$. We recall the definition of $C(J)$, given in Proposition 10.6.13. The complex $C(J)$ is the tensor product

$$C(J) \quad = \quad \bigotimes_{i=0}^{n}{}_{\mathbb{C}} \left\{ A_i^J \longrightarrow B_i^J \right\} ,$$

where $A_i^J \longrightarrow B_i^J$ is given by the rule

(i) If $i \notin J$, then $A_i^J \longrightarrow B_i^J$ is equal to $1 : \mathbb{C}[x_i] \longrightarrow \mathbb{C}[x_i]$.
(ii) If $i \in J$, then $A_i^J \longrightarrow B_i^J$ is equal to $0 \longrightarrow x_i^{-1}\mathbb{C}[x_i^{-1}]$.

The rule for forming the tensor product may be found in Reminder 10.6.7. First of all, for any subset $J' \subset \{0, 1, \ldots, n\}$ we form a vector space which we called $\mathbb{C}(J)_{J'}$. The general formula yields

$$C(J)_{J'} \quad = \quad \bigotimes_{i=0}^{n} C(J)_i^{J'} ,$$

where

$$C(J)_i^{J'} \quad = \quad \begin{cases} A_i^J & \text{if } i \notin J', \\ B_i^J & \text{if } i \in J'. \end{cases}$$

In our special case, if $i \in J$ then $A_i^J = 0$. If there exists an $i \in J - J'$ then $C(J)_i^{J'} = 0$, making the tensor product vanish. Hence the tensor product will vanish unless $J \subset J'$. And in the case when $J \subset J'$ this tensor product is independent of J'; it is

$$F(J) \quad = \quad \bigotimes_{i=1}^{n} F_i^J$$

where

$$F_i^J \quad = \quad \begin{cases} \mathbb{C}[x_i] & \text{if } i \notin J, \\ x_i^{-1}\mathbb{C}[x_i^{-1}] & \text{if } i \in J. \end{cases}$$

Furthermore, if $J \subset J' \subset J''$, then the map $\psi^{J'J''} : C(J)_{J'} \longrightarrow C(J)_{J''}$ is simply $1 : F(J) \longrightarrow F(J)$. And most important of all: in each degree m, the Fréchet topology on $C(J)_{J'} \subset C_{J'}$ is the same as the Fréchet topology on $C(J)_{J''} \subset C_{J''}$. Remark 10.2.15 tells us that the homomorphism $\psi^{J'J''} : C_{J'} \longrightarrow C_{J''}$ is an isometry. We can unambiguously speak of the Fréchet topology of $F(J) = C(J)_{J'} = C(J)_{J''}$.

Perhaps a simpler way to state it is the following. We could look at the complexes

$$\cdots \longrightarrow 0 \longrightarrow \widetilde{A}_i \longrightarrow \widetilde{B}_i \longrightarrow 0 \longrightarrow \cdots$$

where

(i) If $i \notin J$, then $\widetilde{A}_i \longrightarrow \widetilde{B}_i$ is equal to $1 : \mathbb{C} \longrightarrow \mathbb{C}$.

(ii) If $i \in J$, then $\widetilde{A}_i \longrightarrow \widetilde{B}_i$ is equal to $0 \longrightarrow \mathbb{C}$.

Form the complex

$$D(J) \quad = \quad \bigotimes_{i=0}^{n} \{\widetilde{A}_i \longrightarrow \widetilde{B}_i\} \quad .$$

Then $C(J)$ is nothing more nor less that $F(J) \otimes D(J)$; we tensor the complex $D(J)$, term-by-term, with the single vector space $F(J)$.

But now the completion is straightforward; the completion $\widehat{C(J)}$ is the tensor product of $\widehat{F(J)}$ with the complex $D(J)$. And the complex $D(J)$ is contractible. The reason is that J is a proper subset of $\{0, 1, \ldots, n\}$, so there exists $i \notin J$. This makes $\widetilde{A}_i \longrightarrow \widetilde{B}_i$ equal to $1 : \mathbb{C} \longrightarrow \mathbb{C}$, and Lemma 10.6.10 tells us that the tensor product

$$D(J) \quad = \quad \bigotimes_{i=0}^{n} \{\widetilde{A}_i \longrightarrow \widetilde{B}_i\} \quad .$$

must be contractible. □

10.7 The first half of GAGA

In this section we prove Theorem 7.9.8, which corresponds to half of GAGA. The proof, as we have already said several times, is by chasing around long exact sequences in cohomology which arise from short exact sequences of sheaves. Now let us state precisely what we will be proving.

Theorem 10.7.1. *Let $\mathscr{S}$ be a coherent algebraic sheaf. We assert three things:*

(i) *For every i the map $H^i(\lambda^*) : H^i(\mathscr{S}) \longrightarrow H^i(\mathscr{S}^{\text{an}})$ is an isomorphism.*

(ii) *There exists an integer N so that, for all $i > 0$ and all $m \geq N$, the vector spaces $H^i\big(\mathscr{S}(m)\big)$ vanish. Here, $\mathscr{S}(m)$ is as in Definition 10.4.1; that is, $\mathscr{S}(m) = \mathscr{H}om\big(\mathcal{O}(-m)\,,\,\mathscr{S}\big)$.*

(iii) *The vector spaces $H^i(\mathscr{S})$ are all finite dimensional.*

Remark 10.7.2. Observe that Theorem 10.7.1(i) is just exactly Theorem 10.6.1. We are, as promised, proving GAGA. Part (ii) is a technical lemma, which we will need in the proof of the second part of GAGA. Part (iii) is included just for fun.

Proof. First we note that all three assertions are true for the sheaves $\mathscr{S} = \mathcal{O}(m)$, and follow from Lemma 10.6.4. Parts (i) and (iii) follow, respectively, from (i) and (ii) of the Lemma. As for assertion (ii), it comes as follows: from Corollary 10.3.14 we know that, if $\mathscr{S} = \mathcal{O}(m)$, then

$$\mathscr{S}(m') \quad = \quad \mathscr{H}om\big(\mathcal{O}(-m'), \mathcal{O}(m)\big) \quad \cong \quad \mathcal{O}(m+m') \; .$$

Lemma 10.6.4(iii) says that $H^i\big(\mathcal{O}(m+m')\big) = 0$ for all $i > 0$ and all $m+m' \geq -n$. In other words $N = -m-n$ will work. Next observe that all three statements must also be true for finite direct sums of sheaves $\mathcal{O}(m_j)$.

Now we prove the following four assertions, by descending induction on i. For part (i) below, we will also need to assume $i \geq 1$.

(i) For any coherent algebraic sheaf $\mathscr{S}$, on projective space $\mathbb{P}^n$, there exists an integer N so that the vector space $H^j\big(\mathscr{S}(m)\big)$ vanishes, whenever $j \geq i \geq 1$ and $m \geq N$.

(ii) For any coherent algebraic sheaf $\mathscr{S}$, on projective space $\mathbb{P}^n$, one has that $H^j(\mathscr{S})$ is finite dimensional for all $j \geq i$.

(iii) For any coherent algebraic sheaf $\mathscr{S}$, on projective space $\mathbb{P}^n$, the maps $H^j(\lambda^*) : H^j(\mathscr{S}) \longrightarrow H^j(\mathscr{S}^{\mathrm{an}})$ are isomorphisms if $j \geq i$.

(iv) For any coherent algebraic sheaf $\mathscr{S}$, on projective space $\mathbb{P}^n$, the maps $H^j(\lambda^*) : H^j(\mathscr{S}) \longrightarrow H^j(\mathscr{S}^{\mathrm{an}})$ are isomorphisms if $j \geq i$, and the map $H^{i-1}(\lambda^*) : H^{i-1}(\mathscr{S}) \longrightarrow H^{i-1}(\mathscr{S}^{\mathrm{an}})$ is an epimorphism.

All the assertions are true for $i = n+2$; this is because $i - 1 = n+1$, and Fact 10.5.14 tells us that $H^j(\mathscr{S}) = 0 = H^j(\mathscr{S}^{\mathrm{an}})$ whenever $\mathscr{S}$ is a coherent algebraic sheaf and $j \geq n+1$. We have to prove the inductive step. All the proofs depend on the same short exact sequence of coherent sheaves. Let $\mathscr{T}$ be any coherent sheaf on $\mathbb{P}^n$. Lemma 10.1.6 informs us that there is a surjection $\mathscr{S} \longrightarrow \mathscr{T}$, with $\mathscr{S}$ a finite direct sum of $\mathcal{O}(m_k)$'s. Let $\mathscr{R}$ be the kernel; we have an exact sequence of sheaves on $\mathbb{P}^n$

$$0 \longrightarrow \mathscr{R} \longrightarrow \mathscr{S} \longrightarrow \mathscr{T} \longrightarrow 0 \; .$$

The idea of the proof will be to use this exact sequence in all three inductions.

Let us first prove that, if (ii) is true for i, then it must also be true for $i-1$. Our assumption is that, for any coherent sheaf $\mathscr{K}$ and any $j \geq i$, the vector space $H^j(\mathscr{K})$ is finite dimensional. Take a coherent sheaf $\mathscr{T}$; we need to prove $H^{i-1}(\mathscr{T})$ finite dimensional. Now consider our favorite short exact sequence

$$0 \longrightarrow \mathscr{R} \longrightarrow \mathscr{S} \longrightarrow \mathscr{T} \longrightarrow 0 \; .$$

We deduce a long exact sequence in cohomology; in particular, the following bit is exact

$$H^{i-1}(\mathscr{S}) \longrightarrow H^{i-1}(\mathscr{T}) \longrightarrow H^i(\mathscr{R}) \quad .$$

Now $H^i(\mathscr{R})$ is finite dimensional by the induction hypothesis. The sheaf $\mathscr{S}$ is a finite direct sum of $\mathcal{O}(m_k)$'s, and Lemma 10.6.4(iii) guarantees the finite dimensionality of $H^{i-1}(\mathscr{S})$. The exact sequence therefore forces $H^{i-1}(\mathscr{T})$ to be finite dimensional.

Let us next prove that, if (i) is true for $i \geq 2$, then it is must also be true for $i-1$. We are assuming that, for any coherent sheaf $\mathscr{K}$, there is a number N so that $H^j\big(\mathscr{K}(m)\big)$ vanishes, whenever $j \geq i$ and $m \geq N$. Let $\mathscr{T}$ be a coherent sheaf. We need to produce a number N that works also for $j = i-1$. Now take our exact sequence above and apply to it $\mathscr{H}om\big(\mathcal{O}(-m)\ ,\ -\big)$. We deduce a sequence

$$0 \longrightarrow \mathscr{R}(m) \longrightarrow \mathscr{S}(m) \longrightarrow \mathscr{T}(m) \longrightarrow 0\ ,$$

which is exact by Lemma 10.4.7. From Fact 10.5.14(iv) we have a long exact sequence in cohomology, and in particular

$$H^{i-1}\big(\mathscr{S}(m)\big) \longrightarrow H^{i-1}\big(\mathscr{T}(m)\big) \longrightarrow H^i\big(\mathscr{R}(m)\big)$$

must be exact. Induction tells us that there will be an integer N_1 so that $m \geq N_1$ means $H^i\big(\mathscr{R}(m)\big) = 0$. Now $i \geq 2$ means $i-1 \geq 1$, and since the sheaf $\mathscr{S}$ is a finite direct sum of $\mathcal{O}(m_k)$'s there is an integer N_2 so that $H^{i-1}\big(\mathscr{S}(m)\big) = 0$ for all $m \geq N_2$. Put $N = \max(N_1, N_2)$ and the exact sequence implies that $H^{i-1}\big(\mathscr{T}(m)\big) = 0$ for all $m \geq N$.

The way we will prove (iii) and (iv) is by showing that if (iii) is true for i then (iv) is true for i, and if (iv) is true for i then (iii) is true for $i-1$. Begin by assuming (iii) true for i. Let $\mathscr{T}$ be any coherent algebraic sheaf; we need to show that the map $H^{i-1}(\lambda^*) : H^{i-1}(\mathscr{T}) \longrightarrow H^{i-1}(\mathscr{T}^{\text{an}})$ is an epimorphism. Above we found the exact sequence of coherent algebraic sheaves $0 \longrightarrow \mathscr{R} \longrightarrow \mathscr{S} \longrightarrow \mathscr{T} \longrightarrow 0$, with $\mathscr{S}$ a direct sum of $\mathcal{O}(m_k)$'s. Fact 10.5.14(vii) gives a commutative diagram with exact rows

$$\begin{array}{ccccccc}
H^{i-1}(\mathscr{S}) & \longrightarrow & H^{i-1}(\mathscr{T}) & \longrightarrow & H^i(\mathscr{R}) & \longrightarrow & H^i(\mathscr{S}) \\
{\scriptstyle\beta}\big\downarrow & & {\scriptstyle\gamma}\big\downarrow & & {\scriptstyle\delta}\big\downarrow & & {\scriptstyle\varepsilon}\big\downarrow \\
H^{i-1}(\mathscr{S}^{\text{an}}) & \longrightarrow & H^{i-1}(\mathscr{T}^{\text{an}}) & \longrightarrow & H^i(\mathscr{R}^{\text{an}}) & \longrightarrow & H^i(\mathscr{S}^{\text{an}})
\end{array} \quad .$$

By (iii) for i we know that δ and ε are isomorphisms, and because $\mathscr{S}$ is a finite direct sum of $\mathcal{O}(m_k)$'s we also know that β is an isomorphism. It follows that γ is surjective.

Now assume we know (iv) for i and want to deduce (iii) for $(i-1)$.

Let $\mathscr{T}$ be any coherent algebraic sheaf, choose a surjection $\mathscr{S} \longrightarrow \mathscr{T}$ with $\mathscr{S}$ a finite direct sum of $\mathcal{O}(m_k)$'s, and look at our exact sequence of sheaves $0 \longrightarrow \mathscr{R} \longrightarrow \mathscr{S} \longrightarrow \mathscr{T} \longrightarrow 0$. Fact 10.5.14(vii) gives a commutative diagram with exact rows

$$\begin{array}{ccccccc}
H^{i-1}(\mathscr{R}) & \longrightarrow & H^{i-1}(\mathscr{S}) & \longrightarrow & H^{i-1}(\mathscr{T}) & \longrightarrow & H^{i}(\mathscr{R}) \\
{\scriptstyle\alpha}\downarrow & & {\scriptstyle\beta}\downarrow & & {\scriptstyle\gamma}\downarrow & & {\scriptstyle\delta}\downarrow \\
H^{i-1}(\mathscr{R}^{\mathrm{an}}) & \longrightarrow & H^{i-1}(\mathscr{S}^{\mathrm{an}}) & \longrightarrow & H^{i-1}(\mathscr{T}^{\mathrm{an}}) & \longrightarrow & H^{i}(\mathscr{R}^{\mathrm{an}})\,.
\end{array}$$

Because (iv) is true for i we know that δ is an isomorphism and α and γ are surjective. The fact that $\mathscr{S}$ is a finite direct sum of $\mathcal{O}(m_k)$'s means that β is an isomorphism. The fine 5–lemma says that γ must be an isomorphism. □

I have said that we proved Theorem 10.7.1(iii) just for fun. The reason is that we will need the stronger statement:

Theorem 10.7.3. *Let X be a complex analytic space, and assume X is compact and Hausdorff. Let $\mathscr{K}$ be a coherent analytic sheaf on X. Then the vector spaces $H^i(\mathscr{K})$ are all finite dimensional.*

Proof. See [3, page 245, Theorem 19]. □

Remark 10.7.4. In Lemma 8.9.27 we proved that $\{\mathbb{P}^n\}^{\mathrm{an}}$ is compact and Hausdorff, so Theorem 10.7.3 applies to it. What we have done, in Theorem 10.7.1(iii), is given a direct proof of the special case of Theorem 10.7.3, where $X = \{\mathbb{P}^n\}^{\mathrm{an}}$, and where the sheaf $\mathscr{K}$ is isomorphic to $\mathscr{S}^{\mathrm{an}}$, for some coherent algebraic sheaf $\mathscr{S}$. Eventually we will know that every coherent analytic sheaf $\mathscr{K}$ on $\{\mathbb{P}^n\}^{\mathrm{an}}$ is isomorphic to an $\mathscr{S}^{\mathrm{an}}$, but using this would be circular. In the proof, of the existence of an isomorphism $\mathscr{K} \cong \mathscr{S}^{\mathrm{an}}$, we will use the finite dimensionality of $H^i(\mathscr{K})$; see page 385.

10.8 Skyscraper sheaves

It remains to prove Theorem 7.9.1(ii). We must show that any coherent analytic sheaf is isomorphic to $\mathscr{S}^{\mathrm{an}}$, for some coherent algebraic sheaf $\mathscr{S}$. The idea of the proof is to reduce the general statement to the trivial case of skyscraper sheaves. We therefore begin by carefully pondering trivialities. We must make sure we truly understand skyscraper sheaves.

The 1-point space $\left(\mathrm{Spec}(\mathbb{C}), \widetilde{\mathbb{C}}\right)$ is clearly isomorphic to its analytification. Let $(Y, \mathcal{O}_Y)$ be a scheme locally of finite type over $\mathbb{C}$ and let

$p \in Y$ be a closed point. Lemma 4.2.6 tells us that there is a unique morphism of ringed spaces over $\mathbb{C}$

$$(\Phi, \Phi^*) : \left(\operatorname{Spec}(\mathbb{C}), \widetilde{\mathbb{C}}\right) \longrightarrow (Y, \mathcal{O}_Y)$$

such that $\Phi : \operatorname{Spec}(\mathbb{C}) \longrightarrow Y$ takes the only point of $\operatorname{Spec}(\mathbb{C})$ to $p \in Y$. In 5.1.2 we learned that any morphism $(\Phi, \Phi^*) : (X, \mathcal{O}_X) \longrightarrow (Y, \mathcal{O}_Y)$, of schemes locally of finite type over $\mathbb{C}$, gives a commutative square

$$\begin{array}{ccc}
(X^{\mathrm{an}}, \mathcal{O}_X^{\mathrm{an}}) & \xrightarrow{\left(\Phi^{\mathrm{an}}, \{\Phi^*\}^{\mathrm{an}}\right)} & (Y^{\mathrm{an}}, \mathcal{O}_Y^{\mathrm{an}}) \\
{\scriptstyle (\lambda_X, \lambda_X^*)} \downarrow & & \downarrow {\scriptstyle (\lambda_Y, \lambda_Y^*)} \\
(X, \mathcal{O}_X) & \xrightarrow[(\Phi, \Phi^*)]{} & (Y, \mathcal{O}_Y)
\end{array} ,$$

where (λ, λ^*) is the map from the analytification of a scheme to the scheme. We note:

Remark 10.8.1. In the case where $(X, \mathcal{O}_X)$ happens to be the 1–point space $\left(\operatorname{Spec}(\mathbb{C}), \widetilde{\mathbb{C}}\right)$ the map (λ_X, λ_X^*) is an isomorphism, and the diagram becomes

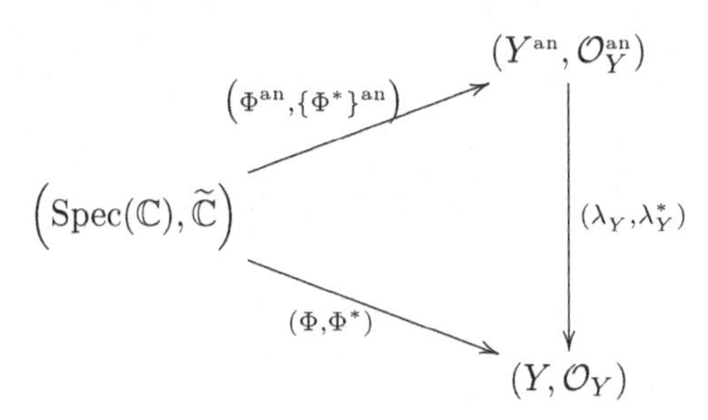

The commutative triangle of continuous maps is just

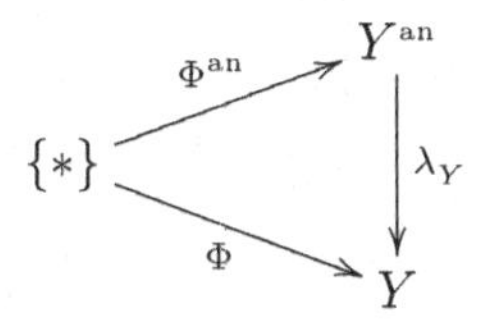

where $\lambda = \lambda_Y$ is the inclusion $Y^{\mathrm{an}} \subset Y$, and Φ^{an} and Φ takes the one point $*$ to $p \in Y^{\mathrm{an}} \subset Y$. Except for the trivial $\{*\}$ the space Y is the only topological space we will consider in the section, and we therefore allow ourselves to abbreviate λ_Y to λ.

Remark 10.8.2. It is the corresponding maps of sheaves that we are after. We have, on Y^{an}, a homomorphism of sheaves of rings $\{\Phi^*\}^{\mathrm{an}}$:

$\mathcal{O}_Y^{\mathrm{an}} \longrightarrow \Phi_*^{\mathrm{an}}\widetilde{\mathbb{C}}$, and on Y another homomorphism of sheaves of rings $\Phi^* : \mathcal{O}_Y \longrightarrow \Phi_*\widetilde{\mathbb{C}}$. The commutativity of the diagram means that $\lambda_*\Phi_*^{\mathrm{an}}\widetilde{\mathbb{C}} = \Phi_*\widetilde{\mathbb{C}}$, where the isomorphism is as sheaves of algebras over the sheaf of rings $\mathcal{O}_Y$. What does this all mean?

Remark 10.8.3. For any continuous map $\alpha : X \longrightarrow Y$, and for any sheaf $\mathscr{S}$ on X, the formula for the sheaf $\alpha_*\mathscr{S}$ on Y is

$$\Gamma(U, \alpha_*\mathscr{S}) \quad = \quad \Gamma(\alpha^{-1}U, \mathscr{S}) \,.$$

In the case where $X = \{*\}$ is the 1–point space, this is particularly simple. The two sheaves $\Phi_*\widetilde{\mathbb{C}}$ and $\Phi_*^{\mathrm{an}}\widetilde{\mathbb{C}}$ are defined on different spaces, but both are given by the very simple formulas

$$\Gamma(U, \Phi_*\widetilde{\mathbb{C}}) = \begin{cases} \mathbb{C} & \text{if } p \in U \\ 0 & \text{otherwise,} \end{cases} \qquad \Gamma(V, \Phi_*^{\mathrm{an}}\widetilde{\mathbb{C}}) = \begin{cases} \mathbb{C} & \text{if } p \in V \\ 0 & \text{otherwise,} \end{cases}$$

where the restriction maps are, for both sheaves,

$$\mathrm{res}_V^{V'} \quad = \quad \begin{cases} 1 : \mathbb{C} \longrightarrow \mathbb{C} & \text{if } p \in V \subset V' \\ 0 & \text{otherwise.} \end{cases}$$

What the commutative triangle of maps of ringed spaces buys us is that, on Y and Y^{an} respectively, there are morphisms of sheaves of rings

$$\Phi^* : \mathcal{O}_Y \longrightarrow \Phi_*\widetilde{\mathbb{C}}, \qquad\qquad \{\Phi^*\}^{\mathrm{an}} : \mathcal{O}_Y^{\mathrm{an}} \longrightarrow \Phi_*^{\mathrm{an}}\widetilde{\mathbb{C}}.$$

That is the sheaf $\Phi_*\widetilde{\mathbb{C}}$ can be viewed as a sheaf of $\mathcal{O}_Y$–modules on Y, and the sheaf $\Phi_*^{\mathrm{an}}\widetilde{\mathbb{C}}$ can be viewed as a sheaf of $\mathcal{O}_Y^{\mathrm{an}}$–modules on Y^{an}. Let us observe

Lemma 10.8.4. *The sheaf $\Phi_*\widetilde{\mathbb{C}}$ is a coherent algebraic sheaf on $(Y, \mathcal{O}_Y)$.*

Proof. The question is local, so we may restrict to an affine open subset $U \subset Y$. If $p \notin U$ then the restriction of $\Phi_*\widetilde{\mathbb{C}}$ to U is trivial, hence certainly coherent. Suppose therefore that $(U, \{\mathcal{O}_Y\}|_U) = \left(\mathrm{Spec}(S), \widetilde{S}\right)$ is affine, and $p \in U$. Then $\Gamma(U, \Phi_*\widetilde{\mathbb{C}}) = \mathbb{C}$ is a finite module over the ring $\Gamma(U, \mathcal{O}_Y) = S$, and the identity gives a homomorphism of S–modules

$$\mathbb{C} \longrightarrow \Gamma(U, \Phi_*\widetilde{\mathbb{C}}) \,.$$

Proposition 7.5.1 gives an induced map of sheaves $\theta : \widetilde{\mathbb{C}} \longrightarrow \Phi_*\widetilde{\mathbb{C}}|_U$, and we want to prove that θ is an isomorphism. We therefore need to understand the structure of $\Gamma(U, \Phi_*\widetilde{\mathbb{C}}) = \mathbb{C}$ as a module over S.

Because the open set U contains the image of Φ, the map of ringed spaces (Φ, Φ^*) must factorize as

$$\left(\mathrm{Spec}(\mathbb{C}), \widetilde{\mathbb{C}}\right) \xrightarrow{(\Psi,\Psi^*)} \left(\mathrm{Spec}(S), \widetilde{S}\right) \xrightarrow{(i,i^*)} (Y, \mathcal{O}) \,,$$

with (i, i^*) the inclusion. Theorem 3.9.4 says that $(\Psi, \Psi^*) = (\mathrm{Spec}(\varphi), \widetilde{\varphi})$ for some $\mathbb{C}$–algebra homomorphism $\varphi : S \longrightarrow \mathbb{C}$, and as Ψ takes the one point in $\mathrm{Spec}(\mathbb{C})$ to $p \in \mathrm{Spec}(S)$, the $\mathbb{C}$–algebra homomorphism $\varphi : S \longrightarrow \mathbb{C}$ is the unique map corresponding to p in Proposition 4.2.4. This means that the ring homomorphism $\Gamma(U, \mathcal{O}_Y) \longrightarrow \Gamma(\Phi^{-1}U, \widetilde{\mathbb{C}})$ is just $\varphi : S \longrightarrow \mathbb{C}$. The structure of $\mathbb{C} = \Gamma(U, \Phi_*\widetilde{\mathbb{C}})$, as a module over $S = \Gamma(U, \mathcal{O}_Y)$, is that $f \in S$ acts on $v \in \mathbb{C}$ to form $\varphi(f) \cdot v$. For the basic open sets $U_f \subset U = \mathrm{Spec}(S)$, this means that $\Gamma(U_f, \widetilde{\mathbb{C}}) = \mathbb{C}[1/f]$ is zero if $f \in \ker(\varphi)$, and $\mathbb{C}$ otherwise. Lemma 4.4.3 establishes that $f \in \ker(\varphi)$ if and only if $p \notin U_f$. We deduce

$$\Gamma(U_f, \widetilde{\mathbb{C}}) \;=\; \begin{cases} \mathbb{C} & \text{if } p \in U_f \\ 0 & \text{otherwise}, \end{cases}$$

which agrees with the formula for $\Phi_*\widetilde{\mathbb{C}}$ in Remark 10.8.3. This means the modules $\Gamma(U_f, \widetilde{\mathbb{C}})$ and $\Gamma\big(U_f, \Phi_*\widetilde{\mathbb{C}}|_U\big)$ are isomorphic, but we need to show that the map

$$\Gamma(U_f, \theta) \;:\; \Gamma(U_f, \widetilde{\mathbb{C}}) \longrightarrow \Gamma\big(U_f, \Phi_*\widetilde{\mathbb{C}}|_U\big)$$

is an isomorphism. If $p \notin U_f$ then the $\Gamma(U_f, \widetilde{\mathbb{C}}) = 0 = \Gamma\big(U_f, \Phi_*\widetilde{\mathbb{C}}|_U\big)$, and the map $\Gamma(U_f, \theta)$ has no choice but to be an isomorphism. If $p \in U_f$ we consider the commutative square

$$\begin{array}{ccccccc}
\mathbb{C} & = & \Gamma(U, \widetilde{\mathbb{C}}) & \xrightarrow{\Gamma(U,\theta)} & \Gamma\big(U, \Phi_*\widetilde{\mathbb{C}}|_U\big) & = & \mathbb{C} \\
{\scriptstyle 1}\downarrow & & {\scriptstyle \mathrm{res}^U_{U_f}}\downarrow & & \downarrow{\scriptstyle \mathrm{res}^U_{U_f}} & & \downarrow{\scriptstyle 1} \\
\mathbb{C} & = & \Gamma(U_f, \widetilde{\mathbb{C}}) & \xrightarrow[\Gamma(U_f,\theta)]{} & \Gamma\big(U_f, \Phi_*\widetilde{\mathbb{C}}|_U\big) & = & \mathbb{C} \;.
\end{array}$$

The map $\mathbb{C} = \Gamma(U, \widetilde{\mathbb{C}}) \longrightarrow \Gamma\big(U, \Phi_*\widetilde{\mathbb{C}}|_U\big)$ is equal to $1 : \mathbb{C} \longrightarrow \mathbb{C}$ by construction, forcing the map $\Gamma(U_f, \theta)$ to be an isomorphism. □

Remark 10.8.5. We now know that $\Phi_*\widetilde{\mathbb{C}}$ is a coherent algebraic sheaf on $(Y, \mathcal{O}_Y)$. The commutative triangle of Remark 10.8.1 means that we have an isomorphism $\Phi_*\widetilde{\mathbb{C}} = \lambda_*\Phi_*^{\mathrm{an}}\widetilde{\mathbb{C}}$ of sheaves of $\mathcal{O}_Y$–modules on Y. Now $\Phi_*^{\mathrm{an}}\widetilde{\mathbb{C}}$ is a sheaf of $\mathcal{O}_Y^{\mathrm{an}}$–modules on Y^{an}. Fact 7.8.9(v) tells us that there is a morphism of sheaves of $\mathcal{O}_Y^{\mathrm{an}}$–modules $\Lambda^* : \{\Phi_*\widetilde{\mathbb{C}}\}^{\mathrm{an}} \longrightarrow \Phi_*^{\mathrm{an}}\widetilde{\mathbb{C}}$ on Y^{an}, so that the isomorphism above factors as

$$\Phi_*\widetilde{\mathbb{C}} \longrightarrow \lambda_*\{\Phi_*\widetilde{\mathbb{C}}\}^{\mathrm{an}} \xrightarrow{\lambda_*\Lambda^*} \lambda_*\Phi_*^{\mathrm{an}}\widetilde{\mathbb{C}} \,.$$

We assert:

Lemma 10.8.6. *The map $\Lambda^* : \{\Phi_*\widetilde{\mathbb{C}}\}^{\mathrm{an}} \longrightarrow \Phi_*^{\mathrm{an}}\widetilde{\mathbb{C}}$ is an isomorphism.*

Proof. On the open set $Y - \{p\}$ the map is clearly an isomorphism, since both sheaves vanish. It therefore suffices to prove it an isomorphism on some polydisc V containing p. Choose therefore an open affine $U \subset Y$, and a polydisc $V \subset U^{\mathrm{an}}$ containing p. For every polydisc $W \subset V$ we must show that the map

$$\Gamma\big(W, \{\Phi_*\widetilde{\mathbb{C}}\}^{\mathrm{an}}\big) \xrightarrow{\ \Lambda_W^*\ } \Gamma\big(W, \lambda_*\Phi_*^{\mathrm{an}}\widetilde{\mathbb{C}}\big)$$

is an isomorphism. If $p \notin W$ this is clear, as

$$\Gamma\big(W, \{\Phi_*\widetilde{\mathbb{C}}\}^{\mathrm{an}}\big) \quad = \quad 0 \quad = \quad \Gamma\big(W, \lambda_*\Phi_*^{\mathrm{an}}\widetilde{\mathbb{C}}\big) \ .$$

If $p \in W$ we consider the composite

$$\Gamma(U, \Phi_*\widetilde{\mathbb{C}}) \xrightarrow{\ \alpha\ } \Gamma\big(W, \{\Phi_*\widetilde{\mathbb{C}}\}^{\mathrm{an}}\big) \xrightarrow{\ \Lambda_W^*\ } \Gamma\big(W, \lambda_*\Phi_*^{\mathrm{an}}\widetilde{\mathbb{C}}\big) \ .$$

This composite $\Lambda_W^*\alpha$ is the identity map $1 : \mathbb{C} \longrightarrow \mathbb{C}$. We want to show that Λ_W^* is an isomorphism; it clearly suffices to prove that α is. The injectivity of α follows from the fact that the composite $\Lambda_W^*\alpha$ is injective; we are reduced to proving α surjective.

Now Fact 7.8.9(iii)(a) tells us that the image of α is dense in the Fréchet topology on $\Gamma\big(W, \{\Phi_*\widetilde{\mathbb{C}}\}^{\mathrm{an}}\big)$. The image of α is $\mathbb{C}$, and the density makes $\Gamma\big(W, \{\Phi_*\widetilde{\mathbb{C}}\}^{\mathrm{an}}\big)$ the completion of the pre-Fréchet subspace $\mathbb{C}$. Definition 5.4.1(ii) guaranteed that there can only be one non-trivial pre-Fréchet topology on $\mathbb{C}$, and in this topology $\mathbb{C}$ is complete. □

Let $(Y, \mathcal{O}_Y)$ be a scheme locally of finite type over $\mathbb{C}$ and let $p \in Y$ be a closed point. We have constructed a sheaf $\Phi_*\widetilde{\mathbb{C}}$ for which we completely understand the analytification; it is canonically isomorphic to $\Phi_*^{\mathrm{an}}\widetilde{\mathbb{C}}$. The next little lemmas help us recognize when we are confronted with a sheaf $\Phi_*\widetilde{\mathbb{C}}$.

Lemma 10.8.7. *Let* $(U, \mathcal{O}) = \big(\mathrm{Spec}(S), \widetilde{S}\big)$ *be an affine scheme of finite type over* $\mathbb{C}$. *If* $\mathscr{S}$ *is a coherent sheaf, and* $\Gamma(U, \mathscr{S}) = \mathbb{C}$, *then* $\mathscr{S}$ *is isomorphic to a sheaf* $\Phi_*\widetilde{\mathbb{C}}$ *as above, for some* $(\Phi, \Phi^*) : \big(\mathrm{Spec}(\mathbb{C}), \widetilde{\mathbb{C}}\big) \longrightarrow (U, \mathcal{O})$.

Proof. Corollary 9.9.1 tells us that $\mathscr{S}$ must be $\widetilde{M}$ for some S–module M, Remark 7.4.2 says $M = \Gamma(U, \widetilde{M}) = \Gamma(U, \mathscr{S})$, and in our case this is $\Gamma(U, \mathscr{S}) = \mathbb{C}$. We only need to understand the structure of $\mathbb{C}$ as an S–module. The fact that $\mathbb{C}$ is a module, over the $\mathbb{C}$–algebra S, gives us a $\mathbb{C}$–algebra homomorphism $\varphi : S \longrightarrow \mathrm{End}(\mathbb{C}) = \mathbb{C}$. By Proposition 4.2.4 this φ corresponds to a unique closed point $p \in U = \mathrm{Spec}(S)$, and the proof of Lemma 10.8.4 identifies the sheaf $\mathscr{S} = \widetilde{\mathbb{C}}$ with the pushforward, of the sheaf $\widetilde{\mathbb{C}}$ on the 1–point space $\operatorname{Spec}(\mathbb{C})$, by the map

$$(\operatorname{Spec}(\varphi), \widetilde{\varphi}) : \left(\operatorname{Spec}(\mathbb{C}), \widetilde{\mathbb{C}}\right) \longrightarrow \left(\operatorname{Spec}(S), \widetilde{S}\right) .$$

□

Lemma 10.8.8. *Let $(Y, \mathcal{O})$ be a scheme locally of finite type over $\mathbb{C}$. Suppose $\mathscr{S}$ is a coherent sheaf on Y. Suppose for some closed point $p \in Y$ we have*

(i) *The restriction of $\mathscr{S}$ to the open set $Y - \{p\}$ vanishes.*
(ii) *There is an open affine subset $U \subset Y$ with $\Gamma(U, \mathscr{S}) = \mathbb{C}$.*

Then $\mathscr{S}$ is isomorphic to the sheaf $\Phi_* \widetilde{\mathbb{C}}$, where the morphism (Φ, Φ^*) : $\left(\operatorname{Spec}(\mathbb{C}), \widetilde{\mathbb{C}}\right) \longrightarrow (Y, \mathcal{O})$ is the unique map, of ringed spaces over $\mathbb{C}$, such that the image of Φ is $p \in Y$.

Proof. By Lemma 10.8.7 we know that the restriction $\mathscr{S}|_U$, of the coherent sheaf $\mathscr{S}$ to the open affine $U \subset Y$, must be isomorphic to the pushforward, of the sheaf $\widetilde{\mathbb{C}}$ on the 1–point space, by a map (Ψ, Ψ^*) : $\left(\operatorname{Spec}(\mathbb{C}), \widetilde{\mathbb{C}}\right) \longrightarrow (U, \mathcal{O}|_U)$. Since $\Gamma(U - \{p\}, \mathscr{S}) = 0$ we have that

$$\Gamma\left(\Psi^{-1}\{U - \{p\}\} \ , \ \widetilde{\mathbb{C}}\right) \quad = \quad \Gamma(U - \{p\} \ , \ \Psi_* \widetilde{\mathbb{C}}) \quad = \quad 0 \ ,$$

which means that $\Psi^{-1}\{U - \{p\}\}$ must be empty; that is the image of the map Ψ must be the point $p \in U$. The composite

$$\left(\operatorname{Spec}(\mathbb{C}), \widetilde{\mathbb{C}}\right) \xrightarrow{(\Psi, \Psi^*)} (U, \mathcal{O}|_U) \xrightarrow{(i, i^*)} (Y, \mathcal{O})$$

gives a map (Φ, Φ^*) : $\left(\operatorname{Spec}(\mathbb{C}), \widetilde{\mathbb{C}}\right) \longrightarrow (Y, \mathcal{O})$. The sheaves $\mathscr{S}$ and $\Phi_* \widetilde{\mathbb{C}}$ agree on the open set U, and agree on $Y - \{p\}$ because both vanish there. This makes them isomorphic, with isomorphisms compatible with restriction maps, on all open sets contained in either U or $Y - \{p\}$. This collection of open sets forms a basis for the topology of Y, and two sheaves, agreeing on a basis of the open sets, are isomorphic. □

We want to have a symbol for finite direct sums of sheaves of $\Phi_* \widetilde{\mathbb{C}}$. Let us therefore introduce

Notation 10.8.9. Let $(Y, \mathcal{O}_Y)$ be a scheme locally of finite type over $\mathbb{C}$, let $p \in Y$ be a closed point, and let W be a finite dimensional vector space over $\mathbb{C}$. Then we will denote by $\mathscr{W}_p$ the sheaf

$$\mathscr{W}_p \quad = \quad W \otimes_{\mathbb{C}} \Phi_* \widetilde{\mathbb{C}} \ .$$

The sheaves $\mathscr{W}_p$ are called the *skyscraper sheaves* concentrated at the closed point $p \in Y$. The analytification $\mathscr{W}_p^{\mathrm{an}}$ will be called a skyscraper sheaf on Y^{an}; we will let the reader figure out, from the context, whether the skyscraper sheaf is algebraic or analytic.

Remark 10.8.10. What we like about the skyscraper sheaves $\mathscr{W}_p$ is that they are all coherent algebraic sheaves, and we completely understand their analytification. If $\mathscr{W}_p = W \otimes_{\mathbb{C}} \Phi_* \widetilde{\mathbb{C}}$ then $\mathscr{W}_p^{\mathrm{an}} = W \otimes_{\mathbb{C}} \Phi_*^{\mathrm{an}} \widetilde{\mathbb{C}}$. When W is 1–dimensional the coherence of the algebraic sheaf is by Lemma 10.8.4, and the statement about the analytification comes from Lemma 10.8.6. Both assertions are stable under taking finite direct sums.

Most readers will undoubtedly agree that the skyscraper sheaves are a very trivial, stupid example. It nevertheless helps to understand trivialities carefully. Our proof, of the general case of GAGA, will be by reduction to trivial cases, more explicitly by reduction to the case of skyscraper sheaves.

10.9 Skyscraper sheaves on $\mathbb{P}^n$

In Section 10.8 we dealt with skyscraper sheaves, on a general scheme $(Y, \mathcal{O})$ locally of finite type over $\mathbb{C}$. Next we want to specialize the knowledge we have gained to the case we care about, namely where $(Y, \mathcal{O}) = (\mathbb{P}^n, \mathcal{O}_{\mathbb{P}^n})$. We return therefore to Notation 10.1.1: We fix the Hopf algebra $R = \mathbb{C}[t, t^{-1}]$, the group will be $(G, \mathcal{O}_G) = \big(\mathrm{Spec}(R), \widetilde{R}\big)$, the G–ring S will be the polynomial ring $S = \mathbb{C}[x_0, x_1, \ldots, x_n]$ in $(n+1)$ variables, the action of $(G, \mathcal{O}_G)$ on S is by the map $a' : S \longrightarrow R \otimes_{\mathbb{C}} S$ with $a'(x_i) = t^{-1} \otimes x_i$, the scheme $(X, \mathcal{O}_X) = \big(\mathrm{Spec}(S), \widetilde{S}\big)$ admits an induced action of $(G, \mathcal{O}_G)$, the quotient map

$$(\pi, \pi^*) : (X, \mathcal{O}_X) \longrightarrow (X/G, \mathcal{O}_{X/G})$$

is as in Definitions 8.6.2 and 8.7.6, the G–invariant open subset we wish to consider is $U = X - \{0\}$, and $\mathbb{P}^n \subset X/G$ is the open subset, whose inverse image $\pi^{-1}\mathbb{P}^n$ is the G–invariant open set $U \subset X$.

The first problem is to understand which G–modules for the G–ring $S = \mathbb{C}[x_0, x_1, \ldots, x_n]$ give rise to skyscraper sheaves. We want to give a basis-free way of saying this. Let V be a vector space of dimension $(n+1)$. Recall that S may be viewed as

$$S \quad = \quad \mathbb{C}[V] \quad = \quad \sum_{i=0}^{\infty} \mathrm{Sym}^i V \; ;$$

Observation 9.2.1 gives a discussion. If we choose a basis $\{x_0, x_1, \ldots, x_n\}$ for V then the basis gives an isomorphism of $\mathbb{C}[V]$ with $\mathbb{C}[x_0, x_1, \ldots, x_n]$; in what follows we want to have the liberty to choose any basis we wish.

Lemma 10.9.1. *Let $\{y_0, y_1, \ldots, y_n\}$ be a basis for V. Let M be the G–module*

$$M \quad = \quad S/\langle y_1, y_2, \ldots, y_n\rangle \quad = \quad \mathbb{C}[y_0]\ .$$

Then the corresponding sheaf $\mathscr{S} = \{\pi_*\widetilde{M}\}^{G^{\text{an}}}$ on $\mathbb{P}^n$ is a skyscraper sheaf $\Phi_*\widetilde{\mathbb{C}}$, supported at the closed point, which is the image of $(1, 0, \ldots, 0) \in \mathbb{C}^{n+1} - \{0\}$ under the map $\pi : X \longrightarrow X/G$.

Proof. We first note that, if $j > 0$, then the sheaf $\mathscr{S}$ vanishes on the open set $\mathbb{P}^n_{y_j} \subset \mathbb{P}^n$. The reason is that $M[1/y_j]$ vanishes as an $S[1/y_j]$–module, and hence so does $\{\pi_*\widetilde{M[1/y_j]}\}^{G^{\text{an}}} = \mathscr{S}|_{\mathbb{P}^n_{y_j}}$. We want to show that this already implies there is a point $p \in \mathbb{P}^n$ so that $\mathscr{S}$ vanishes on $\mathbb{P}^n - \{p\}$. More specifically $p = \pi(a)$, where a is the point $a = (1, 0, \ldots, 0) \in \mathbb{C}^{n+1} = X^{\text{an}}$.

Let therefore $V' = \cup_{j>0}\mathbb{P}^n_{y_j}$. We know that $\mathscr{S}$ vanishes on the open set $V' \subset \mathbb{P}^n$, and we will prove that $V' = \mathbb{P}^n - \{p\}$. Its inverse image in $X = \text{Spec}(S)$ is the union $U' = \cup_{j>0}X_{y_j}$. The closed set $Z = U - U'$ is the intersection of all the closed sets $U - X_{y_j}$. This makes $Z \cap U^{\text{an}} = Z \cap (\mathbb{C}^{n+1} - \{0\})$ the set of all $b \in \mathbb{C}^{n+1} - \{0\}$ not contained in $\mathbb{C}^{n+1}_{y_j}$ for any $j > 0$. Lemma 4.4.3 says that $b \notin \mathbb{C}^{n+1}_{y_j}$ means $y_j(b) = 0$. The set $Z \cap (\mathbb{C}^{n+1} - \{0\})$ is precisely the subset of $U^{\text{an}} = \mathbb{C}^{n+1} - \{0\}$ on which $\{y_1, y_2, \ldots, y_n\}$ all vanish. Put very concretely this is the set of points

$$Z \cap (\mathbb{C}^{n+1} - \{0\}) \quad = \quad \{(b, 0, \ldots, 0) \in \mathbb{C}^{n+1} \mid b \neq 0\}\ .$$

It is clearly a single orbit for the group $G^{\text{an}} = \mathbb{C}^*$. The point $a = (1, 0, \ldots, 0)$ lies in the orbit, and Corollary 8.9.24 says that $Z \cap U^{\text{an}}$ is equal to $\pi^{-1}\pi(a) \cap U^{\text{an}}$. Now Z and $\pi^{-1}\pi(a)$ are two Zariski closed subsets of U, and $Z \cap U^{\text{an}} = \pi^{-1}\pi(a) \cap U^{\text{an}}$. By 9.5.5.1 we know that $Z = \pi^{-1}\pi(a)$. This means that

$$U - \pi^{-1}\pi(a) \quad = \quad U - Z \quad = \quad U' \quad = \quad \pi^{-1}V'\ .$$

Since $\pi : U \longrightarrow \mathbb{P}^n$ is surjective we conclude $V' = \mathbb{P}^n - \pi(a)$.

Now we consider the sheaf $\mathscr{S}$ on the open affine set $\mathbb{P}^n_{y_0} \subset \mathbb{P}^n$. We know that $\Gamma(\mathbb{P}^n_{y_0}, \mathscr{S})$ is computed by taking the G^{an}–invariant elements of $\Gamma(X_{y_0}, \widetilde{M}) = M[1/y_0]$. But $M = \mathbb{C}[y_0]$, hence $M[1/y_0] = \mathbb{C}[y_0, y_0^{-1}]$, and the action is given by $a'(y_0) = t^{-1} \otimes y_0$. More concretely, the way $g \in \mathbb{C}^*$ acts on y_0^m is to take it to $g^m y_0^m$. The only G^{an}–invariant elements for this action are the constants; that is $\Gamma(\mathbb{P}^n_{y_0}, \mathscr{S}) = \mathbb{C}$.

We can now apply Lemma 10.8.8. The sheaf $\mathscr{S}$ is coherent on $\mathbb{P}^n$, its restriction to $\mathbb{P}^n - \{p\}$ vanishes, and on the open affine $\mathbb{P}^n_{y_0} \subset \mathbb{P}^n$ we have $\Gamma(\mathbb{P}^n_{y_0}, \mathscr{S}) = \mathbb{C}$. The sheaf $\mathscr{S}$ must be a skyscraper sheaf $\Phi_*\widetilde{\mathbb{C}}$. □

Construction 10.9.2. Next we want to consider an exact sequence of G–modules for the G–ring $S = \mathbb{C}[V]$. Let $\{y_0, y_1, \ldots, y_n\}$ be a basis for V. The G–module M of Lemma 10.9.1 fits in an exact sequence

$$\bigoplus_{j=1}^{n} \{S \otimes \mathbb{C}e_{-1}\} \xrightarrow{(y_1, y_2, \ldots, y_n)} S \longrightarrow M \longrightarrow 0 \; ,$$

where the map $y_j : S \otimes \mathbb{C}e_{-1} \longrightarrow S$ takes $1 \otimes e_{-1}$ to $y_j \in S$. If we forget the G–action this is just the trivial statement, that the kernel of the map

$$S \longrightarrow M \quad = \quad S/\langle y_1, y_2, \ldots, y_n \rangle$$

is generated by the image of the map $S^n \longrightarrow S$ taking $(s_1, s_2, \ldots, s_n) \in S^n$ to $\sum_{j=1}^n s_j y_j$. The argument above amounted to keeping track of the G–action. In any case we may descend to maps of coherent sheaves on $\mathbb{P}^n$. We obtain an exact sequence

$$\bigoplus_{j=1}^{n} \mathscr{O}(-1) \xrightarrow{(y_1, y_2, \ldots, y_n)} \mathscr{O} \longrightarrow \{\pi_* \widetilde{M}\}^{G^{\mathrm{an}}} \longrightarrow 0 \; ,$$

and Lemma 10.9.1 tells us that $\{\pi_* \widetilde{M}\}^{G^{\mathrm{an}}} \cong \Phi_* \widetilde{\mathbb{C}}$ is a skyscraper sheaf. We want to record the analytification of this exact sequence. Before we state the lemma let us give the notation.

Notation 10.9.3. Let $V \subset S$ be the vector space of all homogeneous elements of degree 1 in the polynomial ring $S = \mathbb{C}[V] = \mathbb{C}[x_0, x_1, \ldots, x_n]$. If $p \in \mathbb{P}^n$ is a closed point then p is the image of a $\mathbb{C}^*$–orbit in $\mathbb{C}^{n+1} - \{0\}$. That is $\pi^{-1}(p) \subset \mathbb{C}^{n+1} - \{0\}$ is a line. Let $V(p) \subset V$ be the vector space of all linear polynomials vanishing on the line $\pi^{-1}(p)$.

Remark 10.9.4. Choose a basis $\{y_0, y_1, \ldots, y_n\} \subset V$, in such a way that $\{y_1, y_2, \ldots, y_n\}$ is a basis for the subspace $V(p) \subset V$. In the coordinates $\{y_0, y_1, \ldots, y_n\}$ the set $\pi^{-1}(p)$ becomes

$$\pi^{-1}(p) \quad = \quad \{(b, 0, \ldots, 0) \in \mathbb{C}^{n+1} \mid b \neq 0\} \; .$$

There is a bijective correspondence between closed points $p \in \mathbb{P}^n$ and subspaces $V(p) \subset V$ of codimension 1; this merely says that every line

$\pi^{-1}(p) \subset \mathbb{C}^{n+1} - \{0\}$ corresponds, bijectively, to the n–dimensional space of linear functions vanishing on it.

Lemma 10.9.5. *Let $p \in \mathbb{P}^n$ be any closed point, and let $V(p) \subset V$ be the space of linear functions vanishing on $\pi^{-1}(p)$, as in Notation 10.9.3. Let $\{y_1, y_2, \ldots, y_n\}$ be any basis for $V(p)$. With the maps $y_j : \mathcal{O}(-1) \longrightarrow \mathcal{O}$ as in Definition 10.3.7, we have an exact sequence*

$$\bigoplus_{j=1}^{n} \mathcal{O}(-1)^{\text{an}} \xrightarrow{(y_1^{\text{an}}, y_2^{\text{an}}, \ldots, y_n^{\text{an}})} \mathcal{O}^{\text{an}} \longrightarrow \Phi_*^{\text{an}}\widetilde{\mathbb{C}} \longrightarrow 0\,.$$

Proof. In Construction 10.9.2 we produced the exact sequence of coherent algebraic sheaves

$$\bigoplus_{j=1}^{n} \mathcal{O}(-1) \xrightarrow{(y_1, y_2, \ldots, y_n)} \mathcal{O} \longrightarrow \Phi_*\widetilde{\mathbb{C}} \longrightarrow 0\,.$$

Analytification is exact by Fact 7.8.9(iv), and hence the sequence

$$\bigoplus_{j=1}^{n} \mathcal{O}(-1)^{\text{an}} \xrightarrow{(y_1^{\text{an}}, y_2^{\text{an}}, \ldots, y_n^{\text{an}})} \mathcal{O}^{\text{an}} \longrightarrow \left\{\Phi_*\widetilde{\mathbb{C}}\right\}^{\text{an}} \longrightarrow 0$$

must be exact; but Lemma 10.8.6 permits us to identify $\left\{\Phi_*\widetilde{\mathbb{C}}\right\}^{\text{an}}$ with $\Phi_*^{\text{an}}\widetilde{\mathbb{C}}$. □

Remark 10.9.6. Choose an open set $\mathbb{P}^n_{x_i} \subset \mathbb{P}^n$; we know that the sheaves $\mathcal{O}(m)$ all trivialize on $\mathbb{P}^n_{x_i}$. Take the sequence of Lemma 10.9.5, and restrict it to the open set $\{\mathbb{P}^n_{x_i}\}^{\text{an}} \subset \{\mathbb{P}^n\}^{\text{an}}$. The sequence becomes

$$\bigoplus_{j=1}^{n} \mathcal{O}^{\text{an}} \xrightarrow{(y_1^{\text{an}}, y_2^{\text{an}}, \ldots, y_n^{\text{an}})} \mathcal{O}^{\text{an}} \longrightarrow \Phi_*^{\text{an}}\widetilde{\mathbb{C}} \longrightarrow 0\,,$$

where we have identified $\mathcal{O}^{\text{an}}|_{\mathbb{P}^n_{x_i}} \cong \mathcal{O}(-1)^{\text{an}}|_{\mathbb{P}^n_{x_i}}$. By Exercise 7.1.16(iii) we know that each map $y_j^{\text{an}} : \mathcal{O}^{\text{an}} \longrightarrow \mathcal{O}^{\text{an}}$ is multiplication by some element $a_j \in \Gamma\big(\{\mathbb{P}^n_{x_i}\}^{\text{an}}, \mathcal{O}^{\text{an}}\big)$, that is by a holomorphic function on $\{\mathbb{P}^n_{x_i}\}^{\text{an}}$. The exact sequence, coupled with the fact that $\Phi_*^{\text{an}}\widetilde{\mathbb{C}}$ does not vanish at p, assures us that none of the holomorphic functions a_j, $1 \leq j \leq n$ can be invertible near p; they all must vanish at p.

Suppose we fix j, with $1 \leq j \leq n$. Lemma 10.3.12 informs us that the maps $y_j^{\text{an}} : \mathcal{O}(m-1)^{\text{an}} \longrightarrow \mathcal{O}(m)^{\text{an}}$ all agree, up to units, on the open set $\{\mathbb{P}^n_{x_i}\}^{\text{an}}$; they are independent of m. Hence they all come down to multiplication by holomorphic functions vanishing at p.

Reminder 10.9.7. Before the next lemma we should remind the reader of Section 10.4. For any coherent analytic sheaf $\mathscr{S}'$ we gave, in Definition 10.4.9, a vector space $\mathrm{ann}(\mathscr{S}') \subset V$. The definition was that $y \in V$ belongs to $\mathrm{ann}(\mathscr{S}')$ if the map

$$\mathscr{H}om(y^{\mathrm{an}},1) \;:\; \mathscr{H}om(\mathcal{O}^{\mathrm{an}}\,,\,\mathscr{S}') \longrightarrow \mathscr{H}om(\mathcal{O}(-1)^{\mathrm{an}}\,,\,\mathscr{S}')$$

vanishes.

Lemma 10.9.8. *Let $\mathscr{S}'$ be a coherent analytic sheaf on projective space $\{\mathbb{P}^n\}^{\mathrm{an}}$. If the annihilator of $\mathscr{S}'$ is of dimension n then $\mathscr{S}'$ is a skyscraper sheaf. More precisely, suppose the n–dimensional subspace $\mathrm{ann}(\mathscr{S}') \subset V$ is $V(p)$, for p a closed point in $\mathbb{P}^n$. Then the sheaf $\mathscr{S}'$ is a skyscraper sheaf supported at $p \in \mathbb{P}^n$.*

Proof. Let us choose a basis $\{y_0, y_1, \ldots, y_n\} \subset V$ so that $\{y_1, y_2, \ldots, y_n\}$ is a basis for $V(p)$. We know that $y_j \in \mathrm{ann}(\mathscr{S}')$ for $1 \le j \le n$, and Lemma 10.4.13 tells us that $\mathscr{S}'$ must vanish on the open sets $\{\mathbb{P}^n_{y_j}\}^{\mathrm{an}}$. Hence $\mathscr{S}'$ vanishes on the union $\cup_{j=1}^n \{\mathbb{P}^n_{y_j}\}^{\mathrm{an}}$, and in the proof of Lemma 10.9.1 we saw that the union of these open sets is $\{\mathbb{P}^n\}^{\mathrm{an}} - \{p\}$. Thus $\mathscr{S}'$ vanishes on $\{\mathbb{P}^n\}^{\mathrm{an}} - \{p\}$.

Now let Δ be a sufficiently small polydisc containing p. The coherence of $\mathscr{S}'$ means that there will be a surjection $\{\mathcal{O}^{\mathrm{an}}\}^m|_\Delta \longrightarrow \mathscr{S}'|_\Delta$ of coherent analytic sheaves on Δ. Let W be an m–dimensional vector space over $\mathbb{C}$, and write $\{\mathcal{O}^{\mathrm{an}}\}^m = W \otimes_{\mathbb{C}} \mathcal{O}^{\mathrm{an}}$. We have a surjection $\{W \otimes_{\mathbb{C}} \mathcal{O}^{\mathrm{an}}\}|_\Delta \longrightarrow \mathscr{S}'|_\Delta$. But now observe that $y_j \in \mathrm{ann}(\mathscr{S}')$ if $1 \le j \le n$, and hence the map

$$\mathscr{H}om(y_j^{\mathrm{an}},1) \;:\; \mathscr{H}om(\mathcal{O}^{\mathrm{an}}\,,\,\mathscr{S}') \longrightarrow \mathscr{H}om(\mathcal{O}(-1)^{\mathrm{an}}\,,\,\mathscr{S}')$$

vanishes. On the open set Δ this means that any composite

$$\mathcal{O}(-1)^{\mathrm{an}}|_\Delta \xrightarrow{\;y_j^{\mathrm{an}}\;} \mathcal{O}^{\mathrm{an}}|_\Delta \longrightarrow \mathscr{S}'|_\Delta$$

must vanish. Any map $\mathcal{O}^{\mathrm{an}}|_\Delta \longrightarrow \mathscr{S}'|_\Delta$ factors through the cokernel of the map $(y_1^{\mathrm{an}}, y_2^{\mathrm{an}}, \ldots, y_n^{\mathrm{an}})$, in the exact sequence

$$\bigoplus_{j=1}^{n} \mathcal{O}(-1)^{\mathrm{an}} \xrightarrow{(y_1^{\mathrm{an}}, y_2^{\mathrm{an}}, \ldots, y_n^{\mathrm{an}})} \mathcal{O}^{\mathrm{an}} \longrightarrow \Phi_*^{\mathrm{an}}\widetilde{\mathbb{C}} \longrightarrow 0$$

of Lemma 10.9.5. That is our surjective map $\{W \otimes_{\mathbb{C}} \mathcal{O}^{\mathrm{an}}\}|_\Delta \longrightarrow \mathscr{S}'|_\Delta$ must factor through a surjection $\{W \otimes_{\mathbb{C}} \Phi_*^{\mathrm{an}}\widetilde{\mathbb{C}}\}|_\Delta \longrightarrow \mathscr{S}'|_\Delta$. Taking sections of this map on Δ we have a surjection

$$W \;=\; \Gamma(\Delta\,,\, W \otimes_{\mathbb{C}} \Phi_*^{\mathrm{an}}\widetilde{\mathbb{C}}) \longrightarrow \Gamma(\Delta, \mathscr{S}')\,.$$

Replacing W by a subspace $W' \subset W$ if necessary, we get a map of sheaves on Δ

$$\{W' \otimes_{\mathbb{C}} \Phi_*^{\mathrm{an}}\widetilde{\mathbb{C}}\}|_\Delta \longrightarrow \{W \otimes_{\mathbb{C}} \Phi_*^{\mathrm{an}}\widetilde{\mathbb{C}}\}|_\Delta \longrightarrow \mathscr{S}'|_\Delta .$$

which induces an isomorphism when we apply $\Gamma(\Delta, -)$. Since Δ is a polydisc, and both sheaves are coherent, the map must be an isomorphism of sheaves on Δ.

The sheaves $\mathscr{W}_p^{\mathrm{an}} = W' \otimes_{\mathbb{C}} \Phi_*^{\mathrm{an}}\widetilde{\mathbb{C}}$ and $\mathscr{S}'$ are isomorphic on Δ and vanish on $\{\mathbb{P}^n\}^{\mathrm{an}} - \{p\}$, hence agree on all open sets contained either in Δ or in $\{\mathbb{P}^n\}^{\mathrm{an}} - \{p\}$. These open sets form a basis for the topology of $\{\mathbb{P}^n\}^{\mathrm{an}}$, and the sheaves must be isomorphic. □

Remark 10.9.9. If the dimension of $\mathrm{ann}(\mathscr{S}')$ is $(n+1)$ then $\mathrm{ann}(\mathscr{S}') = V$. In that case Lemma 10.4.13 says that the restriction of $\mathscr{S}'$ to $\{\mathbb{P}^n_y\}^{\mathrm{an}}$ vanishes, for every $y \in V$. Lemma 8.9.8 tells us that $\{\mathbb{P}^n\}^{\mathrm{an}}$ is covered by its open subsets $\{\mathbb{P}^n_y\}^{\mathrm{an}}$, and hence $\mathscr{S}'$ must vanish. The trivial sheaf is a skyscraper sheaf, so we could say that $\mathscr{S}'$ is a skyscraper sheaf whenever the dimension of $\mathrm{ann}(\mathscr{S}')$ is $\geq n$.

10.10 The second half of GAGA

This section is devoted to the proof of Theorem 7.9.1(ii). We remind the reader: we must prove that every coherent analytic sheaf $\mathscr{S}'$ on $\{\mathbb{P}^n\}^{\mathrm{an}}$ is isomorphic to $\mathscr{S}^{\mathrm{an}}$, for some coherent algebraic sheaf $\mathscr{S}$. We already know this for skyscraper sheaves; the sheaf $\mathscr{W}_p^{\mathrm{an}} = W \otimes_{\mathbb{C}} \Phi_*^{\mathrm{an}}\widetilde{\mathbb{C}}$ is the analytification of $\mathscr{W}_p = W \otimes_{\mathbb{C}} \Phi_*\widetilde{\mathbb{C}}$. The idea of the proof is to reduce to the case of the skyscraper sheaves. Let us make a definition to explain this reduction.

Definition 10.10.1. *Let $\mathscr{S}'$ be a coherent analytic sheaf on $\{\mathbb{P}^n_y\}^{\mathrm{an}}$. The number* $\dim(\mathscr{S}')$ *is the dimension of the vector space* $\mathrm{ann}(\mathscr{S}') \subset V$.

Remark 10.10.2. From Lemma 10.9.8 and Remark 10.9.9 we learnt that, if $\dim(\mathscr{S}') \geq n$, then $\mathscr{S}'$ must be a skyscraper sheaf, and is the analytification of a coherent algebraic sheaf.

Discussion 10.10.3. The idea of the proof of Theorem 7.9.1(ii) will be to use descending induction on $\dim(\mathscr{S}')$. For each i we will consider the statement:

(i) Every coherent analytic sheaf $\mathscr{S}'$, of dimension $\dim(\mathscr{S}') \geq i$, is isomorphic to $\mathscr{S}^{\mathrm{an}}$ for some coherent algebraic sheaf $\mathscr{S}$.

We know (i) for $i = n$, by Remark 10.10.2. Theorem 7.9.1(ii) is the assertion that (i) holds for $i = 0$. The induction step will be to prove that if (i) holds for $(i+1)$ then it also holds for i. The entire section is devoted to the proof of this induction step.

Construction 10.10.4. For any coherent analytic sheaf $\mathscr{S}'$ and any $y \in V$ we gave, in Construction 10.4.15, an exact sequence

$$0 \longrightarrow \mathscr{K}' \longrightarrow \mathscr{S}'(-1) \xrightarrow{\ y\ } \mathscr{S}' \longrightarrow \mathscr{Q}' \longrightarrow 0 .$$

Let us write $\mathscr{S}'/y$ for $\mathscr{Q}'$. We will consider the sheaves $\mathscr{S}'/y$ a great deal.

The proof now proceeds by a sequence of lemmas.

Lemma 10.10.5. *Suppose that every coherent analytic sheaf of dimension $> i$ is the analytification of a coherent algebraic sheaf. Let $\mathscr{S}'$ be a coherent analytic sheaf of dimension i, and let $y \in V$ be any element. Then there exists an integer $N = N(\mathscr{S}', y)$, depending on $\mathscr{S}'$ and y, so that*

$$H^0\big(\mathscr{S}'(m)\big) \longrightarrow H^0\big(\{\mathscr{S}'/y\}(m)\big)$$

is surjective for every $m \geq N$.

Proof. First of all note that if $y \in \text{ann}(\mathscr{S}')$ then there is nothing to prove. We have an exact sequence

$$0 \longrightarrow \mathscr{K}' \longrightarrow \mathscr{S}'(-1) \xrightarrow{\ y\ } \mathscr{S}' \longrightarrow \mathscr{Q}' \longrightarrow 0 ,$$

and Lemma 10.4.11 tells us that the map y vanishes. The exactness now forces the map $\mathscr{S}' \longrightarrow \mathscr{Q}' = \mathscr{S}'/y$ to be an isomorphism, so of course $H^0\big(\mathscr{S}'(m)\big) \longrightarrow H^0\big(\{\mathscr{S}'/y\}(m)\big)$ will be surjective.

We may therefore assume that $y \notin \text{ann}(\mathscr{S}')$. Lemma 10.4.17 tells us that $\text{ann}(\mathscr{K}')$, $\text{ann}(\mathscr{Q}')$ contain both y and $\text{ann}(\mathscr{S}')$. It follows that $\dim(\mathscr{K}')$ and $\dim(\mathscr{Q}')$ must be at least $(i+1)$, and the hypothesis of the lemma says that $\mathscr{K}'$ and $\mathscr{Q}'$ must be isomorphic, respectively, to $\mathscr{K}^{\text{an}}$ and $\mathscr{Q}^{\text{an}}$, for some coherent algebraic sheaves $\mathscr{K}$ and $\mathscr{Q}$. Theorem 10.7.1(ii) tells us that there exists an integer N_1 so that, if $m \geq N_1$ and $j \geq 1$, then the vector spaces $H^j\big(\mathscr{K}(m)\big)$ and $H^j\big(\mathscr{Q}(m)\big)$ all vanish. Theorem 10.7.1(i) says that

$$H^j\big(\mathscr{K}(m)\big) = H^j\big(\{\mathscr{K}(m)\}^{\text{an}}\big) \quad \text{and} \quad H^j\big(\mathscr{Q}(m)\big) = H^j\big(\{\mathscr{Q}(m)\}^{\text{an}}\big) ,$$

while Lemma 10.4.5 asserts

$$\{\mathscr{K}(m)\}^{\text{an}} = \mathscr{K}^{\text{an}}(m) = \mathscr{K}'(m) \quad \text{and} \quad \{\mathscr{Q}(m)\}^{\text{an}} = \mathscr{Q}^{\text{an}}(m) = \mathscr{Q}'(m).$$

In summary we have that, as long as $m \geq N_1$ and $j \geq 1$,

$$H^j\big(\mathscr{K}'(m)\big) \quad = \quad 0 \quad = \quad H^j\big(\mathscr{Q}'(m)\big) .$$

Next we combine this with some exact sequences.

The exact sequence

$$0 \longrightarrow \mathscr{K}' \longrightarrow \mathscr{S}'(-1) \xrightarrow{\ y\ } \mathscr{S}' \longrightarrow \mathscr{Q}' \longrightarrow 0$$

can be split into two short exact sequences

$$0 \longrightarrow \mathscr{K}' \longrightarrow \mathscr{S}'(-1) \longrightarrow \mathscr{R}' \longrightarrow 0$$

$$0 \longrightarrow \mathscr{R}' \longrightarrow \mathscr{S}' \longrightarrow \mathscr{Q}' \longrightarrow 0 \quad .$$

By Lemma 10.4.7 the sequences

$$0 \longrightarrow \mathscr{K}'(m) \longrightarrow \{\mathscr{S}'(-1)\}(m) \longrightarrow \mathscr{R}'(m) \longrightarrow 0$$

$$0 \longrightarrow \mathscr{R}'(m) \longrightarrow \mathscr{S}'(m) \longrightarrow \mathscr{Q}'(m) \longrightarrow 0$$

are also exact, and Lemma 10.4.8 tells us that there is an isomorphism $\{\mathscr{S}'(-1)\}(m) \cong \mathscr{S}'(m-1)$. There are long exact sequences in cohomology induced by the short exact sequences of sheaves, and in particular the following are exact

$$H^1\big(\mathscr{K}'(m)\big) \longrightarrow H^1\big(\mathscr{S}'(m-1)\big) \longrightarrow H^1\big(\mathscr{R}'(m)\big) \longrightarrow H^2\big(\mathscr{K}'(m)\big)$$

$$H^1\big(\mathscr{R}'(m)\big) \longrightarrow H^1\big(\mathscr{S}'(m)\big) \longrightarrow H^1\big(\mathscr{Q}'(m)\big) \quad .$$

If we take $m > N_1$ then $H^1\big(\mathscr{K}'(m)\big) = H^2\big(\mathscr{K}'(m)\big) = H^1\big(\mathscr{Q}'(m)\big) = 0$, and the sequences simplify to

$$0 \longrightarrow H^1\big(\mathscr{S}'(m-1)\big) \xrightarrow{\ \alpha\ } H^1\big(\mathscr{R}'(m)\big) \longrightarrow 0$$

$$H^1\big(\mathscr{R}'(m)\big) \xrightarrow{\ \beta\ } H^1\big(\mathscr{S}'(m)\big) \longrightarrow 0 \ .$$

The map α is an isomorphism and the map β an epimorphism, and so the composite $\beta\alpha$ is an epimorphism

$$H^1\big(\mathscr{S}'(m-1)\big) \xrightarrow{\ \beta\alpha\ } H^1\big(\mathscr{S}'(m)\big) \longrightarrow 0 \quad .$$

Now projective space is a compact, Hausdorff space by Lemma 8.9.27. Using Theorem 10.7.3 the vector spaces $H^1\big(\mathscr{S}'(m)\big)$ are all finite dimensional. In the sequence of finite dimensional vector spaces $H^1\big(\mathscr{S}'(m)\big)$, the dimensions decrease once $m \geq N_1$. The dimensions must therefore stabilize. There must exist an integer $N_2 > N_1$ so that the dimension of $H^1\big(\mathscr{S}'(m)\big)$ is independent of m once $m \geq N_2 - 1$. If we choose $m \geq N_2$, then the composite $\beta\alpha$ below

$$0 \longrightarrow H^1\big(\mathscr{S}'(m-1)\big) \xrightarrow{\ \alpha\ } H^1\big(\mathscr{R}'(m)\big) \longrightarrow 0$$

$$H^1\big(\mathscr{R}'(m)\big) \xrightarrow{\ \beta\ } H^1\big(\mathscr{S}'(m)\big) \longrightarrow 0$$

is a surjective map between vector spaces of the same finite dimension, and must be an isomorphism. Since α is an isomorphism so is β.

Now take $m \geq N_2$ and consider the exact sequence

$$H^0\big(\mathscr{S}'(m)\big) \longrightarrow H^0\big(\mathscr{Q}'(m)\big) \longrightarrow H^1\big(\mathscr{R}'(m)\big) \xrightarrow{\ \beta\ } H^1\big(\mathscr{S}'(m)\big)\ .$$

We have shown above that β is an isomorphism, and the exactness tells us that the map

$$H^0\big(\mathscr{S}'(m)\big) \longrightarrow H^0\big(\mathscr{Q}'(m)\big) \quad = \quad H^0\big(\{\mathscr{S}'/y\}(m)\big)$$

is surjective. □

The next trick is to apply Lemma 10.10.5 to a string of maps. We write down the string we will wish to consider.

Construction 10.10.6. Suppose we are given a closed point $p \in \mathbb{P}^n$, and a coherent analytic sheaf $\mathscr{T}'$ with $\dim(\mathscr{T}') \geq i$. We are about to construct a sequence of maps we want to study.

For our closed point $p \in \mathbb{P}^n$, let $V(p) \subset V$ be the n–dimensional vector space of linear functions vanishing on $\pi^{-1}(p) \subset \mathbb{C}^{n+1} - \{0\}$, as in Notation 10.9.3. We can choose a basis $\{y_1, y_2, \ldots, y_n\}$ for $V(p)$. We will study the sequence of sheaves

$$\mathscr{T}' \longrightarrow \frac{\mathscr{T}'}{y_1} \longrightarrow \frac{\mathscr{T}'}{\langle y_1, y_2\rangle} \longrightarrow \cdots \longrightarrow \frac{\mathscr{T}'}{\langle y_1, y_2, \ldots, y_n\rangle}\ .$$

That is, we put $\mathscr{T}_0' = \mathscr{T}'$, and inductively define $\mathscr{T}_{j+1}' = \mathscr{T}_j'/y_{j+1}$. Lemma 10.4.17 tells us that $\operatorname{ann}(\mathscr{T}_{j+1}')$ contains both $\operatorname{ann}(\mathscr{T}_j')$ and y_{j+1}. It follows that all the annihilators contain $\operatorname{ann}(\mathscr{T}') = \operatorname{ann}(\mathscr{T}_0')$, and $\operatorname{ann}(\mathscr{T}_n')$ contains $V(p)$. What we deduce is:

10.10.6.1. The sheaves $\mathscr{T}_j'$ all satisfy $\dim(\mathscr{T}_j') \geq i$.

10.10.6.2. In the case of $\mathscr{T}_n'$ we know that $\operatorname{ann}(\mathscr{T}_n')$ contains $V(p)$. Therefore either $\operatorname{ann}(\mathscr{T}_n') = V$ or $\operatorname{ann}(\mathscr{T}_n') = V(p)$. In the first case, Remark 10.9.9 informs us that $\mathscr{T}_n'$ vanishes, while in the second case, Lemma 10.9.8 says that $\mathscr{T}_n'$ must be a skyscraper sheaf supported at p.

Discussion 10.10.7. Lemma 10.1.7 produced for us an affine cover of projective space; we have $\mathbb{P}^n = \cup_i \mathbb{P}^n_{x_i}$. Lemma 10.1.7 tells us that, on the open sets $\mathbb{P}^n_{x_i}$, the sheaves $\mathcal{O}(m)$ all trivialize. We may choose an isomorphism of $\mathcal{O}(1)|_{\mathbb{P}^n_{x_i}}$ with $\mathcal{O}|_{\mathbb{P}^n_{x_i}}$. For each i, choose and fix such an isomorphism.

We may now view the map $y_j^{\text{an}} : \mathcal{O}^{\text{an}} \longrightarrow \mathcal{O}(1)^{\text{an}}$, restricted to the open set $\{\mathbb{P}^n_{x_i}\}^{\text{an}} \subset \{\mathbb{P}^n\}^{\text{an}}$, as a map $y_j^{\text{an}} : \mathcal{O}^{\text{an}} \longrightarrow \mathcal{O}^{\text{an}}$. By Exercise 7.1.16(iii) there is an element $a_j \in \Gamma\big(\{\mathbb{P}^n_{x_i}\}^{\text{an}}, \mathcal{O}^{\text{an}}\big)$, that is a holomorphic function on $\{\mathbb{P}^n_{x_i}\}^{\text{an}}$, so that, for all open sets $\Delta \subset \{\mathbb{P}^n_{x_i}\}^{\text{an}}$, the map

$$\Gamma(\Delta, y_j^{\text{an}}) : \Gamma(\Delta, \mathcal{O}^{\text{an}}) \longrightarrow \Gamma\big(\Delta, \mathcal{O}(1)^{\text{an}}\big) \quad \cong \quad \Gamma(\Delta, \mathcal{O}^{\text{an}})$$

is multiplication by the holomorphic function a_j. If $\mathscr{S}'$ is any coherent analytic sheaf and $\Delta \subset \{\mathbb{P}^n_{x_i}\}^{\text{an}}$ is an open subset, then the map $y_j : \mathscr{S}'(-1) \longrightarrow \mathscr{S}'$ was given, in Definition 10.4.3(iii), as

$$\mathscr{H}om(y_j^{\text{an}}, 1) : \mathscr{H}om\big(\mathcal{O}(1)^{\text{an}}, \mathscr{S}')\big) \longrightarrow \mathscr{H}om\big(\mathcal{O}^{\text{an}}, \mathscr{S}'\big) \; .$$

By the above, on the open subset $\Delta \subset \{\mathbb{P}^n_{x_i}\}^{\text{an}}$ we deduce that

$$\Gamma(\Delta, y_j) : \Gamma\big(\Delta, \mathscr{S}'(-1)\big) \longrightarrow \Gamma\big(\Delta, \mathscr{S}'\big)$$

identifies with multiplication by a_j. The sheaf $\mathscr{S}'/y_j$ was defined as the cokernel in the exact sequence of sheaves

$$\mathscr{S}'(-1) \xrightarrow{\;y_j\;} \mathscr{S}' \longrightarrow \mathscr{S}'/y_j \longrightarrow 0 \; .$$

If $\Delta \subset \{\mathbb{P}^n_{x_i}\}^{\text{an}}$ is a polydisc then the sequence

$$\Gamma\big(\Delta, \mathscr{S}'(-1)\big) \xrightarrow{\;\Gamma(\Delta, y_j)\;} \Gamma\big(\Delta, \mathscr{S}'\big) \longrightarrow \Gamma\big(\Delta, \mathscr{S}'/y_j\big) \longrightarrow 0$$

is exact, and we learn that $\Gamma(\Delta, \mathscr{S}'/y_j)$ identifies with

$$\Gamma(\Delta, \mathscr{S}'/y_j) \quad = \quad \frac{\Gamma(\Delta, \mathscr{S}')}{a_j\Gamma(\Delta, \mathscr{S}')} \quad .$$

Applying this to the sequence

$$\mathscr{T}' \longrightarrow \frac{\mathscr{T}'}{y_1} \longrightarrow \frac{\mathscr{T}'}{\langle y_1, y_2 \rangle} \longrightarrow \cdots \longrightarrow \frac{\mathscr{T}'}{\langle y_1, y_2, \dots, y_n \rangle}$$

of Construction 10.10.6, we have that, for each j,

$$\Gamma(\Delta, \mathscr{T}'_{j+1}) \quad = \quad \Gamma(\Delta, \mathscr{T}'_j/y_{j+1}) \quad = \quad \frac{\Gamma(\Delta, \mathscr{T}'_j)}{a_{j+1}\Gamma(\Delta, \mathscr{T}'_j)} \quad .$$

Combining everything we have said in Discussion 10.10.7, we observe:

Remark 10.10.8. In Construction 10.10.6 we produced a morphism of sheaves $\mathscr{T}' \longrightarrow \mathscr{T}'_n$. For any polydisc $\Delta \subset \{\mathbb{P}^n_{x_i}\}^{\text{an}}$, the homomorphism $\Gamma(\Delta, \mathscr{T}') \longrightarrow \Gamma(\Delta, \mathscr{T}'_n)$ identifies with the quotient map

$$\Gamma(\Delta, \mathscr{T}') \longrightarrow \frac{\Gamma(\Delta, \mathscr{T}')}{I \cdot \Gamma(\Delta, \mathscr{T}')} \quad ,$$

where $I \subset \Gamma(\Delta, \mathcal{O}^{\text{an}})$ is the ideal generated by $\{a_1, a_2, \cdots, a_n\}$.

Now recall that Construction 10.10.6 began with a closed point $p \in \mathbb{P}^n$, and a basis $\{y_1, y_2, \dots, y_n\}$ for $V(p)$. Suppose it so happens that the point p lies in $\Delta \subset \{\mathbb{P}^n_{x_i}\}^{\mathrm{an}}$. Then Remark 10.9.6 assures us that the holomorphic functions $a_j \in \Gamma(\Delta, \mathcal{O}^{\mathrm{an}})$ all vanish at $p \in \Delta$, for $1 \le j \le n$. The ideal I therefore consists of holomorphic functions on Δ vanishing at p.

Lemma 10.10.9. *Suppose that every coherent analytic sheaf of dimension $> i$ is isomorphic to an $\mathscr{R}^{\mathrm{an}}$. Let $\mathscr{T}'$ be a coherent analytic sheaf of dimension* $\dim(\mathscr{T}') \ge i$, *and let p be a closed point in $\mathbb{P}^n$. Let the map $\mathscr{T}' \longrightarrow \mathscr{T}'_n$ be as in Construction 10.10.6.*

There exists a coherent algebraic sheaf $\mathscr{S}$, and a map $\varphi : \mathscr{S}^{\mathrm{an}} \longrightarrow \mathscr{T}'$, so that the composite

$$\mathscr{S}^{\mathrm{an}} \xrightarrow{\ \varphi\ } \mathscr{T}' \longrightarrow \mathscr{T}'_n$$

is surjective.

Proof. In Construction 10.10.6 we put $\mathscr{T}'_0 = \mathscr{T}'$, and inductively defined $\mathscr{T}'_{j+1} = \mathscr{T}'_j / y_{j+1}$. From 10.10.6.1 we know that the sheaves $\mathscr{T}'_j$ all satisfy $\dim(\mathscr{T}'_j) \ge i$. By Lemma 10.10.5 it follows that for each j we can find an integer N_j so that, whenever $m \ge N_j$, the map

$$H^0\big(\mathscr{T}'_j(m)\big) \longrightarrow H^0\big(\mathscr{T}'_{j+1}(m)\big)$$

is surjective. Choose and fix an m larger than the maximum of the N_j. Then, in the composite

$$H^0\big(\mathscr{T}'_0(m)\big) \longrightarrow H^0\big(\mathscr{T}'_1(m)\big) \longrightarrow \cdots \longrightarrow H^0\big(\mathscr{T}'_n(m)\big) ,$$

each map is an epimorphism. We conclude that, for our chosen large m, the composite $H^0\big(\mathscr{T}'(m)\big) \longrightarrow H^0\big(\mathscr{T}'_n(m)\big)$ must be an epimorphism.

Now recall that, for any coherent analytic sheaf $\mathscr{R}'$, the sheaf $\mathscr{R}'(m)$ was given, in Definition 10.4.3(i), as

$$\mathscr{R}'(m) \quad = \quad \mathscr{H}om\big(\mathcal{O}(-m)^{\mathrm{an}} , \mathscr{R}'\big) .$$

The surjectivity of the map $H^0\big(\mathscr{T}'(m)\big) \longrightarrow H^0\big(\mathscr{T}'_n(m)\big)$ means that

$$\mathrm{Hom}\big(\mathcal{O}(-m)^{\mathrm{an}} , \mathscr{T}'\big) \longrightarrow \mathrm{Hom}\big(\mathcal{O}(-m)^{\mathrm{an}} , \mathscr{T}'_n\big)$$

is surjective, which very concretely says that any map $\mathcal{O}(-m)^{\mathrm{an}} \longrightarrow \mathscr{T}'_n$ factors through $\mathscr{T}'$. If we can produce a surjective $\mathscr{S}^{\mathrm{an}} \longrightarrow \mathscr{T}'_n$, where $\mathscr{S}$ is a finite direct sum of $\mathcal{O}(-m)$'s, then we would be done; the map must factor through $\mathscr{T}' \longrightarrow \mathscr{T}'_n$. Now we will show how to produce such a surjection $\mathscr{S}^{\mathrm{an}} \longrightarrow \mathscr{T}'_n$.

Choose any small polydisc $\Delta \subset \{\mathbb{P}^n_{x_i}\}^{\mathrm{an}} \subset \{\mathbb{P}^n\}^{\mathrm{an}}$ containing the point

p. Because $\mathscr{T}'_n$ is a coherent analytic sheaf, there is an epimorphism of sheaves on Δ

$$\{W \otimes_{\mathbb{C}} \mathcal{O}^{\mathrm{an}}\}|_{\Delta} \longrightarrow \mathscr{T}'_n|_{\Delta} ,$$

where W is a finite dimensional $\mathbb{C}$–vector space. The polydisc Δ is contained in $\{\mathbb{P}^n_{x_i}\}^{\mathrm{an}}$, and Lemma 10.1.7 says the sheaf $\mathcal{O}(-m)$ trivializes on $\mathbb{P}^n_{x_i}$; there is an isomorphism between $\mathcal{O}|_{\mathbb{P}^n_{x_i}}$ and $\mathcal{O}(-m)|_{\mathbb{P}^n_{x_i}}$. Hence we may view the epimorphism as a map

$$\{W \otimes_{\mathbb{C}} \mathcal{O}(-m)^{\mathrm{an}}\}|_{\Delta} \longrightarrow \mathscr{T}'_n|_{\Delta} .$$

10.10.6.2 tells us that the sheaf $\mathscr{T}'_n$ is a skyscraper sheaf, which vanishes outside the point $p \in \{\mathbb{P}^n\}^{\mathrm{an}}$. On the open set $O = \{\mathbb{P}^n\}^{\mathrm{an}} - \{p\}$ we define the map

$$\{W \otimes_{\mathbb{C}} \mathcal{O}(-m)^{\mathrm{an}}\}|_{O} \longrightarrow \mathscr{T}'_n|_{O}$$

to be the zero map. On the intersection $\Delta - \{p\}$ the two morphisms of sheaves agree, by the vanishing of $\mathscr{T}'_n$ on $\Delta - \{p\}$. They glue to give a global morphism of sheaves

$$W \otimes_{\mathbb{C}} \mathcal{O}(-m)^{\mathrm{an}} \longrightarrow \mathscr{T}'_n .$$

This morphism is surjective on Δ by assumption, while on $\{\mathbb{P}^n\}^{\mathrm{an}} - \{p\}$ it is surjective because the sheaf $\mathscr{T}'_n$ vanishes. Hence the morphism is a global epimorphism of sheaves. □

Lemma 10.10.10. *Suppose that every coherent analytic sheaf of dimension $> i$ is isomorphic to an $\mathscr{R}^{\mathrm{an}}$. Let $\mathscr{T}'$ be a coherent analytic sheaf of dimension $\dim(\mathscr{T}') \geq i$, and let p be a closed point in $\mathbb{P}^n$.*

There exists a coherent algebraic sheaf $\mathscr{S}$, a map $\varphi : \mathscr{S}^{\mathrm{an}} \longrightarrow \mathscr{T}'$, and a polydisc $\Delta \subset \{\mathbb{P}^n_{x_i}\}^{\mathrm{an}} \subset \{\mathbb{P}^n\}^{\mathrm{an}}$ containing the point p, so that the restriction to Δ of the map φ is an epimorphism.

Proof. In Lemma 10.10.9 we produced a map $\varphi : \mathscr{S}^{\mathrm{an}} \longrightarrow \mathscr{T}'$, so that the composite

$$\mathscr{S}^{\mathrm{an}} \xrightarrow{\ \varphi\ } \mathscr{T}' \xrightarrow{\ \alpha\ } \mathscr{T}'_n$$

is surjective. We will now show that there exists a polydisc $\Delta \subset \{\mathbb{P}^n_{x_i}\}^{\mathrm{an}}$, containing p, so that $\varphi|_{\Delta} : \mathscr{S}^{\mathrm{an}}|_{\Delta} \longrightarrow \mathscr{T}'|_{\Delta}$ is an epimorphism.

The sheaf $\mathscr{T}'$ is coherent analytic. Therefore the point p is contained in a polydisc $\Delta \subset \{\mathbb{P}^n_{x_i}\}^{\mathrm{an}}$, where the module $N = \Gamma(\Delta, \mathscr{T}')$ is finitely generated over the ring $A = \Gamma(\Delta, \mathcal{O}^{\mathrm{an}})$. Put $M = \Gamma(\Delta, \mathscr{S}^{\mathrm{an}})$. Remark 10.10.8 tells us that the map $\Gamma(\Delta, \mathscr{T}') \longrightarrow \Gamma(\Delta, \mathscr{T}'_n)$ can be rewritten as $N \longrightarrow N/IN$, where $I \subset A$ is an ideal of holomorphic functions vanishing at p. The composite

$$\Gamma(\Delta, \mathscr{S}^{\mathrm{an}}) \xrightarrow{\Gamma(\Delta,\varphi)} \Gamma(\Delta, \mathscr{T}') \xrightarrow{\Gamma(\Delta,\alpha)} \Gamma(\Delta, \mathscr{T}'_n)$$

rewrites as

$$M \xrightarrow{\varphi'} N \xrightarrow{\alpha'} \frac{N}{IN} ,$$

and $\alpha'\varphi'$ must be a surjection; the map $\alpha\varphi : \mathscr{S}^{\mathrm{an}} \longrightarrow \mathscr{T}'_n$ is an epimorphism of coherent analytic sheaves, and Δ is a polydisc. Nakayama's lemma says there is an element f in the ring $A = \Gamma(\Delta, \mathcal{O}^{\mathrm{an}})$, congruent to 1 modulo the ideal $I \subset A$, so that the map $M[1/f] \longrightarrow N[1/f]$ is surjective; for a proof see [1, page 21, Corollary 2.5]. Since every element of the ideal I vanishes at p we have that $f(p)$ must be 1; in particular f does not vanish at p. If we shrink Δ, to a polydisc $\Delta' \subset \Delta$, we may assume that f is invertible on Δ'. For any smaller polydisc $\Delta'' \subset \Delta'$, the map

$$\Gamma(\Delta'', \mathcal{O}^{\mathrm{an}}) \otimes_{\Gamma(\Delta,\mathcal{O}^{\mathrm{an}})} M \longrightarrow \Gamma(\Delta'', \mathcal{O}^{\mathrm{an}}) \otimes_{\Gamma(\Delta,\mathcal{O}^{\mathrm{an}})} N$$

is therefore surjective, but this identifies as

$$\Gamma(\Delta'', \alpha) : \Gamma(\Delta'', \mathscr{S}^{\mathrm{an}}) \longrightarrow \Gamma(\Delta'', \mathscr{T}') .$$

We conclude that the restriction to Δ', of the sheaf map $\varphi : \mathscr{S}^{\mathrm{an}} \longrightarrow \mathscr{T}'$, must be an epimorphism. □

Lemma 10.10.11. *Suppose that every coherent analytic sheaf $\mathscr{R}'$ of dimension $> i$ is isomorphic to $\mathscr{R}^{\mathrm{an}}$, for some coherent algebraic sheaf $\mathscr{R}$. Let $\mathscr{T}'$ be a coherent analytic sheaf of dimension* $\dim(\mathscr{T}') \geq i$. *Then there exists a surjection of coherent analytic sheaves $\mathscr{S}^{\mathrm{an}} \longrightarrow \mathscr{T}'$, where $\mathscr{S}$ is a coherent algebraic sheaf.*

Proof. If $p \in \mathbb{P}^n$ is a closed point, Lemma 10.10.10 produced for us a coherent algebraic sheaf $\mathscr{S}_p$, a morphism of sheaves $\varphi_p : \mathscr{S}_p^{\mathrm{an}} \longrightarrow \mathscr{T}'$, and an open set $\Delta_p \subset \{\mathbb{P}^n\}^{\mathrm{an}}$ so that $\varphi_p|_{\Delta_p} : \mathscr{S}_p^{\mathrm{an}}|_{\Delta_p} \longrightarrow \mathscr{T}'|_{\Delta_p}$ is an epimorphism. Choose for every p such data $\{\varphi_p : \mathscr{S}_p^{\mathrm{an}} \longrightarrow \mathscr{T}',\ \Delta_p\}$. The open sets Δ_p cover $\{\mathbb{P}^n\}^{\mathrm{an}}$, which is compact by Lemma 8.9.27. There is a finite subcover. That is $\{\mathbb{P}^n\}^{\mathrm{an}} = \cup_{j=1}^k \Delta_{p_j}$. Taking the sum of the maps $\mathscr{S}_{p_j}^{\mathrm{an}} \longrightarrow \mathscr{T}'$, we obtain a morphism

$$\bigoplus_{j=1}^{k} \mathscr{S}_{p_j}^{\mathrm{an}} \longrightarrow \mathscr{T}' ,$$

which must be an epimorphism everywhere, because each of the maps $\mathscr{S}_{p_j}^{\mathrm{an}} \longrightarrow \mathscr{T}'$ is an epimorphism on Δ_{p_j}. □

We are ready for the final punchline:

Lemma 10.10.12. *Assume that every coherent analytic sheaf* $\mathscr{R}'$ *of dimension* $\dim(\mathscr{R}') > i$ *is isomorphic to* $\mathscr{R}^{\mathrm{an}}$*, for some coherent algebraic sheaf* $\mathscr{R}$*. Let* $\mathscr{T}'$ *be a coherent analytic sheaf with* $\dim(\mathscr{T}') = i$*. Then there exists a coherent algebraic sheaf* $\mathscr{T}$ *and an isomorphism* $\mathscr{T}' \cong \mathscr{T}^{\mathrm{an}}$*.*

Proof. Let $\mathscr{T}'$ be a coherent analytic sheaf with $\dim(\mathscr{T}') = i$. From Lemma 10.10.11 we have a coherent algebraic sheaf $\mathscr{K}$ and an epimorphism $\varphi : \mathscr{K}^{\mathrm{an}} \longrightarrow \mathscr{T}'$. Lemma 10.4.18 informs us that the map φ factors as $\mathscr{K}^{\mathrm{an}} \longrightarrow \mathscr{S}^{\mathrm{an}} \longrightarrow \mathscr{T}'$, where $\mathscr{S}$ is also a coherent algebraic sheaf, but now we know further that $\dim(\mathscr{S}^{\mathrm{an}}) \geq i$. Let us complete to an exact sequence of coherent analytic sheaves

$$0 \longrightarrow \mathscr{L}' \longrightarrow \mathscr{S}^{\mathrm{an}} \longrightarrow \mathscr{T}' \longrightarrow 0\,.$$

Now $\mathscr{L}'$ is a coherent analytic subsheaf of $\mathscr{S}^{\mathrm{an}}$, and we know that $\dim(\mathscr{S}^{\mathrm{an}}) \geq i$. Lemma 10.4.14 says that $\dim(\mathscr{L}') \geq i$. We can therefore, again by Lemma 10.10.11, find a coherent algebraic sheaf $\mathscr{R}$ and an epimorphism $\mathscr{R}^{\mathrm{an}} \longrightarrow \mathscr{L}'$. Assembling this we have an exact sequence of coherent analytic sheaves

$$\mathscr{R}^{\mathrm{an}} \xrightarrow{\ f\ } \mathscr{S}^{\mathrm{an}} \longrightarrow \mathscr{T}' \longrightarrow 0\,.$$

The map $f : \mathscr{R}^{\mathrm{an}} \longrightarrow \mathscr{S}^{\mathrm{an}}$ is a morphism of sheaves of $\mathcal{O}^{\mathrm{an}}$–modules on $\{\mathbb{P}^n\}^{\mathrm{an}}$, where $\mathscr{R}$ and $\mathscr{S}$ are coherent algebraic sheaves. By Theorem 7.9.1(i), which we have already proved, there exists a morphism $\varphi : \mathscr{R} \longrightarrow \mathscr{S}$, of coherent algebraic sheaves on $\mathbb{P}^n$, with $f = \varphi^{\mathrm{an}}$. Now complete $\varphi : \mathscr{R} \longrightarrow \mathscr{S}$ to an exact sequence

$$\mathscr{R} \xrightarrow{\ \varphi\ } \mathscr{S} \longrightarrow \mathscr{T} \longrightarrow 0\,.$$

Because analytification is exact the sequence

$$\mathscr{R}^{\mathrm{an}} \xrightarrow{\ \varphi^{\mathrm{an}}\ } \mathscr{S}^{\mathrm{an}} \longrightarrow \mathscr{T}^{\mathrm{an}} \longrightarrow 0$$

is also exact, and $\mathscr{T}^{\mathrm{an}}$ is isomorphic to $\mathscr{T}'$. $\square$

Appendix 1
The proofs concerning analytification

Let $(X, \mathcal{O})$ be a scheme locally of finite type over $\mathbb{C}$. In Chapters 4 and 5 we constructed an analytic space $(X^{\text{an}}, \mathcal{O}^{\text{an}})$ and a map of ringed spaces

$$(\lambda, \lambda^*) : (X^{\text{an}}, \mathcal{O}^{\text{an}}) \longrightarrow (X, \mathcal{O}) .$$

In Chapter 7 we extended this construction to coherent sheaves. Given a coherent algebraic sheaf $\mathscr{S}$ on $(X, \mathcal{O})$, we produced a coherent analytic sheaf $\mathscr{S}^{\text{an}}$ on $(X^{\text{an}}, \mathcal{O}^{\text{an}})$ and a natural comparison map of sheaves on X

$$\lambda^* : \mathscr{S} \longrightarrow \lambda_* \mathscr{S}^{\text{an}} .$$

Very concretely this means the following. The continuous map $\lambda : X^{\text{an}} \longrightarrow X$ is just the inclusion of the set of closed points. We will identify subsets of X^{an} with their images by the inclusions. Given an open set $U \subset X$ we have a ring $\Gamma(U, \mathcal{O})$ and a module over it $\Gamma(U, \mathscr{S})$. Given an open set $V \subset X^{\text{an}}$ we have a ring $\Gamma(V, \mathcal{O}^{\text{an}})$ and a module over it $\Gamma(V, \mathscr{S}^{\text{an}})$. If V happens to be contained in U then there is a ring homomorphism and a homomorphism of modules

$$\Gamma(U, \mathcal{O}) \longrightarrow \Gamma(V, \mathcal{O}^{\text{an}}) , \qquad \Gamma(U, \mathscr{S}) \longrightarrow \Gamma(V, \mathscr{S}^{\text{an}}) .$$

Maybe we should remind the reader: these maps are the composites

$$\Gamma(U, \mathcal{O}) \xrightarrow{\lambda_U^*} \Gamma(U^{\text{an}}, \mathcal{O}^{\text{an}}) \xrightarrow{\text{res}_V^{U^{\text{an}}}} \Gamma(V, \mathcal{O}^{\text{an}}) \quad ,$$
$$\Gamma(U, \mathscr{S}) \xrightarrow{\lambda_U^*} \Gamma(U^{\text{an}}, \mathscr{S}^{\text{an}}) \xrightarrow{\text{res}_V^{U^{\text{an}}}} \Gamma(V, \mathscr{S}^{\text{an}}) \quad .$$

Both $\Gamma(U, \mathscr{S})$ and $\Gamma(V, \mathscr{S}^{\text{an}})$ are modules over $\Gamma(U, \mathcal{O})$, and the homomorphism $\Gamma(U, \mathscr{S}) \longrightarrow \Gamma(V, \mathscr{S}^{\text{an}})$ is a homomorphism of $\Gamma(U, \mathcal{O})$–modules. But $\Gamma(V, \mathscr{S}^{\text{an}})$ is a module over the $\Gamma(U, \mathcal{O})$–algebra $\Gamma(V, \mathcal{O}^{\text{an}})$. We define

Definition A1.0.1. *The map*

$$\Sigma_{\mathscr{S}} : \Gamma(V, \mathcal{O}^{\text{an}}) \otimes_{\Gamma(U,\mathcal{O})} \Gamma(U, \mathscr{S}) \longrightarrow \Gamma(V, \mathscr{S}^{\text{an}})$$

is the unique homomorphism of $\Gamma(V,\mathcal{O}^{\mathrm{an}})$–modules through which the map $\Gamma(U,\mathscr{S}) \longrightarrow \Gamma(V,\mathscr{S}^{\mathrm{an}})$ naturally factors.

Note that the map $\Sigma_{\mathscr{S}}$ is natural in $\mathscr{S}$; if we have a map of coherent algebraic sheaves $\eta : \mathscr{S} \longrightarrow \mathscr{T}$, Exercise 7.8.8(i) asserts that the following square of homomorphisms of commutes

$$\begin{array}{ccc} \Gamma(U,\mathscr{S}) & \xrightarrow{\lambda^*} & \Gamma(V,\mathscr{S}^{\mathrm{an}}) \\ {\scriptstyle \Gamma(U,\eta)}\downarrow & & \downarrow{\scriptstyle \Gamma(V,\eta^{\mathrm{an}})} \\ \Gamma(U,\mathscr{T}) & \xrightarrow[\lambda^*]{} & \Gamma(V,\mathscr{T}^{\mathrm{an}}) \end{array} .$$

But this immediately implies the commutativity of

$$\begin{array}{ccc} \Gamma(V,\mathcal{O}^{\mathrm{an}}) \otimes_{\Gamma(U,\mathcal{O})} \Gamma(U,\mathscr{S}) & \xrightarrow{\Sigma_{\mathscr{S}}} & \Gamma(V,\mathscr{S}^{\mathrm{an}}) \\ {\scriptstyle \Gamma(V,\mathcal{O}^{\mathrm{an}})\otimes_{\Gamma(U,\mathcal{O})}\Gamma(U,\eta)}\downarrow & & \downarrow{\scriptstyle \Gamma(V,\eta^{\mathrm{an}})} \\ \Gamma(V,\mathcal{O}^{\mathrm{an}}) \otimes_{\Gamma(U,\mathcal{O})} \Gamma(U,\mathscr{T}) & \xrightarrow[\Sigma_{\mathscr{T}}]{} & \Gamma(V,\mathscr{T}^{\mathrm{an}}) \end{array} .$$

We will prove most of the facts we stated by understanding the natural map $\Sigma_{\mathscr{S}}$ better.

A1.1 The first facts

We begin with a reminder:

Reminder A1.1.1. Let S be a finitely generated algebra, and let $a_1, a_2, \ldots, a_k$ be generators for the ring S. Let $R = \mathbb{C}[x_1, x_2, \ldots, x_k]$ be the polynomial ring, and let $\theta : R \longrightarrow S$ be the homomorphism with $\theta(x_i) = a_i$. Let I be the kernel of θ. The map θ induces an embedding

$$\left\{\operatorname{Spec}(\theta)\right\}^{\mathrm{an}} : \left\{\operatorname{Spec}(S)\right\}^{\mathrm{an}} \longrightarrow \left\{\operatorname{Spec}(R)\right\}^{\mathrm{an}} \quad = \quad \mathbb{C}^k .$$

Let $W = \Delta(g; w; r) \subset \mathbb{C}^k$ be a polydisc, and let $V = W \cap \left\{\operatorname{Spec}(S)\right\}^{\mathrm{an}}$. The way Section 5.3 defined the sheaf $\mathcal{O}^{\mathrm{an}}$ on $\left\{\operatorname{Spec}(S)\right\}^{\mathrm{an}}$ was by the formula

$$\Gamma(V,\mathcal{O}^{\mathrm{an}}) \quad = \quad \frac{\Gamma\big(W,\mathcal{O}^{\mathrm{an}}_{\mathbb{C}^k}\big)}{I\cdot\Gamma\big(W,\mathcal{O}^{\mathrm{an}}_{\mathbb{C}^k}\big)} .$$

For our purposes now it is helpful to note that this is nothing other than

$$\Gamma(V,\mathcal{O}^{\mathrm{an}}) \quad = \quad \Gamma\big(W,\mathcal{O}^{\mathrm{an}}_{\mathbb{C}^k}\big) \otimes_R S .$$

After all, $S = R/I$ and the tensor product respects quotients.

Lemma A1.1.2. *With the notation as in Reminder A1.1.1, let M be a finite S–module. Then $\widetilde{M}$ is a coherent algebraic sheaf over $\Big(\mathrm{Spec}(S), \widetilde{S}\Big)$, and we have learned how to form the sheaf $\widetilde{M}^{\mathrm{an}}$ on $\big\{\mathrm{Spec}(S)\big\}^{\mathrm{an}}$. I assert that there is for it the simple formula*

$$\Gamma(V, \widetilde{M}^{\mathrm{an}}) \quad = \quad \Gamma\big(W, \mathscr{O}^{\mathrm{an}}_{\mathbb{C}^k}\big) \otimes_R M \ .$$

Proof. In Reminder A1.1.1 we already chose generators $a_1, a_2, \ldots, a_k$ for the ring S as an algebra over $\mathbb{C}$. Now choose generators $m_1, m_2, \ldots, m_\ell$ for M as a module over S. Let R and R' be polynomial rings,

$$R = \mathbb{C}[x_1, x_2, \ldots, x_k] \,, \qquad R' = \mathbb{C}[x_1, x_2, \ldots, x_k, y_1, y_2, \ldots, y_\ell] \,,$$

let $i : R \longrightarrow R'$ be the obvious inclusion, and let $p : R' \longrightarrow R$ be the projection with $p(x_i) = x_i$, $p(y_j) = 0$. As in Reminder A1.1.1 the map $\theta : R \longrightarrow S$ will be the homomorphism with $\theta(x_i) = a_i$, and we set $\theta' : R' \longrightarrow S \oplus M$ to be the homomorphism with $\theta'(x_i) = a_i$, $\theta'(y_j) = m_j$. Let I be the kernel of θ and let I' be the kernel of θ'.

We have a commutative diagram of ring homomorphisms

$$\begin{array}{ccccc} R & \xrightarrow{\ i\ } & R' & \xrightarrow{\ p\ } & R \\ {\scriptstyle\theta}\Big\downarrow & & {\scriptstyle\theta'}\Big\downarrow & & \Big\downarrow{\scriptstyle\theta} \\ S & \xrightarrow[\psi]{} & S \oplus M & \xrightarrow[\varphi]{} & S \end{array}$$

and we want to study the maps of sheaves of rings on $\big\{\mathrm{Spec}(S)\big\}^{\mathrm{an}}$

$$\widetilde{S}^{\mathrm{an}} \xrightarrow{\ \psi^{\mathrm{an}}\ } \{\widetilde{S \oplus M}\}^{\mathrm{an}} \xrightarrow{\ \varphi^{\mathrm{an}}\ } \widetilde{S}^{\mathrm{an}}$$

when evaluated on the polydisc $V = W \cap \big\{\mathrm{Spec}(S)\big\}^{\mathrm{an}}$; after all the sheaf $\widetilde{M}^{\mathrm{an}}$ is defined as the kernel of the map $\varphi^{\mathrm{an}} : \{\widetilde{S \oplus M}\}^{\mathrm{an}} \longrightarrow \widetilde{S}^{\mathrm{an}}$. Now the $\mathbb{C}$–algebra homomorphisms

$$R \xrightarrow{\ i\ } R' \xrightarrow{\ p\ } R$$

induce, when we apply $\big\{\mathrm{Spec}(-)\big\}^{\mathrm{an}}$ to them, the continuous maps of topological spaces

$$\mathbb{C}^k \xrightarrow{\ \{\mathrm{Spec}(p)\}^{\mathrm{an}}\ } \mathbb{C}^{k+\ell} \xrightarrow{\ \{\mathrm{Spec}(i)\}^{\mathrm{an}}\ } \mathbb{C}^k \quad ,$$

where $\big\{\mathrm{Spec}(i)\big\}^{\mathrm{an}}$ is the projection and $\big\{\mathrm{Spec}(p)\big\}^{\mathrm{an}}$ the inclusion. If we pull back to the polydisc W we have maps

$$W \xrightarrow{\ \{\mathrm{Spec}(p)\}^{\mathrm{an}}\ } W \times \mathbb{C}^\ell \xrightarrow{\ \{\mathrm{Spec}(i)\}^{\mathrm{an}}\ } W \ ,$$

and the ring homomorphisms we want to study, that is

$$\Gamma\left(V, \{\widetilde{S}\}^{\text{an}}\right) \longrightarrow \Gamma\left(V, \{\widetilde{S \oplus M}\}^{\text{an}}\right) \longrightarrow \Gamma\left(V, \{\widetilde{S}\}^{\text{an}}\right) ,$$

reduce, by Reminder A1.1.1, to

$$\begin{array}{ccc} \Gamma(W, \mathcal{O}^{\text{an}}_{\mathbb{C}^k}) \otimes_R S & & \Gamma(W, \mathcal{O}^{\text{an}}_{\mathbb{C}^k}) \otimes_R S \\ & \searrow \quad \nearrow & \\ & \Gamma(W \times \mathbb{C}^\ell, \mathcal{O}^{\text{an}}_{\mathbb{C}^{k+\ell}}) \otimes_{R'} (S \oplus M) & \end{array} .$$

And the key point is that the map

$$\Gamma(W, \mathcal{O}^{\text{an}}_{\mathbb{C}^k}) \otimes_R R' \longrightarrow \Gamma(W \times \mathbb{C}^\ell, \mathcal{O}^{\text{an}}_{\mathbb{C}^{k+\ell}}) ,$$

which most definitely is not an isomorphism, becomes an isomorphism after tensor product with $S \oplus M$. That is

$$\Gamma(W, \mathcal{O}^{\text{an}}_{\mathbb{C}^k}) \otimes_R R' \otimes_{R'} (S \oplus M) \quad = \quad \Gamma(W \times \mathbb{C}^\ell, \mathcal{O}^{\text{an}}_{\mathbb{C}^{k+\ell}}) \otimes_{R'} (S \oplus M) .$$

Granting this, the term on the left simplifies to $\Gamma(W, \mathcal{O}^{\text{an}}_{\mathbb{C}^k}) \otimes_R (S \oplus M)$, the sequence of maps defining $\Gamma(V, \widetilde{M}^{\text{an}})$ becomes

$$\Gamma(W, \mathcal{O}^{\text{an}}_{\mathbb{C}^k}) \otimes_R S \xrightarrow{\psi^{\text{an}}} \Gamma(W, \mathcal{O}^{\text{an}}_{\mathbb{C}^k}) \otimes_R (S \oplus M) \xrightarrow{\varphi^{\text{an}}} \Gamma(W, \mathcal{O}^{\text{an}}_{\mathbb{C}^k}) \otimes_R S$$

and it is now transparent that $\Gamma(V, \widetilde{M}^{\text{an}})$, which is the kernel of φ^{an}, is equal to $\Gamma(W, \mathcal{O}^{\text{an}}_{\mathbb{C}^k}) \otimes_R M$.

So it remains to prove the key claim: we need to show that the map

$$\Gamma(W, \mathcal{O}^{\text{an}}_{\mathbb{C}^k}) \otimes_R R' \longrightarrow \Gamma(W \times \mathbb{C}^\ell, \mathcal{O}^{\text{an}}_{\mathbb{C}^{k+\ell}})$$

induces an isomorphism after tensor with $S \oplus M$. Observe that the ring $\Gamma(W, \mathcal{O}^{\text{an}}_{\mathbb{C}^k}) \otimes_R R'$ is the ring of polynomials in $y_1, y_2, \dots, y_\ell$ whose coefficients are holomorphic functions on the polydisc $W \subset \mathbb{C}^k$. The ring $\Gamma(W \times \mathbb{C}^\ell, \mathcal{O}^{\text{an}}_{\mathbb{C}^{k+\ell}})$ is the ring of holomorphic functions on $W \times \mathbb{C}^\ell$. The map is just the natural inclusion, and as we have already said it is *not* an isomorphism. But when we tensor with $S \oplus M$ over R' we divide by a large ideal I', the kernel of the homomorphism $\theta' : R' \longrightarrow S \oplus M$. The ideal I' contains all elements $\{y_i y_j \mid 1 \leq i \leq j \leq \ell\}$. Modulo just the generators $y_i y_j$ we have that both rings become isomorphic; in both cases we have just the free module over $\Gamma(W, \mathcal{O}^{\text{an}}_{\mathbb{C}^k})$ with basis $\{1, y_1, y_2, \dots, y_\ell\}$; all higher degree terms in y_i have been killed. Further dividing, by other elements of I', still gives an isomorphism. □

Making our result more coordinate-free, we can say

Lemma A1.1.3. *Let S be a finitely generated $\mathbb{C}$–algebra and let M be a finite S–module. Put $(X,\mathcal{O}) = \left(\operatorname{Spec}(S), \widetilde{S}\right)$. Let V be a polydisc in $X^{\text{an}} = \left\{\operatorname{Spec}(S)\right\}^{\text{an}}$. Then*

$$\Gamma(V, \widetilde{M}^{\text{an}}) \quad = \quad \Gamma(V, \mathcal{O}^{\text{an}}) \otimes_S M \ .$$

Proof. After choosing coordinates, Reminder A1.1.1 informs us that $\Gamma(V,\mathcal{O}^{\text{an}}) = \Gamma(W, \mathcal{O}^{\text{an}}_{\mathbb{C}^k}) \otimes_R S$. From Lemma A1.1.2 we learn further that $\Gamma(V, \widetilde{M}^{\text{an}}) = \Gamma(W, \mathcal{O}^{\text{an}}_{\mathbb{C}^k}) \otimes_R M$. Combining these two facts, we have

$$\Gamma(V, \widetilde{M}^{\text{an}}) \quad = \quad \Gamma(W, \mathcal{O}^{\text{an}}_{\mathbb{C}^k}) \otimes_R S \otimes_S M \quad = \quad \Gamma(V, \mathcal{O}^{\text{an}}) \otimes_S M \, .$$

□

An even more coordinate-free way to say the same thing is

Lemma A1.1.4. *Let $(X,\mathcal{O})$ be a scheme locally of finite type over $\mathbb{C}$. Let $U \subset X$ be an affine open subset, that is $(U, \mathcal{O}|_U) \cong \left(\operatorname{Spec}(S), \widetilde{S}\right)$ for some finitely generated $\mathbb{C}$–algebra S. Let $\mathscr{S}$ be a coherent algebraic sheaf over $(X,\mathcal{O})$ and assume that, over the open set U, we have $\mathscr{S}|_U \cong \widetilde{M}$, for a finite S–module M. Let $V \subset U^{\text{an}}$ be a polydisc. Then the map of Definition A1.0.1, that is*

$$\Sigma_{\mathscr{S}} : \Gamma(V, \mathcal{O}^{\text{an}}) \otimes_{\Gamma(U,\mathcal{O})} \Gamma(U, \mathscr{S}) \longrightarrow \Gamma(V, \mathscr{S}^{\text{an}}) \ ,$$

is an isomorphism.

Proof. If $(X,\mathcal{O}) = (U, \mathcal{O}|_U) = \left(\operatorname{Spec}(S), \widetilde{S}\right)$ then the map of the lemma is precisely the homomorphism $\Gamma(V, \mathcal{O}^{\text{an}}) \otimes_S M \longrightarrow \Gamma(V, \widetilde{M}^{\text{an}})$, which we proved to be an isomorphism in Lemma A1.1.3. We only need to convince ourselves that the problem is local. But Exercise 7.8.8(i) does precisely that; it establishes that analytification commutes with restriction. That is, $\mathscr{S}^{\text{an}}|_{U^{\text{an}}} = \left\{\mathscr{S}|_U\right\}^{\text{an}}$ and $\mathcal{O}^{\text{an}}|_{U^{\text{an}}} = \left\{\mathcal{O}|_U\right\}^{\text{an}}$ □

We can already conclude

Corollary A1.1.5. *Facts 7.8.9 (i), (iii) and (v) are true.*

Proof. Let us begin with Fact 7.8.9(i); we need to show that, if $\mathscr{S}$ is a coherent algebraic sheaf on $(X,\mathcal{O})$, then $\mathscr{S}^{\text{an}}$ is a coherent analytic sheaf on $(X^{\text{an}}, \mathcal{O}^{\text{an}})$. The problem is local so we may assume $(X,\mathcal{O}) = \left(\operatorname{Spec}(S), \widetilde{S}\right)$ and $\mathscr{S} = \widetilde{M}$, for some finitely generated $\mathbb{C}$–algebra S and some finite S–module M. But then we may choose a finite presentation of M, that is an exact sequence of S–modules

$$S^m \longrightarrow S^n \longrightarrow M \longrightarrow 0 \ .$$

On X^{an} this induces a sequence of sheaves $\{\widetilde{S}^m\}^{\mathrm{an}} \longrightarrow \{\widetilde{S}^n\}^{\mathrm{an}} \longrightarrow \widetilde{M}^{\mathrm{an}}$, which can also be denoted $\{\mathcal{O}^{\mathrm{an}}\}^m \longrightarrow \{\mathcal{O}^{\mathrm{an}}\}^n \longrightarrow \widetilde{M}^{\mathrm{an}}$. On every polydisc $V \subset X^{\mathrm{an}}$, Lemma A1.1.3 identifies the sequence

$$\Gamma\Big(V, \{\widetilde{S}^m\}^{\mathrm{an}}\Big) \longrightarrow \Gamma\Big(V, \{\widetilde{S}^n\}^{\mathrm{an}}\Big) \longrightarrow \Gamma(V, \widetilde{M}^{\mathrm{an}}) \longrightarrow 0$$

with

$$\Gamma(V, \mathcal{O}) \otimes_S S^m \longrightarrow \Gamma(V, \mathcal{O}) \otimes_S S^n \longrightarrow \Gamma(V, \mathcal{O}) \otimes_S M \longrightarrow 0 \,.$$

This is exact since the tensor is right exact. Hence $\widetilde{M}^{\mathrm{an}}$ is a coherent analytic sheaf; see Section 7.7.

Fact 7.8.9(iii) has two parts. Part (a) is an analytic statement, asserting that the image of the map $\Gamma(U, \widetilde{M}) \longrightarrow \Gamma(V, \widetilde{M}^{\mathrm{an}})$ is dense in the Fréchet topology. By Proposition 5.5.7 we know that the ring homomorphism

$$\mu_V : \Gamma(U, \widetilde{S} \oplus \widetilde{M}) \longrightarrow \Gamma(V, \widetilde{S}^{\mathrm{an}} \oplus \widetilde{M}^{\mathrm{an}})$$

has a dense image; we now deduce that the image of the restriction to $\Gamma(U, \widetilde{M})$ is dense in $\Gamma(V, \widetilde{M}^{\mathrm{an}})$. The fact that the continuous ring homomorphisms

$$\Gamma(V, \widetilde{S}^{\mathrm{an}}) \xrightarrow{\ \psi^{\mathrm{an}}\ } \Gamma(V, \widetilde{S}^{\mathrm{an}} \oplus \widetilde{M}^{\mathrm{an}}) \xrightarrow{\ \varphi^{\mathrm{an}}\ } \Gamma(V, \widetilde{S}^{\mathrm{an}})$$

compose to the identity means that the continuous map

$$1 - \psi^{\mathrm{an}}\varphi^{\mathrm{an}} \ : \ \Gamma(V, \widetilde{S}^{\mathrm{an}} \oplus \widetilde{M}^{\mathrm{an}}) \longrightarrow \Gamma(V, \widetilde{S}^{\mathrm{an}} \oplus \widetilde{M}^{\mathrm{an}})$$

has its image in the closed subspace $\Gamma(V, \widetilde{M}^{\mathrm{an}}) \subset \Gamma(V, \widetilde{S}^{\mathrm{an}} \oplus \widetilde{M}^{\mathrm{an}})$. In the commutative square

$$\begin{array}{ccc} \Gamma(U, \widetilde{S} \oplus \widetilde{M}) & \xrightarrow{\ \mu_V\ } & \Gamma(V, \widetilde{S}^{\mathrm{an}} \oplus \widetilde{M}^{\mathrm{an}}) \\ {\scriptstyle 1-\psi\varphi}\big\downarrow & & \big\downarrow{\scriptstyle 1-\psi^{\mathrm{an}}\varphi^{\mathrm{an}}} \\ \Gamma(U, \widetilde{M}) & \xrightarrow[\{\mu_V\}|_M]{} & \Gamma(V, \widetilde{M}^{\mathrm{an}}) \end{array}$$

the map μ_V has a dense image and the map $1 - \psi^{\mathrm{an}}\varphi^{\mathrm{an}}$ is surjective and continuous. The image of $\{\mu_V\}|_M$ must therefore be dense.

Next we prove Fact 7.8.9(iii)(b). Lemma A1.1.4 asserts that the map

$$\Sigma_{\mathscr{S}} : \Gamma(V, \mathcal{O}^{\mathrm{an}}) \otimes_{\Gamma(U,\mathcal{O})} \Gamma(U, \mathscr{S}) \longrightarrow \Gamma(V, \mathscr{S}^{\mathrm{an}}) \,,$$

is an isomorphism, which certainly implies that the image of $\Gamma(U, \mathscr{S})$ in $\Gamma(V, \mathscr{S}^{\mathrm{an}})$ generates the module over the ring $\Gamma(V, \mathcal{O}^{\mathrm{an}})$.

Finally we prove Fact 7.8.9(v). We are given a coherent algebraic sheaf $\mathscr{S}$, a sheaf $\mathscr{T}$ of $\mathcal{O}^{\mathrm{an}}$–modules on X^{an}, and a map $\eta : \mathscr{S} \longrightarrow \lambda_* \mathscr{T}$. This means that, for every open set $U \subset X$, we have a homomorphism of $\Gamma(U, \mathcal{O})$–modules

$$\eta_U : \Gamma(U, \mathscr{S}) \longrightarrow \Gamma(U^{\mathrm{an}}, \mathscr{T}) .$$

If $V \subset U^{\mathrm{an}}$ is any open subset we have a composite

$$\Gamma(U, \mathscr{S}) \xrightarrow{\ \eta_U\ } \Gamma(U^{\mathrm{an}}, \mathscr{T}) \xrightarrow{\ \mathrm{res}_V^{U^{\mathrm{an}}}\ } \Gamma(V, \mathscr{T}) .$$

This is a homomorphism of $\Gamma(U, \mathcal{O})$–modules, but $\Gamma(V, \mathscr{T})$ is a module over that $\Gamma(U, \mathcal{O})$–algebra $\Gamma(V, \mathcal{O}^{\mathrm{an}})$. There is a unique factorization through

$$\Gamma(V, \mathcal{O}^{\mathrm{an}}) \otimes_{\Gamma(U,\mathcal{O})} \Gamma(U, \mathscr{S}) \longrightarrow \Gamma(V, \mathscr{T}) .$$

If U happens to be affine, if $V \subset U^{\mathrm{an}}$ is a polydisc, and if the coherent sheaf $\mathscr{S}$ restricts on the open affine subset $U \subset X$ to $\mathscr{S}|_U \cong \widetilde{M}$, then Lemma A1.1.4 tells us that

$$\Gamma(V, \mathcal{O}^{\mathrm{an}}) \otimes_{\Gamma(U,\mathcal{O})} \Gamma(U, \mathscr{S}) \quad = \quad \Gamma(V, \mathscr{S}^{\mathrm{an}}) .$$

That is we have found, on the polydisc $V \subset U$, a unique factorization

$$\Gamma(U, \mathscr{S}) \longrightarrow \Gamma(V, \mathscr{S}^{\mathrm{an}}) \xrightarrow{\ E_V\ } \Gamma(V, \mathscr{T}) .$$

The uniqueness of the factorization makes clear that it will be compatible with restrictions, both in U and in V. We have therefore defined, on sufficiently small polydiscs V, maps E_V compatible with restriction. Sufficiently small polydiscs form a subbasis for the topology of X^{an}, and there is a unique extension to a morphism of sheaves defined on all open sets in X^{an}. □

A1.2 The flatness

We still need to prove Facts 7.8.9 (ii) and (iv) and (vi). These questions are more subtle. Let us take Fact 7.8.9(iv); it asserts that analytification takes exact sequences of coherent algebraic sheaves to exact sequences of coherent analytic sheaves. The question is local, so we may assume that $(X, \mathcal{O}) = \left(\mathrm{Spec}(S), \widetilde{S}\right)$, and that all the finitely many coherent algebraic sheaves we will deal with on X are of the form $\widetilde{M}$. Now we should explain the subtlety involved in proving this fact. By Lemma 7.6.14, the sequence of sheaves

$$\widetilde{L} \longrightarrow \widetilde{M} \longrightarrow \widetilde{N}$$

will be exact precisely when the ordinary sequence of S–modules

$$L \longrightarrow M \longrightarrow N$$

is exact. Let us choose coordinates; in other words choose a surjective homomorphism $\theta : R \longrightarrow S$, with $R = \mathbb{C}[x_1, x_2, \ldots, x_k]$ the polynomial ring. By Lemma A1.1.2 the maps of the analytic sheaves $\widetilde{L}^{\mathrm{an}} \longrightarrow \widetilde{M}^{\mathrm{an}} \longrightarrow \widetilde{N}^{\mathrm{an}}$, evaluated on a polydisc V, are nothing other than

$$\Gamma(W, \mathcal{O}^{\mathrm{an}}_{\mathbb{C}^k}) \otimes_R L \longrightarrow \Gamma(W, \mathcal{O}^{\mathrm{an}}_{\mathbb{C}^k}) \otimes_R M \longrightarrow \Gamma(W, \mathcal{O}^{\mathrm{an}}_{\mathbb{C}^k}) \otimes_R N .$$

Fact 7.8.9(iv) asserts that this is exact. In other words, what we have to prove is

Lemma A1.2.1. *Let R be the polynomial ring in k variables, and let $W = \Delta(g; w; r) \subset \mathbb{C}^k$ be a polydisc. Then the ring of holomorphic functions $\Gamma(W, \mathcal{O}^{\mathrm{an}}_{\mathbb{C}^k})$ is flat over R.*

Proof. Let A be the ring $A = \Gamma(W, \mathcal{O}^{\mathrm{an}}_{\mathbb{C}^k})$. We need to show that, for any ideal $\mathfrak{a} \subset R$, the group $\mathrm{Tor}_1^R(R/\mathfrak{a}, A)$ vanishes; see [1, page 35, Exercise 26]. Concretely this means we need to show that, for any exact sequence $R^m \longrightarrow R^n \longrightarrow R$, the induced sequence

$$\Gamma(W, \mathcal{O}^{\mathrm{an}}_{\mathbb{C}^k})^m \longrightarrow \Gamma(W, \mathcal{O}^{\mathrm{an}}_{\mathbb{C}^k})^n \longrightarrow \Gamma(W, \mathcal{O}^{\mathrm{an}}_{\mathbb{C}^k})$$

is exact. In any case we have maps of coherent sheaves on $\mathbb{C}^k$

$$\{\mathcal{O}^{\mathrm{an}}_{\mathbb{C}^k}\}^m \xrightarrow{\ \alpha\ } \{\mathcal{O}^{\mathrm{an}}_{\mathbb{C}^k}\}^n \xrightarrow{\ \beta\ } \mathcal{O}^{\mathrm{an}}_{\mathbb{C}^k} .$$

The kernel $\mathscr{I}$ of β is a coherent analytic sheaf on $\mathbb{C}^k$, and α factors uniquely through a homomorphism into $\mathscr{I}$; we deduce a longer sequence of maps

$$\{\mathcal{O}^{\mathrm{an}}_{\mathbb{C}^k}\}^m \longrightarrow \mathscr{I} \longrightarrow \{\mathcal{O}^{\mathrm{an}}_{\mathbb{C}^k}\}^n \xrightarrow{\ \beta\ } \mathcal{O}^{\mathrm{an}}_{\mathbb{C}^k} .$$

Because $\mathscr{I}$ is the kernel sheaf, the sequence

$$\Gamma(W, \mathscr{I}) \longrightarrow \Gamma\big(W, \{\mathcal{O}^{\mathrm{an}}_{\mathbb{C}^k}\}^n\big) \xrightarrow{\ \beta\ } \Gamma(W, \mathcal{O}^{\mathrm{an}}_{\mathbb{C}^k})$$

must be exact. It would be enough to prove that the homomorphism $\Gamma(W, \{\mathcal{O}^{\mathrm{an}}_{\mathbb{C}^k}\}^m) \longrightarrow \Gamma(W, \mathscr{I})$ is surjective for every polydisc W. The first reduction is that it suffices to prove this in the special case where the polydisc W is all of $\mathbb{C}^k$. The reason is the following: for any polydisc W, we have a commutative square

$$\begin{array}{ccc}
\Gamma\big(\mathbb{C}^k, \{\mathcal{O}^{\mathrm{an}}_{\mathbb{C}^k}\}^m\big) & \xrightarrow{\ \gamma\ } & \Gamma(\mathbb{C}^k, \mathscr{I}) \\
{\scriptstyle \mathrm{res}^{\mathbb{C}^k}_W}\Big\downarrow & & \Big\downarrow{\scriptstyle \mathrm{res}^{\mathbb{C}^k}_W} \\
\Gamma\big(W, \{\mathcal{O}^{\mathrm{an}}_{\mathbb{C}^k}\}^m\big) & \xrightarrow[\ \gamma'\ ]{} & \Gamma(W, \mathscr{I})
\end{array} .$$

We are assuming that the map γ is surjective, and want to deduce that so is the map γ'. We know that the maps $\mathrm{res}_W^{\mathbb{C}^k}$ have dense images in the Fréchet topology; see [3, page 241, Theorem 11]. Hence the image of the composite

$$\begin{array}{ccc} \Gamma\big(\mathbb{C}^k, \{\mathscr{O}_{\mathbb{C}^k}^{\mathrm{an}}\}^m\big) & \xrightarrow{\ \gamma\ } & \Gamma(\mathbb{C}^k, \mathscr{I}) \\ & & \big\downarrow \mathrm{res}_W^{\mathbb{C}^k} \\ & & \Gamma(W, \mathscr{I}) \end{array}$$

is dense in $\Gamma(W, \mathscr{I})$. But the image of $\gamma' : \Gamma\big(W, \{\mathscr{O}_{\mathbb{C}^k}^{\mathrm{an}}\}^m\big) \longrightarrow \Gamma(W, \mathscr{I})$ is closed in $\Gamma\big(W, \mathscr{I}\big)$ by [3, page 243, Theorem 14], and contains a dense set. Hence the map γ' must be surjective.

Thus we are left with proving that the ring of holomorphic functions, over the entire $\mathbb{C}^k$, is flat over the ring R of polynomials. We will prove this by induction on k, the case $k = 0$ being obvious. Assume therefore that we know the theorem for $\ell < k$, and want to prove it for k. Let A be the ring $A = \Gamma(\mathbb{C}^k, \mathscr{O}_{\mathbb{C}^k}^{\mathrm{an}})$. We need to show that, for any exact sequence over the ring R of the form $R^m \longrightarrow R^n \longrightarrow R$, the sequence $A^m \longrightarrow A^n \longrightarrow A$ is also exact.

The sequence $R^m \longrightarrow R^n \longrightarrow R$ is given by two matrices over the polynomial ring $R = \mathbb{C}[x_1, x_2, \ldots, x_k]$; the map $R^m \longrightarrow R^n$ is given by an $n \times m$ matrix, while the map $R^n \longrightarrow R$ by a $1 \times n$ matrix, and we can assume that all the entries of the $1 \times n$ matrix are non-zero. Anyway we have a finite number of polynomials over $\mathbb{C}$. The product of all the non-zero matrix entries is a polynomial over $\mathbb{C}$. The Weierstrass Preparation Theorem tells us that, after a linear change of variables, we may assume the polynomial to be a monic polynomial in x_k with coefficients in $\mathbb{C}[x_1, x_2, \ldots, x_{k-1}]$; for a proof see the hint to [1, page 69, Exercise 16]. Of course if a product of finitely many polynomials is monic then each of the factors must be monic. We may therefore assume that all the entries in the matrices of the maps $R^m \longrightarrow R^n \longrightarrow R$ are monic polynomials in x_k with coefficients in $\mathbb{C}[x_1, x_2, \ldots, x_{k-1}]$. Let N be some integer greater than any of the degrees of the matrix entries in x_k.

Given a vector $a = (a_1, a_2, \ldots, a_n) \in R^n$, the map $\varphi : R^n \longrightarrow R$ takes it to

$$\varphi(a) \quad = \quad a_1 p_1 + a_2 p_2 + \cdots + a_n p_n \quad \in \quad R\ ,$$

where p_j are monic polynomials in x_k of degree $< N$, with coefficients in the smaller polynomial ring $\mathbb{C}[x_1, x_2, \ldots, x_{k-1}]$. The obvious elements of the kernel are the vectors

$$r_j \quad = \quad (-p_j, 0, \ldots, 0, p_1, 0, \ldots, 0)$$

with a p_1 in the jth position; of course all such elements are in the image of the map $\psi : R^m \longrightarrow R^n$, since ψ surjects to the kernel. Suppose we are given an arbitrary element in the kernel of the map $A \otimes_R \varphi : A^n \longrightarrow A$, that is an n–tuple of holomorphic functions $a = (a_1, a_2, \ldots, a_n) \in A^n$ with

$$a_1 p_1 + a_2 p_2 + \cdots + a_n p_n \quad = \quad 0 \, .$$

We can write each a_i, with $i \geq 2$, as $a_i = \lambda_i p_1 + s_i$, with s_i a polynomial in x_k of degree $< N$, and with coefficients holomorphic functions in $x_1, x_2, \ldots, x_{k-1}$. Replacing a with $a - \sum_{i=2}^k \lambda_i r_i$ we discover that, modulo the image of $A \otimes_R \psi : A^m \longrightarrow A^n$, we can reduce a to a vector $s = (s_1, s_2, \ldots, s_n) \in A^n$, with each of $s_i, i \geq 2$ a polynomial in x_k of degree $< N$, with coefficients holomorphic in $x_1, x_2, \ldots, x_{k-1}$. But the identity

$$s_1 p_1 + s_2 p_2 + \cdots + s_n p_n \quad = \quad 0$$

immediately forces the degree of s_1 in x_k to be bounded by $2N$. Anyway, all the degrees in x_k are bounded.

We have now reduced to a problem in the variables $x_1, x_2, \ldots, x_{k-1}$. We are given a free module of finite rank over $R' = \mathbb{C}[x_1, x_2, \ldots, x_{k-1}]$, namely the submodule of R^n of polynomials of degree $< 2N$ in x_k. It maps to a free module over R', namely the submodule of R of polynomials of degree $< 3N$ in x_k. The kernel is generated by some free module over R' of finite rank, which we can express as mapping through a submodule of R^m. And the question is whether the sequence remains exact after extending scalars to the holomorphic functions in $(k-1)$ variables. Induction now says all is well. □

This completes the proof of Fact 7.8.9(iv). Let us make the following observation, which renders Lemma A1.2.1 more coordinate-free.

Corollary A1.2.2. *Let S be a finitely generated $\mathbb{C}$–algebra, denote $(X, \mathcal{O}) = \big(\mathrm{Spec}(S), \widetilde{S}\big)$ and let V be a polydisc in $\big\{\mathrm{Spec}(S)\big\}^{\mathrm{an}}$. Then the ring $\Gamma(V, \mathcal{O}^{\mathrm{an}})$ is flat over S.*

Proof. If we choose generators for S, they give a surjective homomorphism $\theta : R \longrightarrow S$ with R the polynomial ring. Let W be a polydisc in $\mathbb{C}^k$ with $V = W \cap \big\{\mathrm{Spec}(S)\big\}^{\mathrm{an}}$. Lemma A1.2.1 proves that $\Gamma(W, \mathcal{O}^{\mathrm{an}}_{\mathbb{C}^k})$ is flat over R, and Reminder A1.1.1 tells us that $\Gamma(V, \mathcal{O}^{\mathrm{an}}) = \Gamma(W, \mathcal{O}^{\mathrm{an}}_{\mathbb{C}^k}) \otimes_R S$. Hence the corollary follows from [1, page 29, Exercise 2.20]. □

A1.3 The faithful flatness

Next we will prove Fact 7.8.9(ii). We need to show that the sheaf homomorphism $\lambda^* : \mathscr{S} \longrightarrow \lambda_* \mathscr{S}^{\mathrm{an}}$ is injective. Lemma 7.1.6 tells us that there is a sheaf $\ker(\lambda^*)$ with

$$\Gamma\big(U, \ker(\lambda^*)\big) \quad = \quad \ker\left\{\lambda_U^* : \Gamma(U, \mathscr{S}) \longrightarrow \Gamma(U^{\mathrm{an}}, \mathscr{S}^{\mathrm{an}})\right\}$$

We need to prove that this is the zero sheaf. Since we know it is a sheaf, it is enough to show that $\Gamma\big(U, \ker(\lambda^*)\big)$ vanishes for all U in some subbasis of the open sets; a sheaf is uniquely determined by a subbasis. We will verify this claim on affine open sets $(U, \mathcal{O}|_U) \cong \big(\mathrm{Spec}(S), \widetilde{S}\big)$ with $\mathscr{S}|_U \cong \widetilde{M}$.

Choosing generators for the ring S we can produce a surjective homomorphism $R \longrightarrow S$, where R is a polynomial ring. Lemma A1.1.2 allows us to identify the homomorphism

$$\lambda^* : \Gamma(U, \mathscr{S}) \longrightarrow \Gamma(U^{\mathrm{an}}, \mathscr{S}^{\mathrm{an}})$$

with the homomorphism

$$M \longrightarrow \Gamma(\mathbb{C}^k, \mathcal{O}_{\mathbb{C}}^{\mathrm{an}}) \otimes_R M \, .$$

We need to prove this map injective. By Lemma A1.2.1 we already know that $\Gamma(\mathbb{C}^k, \mathcal{O}_{\mathbb{C}}^{\mathrm{an}})$ is flat over R; the point is that we will now prove it faithfully flat. For a discussion of faithful flatness the reader is referred to [1, pages 45-46, Exercise 16]. In the exercise one shows that five statements are equivalent, and a ring homomorphism $R \longrightarrow A$ possessing these five equivalent properties is called faithfully flat. If we put $A = \Gamma(\mathbb{C}^k, \mathcal{O}_{\mathbb{C}}^{\mathrm{an}})$, the injectivity of the homomorphisms $M \longrightarrow A \otimes_R M$ is part (v) of the five equivalent statements. Since the five conditions are equivalent it suffices to prove (iii). That is we need to prove that, for any maximal ideal $\mathfrak{m} \subset R$, the ideal $A\mathfrak{m} \subset A$ is proper.

But now notice that if $\mathfrak{m}$ is a maximal ideal of R then, by the Nullstellensatz, there is a point $a \in \mathbb{C}^k$ so that $\mathfrak{m}$ is the kernel of the homomorphism φ_a, with $\varphi_a(f) = f(a)$. This homomorphism extends to the ring A of holomorphic functions on $\mathbb{C}^k$. It factors as

$$R \longrightarrow A \xrightarrow{\ \psi\ } \mathbb{C} \, ,$$

where $\psi(f) = f(a)$ for any homolorphic function f on $\mathbb{C}^k$. Now $A\mathfrak{m}$ is in the kernel of ψ, which is a proper ideal in A.

A1.4 The Hom-sheaves

Next we prove Fact 7.8.9(vi). We want to show that a map

$$\Theta : \left\{\mathscr{H}om(\mathscr{S},\mathscr{T})\right\}^{\mathrm{an}} \longrightarrow \mathscr{H}om(\mathscr{S}^{\mathrm{an}},\mathscr{T}^{\mathrm{an}})$$

of sheaves on X^{an} is an isomorphism. The problem is local. Hence we will assume $(X,\mathcal{O}) = \left(\mathrm{Spec}(S),\widetilde{S}\right)$, and we will assume that $\mathscr{S} = \widetilde{M}$ and $\mathscr{T} = \widetilde{N}$.

If V is a polydisc in X^{an}, Lemma A1.1.3 tells us that

$$\Gamma(V,\widetilde{M}^{\mathrm{an}}) \quad = \quad \Gamma(V,\widetilde{\mathcal{O}}^{\mathrm{an}}) \otimes_S M\ .$$

It follows immediately that

Lemma A1.4.1. *For any polydisc $V \subset X^{\mathrm{an}}$, there is a bijective correspondence*

$$\left\{\begin{array}{c}\text{maps of sheaves}\\ \text{of } \mathcal{O}^{\mathrm{an}}\text{–modules}\\ \mathscr{S}^{\mathrm{an}}|_V \longrightarrow \mathscr{T}^{\mathrm{an}}|_V\end{array}\right\} \Longleftrightarrow \left\{\begin{array}{c}\text{maps of } \Gamma(V,\mathcal{O}^{\mathrm{an}})\text{–modules}\\ \Gamma(V,\mathscr{S}^{\mathrm{an}}) \longrightarrow \Gamma(V,\mathscr{T}^{\mathrm{an}})\end{array}\right\}\ .$$

The map in one direction is clear: given a homomorphism of sheaves $\mathscr{S}^{\mathrm{an}}|_V \longrightarrow \mathscr{T}^{\mathrm{an}}|_V$, the corresponding map of modules is obtained by taking global sections.

Proof. We need to prove that the correspondence is bijective. That is, given a map $\Gamma(V,\mathscr{S}^{\mathrm{an}}) \longrightarrow \Gamma(V,\mathscr{T}^{\mathrm{an}})$ we must prove that there is a unique map of sheaves $\mathscr{S}^{\mathrm{an}}|_V \longrightarrow \mathscr{T}^{\mathrm{an}}|_V$ inducing it. Let us therefore be given a map $\varphi : \Gamma(V,\mathscr{S}^{\mathrm{an}}) \longrightarrow \Gamma(V,\mathscr{T}^{\mathrm{an}})$. We need to produce the map of sheaves and prove its uniqueness.

But if there is a map of sheaves $\Phi : \mathscr{S}^{\mathrm{an}}|_V \longrightarrow \mathscr{T}^{\mathrm{an}}|_V$ then, for any polydisc $V' \subset V$ we would have a commutative diagram

$$\begin{array}{ccc}\Gamma(V,\mathscr{S}^{\mathrm{an}}) & \xrightarrow{\Gamma(V,\Phi)} & \Gamma(V,\mathscr{T}^{\mathrm{an}})\\ {\scriptstyle \mathrm{res}^V_{V'}}\downarrow & & \downarrow{\scriptstyle \mathrm{res}^V_{V'}}\\ \Gamma(V',\mathscr{S}^{\mathrm{an}}) & \xrightarrow[\Gamma(V',\Phi)]{} & \Gamma(V',\mathscr{T}^{\mathrm{an}})\end{array}\ .$$

By Lemma A1.1.3 this diagram is isomorphic to

$$\begin{array}{ccc}\Gamma(V,\mathcal{O}^{\mathrm{an}})\otimes_S M & \xrightarrow{\Gamma(V,\Phi)} & \Gamma(V,\mathcal{O}^{\mathrm{an}})\otimes_S N\\ {\scriptstyle \mathrm{res}^V_{V'}}\downarrow & & \downarrow{\scriptstyle \mathrm{res}^V_{V'}}\\ \Gamma(V',\mathcal{O}^{\mathrm{an}})\otimes_S M & \xrightarrow[\Gamma(V',\Phi)]{} & \Gamma(V',\mathcal{O}^{\mathrm{an}})\otimes_S N\end{array}\ .$$

In this diagram the maps

$$\begin{array}{ccc}
\Gamma(V,\mathcal{O}^{\mathrm{an}})\otimes_S M & \xrightarrow{\varphi=\Gamma(V,\Phi)} & \Gamma(V,\mathcal{O}^{\mathrm{an}})\otimes_S N \\
 & & \big\downarrow \mathrm{res}^V_{V'} \\
 & & \Gamma(V',\mathcal{O}^{\mathrm{an}})\otimes_S N
\end{array}$$

are given; the composite is a homomorphism from the $\Gamma(V,\mathcal{O}^{\mathrm{an}})$–module $\Gamma(V,\mathcal{O}^{\mathrm{an}})\otimes_S M$ to the $\Gamma(V',\mathcal{O}^{\mathrm{an}})$–module $\Gamma(V',\mathcal{O}^{\mathrm{an}})\otimes_S N$, and therefore factors uniquely through the module

$$\Gamma(V',\mathcal{O}^{\mathrm{an}})\otimes_{\Gamma(V,\mathcal{O}^{\mathrm{an}})}\Gamma(V,\mathcal{O}^{\mathrm{an}})\otimes_S M \quad = \quad \Gamma(V',\mathcal{O}^{\mathrm{an}})\otimes_S M\ .$$

This gives us the uniqueness of $\Gamma(V',\Phi)$. There is at most one possible factorization.

Proving the existence means showing that, with the unique choice forced on us, certain squares commute. Assume $V''\subset V'\subset V$ are polydiscs. We have a large diagram

$$\begin{array}{ccc}
\Gamma(V,\mathcal{O}^{\mathrm{an}})\otimes_S M & \xrightarrow{\Gamma(V,\Phi)} & \Gamma(V,\mathcal{O}^{\mathrm{an}})\otimes_S N \\
\mathrm{res}^V_{V'}\big\downarrow & & \big\downarrow \mathrm{res}^V_{V'} \\
\Gamma(V',\mathcal{O}^{\mathrm{an}})\otimes_S M & \xrightarrow[\Gamma(V',\Phi)]{} & \Gamma(V',\mathcal{O}^{\mathrm{an}})\otimes_S N \\
\mathrm{res}^{V'}_{V''}\big\downarrow & & \big\downarrow \mathrm{res}^{V'}_{V''} \\
\Gamma(V'',\mathcal{O}^{\mathrm{an}})\otimes_S M & \xrightarrow[\Gamma(V'',\Phi)]{} & \Gamma(V'',\mathcal{O}^{\mathrm{an}})\otimes_S N
\end{array}$$

where we know that two of the squares commute, and hence that the three composites from top left to bottom right agree. This forces the two composites from top left to bottom right

$$\begin{array}{ccc}
\Gamma(V,\mathcal{O}^{\mathrm{an}})\otimes_S M & & \\
\mathrm{res}^V_{V'}\big\downarrow & & \\
\Gamma(V',\mathcal{O}^{\mathrm{an}})\otimes_S M & \xrightarrow{\Gamma(V',\Phi)} & \Gamma(V',\mathcal{O}^{\mathrm{an}})\otimes_S N \\
\mathrm{res}^{V'}_{V''}\big\downarrow & & \big\downarrow \mathrm{res}^{V'}_{V''} \\
\Gamma(V'',\mathcal{O}^{\mathrm{an}})\otimes_S M & \xrightarrow[\Gamma(V'',\Phi)]{} & \Gamma(V'',\mathcal{O}^{\mathrm{an}})\otimes_S N
\end{array}$$

to agree with each other. But then the square

$$\begin{array}{ccc} \Gamma(V', \mathcal{O}^{\mathrm{an}}) \otimes_S M & \xrightarrow{\Gamma(V',\Phi)} & \Gamma(V', \mathcal{O}^{\mathrm{an}}) \otimes_S N \\ {\scriptstyle \mathrm{res}^{V'}_{V''}} \downarrow & & \downarrow {\scriptstyle \mathrm{res}^{V'}_{V''}} \\ \Gamma(V'', \mathcal{O}^{\mathrm{an}}) \otimes_S M & \xrightarrow[\Gamma(V'',\Phi)]{} & \Gamma(V'', \mathcal{O}^{\mathrm{an}}) \otimes_S N \end{array}$$

gives two homomorphisms of $\Gamma(V', \mathcal{O}^{\mathrm{an}})$–modules that agree on a subset which generates the module $\Gamma(V', \mathcal{O}^{\mathrm{an}}) \otimes_S M$. Hence the square must commute, and we obtain our morphism of sheaves. □

Remark A1.4.2. We have the identifications

$$\mathrm{Hom}(\mathscr{S}, \mathscr{T}) \quad = \quad \mathrm{Hom}(\widetilde{M}, \widetilde{N}) \quad = \quad \mathrm{Hom}(M, N) \ ,$$

where the second equality is by Proposition 7.5.1. We also have the identification, by Lemma A1.4.1,

$$\mathrm{Hom}(\mathscr{S}^{\mathrm{an}}|_V, \mathscr{T}^{\mathrm{an}}|_V) \quad = \quad \mathrm{Hom}\big(\Gamma(V, \mathcal{O}^{\mathrm{an}}) \otimes_S M \ , \ \Gamma(V, \mathcal{O}^{\mathrm{an}}) \otimes_S N\big) \ .$$

The analytification functor takes a morphism $\eta : \mathscr{S} \longrightarrow \mathscr{T}$ to the morphism $\eta^{\mathrm{an}}|_V : \mathscr{S}^{\mathrm{an}}|_V \longrightarrow \mathscr{T}^{\mathrm{an}}|_V$, and the reader can easily check that it agrees with the natural map

$$\mathrm{Hom}(M, N) \longrightarrow \mathrm{Hom}\big(\Gamma(V, \mathcal{O}^{\mathrm{an}}) \otimes_S M \ , \ \Gamma(V, \mathcal{O}^{\mathrm{an}}) \otimes_S N\big)$$

which takes $f : M \longrightarrow N$ and tensors it with $\Gamma(V, \mathcal{O}^{\mathrm{an}})$. This natural map is a map from the S–module $\mathrm{Hom}(M, N)$ to a $\Gamma(V, \mathcal{O}^{\mathrm{an}})$–module, and therefore factors uniquely through

$$\Gamma(V, \mathcal{O}^{\mathrm{an}}) \otimes_S \mathrm{Hom}(M, N) \xrightarrow{\Theta_V} \mathrm{Hom}\big(\Gamma(V, \mathcal{O}^{\mathrm{an}}) \otimes_S M \ , \ \Gamma(V, \mathcal{O}^{\mathrm{an}}) \otimes_S N\big).$$

This homomorphism Θ_V is nothing other than $\Gamma(V, \Theta)$, where Θ is the morphism of sheaves of Fact 7.8.9(vi). We need to show this map an isomorphism. We know, by Corollary A1.2.2, that the ring $\Gamma(V, \mathcal{O}^{\mathrm{an}})$ is flat over S. Fact 7.8.9(vi) is therefore an immediate consequence of the following general observation:

Lemma A1.4.3. *Let S be a ring, A an algebra flat over S. Let M and N be two S–modules, with M finitely presented. The natural homomorphism*

$$\Theta : A \otimes_S \mathrm{Hom}_S(M, N) \longrightarrow \mathrm{Hom}_A(A \otimes_S M, A \otimes_S N)$$

is an isomorphism.

Proof. Choose a finite presentation for M, that is an exact sequence

$$S^m \longrightarrow S^n \longrightarrow M \longrightarrow 0\,.$$

Tensoring with A we have the exact sequence

$$A^m \longrightarrow A^n \longrightarrow A \otimes_S M \longrightarrow 0\,.$$

Since $\mathrm{Hom}_S(-, N)$ and $\mathrm{Hom}_A(-, A \otimes_S N)$ are both left exact we have two exact sequences

$$0 \longrightarrow \mathcal{H}om(M, N) \longrightarrow \mathrm{Hom}(S^n, N) \longrightarrow \mathrm{Hom}(S^m, N)$$

$$0 \longrightarrow \mathrm{Hom}(A \otimes_S M, A \otimes_S N) \longrightarrow \mathrm{Hom}(A^n, A \otimes_S N) \longrightarrow \mathrm{Hom}(A^m, A \otimes_S N)$$

and, since A is flat over S, the top row in the commutative diagram below is also exact

$$\begin{array}{ccccccc}
0 \longrightarrow & A \otimes_S \mathrm{Hom}(M, N) & \longrightarrow & A \otimes_S \mathrm{Hom}(S^n, N) & \longrightarrow & A \otimes_S \mathrm{Hom}(S^m, N) \\
 & \downarrow{\scriptstyle \alpha} & & \downarrow{\scriptstyle \beta} & & \downarrow{\scriptstyle \gamma} \\
0 \longrightarrow & \mathrm{Hom}(A \otimes_S M, A \otimes_S N) & \longrightarrow & \mathrm{Hom}(A^n, A \otimes_S N) & \longrightarrow & \mathrm{Hom}(A^m, A \otimes_S N)
\end{array}$$

Since β and γ are clearly isomorphisms, so must be α. □

A1.5 Vector bundles

It only remains to complete the proof of Lemma 7.9.6. We must convince the reader that, if $\mathscr{V}$ is a coherent algebraic sheaf on $(X, \mathcal{O})$ and if $\mathscr{V}^{\mathrm{an}}$ is an analytic vector bundle, then $\mathscr{V}$ is a vector bundle. The question is local: we may suppose $(X, \mathcal{O}) = \left(\mathrm{Spec}(S), \widetilde{S}\right)$ with S a finitely generated $\mathbb{C}$–algebra, and $\mathscr{V} = \widetilde{M}$ where M is a finite S–module. We prove:

Lemma A1.5.1. *Let the notation be as above. Let p be a point in $\left\{\mathrm{Spec}(S)\right\}^{\mathrm{an}}$ and assume that, for some polydisc V containing p, we have that $\Gamma(V, \widetilde{M}^{\mathrm{an}})$ is a free $\Gamma(V, \mathcal{O}^{\mathrm{an}})$–module. Then there is a basic open affine $X_f = \mathrm{Spec}(S[1/f])$ containing p over which $\mathscr{V}$ trivializes; that is, $\mathscr{V}|_{X_f} = \widetilde{M[1/f]}$ with $M[1/f]$ a free $S[1/f]$–module.*

Proof. By Lemma A1.1.3 we know that $\Gamma(V, \widetilde{M}^{\mathrm{an}}) = \Gamma(V, \mathcal{O}^{\mathrm{an}}) \otimes_S M$. The hypothesis of the lemma is that $\Gamma(V, \widetilde{M}^{\mathrm{an}})$ is free. We must prove that $M[1/f]$ is free over $S[1/f]$, for some f not vanishing at p. In any case we have that the closed point p corresponds to a unique homomorphism

$\varphi : S \longrightarrow \mathbb{C}$; see Proposition 4.2.4. The fact that p lies in the polydisc V means that the homomorphism φ, which evaluates a function at p, factors through $\Gamma(V, \mathcal{O}^{\mathrm{an}})$. We have ring homomorphisms

$$S \xrightarrow{\mu_V} \Gamma(V, \mathcal{O}^{\mathrm{an}}) \xrightarrow{\psi} \mathbb{C} .$$

The vector space

$$\mathbb{C} \otimes_S M \quad = \quad \mathbb{C} \otimes_{\Gamma(V,\mathcal{O}^{\mathrm{an}})} \Gamma(V, \mathcal{O}^{\mathrm{an}}) \otimes_S M$$

is, by the right hand side of the above equality, a vector space over $\mathbb{C}$ of dimension the rank of the vector bundle $\mathscr{V}^{\mathrm{an}}|_V$. Let this rank be n. We can choose n elements of M, let us label them $m_1, m_2, \ldots, m_n$, which map to a basis of $\mathbb{C} \otimes_S M$.

This gives us a map of S–modules $S^m \longrightarrow M$. Let the kernel of the map be K and the cokernel Q; we have an exact sequence

$$0 \longrightarrow K \longrightarrow S^n \longrightarrow M \longrightarrow Q \longrightarrow 0 .$$

The ring $A = \Gamma(V, \mathcal{O}^{\mathrm{an}})$ is flat over S by Corollary A1.2.2, and hence the sequence

$$0 \longrightarrow A \otimes_S K \longrightarrow A^n \longrightarrow A \otimes_S M \longrightarrow A \otimes_S Q \longrightarrow 0$$

is also exact. But the module $A \otimes_S M$ is a free module of rank n, the map $A^n \longrightarrow A \otimes_S M$ is an $n \times n$ matrix, and we have chosen the map $S^n \longrightarrow M$ so that at p this matrix in invertible. The map must therefore be invertible in some neighborhood of p. If we shrink the polydisc we have a homomorphism

$$A \quad = \quad \Gamma(V, \mathcal{O}^{\mathrm{an}}) \xrightarrow{\mathrm{res}^V_{V'}} \Gamma(V', \mathcal{O}^{\mathrm{an}}) \quad = \quad B ,$$

so that, in the sequence

$$0 \longrightarrow B \otimes_S K \longrightarrow B^n \xrightarrow{\theta} B \otimes_S M \longrightarrow B \otimes_S Q \longrightarrow 0$$

the map θ is an isomorphism. We conclude that $B \otimes_S K = 0 = B \otimes_S Q$.

But p lies in V', and hence our homomorphism $\varphi : S \longrightarrow \mathbb{C}$ factors through B. Therefore

$$\mathbb{C} \otimes_S K \quad = \quad \mathbb{C} \otimes_B B \otimes_S K \quad = \quad 0 \quad = \quad \mathbb{C} \otimes_B B \otimes_S Q \quad = \quad \mathbb{C} \otimes_S Q .$$

Nakayama's lemma, for the finite modules K and Q, says that there exists an element $f \in S$, $f \notin \ker(\varphi)$, which annihilates the modules K and Q. For a proof see [1, page 21, Corollary 2.5]. Over the ring $S[1/f]$ the exact sequence

$$0 \longrightarrow K[1/f] \longrightarrow \{S[1/f]\}^n \longrightarrow M[1/f] \longrightarrow Q[1/f] \longrightarrow 0$$

becomes

$$0 \longrightarrow \{S[1/f]\}^n \longrightarrow M[1/f] \longrightarrow 0\,,$$

meaning that $M[1/f]$ is free of rank n. □

Bibliography

[1] Michael F. Atiyah and Ian G. Macdonald. *Introduction to commutative algebra.* Addison-Wesley Publishing Co., Reading, Mass.-London-Don Mills, Ont., 1969.

[2] Hans Grauert and Reinhold Remmert. *Coherent analytic sheaves*, volume 265 of *Grundlehren der Mathematischen Wissenschaften [Fundamental Principles of Mathematical Sciences].* Springer-Verlag, Berlin, 1984.

[3] Robert C. Gunning and Hugo Rossi. *Analytic functions of several complex variables.* Prentice-Hall Inc., Englewood Cliffs, N.J., 1965.

[4] Adolf Hurwitz and R. Courant. *Vorlesungen über allgemeine Funktionentheorie und elliptische Funktionen.* Interscience Publishers, Inc., New York, 1944.

[5] David Mumford. *Abelian varieties.* Tata Institute of Fundamental Research Studies in Mathematics, No. 5. Published for the Tata Institute of Fundamental Research, Bombay, 1970.

[6] David Mumford, John Fogarty, and Frances Kirwan. *Geometric invariant theory*, volume 34 of *Ergebnisse der Mathematik und ihrer Grenzgebiete (2) [Results in Mathematics and Related Areas (2)].* Springer-Verlag, Berlin, third edition, 1994.

[7] Jean-Pierre Serre. Géométrie algébrique et géométrie analytique. *Ann. Inst. Fourier, Grenoble*, 6:1–42, 1955–1956.

[8] Joseph H. Silverman. *The arithmetic of elliptic curves*, volume 106 of *Graduate Texts in Mathematics.* Springer-Verlag, New York, 1986.

[9] Charles A. Weibel. *An Introduction to Homological Algebra*, volume 38 of *Cambridge Studies in Advanced Mathematics.* Cambridge University Press, 1994.

Glossary

Cross references direct the reader to entries in the index. When several page numbers are given, the one in boldface is (in the author's opinion) the main reference.

$\operatorname{ann}(\mathscr{S}')$, *see* annihilator of $\mathscr{S}'$
$\langle \mathbf{a} \rangle$, 237
$|\mathbf{a}|$, 146

$\mathbb{C}e_m$, $S \otimes_{\mathbb{C}} \mathbb{C}e_m$, 327
$\mathbb{C}[V]$, 272–273; *see also* polynomial ring $\mathbb{C}[V]$
$\mathbb{C}[x_0, x_1, \ldots, x_n]$ as a G–ring, *see* viewing $S = \mathbb{C}[x_0, x_1, \ldots, x_n]$ as a G–ring
$\mathbb{CP}^n$, 5–6, 211; *see also* projective space $(\mathbb{P}^n, \mathcal{O})$
C_J, 47

$\dim(\mathscr{S}')$, 383

$f_*\mathscr{S}$, 177–178

$(G, \mathcal{O})$, *see* affine group scheme $(G, \mathcal{O})$
$\Gamma(U, \mathcal{O})$, 16

$H^i(\mathscr{S})$, *see* cohomology of coherent sheaves on $\mathbb{P}^n$
$\mathscr{H}om(\mathscr{S}, \mathscr{T})$, 175–177
computation of $\mathscr{H}om(\mathcal{O}, \mathscr{S})$, 178–179

$\ker(f)$, 172–175

$(\lambda_X, \lambda_X^*) : (X^{\mathrm{an}}, \mathcal{O}^{\mathrm{an}}) \longrightarrow (X, \mathcal{O})$, 100, 137–139, 159–161
$\lambda_X : X^{\mathrm{an}} \longrightarrow X$, *see* continuous map $\lambda_X : X^{\mathrm{an}} \longrightarrow X$

$\widetilde{M}$, *see* sheaf of modules $\widetilde{M}$

$\{\pi_*\widetilde{M}\}^{G^{\mathrm{an}}}$, *see* invariant subsheaf $\{\pi_*\widetilde{M}\}^{G^{\mathrm{an}}} \subset \pi_*\widetilde{M}$
$m : C_J \longrightarrow S_J$, 47
$\mathrm{Max}(X)$, *see* topological space $\mathrm{Max}(X)$
$\mathrm{MaxSpec}(R)$, *see* topological space $\mathrm{MaxSpec}(R)$
for R a polynomial ring, *see* space $\mathrm{MaxSpec}(R)$ for polynomial rings
$\mathrm{Max}(\Psi)$, 76
$\mathrm{MaxSpec}(\theta)$, *see* continuous map $\mathrm{MaxSpec}(\theta)$
$\mu : S \to \Gamma(\{\,\mathrm{Spec}(S)\}^{\mathrm{an}}, \mathcal{O}^{\mathrm{an}})$, 117
$\mu_V : S \longrightarrow \Gamma(V, \mathcal{O}^{\mathrm{an}})$, *see* homomorphism $\mu_V : S \longrightarrow \Gamma(V, \mathcal{O}^{\mathrm{an}})$

$\mathcal{O}(m) = \{\pi_*(\widetilde{S \otimes_{\mathbb{C}} \mathbb{C}e_m})\}^{G^{\mathrm{an}}}$, 327; *see also* invertible sheaf $\mathcal{O}(m)$
$\mathcal{O}^{\mathrm{an}}$, *see* analytic sheaf $\mathcal{O}^{\mathrm{an}}$
$\mathcal{O}_{X/G}$, *see* quotient ringed space $(X/G, \mathcal{O}_{X/G})$

$(\mathbb{P}^n, \mathcal{O})$, *see* projective space $(\mathbb{P}^n, \mathcal{O})$
$\mathbb{P}^n_i$, $\mathbb{P}^n_J$ and $\mathbb{P}^n_f \subset \mathbb{P}^n$, *see* projective space $(\mathbb{P}^n, \mathcal{O})$
$\pi : X \to X/G$, *see* quotient space X/G
p_ℓ, **113–114**, 156–157
$\{\pi_*\widetilde{M}\}^{G^{\mathrm{an}}}$, *see* invariant subsheaf $\{\pi_*\widetilde{M}\}^{G^{\mathrm{an}}} \subset \pi_*\widetilde{M}$
$\Phi^{\mathrm{an}} : X^{\mathrm{an}} \longrightarrow Y^{\mathrm{an}}$, *see* continuous map $\Phi^{\mathrm{an}} : X^{\mathrm{an}} \longrightarrow Y^{\mathrm{an}}$
$(\Phi^{\mathrm{an}}; \{\Phi^*\}^{\mathrm{an}}) : (X^{\mathrm{an}}; \mathcal{O}^{\mathrm{an}}_X) \to (Y^{\mathrm{an}}; \mathcal{O}^{\mathrm{an}}_Y)$, *see* morphism of analytic spaces $(\Phi^{\mathrm{an}}; \{\Phi^*\}^{\mathrm{an}})$

q_ℓ, **154–159**, 266–268

$\widetilde{R}$, *see* sheaf of rings $\widetilde{R}$
r_ℓ, **267–268**, 335–336
$\{\widetilde{R}\}^{\mathrm{an}}$, *see* analytic sheaf $\mathcal{O}^{\mathrm{an}}$

$[S, V]$, *see* pre-Fréchet ring $[S, V]$
S'/y, 384
$S = \mathbb{C}[x_0, x_1, \ldots, x_n]$ as a G–ring , *see* viewing $S = \mathbb{C}[x_0, x_1, \ldots, x_n]$ as a G–ring
$\mathscr{S}'(m)$, *see* coherent analytic sheaf on $\{\mathbb{P}^n\}^{\mathrm{an}}$
$\mathscr{S}(m)$, *see* coherent algebraic sheaf on $\mathbb{P}^n$
$\mathscr{S}^{\mathrm{an}}$, *see* analytification of coherent algebraic sheaf
$\mathscr{S}^G$, *see* G–sheaf $\mathscr{S}$, invariant subseaf
$\Sigma_{\mathscr{S}}$, 392–393
$\mathrm{Sym}^r V$, 272
s_ℓ, 335–337

S_J, 47

$S_J(m) \subset S_J$, 331

$S_J^{G^{\text{an}}} \subset S_J$, *see* viewing $S = \mathbb{C}[x_0, x_1, \ldots, x_n]$ as a G–ring

pre-Fréchet topology on $S_J \subset \Gamma(X_J^{\text{an}}, \mathcal{O}^{\text{an}})$, 157–158

$\mathrm{Spec}(R)$, *see* Zariski topology on $\mathrm{Spec}(R)$

$\left(\mathrm{Spec}(R), \widetilde{R}\right)$, *see* Zariski topology on $\mathrm{Spec}(R)$ *and* sheaf of rings $\widetilde{R}$

$\{\mathrm{Spec}(R)\}^{\text{an}}$, *see* complex topology on $\mathrm{MaxSpec}(R)$

$\{\mathrm{Spec}(R)\}^{\text{an}} \subset \mathbb{C}^n$, *see* generators and the embedding $\{\mathrm{Spec}(R)\}^{\text{an}} \to \mathbb{C}^n$

$(\mathrm{Spec}(\varphi), \widetilde{\varphi})$, *see* morphism of ringed spaces $(\mathrm{Spec}(\varphi), \widetilde{\varphi})$

$\mathrm{Spec}(\varphi)$, *see* continuous map $\mathrm{Spec}(\varphi)$

$\{\mathrm{Spec}(\varphi)\}^{\text{an}}$, *see* continuous map $\{\mathrm{Spec}(\varphi)\}^{\text{an}}$

$V(p) \subset V$, for $p \in \{\mathbb{P}^n\}^{\text{an}}$, 380

$(X/G, \mathcal{O}_{X/G})$, *see* quotient ringed space $(X/G, \mathcal{O}_{X/G})$

X/G, *see* quotient space X/G

$X \times Y$, *see* product of affine schemes

$\mathbf{x}^{\mathbf{a}}$, $\mathbf{x}^{\mathbf{a}+\mathbf{1}}$, 146

X_J, X_J^{an}, 144

$(X^{\text{an}}, \mathcal{O}^{\text{an}})$, *see* analytic space $(X^{\text{an}}, \mathcal{O}^{\text{an}})$

X^{an}, *see* topological space X^{an}

Index